CONSTRUCTING
QUARKS

for Lucy

CONSTRUCTING
QUARKS

A SOCIOLOGICAL HISTORY
OF PARTICLE
PHYSICS

ANDREW PICKERING

EDINBURGH
UNIVERSITY PRESS

© Andrew Pickering 1984
Edinburgh University Press
22 George Square
Edinburgh

Set in Lasercomp Times Roman
and printed in Great Britain by
The Alden Press
Oxford

British Library Cataloguing
in Publication Data
Pickering, Andrew
Constructing Quarks
1. Quarks – History
I. Title
539.7′21   QC793.5.Q252
ISBN 0 85224 458 4

CONTENTS

Preface   ix

PART ONE
Introduction, the Prehistory of HEP and its Material Constraints

1.   Introduction   3

2.   Men and Machines   21
      2.1 The HEP Community   21
      2.2 HEP Experiment: Basics   23
      2.3 Accelerator Chronology   32

3.   The Old Physics: HEP, 1945–64   46
      3.1 The Population Explosion   47
      3.2 Conservation Laws and Quantum Numbers:
           From Spin to the Eightfold Way   50
      3.3 Quantum Field Theory   60
      3.4 The S-Matrix   73

PART TWO
Constructing Quarks and Founding the New Physics:
      HEP, 1964–74

4.   The Quark Model   85
      4.1 The Genesis of Quarks   85
      4.2 The Constituent Quark Model   89
      4.3 Quarks and Current Algebra   108
      4.4 The Reality of Quarks   114

5.   Scaling, Hard Scattering and the Quark-Parton Model   125
      5.1 Scaling at SLAC   125
      5.2 The Parton Model   132
      5.3 Partons, Quarks and Electron Scattering   140
      5.4 Neutrino Physics   143

v

5.5   Lepton-Pair Production, Electron–Positron
Annihilation and Hadronic Hard Scattering   147

6.   Gauge Theory, Electroweak Unification
and the Weak Neutral Current   159
6.1   Yang–Mills Gauge Theory   160
6.2   Electroweak Unification and Spontaneous
Symmetry Breaking   165
6.3   The Renormalisation of Gauge Theory   173
6.4   Electroweak Models and the Discovery of
the Weak Neutral Current   180
6.5   Neutral Currents and Neutron Background   187

7.   Quantum Chromodynamics: A Gauge Theory
of the Strong Interactions   207
7.1   From Scale Invariance
to Asymptotic Freedom   207
7.2   Quantum Chromodynamics   215
7.3   The Failings of QCD   224

8.   HEP in 1974: The State of Play   231
8.1   Gauge Theory – Reprise   231
8.2   The End of an Era   235
8.3   Three Transitional Biographies   238

PART THREE
Establishing the New Physics: The November Revolution
and Beyond

9.   Charm: The Lever that Turned the World   253
9.1   The November Revolution   253
9.2   The *R*-crisis   254
9.3   The New Particles   258
9.4   Charm   261
9.5   A New World   270

10.   The Standard Model of Electroweak Interactions   280
10.1   More New Quarks and Leptons   280
10.2   The Slaying of Mutants   290
10.3   The Standard Model Established:
Unification, Social and Conceptual   300

11. QCD in Practice 309

    11.1 Phenomenological QCD – An Overview 310
    11.2 QCD and Spectroscopy 311
    11.3 QCD and Hard Scattering 317

12. Gauge Theory and Experiment: 1970–90 347

    12.1 Experimental Trends 349
    12.2 Theory Incarnate 353
    12.3 Visions 364

13. Grand Unification 383

    13.1 GUTs and HEP 383
    13.2 GUTs and Cosmology 388
    13.3 GUTs and Proton Decay 391
    13.4 Gauge Theory Supreme 396

14. Producing a World 403

    14.1 The Dynamics of Practice 403
    14.2 Traditions and Symbiosis 405
    14.3 Incommensurability 407

Bibliography 416
Index 459

PREFACE

Etched into the history of twentieth-century physics are cycles of atomism. First came atomic physics, then nuclear physics and finally elementary-particle physics. Each was thought of as a journey deeper into the fine structure of matter. In the beginning was the nuclear atom. The early years of this century saw a growing conviction amongst scientists that the chemical elements were made up of atoms, each of which comprised a small, positively charged core or nucleus surrounded by a diffuse cloud of negatively charged particles known as electrons. Atomic physics was the study of the outer layer of the atom, the electron cloud. Nuclear physics concentrated in turn upon the nucleus, which was itself regarded as composite: the atomic nucleus of each chemical element was supposed to be made up of a fixed number of positively charged particles called protons plus a number of electrically neutral neutrons. Protons, neutrons and electrons were the original 'elementary particles' – the building blocks from which physicists considered that all forms of matter were constructed. However, in the post-World War II period, many other particles were discovered which appeared to be just as elementary as the proton, neutron and electron, and a new specialty devoted to their study grew up within physics. The new specialty was known either as elementary-particle physics, after its subject matter, or as high-energy physics (HEP), after its primary experimental tool – the high-energy particle accelerator.

Atomic physics, nuclear physics and even the early years of elementary-particle physics have all been subject to historical scrutiny, but the story of the latest cycle of atomism has yet to be told. Not content with regarding protons, neutrons and the like as truly elementary particles, in the 1960s and 1970s high-energy physicists became increasingly confident that they had plumbed a new stratum of matter: *quarks*. Gross matter was composed of atoms; within the atom was the nucleus; within the nucleus were protons and neutrons; and, finally, within protons and neutrons (and all sorts of other particles) were quarks. The aim of this book is to document and analyse this latest step into the heart of the material world.

The analysis given here is somewhat unconventional in that its thrust is sociological. Rather than treat the quark view of matter as an abstract conceptual system, I seek to analyse its construction, elaboration and use in the context of the developing practice of the HEP community. My explanation of why the reality of quarks came to be accepted relates to the dynamics of that practice, a dynamics which is at once social and conceptual. I try to avoid the circular idiom of naive realism whereby the product of a historical process, in this case the perceived reality of quarks, is held to determine the process itself. The view taken here is that the reality of quarks was the upshot of particle physicists' practice, and not the reverse: hence the title of the book, *Constructing Quarks*.

The sociology of science is often taken to relate purely to the social relations between scientists, and hence to exclude esoteric technical and conceptual matters. That is not the case here. There is no escape from such topics because the emphasis is on practice and, in HEP, practice is irredeemably esoteric. This, of course, creates a communication problem: arcane practices are best described in arcane language, otherwise known as scientific jargon. To ameliorate the problem, I have sought to explain each item of HEP jargon whenever it first enters the text. I have also cited popular accounts of the developments at issue, for the reader who feels he would benefit from more background on particular cases. In this way I have tried to make the text accessible to anyone with a basic scientific education. It is perhaps appropriate to add that the main object of my account is not to explain technical matters *per se* but to explain how knowledge is developed and transformed in the course of scientific practice. My belief is that the processes of transformation are easier to grasp than the full ramifications of, say, a given theoretical structure at a particular time. My hope is to give the reader some feeling for what scientists do and how science develops, not to equip him or her as a particle physicist.

There remain, nevertheless, sections of the account which may prove difficult for outsiders to physics. I have in mind here especially the passages in Part II of the account which deal with the formal development of 'gauge theory'. Gauge theory provided the theoretical framework within which quark physics was eventually set, and the development of gauge theory was intrinsic to the establishment of the quark picture. It would therefore be inappropriate to omit these passages from a historical account. However, the key idea which I use in analysing the development of theories is that of modelling or analogy and, unfortunately in the present context, gauge theory was modelled upon a highly complex, mathematically-sophisticated theory known as quantum electrodynamics (QED). QED was taken for granted by physicists during the period we will be considering, and it is beyond the scope of this work to go at all deeply into its origins. Thus, in discussing the formal development of gauge theory, I have to refer to accepted properties of QED (and quantum field theory in general) which I can only partially explain. This is the origin of the communication gap which remains in the text. On the positive side, I should stress that the more difficult phases of the narrative are non-cumulative in effect. For example, the uses to which particle physicists put gauge theory, discussed in Part III of the account, are much more easily understood than the prior process of theory development, discussed in Part II. The conceptual intricacies of the early discussions of field theory and gauge theory need not, therefore, deter the reader from reading on.

It remains for me to acknowledge some of my many debts: intellectual, material and financial. Taking the latter first, the research on which this book is based has been supported by grants from the UK Social Science Research Council. Without their support, it would have been impossible. The active co-operation of the HEP community was likewise crucial, and I would like to thank the many physicists who found the time for interview and correspondence. In the following pages I argue that research is a social activity, and I am happy to apply the same argument to my own work. The account of particle physics offered here is intended as a contribution to the 'relativist-constructivist' programme in the sociology of scientific knowledge, and I owe a considerable debt to all of those working in this tradition. I am particularly grateful to Harry Collins, Trevor Pinch, Dave Travis and John Law. The Science Studies Unit of the University of Edinburgh has been my base for this work; it has supported me in many ways and I am indebted to everyone there, especially Mike Barfoot, Barry Barnes, David Edge, Bill Harvey, Malcolm Nicolson and Steve Shapin. Thanks also to Peter Higgs and the other members of the Edinburgh HEP group for much information and many useful discussions. Moyra Forrest prepared the Index, and her assistance with library materials has also been invaluable. The typing of Carole Tansley, Margaret Merchant and, especially, Jane Flaxington has been heroic. To Jane F. likewise my thanks for life-support while this book was being written.

Edinburgh   A.P.
August 1983

PART I

INTRODUCTION,
THE PREHISTORY OF HEP AND
ITS MATERIAL CONSTRAINTS

1

INTRODUCTION

The scientist . . . must appear to the systematic epistemologist as an unscrupulous opportunist.   ALBERT EINSTEIN[1]

The historian of modern science has to come to terms with the fact that the scientists have got there first. Very many accounts of the topics to be examined here have already been presented by particle physicists in the popular scientific press, as well as in the professional literature of high-energy physics (HEP).[2] These accounts all have a similar form – they are contributions to a well-established genre of scientific writing – and present a vision of science which is in some ways the mirror image – or reverse – of that developed in the following pages. I want therefore to begin by sketching out the archetypal 'scientist's account' of the history of HEP, in order both to indicate its shortcomings and to introduce the motivations behind my own approach.[3] This sketch will involve the use of some technical terms which will only be explained in later chapters, but the detailed meanings of the terms are not important in the present context.

The scientist's account begins in the early 1960s. At that time, particle physicists recognised four fundamental forces of nature, known, in order of decreasing strength, as the strong, electromagnetic, weak and gravitational interactions. The strong force, responsible for the binding of neutrons and protons in nuclei, was of short range and was the dominant force in elementary-particle interactions. The electromagnetic force, around 10^3 (i.e. 1000) times weaker than the strong force and acting at long range, was responsible for binding nuclei and electrons together in atoms, and also for macroscopic electromagnetic phenomena: light, radio waves and so on. The weak force, of short range and around 10^5 (100,000) times weaker than the strong force, had, except in special circumstances, negligible effects. Those special circumstances pertained to certain radioactive decays of nuclei and elementary particles, and to processes of energy generation in stars. Finally the gravitational force was a long-range force like electromagnetism. It was responsible for macroscopic gravitational phenomena – apples falling from trees, the earth orbiting the sun – but it was also 10^{38} times weaker than the strong force, and its effects were considered to be completely negligible in the world of elementary particles.

Associated with this classification of forces was a classification of elementary particles. Particles which experienced the strong force were called *hadrons*. There were many hadrons, including the proton and neutron, the constituents of atomic nuclei. Particles which were immune to the strong force – the electron and a handful of other particles – were known as *leptons*. The picture began to change in 1964, with the proposal that the constituents had constituents: hadrons were to be seen as made up of more fundamental entities known as *quarks*. Although it left many questions unanswered, the quark model explained certain experimentally-observed regularities of the hadron mass spectrum and of hadronic decay processes. Moreover, in the late 1960s and early 1970s it was seen that quarks could explain the phenomenon known as *scaling*, which had recently been discovered in experiments on the interaction of leptons with hadrons. In the scientist's account quarks thus represented the fundamental entities of a new layer of matter. Initially, though, the existence of quarks was not regarded as firmly established, principally because experimental searches had failed to detect any particles having the distinctive properties postulated for them (i.e. their having fractional electric charges). Leptons were not subject to a parallel ontological transformation; unlike hadrons they continued to be regarded as truly elementary particles.

Early in the 1970s, new theories of the interactions of quarks and leptons began to be formulated. First came the realisation that the weak and electromagnetic interactions could be seen as manifestations of a single *electroweak* force within the context of a theoretical approach known as gauge theory. This unification, reminiscent of Maxwell's nineteenth-century unification of electricity and magnetism, carried with it the prediction of the existence of the *weak neutral current*, which was verified in 1973, and of *charmed particles*, which was verified in 1974. Meanwhile, it had been realised in 1973 that a particular gauge theory, known as *quantum chromodynamics* or QCD, was a possible theory of the strong interactions of quarks. It was found first to explain scaling, and later to explain observed deviations from scaling. It explained certain interesting properties of charmed and other particles, and various other hadronic phenomena. Therefore QCD became the accepted theory of the strong interactions. Quarks had still not been observed in isolation. But both electroweak theory and QCD assumed the validity of the quark picture, and thus the existence of quarks was established simultaneously with the establishment of the gauge-theory description of their interactions. In the late 1970s, particle physicists were agreed that the world of elementary particles was one of quarks and leptons interacting according to the dictates of the twin gauge theories: electroweak theory and QCD. Finally it was noticed that since the unified electroweak theory and QCD were both gauge theories, they could, in their turn, be unified with one another. This last unification brought with it more fascinating predictions, which began to arouse the interests of experimenters in 1979. These predictions were not immediately verified, but many physicists were confident that they would be. Thus, not only was a new and fundamental layer of structure discovered, in the shape of quarks, but three forces previously thought to be quite different from one another – the strong, electromagnetic and weak interactions – stood revealed as but particular manifestations of a single force.

Apart from brevity, this sketch of the scientist's account of the history of quarks has the virtue that it names key developments. For example, it indicates that the quark idea was only one theoretical component in the developments we will discuss. The other component was gauge theory, which eventually provided the framework within which the interactions of quarks and leptons were understood. This is something to keep firmly in mind: it is impossible to understand the establishment of the quark picture without at the same time understanding the perceived virtues of gauge theory. The scientist's account also points to the role of new phenomena – scaling, neutral currents, charmed particles and the like – in supporting the quark–gauge theory view. The observation that the quark–gauge theory picture referred to new phenomena, quite different from those which supported the pre-quark view of the world, will also figure prominently in the following account. However, beyond specifying the key developments in HEP, the scientist's account goes on, either explicitly or implicitly, to specify a relationship between them. I want now to discuss this relationship, in order to highlight where the present approach departs from that of the scientist.

In the scientist's account, experiment is seen as the supreme arbiter of theory. Experimental facts dictate which theories are to be accepted and which rejected. Experimental data on scaling, neutral currents and charmed particles, for example, dictated that the quark–gauge theory picture was to be preferred over alternative descriptions of the world. There are, though, two well-known and forceful philosophical objections to this view, each of which implies that experiment cannot *oblige* scientists to make a particular choice of theories.[4] First, even if one were to accept that experiment produces unequivocal fact, it would remain the case that choice of a theory is underdetermined by any finite set of data. It is always possible to invent an unlimited set of theories, each one capable of explaining a given set of facts. Of course, many of these theories may seem implausible, but to speak of plausibility is to point to a role for scientific *judgment*: the relative plausibility of competing theories cannot be seen as residing in data which are equally well explained by all of them. Such judgments are intrinsic to theory choice, and clearly entail something more than a straightforward comparison of predictions with data. Furthermore, whilst one could in principle imagine that a given theory might be in perfect agreement with all of the relevant facts, historically this seems never to be the case. There are always misfits between theoretical predictions and contemporary experimental data. Again judgments are inevitable: which theories merit elaboration in the face of apparent empirical falsification, and which do not?

The second objection to the scientist's version is that the idea that experiment produces unequivocal fact is deeply problematic. At the heart of the scientist's version is the image of experimental apparatus as a 'closed', perfectly well understood system. Just because the apparatus is closed in this sense, whatever data it produces must command universal assent; if everyone agrees upon how an experiment works and that it has been competently performed, there is no way in which its findings can be disputed. However, it appears that this is not an adequate image of actual experiments. They are better regarded as being performed upon 'open', imperfectly understood systems, and therefore experimental reports are *fallible*. This fallibility arises in two ways. First, scientists' understanding of any experiment is dependent upon theories of how the apparatus performs, and if these theories change then so will the data produced. More far reaching than this, though, is the observation that experimental reports necessarily rest upon incomplete foundations. To give a relevant example, one can note that much of the effort which goes into the performance and interpretation of HEP experiments is devoted to minimising 'background' – physical processes which are uninteresting in themselves, but which can mimic the phenomenon under investigation. Experimenters do their best, of course, to eliminate all possible sources of background, but it is a commonplace of experimental science that this process has to stop somewhere if results are ever to be presented. Again a *judgment* is required, that *enough* has been done by the experimenters to make it probable that background effects cannot explain the reported signal, and such judgments can always, in principle, be called into question. The determined critic can always concoct some possible, if improbable, source of error which has not been ruled out by the experimenters.[5]

Missing from the scientist's account, then, is any apparent reference to the judgments entailed in the production of scientific knowledge – judgments relating to the acceptability of experimental data as facts about natural phenomena, and judgments relating to the plausibility of theories. But this lack is only apparent. The scientist's account avoids any explicit reference to judgments by *retrospectively adjudicating upon their validity*. By this I mean the following. Theoretical entities like quarks, and conceptualisations of natural phenomena like the weak neutral current, are in the first instance *theoretical constructs*: they appear as terms in theories elaborated by scientists. However, scientists typically make the realist identification of these constructs with the contents of nature, and then use this identification retrospectively to legitimate and make unproblematic existing scientific judgments. Thus, for example, the experiments which discovered the weak neutral current are now represented in the scientist's account as closed systems just because the neutral current is seen to be real. Conversely, other observation reports which were once taken to imply the non-existence of the neutral current are now represented as being erroneous: clearly, if one accepts the reality of the neutral current, this must be the case. Similarly, by interpreting quarks and so on as real entities, the choice of quark models and gauge theories is made to seem unproblematic: if quarks really are the fundamental building blocks of the world, why should anyone want to explore alternative theories?

Most scientists think of it as their purpose to explore the underlying structure of material reality, and it therefore seems quite reasonable for them to view their history in this way.[6] But from the perspective of the historian the realist idiom is considerably less attractive. Its most serious shortcoming is that it is retrospective. One can only appeal to the reality of theoretical constructs to legitimate scientific judgments when one has already decided *which* constructs are real. And consensus over the reality of particular constructs is the outcome of a historical process. Thus, if one is interested in the nature of the process itself rather than in simply its conclusion, recourse to the reality of natural phenomena and theoretical entities is self-defeating.

How is one to escape from retrospection in analysing the history of science? To answer this question, it is useful to reformulate the objection to the scientist's account in terms of the location of *agency* in science. In the scientist's account, scientists do not appear as genuine agents. Scientists are represented rather as passive observers of nature: the facts of natural reality are revealed through experiment; the experimenter's duty is simply to report what he sees; the theorist accepts such reports and supplies apparently unproblematic explanations of them. One gets little feeling that scientists actually *do* anything in their day-to-day practice. Inasmuch as agency appears anywhere in the scientist's account it is ascribed to natural phenomena which, by manifesting themselves through the medium of experiment, somehow direct the evolution of science. Seen in this light, there is something odd about the scientist's account. The attribution of agency to inanimate matter rather than to human actors is not a routinely acceptable notion. In this book, the view will be that agency belongs to actors not phenomena: scientists make their own history, they are not the passive mouthpieces of nature. This perspective has two advantages for the historian. First, while it may be the scientist's job to discover the structure of nature, it is certainly not the historian's. The historian deals in texts, which give him access not to natural reality but to the actions of scientists – scientific practice.[7] The historian's methods are appropriate to the exploration of what scientists were *doing* at a given time, but will never lead him to a quark or a neutral current. And, by paying attention to texts as indicators of contemporary scientific practice, the historian can escape from the retrospective idiom of the scientist. He can, in this way, attempt to understand the process of scientific development, and the judgments entailed in it, in contemporary rather than retrospective terms – but only, of course, if he distances himself from the realist identification of theoretical constructs with the contents of nature.[8]

This is where the mirror symmetry arises between the scientist's account and that offered here. The scientist legitimates scientific judgments by reference to the state of nature; I attempt to understand them by reference to the cultural context in which they are made. I put scientific practice, which is accessible to the historian's methods, at the centre of my account, rather than the putative but inaccessible reality of theoretical constructs. My goal is to interpret the historical development of particle physics, including the pattern of scientific judgments entailed in it, in terms of the dynamics of research practice. To explain how I seek to accomplish this, I will sketch out here some of the salient features of the development of HEP, and describe the framework I adopt for their analysis.[9]

The establishment of the quark–gauge theory view of elementary particles did not take place in a single leap. As we shall see, it was a product of the founding and growth of a whole constellation of experimental and theoretical research traditions structured around the exploration and explanation of a circumscribed range of natural phenomena. Traditions within this constellation drew upon different aspects of the quark–gauge theory picture of elementary particles, and, as they grew during the late 1960s and 1970s, they eventually displaced traditions which drew upon alternative images of physical reality. Seen from this perspective, the problem of understanding the establishment of quarks and gauge theory in the practice of the HEP community is equivalent to that of understanding the dynamics of research traditions. To see what is involved here, consider an idealised discovery process.

Suppose that a group of experimenters sets out to investigate some facet of a phenomenon whose existence is taken by the scientific community to be well established. Suppose, further, that when the experimenters analyse their data they find that their results do not conform to prior expectations. They are then faced with one of the problems of scientific judgment noted above, that of the potential fallibility of all experiments. Have they discovered something new about the world or is something amiss with their performance or interpretation of the experiment? From an examination of the details of the experiment alone, it is impossible to answer this question. However thorough the experimenters have been, the possibility of undetected error remains.[10] Now suppose that a theorist enters the scene. He declares that the experimenters' findings are not unexpected to him – they are the manifestation of some novel phenomenon which has a central position in his latest theory. This creates a new set of options for research practice. First, by identifying the unexpected findings with an attribute of nature rather than with the possible inadequacy of a particular experiment, it points the way forward for further experimental investigation. And secondly, since the new phenomenon is conceptualised within a theoretical framework, the field is open for theorists to elaborate further the original proposal.

One can imagine a variety of sequels to this episode, but it is sufficient to outline two extreme cases. Suppose that a second generation of experiments is performed, aimed at further exploration of the new phenomenon, and that they find no trace of it. In this case, one would expect that suspicion will again fall upon the performance of the so-called discovery experiment, and that the theorist's conjectures will once more be seen as pure theory with little or no empirical support. Conversely, suppose that the second-generation experiments do find traces which conform in some degree with expectations deriving from the new theory. In this case, one would expect scientific realism to begin to take over. The new phenomenon would be seen as a real attribute of nature, the original experiment would be regarded as a genuine discovery, and the initial theoretical conjecture would be seen as a genuine basis for the explanation of what had been observed. Furthermore, one would expect further generations of experimentation and theorising to take place, elaborating the founding experimental and theoretical achievements into what I am calling research *traditions*.

This image of the founding and growth of research traditions is very schematic. It typifies, nevertheless, many of the historical developments which we will be examining, and I will therefore discuss it in some detail. In particular, I want to enquire into the conditions of growth of such traditions. I have so far spoken as though traditions are self-moving; as though succeeding generations of research come into being of their own volition. Clearly this image is inadequate as it stands: research traditions prosper only to the extent that scientists decide to work within them. What is lacking is a framework for understanding the dynamics of practice, the structuring of such decisions. I want, therefore, to present a simple model of this dynamics which will inform my historical account. The model can be encapsulated in the slogan '*opportunism in context*'.

To explain what context entails, let me continue with the example of two research traditions, one experimental the other theoretical, devoted to the exploration and explanation of some natural phenomenon. It is, I think, clear that each generation of practice within one tradition provides a context wherein the succeeding generation of practice in the other can find both its *justification* and *subject matter*. Consider the theoretical tradition: to justify his choice to work within it, a theorist has only to cite the existence of a body of experimental data in need of explanation. And fresh data, from succeeding generations of experiment, constitute the subject matter for further elaborations of theory. Conversely, for the experimenter, his decision to investigate the phenomenon in question rather than some other process is justified by its theoretical interest, as manifested by the existence of the theoretical tradition. And each generation of theorising serves to mark out fresh problem areas to be investigated by the next generation of experiment. Thus, through their reference to the same natural phenomenon, theoretical and experimental traditions constitute mutually reinforcing contexts. Without in any way committing oneself to the reality of the phenomenon, then, one can observe that through the medium of the phenomenon the two traditions maintain a *symbiotic* relationship.

This idea of the symbiosis of research practice, wherein the practice of each group of physicists constitutes both justification and subject

10

matter for that of the others, will be central to my analysis of the history of HEP. I have explained it for the simple case of experimental and theoretical traditions structured around a particular phenomenon, because this is the archetype of many of the developments to be discussed. But it will also underlie my treatment of more complex situations – where, for example, many traditions interact with one another, or when the traditions at issue are purely theoretical ones. In itself, though, reference to context is insufficient to explain the cultural dynamics of research traditions. The question remains of why particular scientists contribute to particular traditions in particular ways. Here I find it very useful to refer to 'opportunism'. The point is this. Each scientist has at his disposal a distinctive set of resources for constructive research. These may be material – the experimenter, say, may have access to a particular piece of apparatus – or they may be intangible – expertise in particular branches of experiment or theory acquired in the course of a professional career, for example.[11] The key to my analysis of the dynamics of research traditions will lie in the observation that these resources may be well or ill matched to particular contexts. Research strategies, therefore, are structured in terms of the relative *opportunities* presented by different contexts for the constructive exploitation of the resources available to individual scientists.

Opportunism in context is the theme which runs through my historical account. I seek to explain the dynamics of practice in terms of the contexts within which researchers find themselves, and the resources which they have available for the exploitation of those contexts. It is, of course, impossible to discuss the entire practice of the HEP community at the micro-level of the individual, and when discussing routine developments within already established traditions I confine myself to an aggregated, macro-level analysis in terms of *shared* resources and contexts. As far as experimental traditions are concerned this creates no special problems. Resources for HEP experiment are limited by virtue of their expense; major items of equipment are located at a few centralised laboratories. Those facilities constitute the shared resources of HEP experimenters. The interest of theorists in particular phenomena likewise constitutes a shared context. If the facilities available are adequate to the investigation of questions of theoretical interest, one can readily understand that an experimental programme devoted to that phenomenon should flourish. Similarly, the data generated within experimental traditions constitute a shared context for theorists. But a problem arises when one comes to discuss the nature of shared theoretical resources: what is the theoretical equivalent of shared experimental facilities? One might look here to the shared material resources of theorists – the computing facilities, for example, which have played an increasingly significant role in the history of modern theoretical (as well as experimental) physics. But this would be insufficient to explain why quarks and gauge theory triumphed at the expense of other theoretical orientations within HEP. Instead I focus primarily upon the intangible resource of theoretical expertise. To explain how I use this concept let me briefly preview the main features of theory development which emerge from the historical account.

The most striking feature of the conceptual development of HEP is that it proceeded through a process of *modelling* or *analogy*.[12] Two key analogies were crucial to the establishment of the quark–gauge theory picture. As far as quarks themselves were concerned, the trick was for theorists to learn to see hadrons as quark composites, just as they had already learned to see nuclei as composites of neutrons and protons, and to see atoms as composites of nuclei and electrons. As far as the gauge theories of quark and lepton interactions were concerned, these were explicitly modelled upon the already established theory of electromagnetic interactions known as quantum electrodynamics. The point to note here is that the analysis of composite systems was, and is, part of the training and research experience of all theoretical physicists. Similarly, in the period we will be considering, the methods and techniques of quantum electrodynamics were part of the common theoretical culture of HEP. Thus expertise in the analysis of composite systems and, albeit to a lesser extent, quantum electrodynamics constituted a set of shared resources for particle physicists. And, as we shall see, the establishment of the quark and gauge-theory traditions of theoretical research depended crucially upon the analogical recycling of those resources into the analysis of various experimentally accessible phenomena.

In discussing the development of established traditions, then, my primary explanatory variables are the shared material resources of experimenters and the shared expertise of theorists. However, there remains the problem of accounting for the founding of new traditions and the first steps towards the establishment of new phenomena. Such episodes are not to be understood in terms of the gross distribution of shared material resources or expertise, but they pose no special problems because of that. I will try to show that these episodes are just as comprehensible as routine developments within established traditions, and are similarly to be understood. The only difference between my accounts of the development of traditions and of their founding is that to discuss the latter I move, perforce, from the macro-level of the group to the micro-level of the individual. I aim to show that the founding of new traditions can be understood in terms of the *particular resources and context* of the individuals concerned, just as the elaboration of those traditions can be understood in terms of the shared resources and contexts of the groups involved.[13]

Having outlined the opportunism-in-context model, we can now return to the problem posed earlier of how scientific judgments pertaining to experimental fallibility and theoretical underdetermination can be related to the dynamics of practice. Consider first the fallibility of experiment. Whilst it is possible to argue that all experiments are in principle fallible, experimenters do not enter this as an explicit caveat when reporting their findings: they simply report that, say, a particular quantity has been measured to have a particular value (often within a stated margin of uncertainty). It is then up to their colleagues to decide whether or not to challenge them, *and such decisions are related to decisions over future practice.* If the theorist can find the resources for a constructive analysis of the data, one should not expect him to spend long periods of time in searching for ways to challenge the experiment. If the data, through the medium of theory, raise new problems for experiment to investigate, then neither will experimenters be disposed to probe deeply. In this way, the potential fallibility of experiments is rendered *manageable.*[14] A parallel solution is available to the problem of the underdetermination of theory. While a multiplicity of different theoretical explanations may be conceivable for any set of data, not all of these will be attractive in terms of the dynamics of practice. Theorists may possess the appropriate expertise to articulate one or more in a constructive fashion, raising, say, new questions which experimenters can tackle; other theoretical proposals may be divorced from any significant pool of theoretical expertise and effectively meaningless; and yet others will lead to questions which experimenters cannot investigate with available techniques, and therefore fail to sustain a symbiosis between theory and experiment. A choice which is impossible to make on purely empirical grounds can thus be straightforward when construed in terms of the dynamics of practice.

The outline of the explanatory framework to be adopted here is now complete. The scientific judgments, which in the scientist's account are retrospectively legitimated by reference to the reality of theoretical entities and phenomena, will here be related to the dynamics of contemporary practice. That dynamics will be analysed as a manifestation of opportunism in context.[15] However, before closing this discussion, I want to draw attention to an important point which has been so far left implicit. I have stressed that the fallibility of experiment is rendered manageable through the symbiosis of experimental and theoretical research traditions, and this has a rather striking consequence. To say that all experiments are 'open' and fallible is to note that no experimental technique (or procedure or mode of interpretation) is ever completely unproblematic. Part of the assessment of any experimental technique is thus an assessment of whether it 'works' – of whether it contributes to the production of data which are significant within the framework of contemporary practice. And this implies the possibility, even the inevitability, of the 'tuning' of experimental techniques – their pragmatic adjustment and development according to their success in displaying phenomena of interest. If one takes a realist view of natural phenomena, then such tuning is both unproblematic and uninteresting, being simply a necessary skill of the experimenter, and the scientist's account correspondingly fails to pay any attention to it. But if one abstains from realism, tuning becomes much more interesting. Natural phenomena are then seen to serve a dual purpose. As theoretical constructs they serve to mediate the symbiosis of theoretical and experimental practice (and hence to make realist discourse retrospectively possible); and, at the same time, they sustain and legitimate the particular experimental practices inherent in their own production. To suspend my earlier strictures against the imputation of agency to inanimate objects for the sake of a symmetric formulation, one can speak of a symbiosis between natural phenomena and the techniques entailed in their production, wherein each confers legitimacy upon the other. Such a symbiosis is a far cry from the antagonistic idea of experiment as an independent and absolute arbiter of theory, and does, I think, call for more attention than it has so far received from historians and philosophers of science. Perhaps the best way to explore it is through detailed case-studies of individual experiments. Space permits the inclusion of only one such study here: the set-piece analysis of the discovery of the weak neutral current, in Section 6.5. But throughout the book I point to instances of technical tuning in less detailed analyses of individual experiments, and in the closing chapter I return to a discussion of its implications.[16]

To round off these introductory remarks, let me briefly review the major historical developments discussed in the following chapters. The story to be told is of the founding and growth of research traditions structured around the quark concept. The account accordingly focuses upon the period from 1964, when quarks were first invented, to 1980, when quark-physics traditions had come to dominate research practice in HEP.[17] Within that period it is useful to distinguish between two constellations of symbiotic research traditions which I will refer to as the 'old physics' and the 'new physics'. The old physics dominated practice in HEP throughout the 1960s and was distinguished by a 'common-sense' approach to the subject matter of the field. By this I mean that experimenters concentrated their efforts upon the phenomena most commonly encountered in the laboratory, and theorists sought to explain the data so produced. Amongst the theories developed in this period were early formulations of the quark model, as well as theories related to the so-called 'bootstrap' conjecture which explicitly disavowed the existence of quarks. The old physics thus consisted of traditions of common-sense experiment, devoted to the investigation of the most conspicuous laboratory phenomena, which sustained and were sustained by both quark and non-quark traditions of theorising. It should be noted that gauge theory did not figure in any of the dominant theoretical traditions of the old physics. By the end of the 1970s the old physics had been almost entirely displaced by the new. Experimentally, the new physics was 'theory-oriented': experimenters had come to eschew the common-sense approach of investigating the most conspicuous phenomena, and research traditions focused instead upon certain very rare processes. Theoretically, the new physics was the physics of quarks and gauge theory. Once more, theoretical and experimental research traditions reinforced one another, since the rare phenomena on which experimenters focused their attention were just those for which gauge theorists could offer a constructive explanation.

Thus there were elements of continuity between the old and new physics: the quark concept, for example, was carried over from one to the other. But there were also discontinuities. The non-quark, bootstrap approach to theorising largely disappeared from sight, and the quark concept became embedded in the wider framework of gauge theory. And, more strikingly, experimental practice in HEP was almost entirely restructured in order to explore the rare phenomena which were at the heart of the new physics. Thus the transformation between the old and new physics and the consequent establishment of the quark–gauge theory world-view involved much more besides conceptual innovation. It was intrinsic to the transformation that particle physicists' way of interrogating the world through their experimental practice was transformed too. This is one of the most fascinating features of the history of HEP and I will try to explain how and why it came about. One wonders whether other great conceptual developments in the history of science – the establishment of quantum mechanics, for example – were associated with similar shifts in experimental practice. Unfortunately, much historical writing tends to reflect the scientist's retrospective realism, regarding experiment as unproblematic and therefore uninteresting, and it is hard to answer such questions at present.

The historical account, then, focuses upon the origins of the quark concept in the old physics and upon the establishment of the new physics of quarks and gauge theory, paying particular attention to the transformation between the old and the new. The structure of the narrative reflects these preoccupations. Part I, comprising this and the two following chapters, aims to delineate the context within which the history of quark physics was enacted. Chapter 2 gives some statistics on the size and composition of the HEP community, discusses the general features of HEP experiment, and outlines the post-1945 development of experimental facilities in HEP. Chapter 3 sketches out the growth of the major traditions of HEP theory and experiment between 1945 and 1964, the year in which quarks were invented. It thus sets the scene for the intervention of quarks into the old physics.

Part II begins the story proper. It covers the period from 1964 to 1974, ten years which saw the old quark physics established and the embryonic new physics traditions founded. Chapter 4 discusses the two earliest formulations of the quark model and the relationships of these with common-sense traditions of experiment. Chapter 5 discusses the founding of the first new-physics traditions: the experimental discovery in 1967 of a new phenomenon – scaling – and its explanation in terms of a third variant on the quark theme, the quark–parton model. Chapter 6 reviews the first impact of gauge theory upon the experimental scene. This came with the 1973 discovery of another new phenomenon, the weak neutral current, the existence of which had been predicted on the basis of unified electroweak gauge theory. Chapter 7 covers the development of a gauge theory of the strong interactions. This was quantum chromodynamics, which promised to underwrite the quark–parton model, and hence to make contact with experimental work on scaling phenomena. Chapter 8 concludes Part II and has three objectives. First, it summarises the conceptual bases of electroweak gauge theory and QCD in preparation for Part III. Secondly, it provides an overview of the topics of active HEP research interest in 1974, emphasising that gauge theory had had relatively little impact at this stage: the old physics – of both quark and non-quark varieties – still dominated experimental and theoretical HEP. Thirdly, Chapter 8 discusses three transitional biographies, seeking to analyse why theorists who grew up (professionally) in the old-physics era should have appeared at the forefront of the developments of Part III.

Part III of the account deals with the establishment of the quark–gauge theory world-view in the period from 1974 to 1980. Its subject is the growth and diversification of the experimental and theoretical traditions of the new physics. Chapter 9 discusses a crucial episode in the history of the new physics: the discovery of the 'new particles' and their theoretical explanation in terms of 'charm'. The discovery of the first of the new particles was announced in November 1974 and by mid-1976 the charm explanation had become generally accepted. This was 'the lever that turned the world' in terms of the transformation between the old and the new physics. Theoretically, the existence of quarks came to be regarded as unproblematically established (despite the continuing failure of experimenters to observe isolated quarks in the laboratory), a specific electroweak gauge theory was singled out for intensive investigation, and QCD was brought most forcibly to the attention of particle physicists. On the experimental plane, charmed particles became a prime target for investigation, and the new interest in gauge theory provided a context in which all of the traditions of new-physics experiment could flourish, eventually to the exclusion of all else. Chapter 10 reviews developments concerning electroweak theory in the latter half of the 1970s. More new particles were discovered, pointing to an expanded ontology of quarks and leptons; detailed experimentation on the weak neutral current culminated in consensus in favour of 'the standard model' – the simplest and prototypical electroweak theory. Chapter 11 reviews the theoretical development of QCD in symbiosis with a range of experimental traditions. Chapter 12 then examines the growth of the new physics from the perspective of experimental practice. It aims to show that by the late 1970s the phenomenal world of the new physics had been built into both the present and future of experimental HEP; the world was, in effect, defined by experimenters to be one of quarks and leptons interacting as gauge theorists said they should. Quantitative data on the new-physics takeover of experimental programmes are presented. The transformation of experimental hardware in this takeover is discussed. And the predication of planning for future major facilities on the new-physics world-view is documented.

By the end of the 1970s, the new physics had established a stranglehold upon HEP. Gauge theories of quarks and leptons dominated contemporary practice; the same theories constituted particle physicists' visions of the future and, via the realist assump-

17

tion, their understanding of the past. In this sense, the story is complete with Chapter 12. But gauge theorists did not rest content with electroweak theory and QCD, and one further substantial chapter is included. Chapter 13 discusses the synthesis of electroweak theory with QCD in the so-called Grand Unified Theories or GUTs, and outlines the implications for experiment drawn from GUTs in the late 1970s. Here we shall see gauge theory permeating the cosmos in a symbiosis between HEP theorists, cosmologists and astrophysicists, and we shall see experimenters disappearing down deep mines in search of unstable protons. Finally, Chapter 14 summarises the overall form of the narrative, and suggests that the history of HEP should be seen as one of the social production of a culturally specific world.

NOTES AND REFERENCES

1 Schilpp (1949, 684).

2 A representative selection of popular articles by leading particle physicists are cited in the following chapters. Book length popular accounts have been given by Segrè (1980) and Trefil (1980). Many historical reviews appearing in the professional HEP literature are cited in the text.

3 In referring to the 'scientist's account' I do not wish to suggest that all scientists produce such accounts nor that their production is solely confined to scientists. I mean only to suggest that such accounts are routinely produced and defended by members of the scientific community.

4 For a variety of philosophical perspectives on the problems at issue here, see Duhem (1954), Feyerabend (1975), Hanson (1958), Harding (1976), Hesse (1974), Kuhn (1970) and Quine (1964).

5 This is illustrated in many recent case-studies. See, for example, Collins (1975a, 1981a), Collins and Pinch (1982), Knorr et al. (1981), and Pickering (1981a, 1981b, 1984).

6 The scientific habit of rewriting the history of science from the standpoint of currently accepted knowledge (and the pedagogical function of this activity) was first remarked on by Kuhn (1970). Latour and Woolgar (1979) suggest that such rewriting entails a two-stage process of 'splitting' and 'inversion'. Scientists first argue about the meaning and significance of their research in terms of the nature of reality. If and when a consensus is reached, splitting can take place: particular theoretical constructs become regarded as pre-existing attributes of the natural world, independent of the particular arguments and practices implicit in their establishment. Inversion is then possible: the reality of the theoretical constructs is used to explain the validity of these arguments and practices, and to indicate the invalidity of any arguments or practices which support an alternative construction of reality.

7 By texts, I refer both to standard historical sources – published and

unpublished scientific papers, scientific correspondence, laboratory notebooks and so on – and to interview material, which I have found very useful in constructing this account.

8 For more discussion of these methodological remarks, see Collins (1981b).

9 I should note here that throughout the narrative I will refer to the scientist's account of events as a foil for my own analysis. Philosophers may accuse me of assaulting a straw man. The scientist's version, they might argue, is a form of naive realism – appropriate to the unreflective practice of science, perhaps, but not a view which any philosopher would care to defend. My response would be that the thrust of my analysis is that the development of science should be understood in sociological terms, and almost any philosophical position would serve as a foil for this argument. I choose to take issue with the naive realist view of science which is at least widely held, even if philosophically disreputable. I might add that while most academic observers would acknowledge the failings of naive realism, the habit of putting the phenomena first – the attribution of agency to theoretical constructs – is hardly conspicuous by its absence from contemporary writing in the history and philosophy of science.

10 For interesting reflections on this point from a practising HEP experimenter, see Deutsch (1958).

11 By referring to expertise, I want to point to the fact that science is not purely an articulated body of knowledge. The articulated knowledge of the community is sustained by, and developed in the context of, a wealth of unverbalised theoretical and experimental skills unevenly distributed amongst its members. For more discussion of this point and its consequences, see Kuhn (1970), Polanyi (1973) and Collins (1974, 1975b).

12 The importance of analogy in theory development has been discussed by many authors. For a philosophical discussion, see Hesse (1974). Masterman (1970) has identified Kuhn's (1970) notion of an 'exemplar' with that of a crude analogy. Knorr (1981) gives a sociologically oriented discussion. The utility of analogy in analysing experimental as well as theoretical practice is shown in Schon (1969). For applications to the history of HEP, see Pickering (1980, 1981c).

13 Since such formulations seem conducive to accusations of idealism in certain quarters, I should make my position clear. I have no wish to deny reality – in the shape of experimental data – a role in the development of scientific knowledge. As I hope is clear in the historical account, experimental data are at the very heart of scientific argument over the reality of theoretical constructs. However, these arguments can encompass a variety of positions which call into question the *meaning* and *validity* of the data, and it seems therefore impossible to regard the data *alone* as forcing a particular outcome to the debate. Only when seen within a field of practice can the force ascribed to given data be understood.

14 Let me reiterate that this is a point of principle, and is most certainly not an imputation of theoretical bias or bad faith either to individual experimenters or to the scientific community in general. Given that all experiments are fallible, even those most conscientiously performed, it seems inevitable that experimental adequacy should ultimately be assessed in terms of the data produced.

15 In regarding the dynamics of practice rather than natural phenomena as the motor of scientific change, the present account falls within the tradition of 'constructivist' approaches to the history and sociology of science. For other extended case-studies within this tradition, see Fleck (1979), Knorr-Cetina (1981), Latour and Woolgar (1979) and MacKenzie (1981). For more general discussions of the approach, see Barnes (1977, 1982) and Bloor (1976). Shapin (1982) provides an extensive review of the historical literature from this perspective. Barboni (1977) offers a well documented analysis of the dynamics of practice in experimental HEP, but does not seek to relate his analysis to the content of the knowledge produced.

16 Many of the detailed case studies listed above illustrate the tuning of experimental techniques within a theoretically-defined context, for example, Collins (1975a), Pickering (1981a, 1984) and Fleck (1979). Further studies which focus on this issue are Galison (1982), Gooding (1982), Holton (1978) and Shapin (1979).

17 The designation of 1980 as a cut-off date for this study is somewhat arbitrary. Many of the developments to be covered were essentially complete by 1978, and that will provide a natural point for termination of much of the account. However, certain important developments concerning 'grand unification' only got under way in the late 1970s, and I will follow these into the 1980s.

2

MEN AND MACHINES

This chapter sets the scene for the historical drama to follow. Section 1 discusses the demography of the HEP community. Section 2 reviews the basic hardware of HEP experiment and the distinctively different patterns of its use in the old and the new physics. Section 3 outlines the chronology of accelerator construction, relating this to the fluctuating fortunes of old and new physics traditions of experimental research.

2.1 The HEP Community

HEP has always been Big Science. In comparison with many other branches of pure research its expense is enormous. In 1980, for example, the annual expenditure of CERN, the major European HEP laboratory, was nearly 600 million Swiss Francs, and governmental support for HEP in the United States ran to $343 million.[1] This is a field in which only the rich can compete. The history of experimental HEP has accordingly been dominated by the United States and Europe. Outside those areas, the only significant experimental programme has been that of the Soviet Union, which has, in general, lagged behind that of the West.[2] The story to be told here is thus largely one of developments on either side of the North Atlantic.[3]

HEP emerged as a recognisable scientific specialty shortly after the end of World War II and, with Europe devastated, the USA quickly took the lead. However, the Europeans made determined attempts to overhaul the Americans, and by the early 1960s had achieved rough comparability in manpower. In 1962, for example, there were estimated to be 685 practising elementary-particle physicists in Europe and 850 in the USA.[4] These research communities grew continuously during the 1960s before stabilising in the early 1970s as governments began to resist the physicists' ever-increasing requests for funds.[5] By this stage, Europe had achieved pre-eminence in manpower and in the late 1970s the 3,000 European particle physicists easily outnumbered their 1,650 American colleagues.[6]

Throughout this book, then, we will be dealing with a cast of thousands, concentrated in the economically advanced countries but with representatives sprinkled all over the world. This cast, predom-

21

inantly white and almost entirely male, can be subdivided in various ways. One important distinction is that between theorists and experimenters. Theory and experiment constitute distinct professional roles within HEP. Each form of practice is highly technical, drawing upon quite different forms of expertise, and it is rare to find an individual who successfully engages in both.[7] During the 1970s, there were around two experimenters for every theorist.[8] A second axis of division relates to the subject matter of HEP research. As mentioned in the preceding chapter, particle physicists recognise four fundamental forces of nature: the strong, electromagnetic, weak and gravitational interactions. Amongst these, the electromagnetic interaction has been regarded as well-understood since the 1950s, while the gravitational force is too weak to have a measurable effect upon elementary-particle phenomena. HEP research has therefore primarily aimed at investigation of the strong and weak interactions, and studies of these forces have constituted two relatively distinct fields of practice: different experimental and theoretical strategies were considered appropriate to each. This was especially true in the era of the old physics, when a sizeable majority of particle physicists concentrated their energies upon the strong rather than the weak interactions.[9] In the context of the new physics the distinctions in practice between research into the strong and weak interactions remained significant but became blurred. Theoretically, a unified framework for the strong, weak and electromagnetic interactions was developed; and many new physics experiments produced data relevant to theoretical models of both the strong and weak interactions.

Finally, a comment upon the institutional locus of HEP research is needed. Particle physics is a pure science, with no evident practical application for the knowledge it produces. Its institutional base lies, as one would expect, in the universities, and this is where most theorists live and work. However, the expense entailed in building and running experimental facilities has increased continuously during the history of HEP, and there has been a corresponding tendency to divorce these resources from individual universities in favour of regional, national and international laboratories. Teams of experimenters, usually drawn from several universities, assemble at such laboratories to perform specific experiments, often dispersing again to their home institutions to analyse their data.[10] The laboratories typically also have their own permanent staff of researchers comprising both an experimental group and a theory group.[11] What goes on within the laboratories is the subject of the next section.

2.2 HEP Experiment: Basics

Although technologically extremely complex and sophisticated, experiments in HEP are in principle rather simple. As indicated in Figure 2.1, one takes a beam of particles and fires it at a target. Interactions take place within the target – some of the beam particles are deflected or 'scattered'; often additional particles are produced – and the particles that emerge are registered in detectors of various kinds.

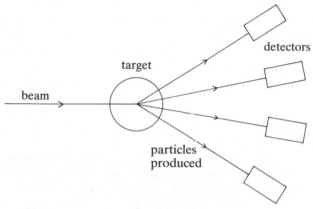

Figure 2.1. Layout of an HEP experiment.

Hardware

Beam, target and detectors are the three important components in an HEP experiment, and we can discuss them in turn. Historically, a variety of methods have been used for producing particle beams, but in the post-war era the workhorse of HEP has been the synchrotron.[12] A synchrotron consists of a continuous evacuated pipe in the form of a circular ring – a stretched-out doughnut – which is encased in a jacket of electromagnets. A bunch of stable particles – either protons or electrons – is injected into the ring at low energy, and the magnetic fields produced by the magnets are so arranged that the particles move in closed orbits around the ring. At a given point in each orbit the particles are given a radio-frequency 'kick' to increase their energy, and this kick is repeated until the particles within the machine attain the energy required. They are then extracted from their orbits and directed as a high-energy beam towards the experimental areas. One might imagine that beams of arbitrarily high energy could be produced by the administration of a sufficient number of kicks, but this is not the case. Every machine has an upper energy limit dictated by its radius and the maximum strength of its electromagnets. As we shall see in Section 3, the trend in accelerator construction has been towards ever-higher maximum energies, attained by machines of ever-increasing size. The big machines of the 1950s had radii around 10 to 20 metres, and could be housed in a covered building; by the 1960s they had grown to around 100 metres, and were buried underground; by the 1970s radii had reached a kilometre; and Europe's projected big machine for the 1980s, LEP, will have a radius of 4 kilometres.[13]

The stream of particles ejected from an accelerator is known as the primary beam. It can be used directly for experiment, or it can be used to generate secondary beams. In the latter case, the primary beam is directed upon a metal target where it creates a shower of different kinds of particles. Secondary beams of the desired species of particle are then singled out from the shower using appropriate combinations of electric and magnetic fields. In this way, beams of particles which are not amenable to direct acceleration – either because they are unstable and would disintegrate during the acceleration process, or because they are electrically neutral and hence immune to the electric and magnetic fields used in acceleration – are made available for experiment. The chosen beam, primary or secondary, is brought to bear upon the experimental target, and the interactions which take place within the target are interpreted as those between the beam particles and the target nuclei. Liquid hydrogen is often the target of choice, since this constitutes a dense aggregate of the simplest nuclei – single protons – and thus facilitates the interpretation of data. When studying rare processes, however, interaction rate rather than simplicity of interpretation may be a prime consideration. In this case a heavy metal target may be used, in order to present as many nucleons (neutrons and protons) to the incoming beam as possible.

In a typical beam–target interaction several elementary particles are produced (the number increases with beam energy) and emerge to pass through detectors. A common characteristic of HEP detectors is that they are only sensitive to electrically charged particles, since in different ways they all register the disruption produced by such particles in their passage through matter. Electrically neutral particles can only be detected indirectly, either by somehow converting them to charged particles and detecting the latter or, inferentially, by consideration of the energy imbalance between the incoming beam and the outgoing charged particles. Many different kinds of detectors have been used in HEP over the years. The details of these need not concern us here, but a few words on the general tactics of particle detection may be useful. Detectors can be classified into two groups – visual and electronic – and we can discuss examples of each in turn. Since its invention in 1953, the pre-eminent visual HEP detector has been the bubble chamber.[14] A bubble chamber is a tank full of superheated liquid held under pressure. When the pressure is released, the liquid begins to boil. Bubbles form preferentially along the trajectories of any charged particles which have recently traversed the chamber. Thus lines of bubbles mark out particle tracks, which can be photographed and recorded for subsequent analysis. Figure 2.2 reproduces a bubble-chamber photograph, where the particle tracks stand out as continuous white lines (the large white blobs are reference marks on the walls of the chamber).

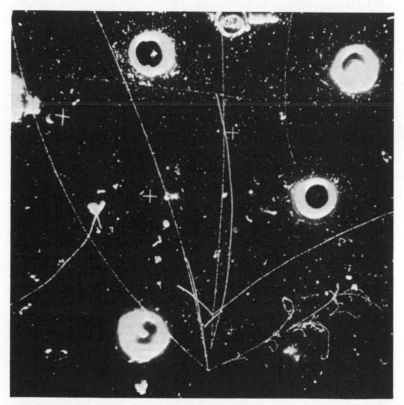

Figure 2.2. Bubble-chamber photograph.

In HEP experiments the bubble chamber serves as both target and detector. The chamber is placed in the path of a particle beam and repeatedly expanded and repressurised. At each expansion, the tracks within the chamber are photographed, yielding, in a typical experiment, many thousands of pictures of beam–target interactions or 'events'. As noted above, the most popular filling for bubble chambers is liquid hydrogen, but when higher interaction rates are sought a heavier liquid (such as freon) may be substituted.

Bubble chambers, and other visual detectors, have two principal advantages for experimental HEP. First, the visible tracks give direct access to elementary particles – if one forgets about philosophical niceties, one can imagine that one actually sees the particles themselves. And secondly, they generate a lot of data. A bubble chamber in a particle beam can record events much faster than they can be analysed. Historically, this was an important factor in the rapid post-war growth and diffusion of HEP in universities around the world. It meant that university scientists with no accelerator at their home institution could visit an accelerator laboratory for a short time, collect a lot of data, and then return to base for months of profitable analysis. However, the productivity of visual detectors is a mixed blessing. They generate a lot of data because they are indiscriminate: they register everything that takes place within them, whether it is interesting or not. Much of the analysis of track data therefore consists of separating wheat from chaff: isolating phenomena of interest from uninteresting background.

Electronic detectors minimise the problem of separating wheat from chaff because they are discriminating. A discriminating detector is one which can be 'triggered' – programmed to decide whether to record what is taking place within it as interesting, or to discard the event as uninteresting. High-energy particles travel at almost the speed of light and traverse any detector array in a tiny fraction of a second. Thus the time-scale for triggering is of the order of nanoseconds (10^{-9} seconds). The only way to process information so quickly is electronically, and the requirement of a discriminatory detector therefore translates into that of a detector whose output is an electrical signal. The first device to satisfy this criterion was the scintillation counter, which came into use in HEP in the late 1940s and has since remained an important weapon in the HEP experimenter's arsenal.[15] Scintillators are materials which emit a flash of light when struck by a charged particle. In his pioneering work on radioactivity, Rutherford had counted alpha-particles (helium nuclei) by eye by observing flashes on a scintillating screen of zinc sulphide. But scintillators only became useful for the purposes of HEP when they were conjoined with the newly developed photo-multiplier tube (PMT). PMTs register flashes of light and amplify them electronically. Their output is an electronic pulse which can be fed directly into fast

electronic logic circuits, and thus arrays of scintillation counters – scintillating materials monitored by PMTs – nicely fulfil the
conditions for discriminating detectors. Spark chambers took their
place alongside scintillation counters in the late 1950s.[16] While the
latter transform an optical signal into an electrical pulse, the former
generate electrical signals directly. Descendants of the pre-war
Geiger-Muller counter, the basic element of spark chambers is a pair
of wires (or metal plates) to which a high voltage is applied. Charged
particles traversing the gaseous medium of the chamber precipitate
sparking between the wires. This generates an electrical pulse which
can be fed into amplification and logic circuits in much the same way
as the output of scintillation counters. During the 1970s highly
sophisticated descendants of the spark chamber – multi-wire proportional chambers, drift chambers and streamer chambers – played an
increasingly important role in experimental HEP.[17] They offered
high-precision measurements conjoined with refined triggering capability, and threatened to make the bubble chamber redundant. Figure
2.3 shows the output from an electronic detector of the early 1980s.
The white dots represent the firing of individual counters, but the eye
readily reconstructs them as continuous particle tracks.

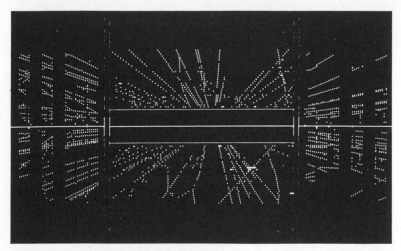

Figure 2.3. Output from an electronic detector array.

One last component of HEP hardware remains to be discussed: the
particle collider. The particle accelerators so far mentioned are
fixed-target machines – their beams are directed upon experimental
targets at rest in the laboratory. Colliders have no such fixed targets.

They store counter-rotating beams of particles and clash them together. In colliders, the target is itself a beam of particles: two counter-rotating beams of particles are stored within a single ring or within two interlaced rings, and interactions take place at the predetermined points where the beams cross one another. The advantage which colliders enjoy over fixed-target machines is the following. In fixed-target experiments only a fraction of the beam energy – known as the centre-of-mass (CM) energy – is available for physically-interesting processes; the remaining energy is effectively locked up in the overall motion of the beam-plus-target system relative to the laboratory. In a collider this is not the case. Because the two beams collide head on, the beam–target system is stationary relative to the laboratory, and all of the beam energy is CM energy. Thus CM energy in colliders grows linearly with beam energy (it is just twice the energy of a single beam) while, according to the theory of special relativity, CM energy in fixed-target experiment grows only as the square-root of beam energies. Maximum beam energy is a major determinant of accelerator cost, and colliders therefore offer a very cost-effective route to high CM energies. As an experimental tool, however, colliders are not without certain disadvantages. In fixed-target experiments, one fires a tenuous beam at a dense target; in colliders one fires a tenuous beam at another tenuous beam, and the reaction rate is correspondingly reduced. Broadly speaking, therefore, the collection of comparable data requires considerably more time and effort at a collider than at a fixed-target machine. Furthermore colliders are limited to the investigation of interactions between stable, electrically charged particles – electrons, protons and their antiparticles – and offer a highly restricted range of experimental possibilities relative to fixed-target machines with their wide variety of secondary beams.[18]

Experiment

We can now turn from the hardware of HEP experiment to its substance. What do HEP experimenters seek to find out, and how do they assemble their resources to do so? A typical HEP experiment measures 'cross-sections'. Cross-sections are quantified in 'barns', units of area ($1 \text{ barn} = 10^{-24} \text{ cm}^2$) which represent the effective area of interaction between beam and target particles. Cross-sections are best thought of as measures of the relative probabilities of different kinds of events: a high cross-section corresponds to a highly probable kind of event, a low cross-section to a rare one. Cross-section data constitute the empirical base of particle physics. For future reference, it will be useful here to introduce some technical distinctions between different types of cross-section. The *total* cross-section (denoted 'σ_{tot}') for the interaction of two species of particles measures the overall probability that they will interact together in any way. In self-evident notation, it refers to the process AB→X where A and B are the beam and target particles, and X represents all conceivable products of their interaction. Total cross-sections can be regarded as sums of various partial cross-sections defined by specifying X. The most basic partial cross-sections are *exclusive* cross-sections, in which X is fully specified. Thus one might measure the exclusive cross-section for the process AB→AABBB, in which one A and two B particles are produced. Exclusive cross-sections can be further divided into *elastic* and *inelastic* cross-sections. Elastic cross-sections refer to processes in which no new particles are produced: AB→AB; inelastic cross-sections refer to processes in which new particles are produced: AB→AAAB, for example. Detailed measurements on inelastic processes in which many particles are produced are very difficult, and in this case experimenters often measure *inclusive* cross-sections. In such measurements X is only partially specified by, say, insisting that it contains at least one A particle: AB→AZ, where no attempt is made to identify the particle or particles comprising Z.

Cross-sections are functions of the CM energy at which they are measured and of the energies and momenta of the particles detected. And, in principle, one could imagine that HEP experimenters would set out to measure *all* conceivable cross-sections. They would take beams of different types of particles at all available energies, fire them at a various targets, and make detailed measurements on every single particle which emerged. One could equally well imagine that this programme of comprehensive enquiry would take an indefinite length of time. It would also be highly inefficient, since an overwhelming proportion of the data would be irrelevant to contemporary theory and effectively meaningless. HEP experiment, like experiment in all branches of science, is therefore systematic: it seeks to measure cross-sections which the experimenters consider to be potentially meaningful. The important parameters which the HEP experimenter has at his disposal in pursuit of meaningful data are the nature of the beam, the target and the detector array. Within technological limits, he can choose a beam of specified energy and intensity containing a specified species of particle; he can choose the target particles – different atomic nuclei or (in the case of colliders), electrons, positrons, protons and antiprotons – and, through choice of detectors, their configuration in space, and electronic logic circuits, he can attempt to register the production of predetermined species of particles emerging from the target at given angles and with given momenta. At different periods in the history of particle physics, different patterns of choice of beams and detectors have dominated experimental practice. These patterns were characteristic of first the old and later the new physics, and I want now to review the gross differences between them (more details will follow in later chapters).

Accelerators capable of providing beams of ever-increasing energy were built in the post-war period, and each new machine opened up new areas for experiment. As one might expect, experimenters concentrated initially upon exploring the processes most commonly encountered in the laboratory – processes having large cross-sections. By definition, such processes are those mediated by the strong interactions, and beams of strongly interacting particles – hadrons – were accordingly favoured in old-physics experiments. Leptons experience only the electromagnetic and weak interactions, and their interaction cross-sections are one thousand or more times smaller than those of hadrons.[19] Lepton-beam experiments therefore played only a minor role in the old physics. In the new physics, the roles were reversed. The earliest new-physics traditions of experiment were based on the use of lepton beams. Experiments with hadron beams did persist into the new physics, but now detectors were redesigned to emphasise rare, low cross-section processes quite different from those explored in the old physics. The old-physics experiments had established that most hadronic reactions were 'soft': most particles emerged from the interaction region with only a small momentum component transverse to the beam axis. It was as though beam particles typically struck the target nuclei a glancing blow, suffering only a slight deflection. Cross-sections were found to fall off exponentially fast with transverse momentum both for the scattering of beam particles and for the production of new particles. Old-physics experiments accordingly concentrated upon exploring the details of high cross-section soft-scattering processes. The new physics, in contrast, largely abandoned this pursuit, and focused instead upon very rare 'hard-scattering' phenomena in which particles emerged with high transverse momenta.

Experimenters sought to isolate such high-transverse-momentum particles through appropriate detector configurations, which are most readily described in connection with proton–proton collider research. In such machines, proton beams collide head-on. The centre-of-mass system is at rest in the laboratory and this makes it particularly easy to distinguish experimentally between hard and soft scattering. Soft scattering corresponds to small transverse momenta, and hence to the emergence of particles from the interaction region at small angles to the beam axis. To investigate soft scattering at colliders, therefore, experimenters set up their detectors as close to the beam axis as possible – as represented by detector A in Figure 2.4. In contrast hard scattering corresponds to the emergence of fast particles at large angles to the beam axis. To investigate hard scattering, experimenters simply moved their detectors to a point at right-angles to the interaction region relative to the beam axis (and arranged only to count high-energy particles) as represented by detector B in Figure 2.4. In time, hard scattering became an important new physics tradition at fixed-target machines as well as at colliders, but at the former it is much more difficult to pick out the relevant events from the enormous soft-scattering background. Because the beam–target system is in motion in fixed-target experiments, all of the produced and scattered particles move in the general direction of the incoming beam. Simple changes in geometry are therefore insufficient to pick out hard-scattering events: more subtle arrangements of detectors, electronic logic and analysis are required which I will not discuss here.

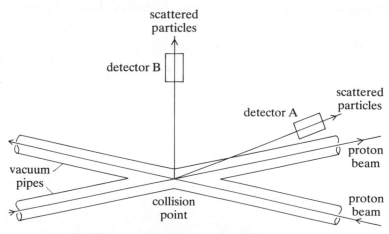

Figure 2.4. Detector configurations in a proton–proton collider experiment: detector A registers low-transverse-momentum soft scatters; detector B registers high-transverse-momentum hard scatters.

From the experimental point of view, then, two key differences between the new physics and the old physics which preceded it were: an emphasis on the use of lepton rather than hadron beams; and, within hadron-beam physics, an emphasis on hard rather than soft scattering achieved through novel combinations of detectors. But these were differences *within* the experimental programmes conducted at fixed-target accelerators and proton–proton colliders, while perhaps the most striking correlate of the transition to the new physics lay in the importance which became attached to experiment at a third class of machines: electron–positron colliders. Electrons and their antiparticles, positrons, are electrically-charged leptons. Their interaction cross-sections have the relatively small values characteristic of electromagnetic processes, and electron–positron experiment played at most a marginal role in the old physics. In contrast, as discussed below, in the latter half of the 1970s electron–positron colliders came to be seen as new physics laboratories *par excellence*.

2.3 Accelerator Chronology

We can turn now from the generalities of HEP experiment to particulars. This section reviews the history of the construction of the big machines of HEP, accelerators and colliders, and discusses how their use was intertwined with the old and new physics. Table 2.1 lists the major particle accelerators built or planned in the post-war era.[20]

Table 2.1. Particle accelerators

Accelerator	Date of first operation	Particles accelerated	Beam energy (GeV)	CM energy (GeV)
Cosmotron, Brookhaven	1952	Protons	3	2.8
Bevatron, Berkeley	1954	Protons	6.2	3.5
Dubna	1957	Protons	10	4.5
CERN PS	1959	Protons	28	7
AGS, Brookhaven	1961	Protons	33	8
CEA, Cambridge, MA	1962	Electrons	6	3.5
ZGS Argonne	1963	Protons	13	5
NIMROD, Rutherford Lab.	1963	Protons	7	3.7
DESY, Hamburg	1964	Electrons	7	3.8
NINA, Daresbury	1966	Electrons	5	3.2
SLAC, Stanford	1966	Electrons	22	7
Yerevan, USSR	1967	Electrons	6	3.5
Cornell	1967	Electrons	12	5
Serpukhov	1967	Protons	76	12
Fermilab, Chicago	1972	Protons	500	32
CERN SPS	1976	Protons	450	30
KEK, Tsukuba, Japan	1977	Protons	12	5
Tevatron, Fermilab	1985	Protons	1000	43
UNK, Serpukhov	Late 1980s	Protons	3000	76

Proton Accelerators

All of the machines listed in Table 2.1 as accelerating protons are proton synchrotrons (PSS). The first major PS came into operation at the first of the national HEP laboratories: the Brookhaven National Laboratory (BNL) on Long Island, USA. BNL was primarily intended to serve the universities of the East Coast, and became a centre for HEP experiment in 1952 with the inception of the Cosmotron.[21] The Cosmotron produced a 3 GeV primary beam,[22] and was the first accelerator to have sufficient energy to produce the so-called 'strange' particles in the laboratory. Strange particles had hitherto only been observed in experiments with cosmic rays (high-energy particles impinging on the earth from space). The Cosmotron beam was more intense than its naturally occurring equivalent, and amenable to human control as cosmic rays are not, and thus investigation of strange particles passed from the province of cosmic-ray physics to what one would now call HEP proper. The transfer of attention from tenuous and uncontrollable cosmic rays to accelerator experiments, begun at Brookhaven, was essentially complete by the early 1960s, and cosmic-ray physics will not figure in the subsequent account.[23]

Accelerator physics truly came into its own in 1954 with the completion of the 6 GeV Bevatron at Berkeley, California – the university laboratory where E.O. Lawrence had led the world in pre-war accelerator development.[24] The Bevatron was designed to have sufficient energy to be the first machine artificially to produce antiprotons (the antiparticles of the proton) and this it dutifully did. To physicists' surprise it did much more besides. Beginning at Berkeley, more and more quite unexpected, unstable strongly interacting particles or 'resonances' were discovered.[25] Before the Bevatron, the known elementary particles could be counted on one's fingers – by the time the 'population explosion' of hadronic resonances had run its course there were hundreds. The population explosion had, by the early 1960s, transformed much of the HEP landscape. From the experimenter's point of view, it revealed strong-interaction physics to be an extraordinarily rich field of enquiry, and confirmed the PS as the prime experimental tool. On the theoretical side, while it had been relatively easy to accept a handful of particles as truly fundamental, it was not easy to maintain such a position when the list got longer every day. The theorists' response to this manifested a kind of social schizophrenia. One group espoused the so-called 'S-matrix' or 'bootstrap' programme, which asserted that there were *no* truly elementary particles. The other group asserted that there were elementary particles, but that these were not

33

the particles observed in experiment: they were the quarks, from which all hadrons were built. The rival positions were often referred to as nuclear democracy and aristocracy respectively – in the former all particles were equal, while in the latter quarks had a privileged ontological position – and together they dominated theoretical HEP in the 1960s. Only in the 1970s, with the rise of the new physics, did the quark programme eclipse the bootstrap.

The Berkeley Bevatron was soon overtaken in energy by machines at several national and international laboratories. An international laboratory serving the Eastern bloc countries was set up at Dubna in the Soviet Union, where a 10 GeV PS began operation in 1957. In 1954 the nations of Western Europe set up a joint HEP laboratory just outside Geneva, Switzerland, and in 1959 a 28 GeV PS came into operation there. The European venture was formally known as the European Organisation for Nuclear Research, but was more often referred to as CERN (an acronym for Conseil Européen pour la Recherche Nucléaire, the body which oversaw the establishment of CERN in the early 1950s). In 1961 the USA regained a narrow lead in energy, with the first operation of a new machine at Brookhaven, the 31 GeV Alternating Gradient Synchrotron (AGS).[26] In 1963, two national machines of intermediate energy appeared: in the USA, the 13 GeV Zero Gradient Synchrotron (ZGS) came into operation at the Argonne National Laboratory, Illinois;[27] and in the UK experiment began at a 7 GeV PS called Nimrod, situated at the Rutherford Laboratory, not far from Oxford.[28]

In the 1960s, with the advent of new accelerators, research in strong-interaction physics split into two rather distinct domains. At low energies, cross-sections were observed to be large and to fluctuate rapidly as a function of beam energy and momentum transfer. These fluctuations were ascribed to the production and decay of hadronic resonances. At higher energies cross-sections remained large in the forward direction, but manifested a smooth non-resonant behaviour. This was ascribed to soft-scattering processes which were routinely analysed in terms of 'Regge' models belonging to the S-matrix programme. Investigations of the properties of resonances at low energies and of soft-scattering phenomena at high energies comprised the two principal strands of the old physics of the 1960s.

Both the CERN PS and the Brookhaven AGS were built on the so-called 'alternating-gradient' principle and with their successful operation it became clear that there should be no special problems involved in the construction of proton accelerators of even higher energies.[29] In 1967 the 76 GeV PS at Serpukhov regained the high-energy record for the Soviet Union. This machine produced data of considerable interest for the Regge tradition but, like the Dubna machine before it, did not have the overall impact one might have expected.[30] USA plans conceived in the mid-1960s came to fruition in 1972 with the first operation of a 500 GeV PS at a new national laboratory just outside Chicago.[31] This laboratory was first called the National Accelerator Laboratory, then restyled the Fermi National Accelerator Laboratory, or Fermilab for short. The European reply to the Fermilab machine was a 400 GeV accelerator called the Super Proton Synchrotron (SPS) which delivered its first beam in 1976.[32]

Both the Fermilab PS and the CERN SPS were conceived during the era of the old physics, and old-physics topics of resonance production and soft scattering constituted the bulk of their initial research programmes. During the late 1970s, however, the old physics was largely displaced at these accelerators by new-physics research. But, as discussed below, the main themes of the new physics – lepton beam and hard-scattering experiments, also the study of charmed particles – all emerged from other machines. In the context of the new physics, proton synchrotrons lost the obvious pre-eminence as a research tool which they had enjoyed in the days of the old physics. They remained unique, however, in their ability to provide a wide variety of secondary beams and thus to support a large and diversified research programme. In the late 1970s plans were in hand to extend the maximum energy of the Fermilab machine to 1 TeV (= 1000 GeV) – the Tevatron project[33] – and the Soviet Union envisaged operating a 3 TeV PS at Serpukhov by the late 1980s.

Electron Accelerators
As electrically charged and stable particles, electrons can be accelerated in machines identical in principle to proton accelerators. However, electrons are much lighter than protons (0.5 MeV and 1 GeV in mass, respectively) and therefore lose energy much faster than protons when accelerated in circular orbits: to achieve a beam of a given energy it is much more efficient to accelerate protons rather than electrons. Historically, too, proton beams have appeared a more direct probe of elementary particle interactions – especially, in a common-sense way, of the strong interactions. Thus the electron machines listed in Table 2.1 have always lagged behind PSs in energy and were for a long time seen as second best. There were several small electron accelerators constructed in the post-war era, but even with the mid-1960s arrival of the 6 GeV Cambridge Electron Accelerator in the USA, the 7 GeV Deutsches Elektronen-Synchrotron (DESY) in Hamburg, W. Germany, the 5 GeV NINA synchrotron at the

Daresbury Laboratory in the UK and a 6 GeV synchrotron at Yerevan in the USSR, such machines remained the poor relations of proton accelerators.[34] Of lower energy than their PS contemporaries, electron machines served to add new data on the old-physics phenomena already explored using proton beams, rather than to point to entirely new research topics. But, in the late 1960s, the first results from the new 22 GeV electron machine at the Stanford Linear Accelerator Center (SLAC) in California thrust electron-beam physics into the HEP limelight. The SLAC machine, alone amongst the machines discussed here, was a linear accelerator (a 'linac') rather than a circular synchrotron. Electrons at SLAC were simply accelerated along a linear path, emerging from the end of the machine as a 22 GeV beam. Stanford was the home of a long tradition of electron linac construction dating back to pre-war years, and the 22 GeV SLAC machine was the biggest (two miles long), best, and most expensive ($114 million, construction; $25 million per annum, running costs) of them all.[35] In the first round of experiments at SLAC a remarkable new phenomenon – scaling – was discovered. Scaling, and its theoretical interpretation in terms of first the quark–parton model and later QCD, constituted the first strand of the new physics to emerge. The discovery of scaling and its impact upon experimental and theoretical practice in HEP is the topic of Chapter 5, and here I want just to comment upon the marked absence from Table 2.1 of any post-SLAC electron accelerators.

The discovery of scaling guaranteed high-energy electron scattering a place in the forefront of HEP interest. However, as mentioned above, to attain a given energy, proton accelerators are much more efficient and economical than electron accelerators. And amongst the various secondary beams which can be generated from the primary proton beam of a PS are beams of leptons: electrons, muons and neutrinos (for a discussion of muons and neutrinos, see Chapter 3). Thus the impact of the discovery of scaling at SLAC was to focus attention on lepton beam experiments at the high-energy proton synchrotrons. In particular, the early 1960s had seen the beginning of programmes of neutrino experiment at both Brookhaven and CERN and, after its discovery at SLAC, scaling was soon observed to hold in neutrino scattering as well. Thus the mantle passed from SLAC not, as one might have expected, to a new generation of electron accelerators, but to the very high-energy neutrino beams available at Fermilab and the CERN SPS. Neutrino physics experiments aiming at the investigation of scaling were one way in which the new physics entered the experimental programmes of those machines. As it happened, neutrino experiments in the 1970s had a dual significance for the new physics. Besides scaling, neutrino beams were also the principal probe of another new phenomenon: the weak neutral current. The neutral current was discovered in 1973, and was regarded as the first piece of empirical evidence in favour of unified electroweak gauge theory. It was initially observed in a neutrino experiment at the CERN PS, but quickly became a prime research topic at Fermilab and the CERN SPS where higher beam energies made it possible to collect more extensive and detailed neutral current data. One identifying characteristic, then, of new-physics PS experiments was an emphasis upon lepton beams. The other principal characteristics were an increasing focus upon hard-scattering phenomena and charmed particles. The emergence of these topics is discussed below.

Colliders

Table 2.2 lists the major colliding beam machines of HEP. The CM-energy advantage which colliders enjoy is clearly evident in comparison with Table 2.1. The CERN ISR proton–proton collider, for example, is fed by 28 GeV proton beams from the CERN PS. After a slight increase in beam energy to 31 GeV, the ISR realises CM energies of 62 GeV ($=2 \times 31$ GeV), while to attain such a CM energy in a fixed-target experiment would require a 2 TeV primary beam.

Table 2.2. Colliders

Collider	Date of first operation	Beams	Energy per beam (GeV)	CM energy (GeV)
ADONE, Frascati	1967	e^+e^-	1.5	3
CEA 'Bypass', Cambridge, MA	1967	e^+e^-	3.5	7
ISR, CERN	1971	pp	31	62
SPEAR, SLAC	1972	e^+e^-	4.2	8.4
DORIS, DESY	1974	e^+e^-	4.5	9
DCI, Orsay, France	1976	e^+e^-	1.6	3.2
PETRA, DESY	1978	e^+e^-	19	38
CESR, Cornell	1979	e^+e^-	8	16
PEP, SLAC	1980	e^+e^-	18	36
VEPP-4, Novosibirsk, USSR	1980	e^+e^-	7	14
SPS pp̄, CERN	1981	pp̄	270	540
Tevatron pp̄, Fermilab	1986	pp̄	1000	2000
Isabelle, Brookhaven Lab.	Late 1980s?	pp	400	800
SLC, SLAC	1986?	e^+e^-	50	100
LEP, CERN	1987	e^+e^-	50	100

The Prehistory of High-Energy Physics

Electron–Positron Colliders

Most of the early colliders (not shown in Table 2.1) were very low energy electron–positron (or electron–electron) machines. Their primary physics interest lay in testing the successful theory of quantum electrodynamics (QED) in a new regime, but they were beset by technical problems and very much the toys of particular, idiosyncratic, experimenter/machine-builders. In the early 1960s a set of hadronic particles known as 'vector mesons' were discovered using conventional accelerators. These particles were found occasionally to decay into electron–positron (e^+e^-) pairs, and it was quickly realised that they should be produced by a converse process in e^+e^- collisions. It therefore appeared possible that beyond the investigation of QED, interesting hadronic physics could be done at e^+e^- colliders. None the less, it was generally felt that above the masses of the vector mesons (1 GeV) experiment would reveal an uninspiring 'desert'.

One of the first machines to put this feeling to the test was the ADONE collider which began operation at the CNEN National Laboratory at Frascati, Italy in 1967. ADONE was the fruit of earlier experience gained at Frascati in the construction of a smaller ring, AdA, and used as its source of electrons and positrons a low energy electron synchrotron already in operation at Frascati.[36] ADONE proved an excellent tool for the investigation of the vector mesons, and in the early 1970s showed that in the energy range above these (1–3 GeV) the desert was far from infertile. Theorists had expected that the cross-section for the production of hadrons would fall quickly towards zero in this region, whereas experiments at ADONE revealed that it stood up rather well. But no striking new phenomena were found, and e^+e^- physics remained of marginal significance in HEP. In the same year as ADONE, the CEA 'Bypass' e^+e^- facility became operational at Cambridge, fed once more from an existing electron synchrotron. By 1972, CM energies at CEA had been extended up to 7 GeV, more than twice those attainable at ADONE, but with very low luminosity. Luminosity is a technical measure of the density of the beams in the interaction region, and CEA's low luminosity implied a very low interaction rate. This in turn implied large statistical uncertainties in data taken in a reasonable length of time, and although observations made at CEA were rather suggestive, it was left to the SPEAR ring at Stanford to mark the real arrival of e^+e^- physics.

The moving force behind the construction of SPEAR was Stanford physicist Burton Richter. Having advocated for many years the construction of an e^+e^- collider to exploit the beam from the SLAC linear accelerator, Richter finally saw his dreams take shape in a cut-price, 8 GeV CM-energy version (SPEAR was funded from the SLAC operational budget) which began to do physics in 1972.[37] In 1974 a watershed in the development of the new physics came with the discovery of the first of the new particles – the J-psi – announced simultaneously by experimenters at SPEAR and the Brookhaven AGS. The AGS proved to have insufficient energy for detailed investigation of the properties of the new particles, and amongst fixed-target machines the Fermilab PS and CERN SPS took over this role. SPEAR, however, was well matched to experimental requirements. Much of the work which contributed to the identification of the new particles as manifestations of 'charm' was done there, and at another electron–positron collider, DORIS, which began operation at DESY in 1974. As indicated in Chapter 1, data on charmed particles were seen as crucial to the theories of the new physics – both electroweak gauge theory and QCD – and in the latter half of the 1970s, e^+e^- interactions came to be seen as perhaps the single most revealing source of information in experimental physics. In the space of a couple of years, electron–positron colliders moved from the margins to the forefront of HEP research.

In 1978, DESY took over the lead in electron–positron physics with the commissioning of the 38 GeV CM-energy collider, PETRA. The equivalent US facility, PEP at SLAC, was delayed by funding problems until 1980. Another US e^+e^- machine – an intermediate energy collider, CESR at Cornell – came into operation in 1979. In Europe, consensus hardened that the major accelerator project for the 1980s should be a giant e^+e^- machine, and the decision to build LEP (the Large Electron–Positron collider) at CERN was ratified by the Member States in 1981.[38] LEP was envisaged as a new-physics machine, a gauge theory laboratory. A whole variety of new physics phenomena could be explored with it, but it had one primary objective: to detect a set of particles known as 'intermediate vector bosons' (IVBS). The IVBS were the key elements of unified electroweak gauge theory, and were predicted to be so heavy that no existing machine had sufficient energy to produce them. A late and novel entry into the field of electron–positron machines was the proposed Stanford Linear Collider (SLC). This machine was envisaged to take 50 GeV beams of electrons and positrons from the SLAC linac (upgraded in energy) and collide them just once before discarding them (in contrast to conventional colliders in which stored beams collide repeatedly). The motivation behind the apparently wasteful SLC was twofold. First, it could be argued that for energies much beyond LEP it would be cheaper to build and operate two linear accelerators and fire them at one another, rather than to go to another generation of circular colliders. The SLC could thus be regarded as a prototype for the machines of the future. More importantly in the short term, it seemed that the SLC could be built quickly, offering US physicists a chance to detect the IVBS before LEP began operating in Europe. Although not officially funded, by 1982 it appeared likely that the SLC would be the next major facility to be built in the USA.[39]

Proton–Proton and Proton–Antiproton Colliders

Electron–positron machines provided the major experimental sensations in HEP in the 1970s, but from 1971 onwards the world's highest CM-energy machine was a proton–proton collider – the Intersecting Storage Rings (ISR) at CERN. The ISR stores counter-rotating proton beams from the CERN PS in two interlaced rings. At the intersections of the rings a total of 62 GeV CM energy is available. Early studies at the ISR concentrated on old-physics topics—in particular, on measurements of soft, low-transverse-momentum processes relevant to the contemporary versions of Regge theory. However, for the reasons outlined in Section 2.2, experimenters found the ISR an especially convenient machine at which to investigate high-momentum transfers, and in the first round of ISR experiments significant steps were taken towards establishing traditions of hard-scattering research. These early ISR experiments observed that while cross-sections were indeed small at high-momentum transfers they were considerably larger than expected from a straightforward extrapolation of soft-scattering data. The excess of high-transverse-momentum events was construed as evidence of hard collisions between the quark constituents of the interacting hadrons, and was interpreted in terms of the quark–parton model originally developed for the analysis of the SLAC electron scattering data. Traditions of hard-scattering experiment grew rapidly at the ISR during the 1970s. Through their relation to the quark–parton model (and later to QCD) they represented the first incursion of the new physics into purely hadronic experiment, and they eventually came to dominate the ISR programme. Despite technical difficulties in detecting high-transverse-momentum particles in fixed-target experiments (see Section 2.2) traditions of hard-scattering research also sprang up at Fermilab and the CERN SPS, where they contributed yet another element to the new physics takeover.

No more proton–proton colliders were constructed during the 1970s, but US plans for a major investment in the 1980s came to centre on an 800 GeV CM-energy pp collider. Called Isabelle, this was to be built at the Brookhaven National Laboratory. Construction began in

1978 and tunnelling for the ring was complete in 1981. Unfortunately, the design for Isabelle relied upon largely untried superconducting-magnet technology, and the development of working magnets encountered many problems. These caused delays and cost escalation, and by the early 1980s it was unclear whether Isabelle would ever be completed.[40] Construction of two proton–antiproton ($p\bar{p}$) colliders also got under way in the 1970s. This followed the 1976 proposal that it might be possible to fill the Fermilab PS and the CERN SPS with counter-rotating beams of protons and anti-protons and thus to operate them in a colliding-beam mode.[41] These $p\bar{p}$ colliders promised to reach the same CM-energy range envisaged for LEP and Isabelle. They could not promise data of the same quantity and quality but, as conversions of existing machines, they could be brought into operation relatively quickly and inexpensively. Their aim was a simple one: the first sighting of the IVBs.

To summarise this section, the history of accelerator physics can be divided into three eras. In the first era, which extended roughly to the end of the 1960s, accelerators were both built and used for old-physics purposes. Hadron-beam experiments at PSs of increasing energy set the pace. The second era spanned the 1970s. Most of the machines which came into operation then had been planned with the old physics in mind, but they were increasingly used for new-physics purposes. PS programmes were reorganised around the investigation of rare phenomena, and colliders came to the fore. The third era lay in the future, with big machines like LEP and the $p\bar{p}$ colliders explicitly conceived as new-physics factories. In the following chapters we can analyse how the transitions between the three eras came about.

NOTES AND REFERENCES

1   For CERN, see Adams (1981); for the US, see Robinson (1981).
2   With the exception of a late entry from Japan (discussed in Hoddeson 1983) and plans in hand in China (Lock 1981) all of the major experimental HEP laboratories have been concentrated in the US, Europe or the Soviet Union. For documentation of the relative lack of impact of the Soviet experimental programme, see Irvine and Martin (1983b). Irvine and Martin discuss a range of factors to which Eastern and Western physicists have attributed the Soviet lack of success, including the lack of fast electronic equipment, insufficient technical support, and problems associated with the bureaucratic management of scientific research. Sullivan, Barboni and White (1981) discuss how technical shortcomings affected the strategy of Soviet experimental research in the mid-1960s.
3   Two factors have fostered the growth of HEP communities outside the major blocs, especially in Japan and India. First, individual laboratories have generally been open to experimenters irrespective of their countries

of origin. Secondly, research in theoretical HEP does not require access to experimental facilities, and is relatively cheap (the primary financial input is researchers' salaries). Despite these mitigating circumstances, it remains the case that the development of HEP has been dominated by the US, Europe and the Soviet Union.

4   'Amaldi Report' (1963, 70, 74).

5   This trend is documented for the overall number of PhDs employed in physics in the USA in Fiske (1979) and Alpher, Fiske and Porter (1980). The 'Bromley Report' confirms that in the early 1970s there were around 1600 practising particle physicists in the USA, essentially the same number as in the late 1970s (Physics Survey Committee 1973, 321, 334). The constant level of experimental manpower in Europe during the 1970s is documented in Martin and Irvine (1983, 17, Table 2).

6   For Europe, see ECFA (European Committee for Future Accelerators) Working Group (1979, 3); for the USA, see High Energy Physics Advisory Panel (1980, 27). Note that all of the figures quoted here are for active HEP researchers, and do not include the large numbers of support staff working at the major laboratories. For example, at CERN the number of support staff grew continually from its foundation in 1954, reaching a peak of nearly 4,000 in 1975 before declining to around 3,500 in 1980 (Adams 1980, 16).

7   For further discussion of the differentiation of professional roles within HEP, see Gaston (1973).

8   ECFA Working Group (1979, 3); High Energy Physics Advisory Panel (1980, 27).

9   A survey of the HEP literature published between 1950 and 1972 located a total of 5000 papers (by 4000 authors) on weak interaction physics. These comprised one-sixth of the total HEP literature of the period (Sullivan, Barboni and White 1981, 165).

10 For more on the collaborative aspect of experimental HEP research, see Morrison (1978). In the 1950s and into the 1960s experiments were typically performed by a group of less than 10 physicists often drawn from a single institution. As experiments increased in complexity and sophistication through the 1960s and 1970s so did the size of collaborations. In the late 1970s a typical collaboration had around 40 members, with a few groups having a complement around 80 (ECFA Working Group 1979, 17). ECFA (1979, 21) envisaged that in the late 1980s experiments at LEP (see below) would be performed by groups of up to 100 physicists. This they regarded as an upper limit on collaboration size, set by difficulties of organisation and communication. Within the HEP community at large, however, a figure of 200 physicists for each LEP experiment was regarded as more realistic. In contrast, the working practice of HEP theorists has continued to resemble the pre-Second-World-War ideal, typical theoretical publications having one or a few authors.

11 Some comment is appropriate on the institutional procedures whereby experimenters gain access to accelerators. With the growth of national laboratories have come advisory and policy committees. Experimental collaborations prepare detailed proposals describing the experiments

Men and Machines

which they wish to perform, and submit these to the management of
whichever laboratory they think appropriate. The proposals are then
vetted by committees, which assess them in terms of their physics
interest, the availability of the appropriate material and financial
resources, the competence of the proposing group, and so on. Successful
groups thus gain access to the machine of their choice; other groups may
reformulate their proposals, change their structure, or disband.
Members of the relevant committees are in general respected practising
particle physicists (experimenters and theorists).

12 For a comprehensive review of accelerator developments from the late
19th century to the late 1970s, see Livingston (1980).

13 See Livingston (1980, 49, 74, 80) and *CERN Courier* (1982c).

14 An important milestone in bubble-chamber development came in 1959
with the first operation of an enormous (72-inch long) chamber designed
and built by Luis Alvarez' group at Berkeley. Many new particles were
discovered with this chamber, and its success sealed the fate of the
bubble chamber's principal rival in the field of visual detectors, namely
the nuclear emulsion method. Nuclear emulsions are basically thick,
high-density photographic films. They came into use in experimental
HEP in the late 1940s. Emulsion stacks were placed in particle beams for
a period of time. After removal, charged particle tracks corresponding
to interactions within the emulsion could be made visible by standard
photographic developing techniques. A great deal of data was produced
using the emulsion method in the 1950s but, relative to bubble
chambers, emulsions had two significant drawbacks: finding and
measuring tracks within emulsions was a difficult, tedious and time-con-
suming business; and it was difficult to prepare large homogeneous
emulsion stacks. In the 1960s and 1970s emulsions were only used for
specialised purposes.

For the development of bubble-chamber technology and analysis
techniques up to the early 1960s, see Alvarez (1970). For later
developments, see Derrick (1970) and Ballam and Watt (1977). For a
popular account of the emulsion method, see Yagoda (1956).

15 For a popular account of scintillation counters, see Collins (1953).

16 For a popular account of spark chambers, see O'Neill (1962).

17 For technical accounts of the development of multi-wire, drift and
streamer chambers, see Charpak (1970) and Rice-Evans (1974). For a
semi-popular account, see Charpak (1978).

18 For an early popular account of collider development, see O'Neill
(1966).

19 Hadronic total cross-sections are typically in the range 10^{-1} to 10^{-2}
barns. Total cross-sections for electron scattering are around 10^{-4}
barns. Neutrinos (electrically neutral leptons which experience only the
weak interactions: see Chapter 3.3) have scattering cross-sections of the
order of 10^{-13} barns.

20 Tables 2.1 and 2.2 are compiled from various sources: principally the
reviews given by Barton (1961), Howard (1967, 1, 18, 43), Panofsky
(1974, xi–xii), ECFA Working Group (1979, Table 7) and Madsen and
Standley (1980), plus the individual sources cited below. It should be

emphasised that these tables do not attempt exhaustive coverage. They refer only to the highest energy machines operating in a particular period, and the many small accelerators built in the 1940s, 50s and 60s have been omitted. For more complete lists see the above reviews and, for machines built prior to 1956, Amaldi (1977, 347).

21 For the history of BNL and the Cosmotron, see Ramsey (1968), Greenberg (1971) and Baggett (1980).

22 An electron-volt (eV) is defined as the energy gained by an electron accelerated through a potential difference of one volt. This is a very small unit for the purposes of HEP and, as accelerators have become more powerful, larger units have received conventional denominations. These are related to the electron volt as follows: 1 keV $= 10^3$ eV; 1 MeV $= 10^6$ eV; 1 GeV $= 10^9$ eV; 1 TeV $= 10^{12}$ eV. According to the theory of special relativity, energy and mass are interconvertible and, to give some idea of the scale of these units, the mass of the electron is around 0.5 MeV, while that of the proton is almost 1 GeV.

23 It should be noted that the cosmic-ray spectrum extends to extremely high energies, well beyond those artificially obtainable, albeit with very low intensity, and particle physicists continue to look to cosmic rays for indications of new phenomena.

24 For the history of Lawrence's Radiation Laboratory (later renamed the Lawrence Berkeley Laboratory) at the University of California, Berkeley, see Kevles (1978), Greenberg (1971) and Heilbron, Seidel and Wheaton (1981).

25 The leading role of Berkeley in the early years of resonance physics is only partly ascribable to the high energy of the Bevatron. In increasing measure, it was due to the pioneering work in bubble-chamber techniques of Alvarez' group (both building the 72-inch chamber', and developing sophisticated methods of film analysis: see note 14 above). For an account of this work and its relation to theoretical developments, see Alvarez (1970). Swatez (1970) gives an interesting discussion of the social organisation of Alvarez' group.

26 For the history of all three institutions and machines, see Jungk (1968). A major study of the history of CERN is currently in progress (Hermann 1980): in the meantime, see Amaldi (1977) and Kowarski (1977a,b). For a discussion of CERN's research performance in comparison with that of the other major HEP laboratories, see Martin and Irvine (1983) and Irvine and Martin (1983a). The current member states of CERN are Austria, Belgium, Denmark, France, Federal Republic of Germany, Greece, Italy, Netherlands, Norway, Spain, Sweden, Switzerland and the United Kingdom.

27 Greenberg (1971).

28 Litt (1979).

29 Livingston (1980, 65–85).

30 See note 2 above.

31 Greenberg (1971), Jachim (1975), Sanford (1976), Hoddeson (1983).

32 Goldsmith and Shaw (1977).

33 CERN Courier (1982d).

34 For details of these machines and a comparative discussion of their

research performance, see Martin and Irvine (1981).

35 Neal (1968), Greenberg (1971).
36 For the history of ADA, see Bernadini (1978) and Amaldi (1981).
37 Richter (1977).
38 *CERN Courier* (1982b).
39 Robinson (1982).
40 Broad (1982).
41 Cline and Rubbia (1980).

3

THE OLD PHYSICS: HEP, 1945-64

This chapter traces out the broad features of the post-war development of HEP.[1] It aims to delimit the principal preoccupations of the old physics in the early 1960s, the immediate pre-quark era. The old physics was characterised by its commonsense approach to elementary-particle phenomena. Experimenters explored high cross-section processes, and theorists constructed models of what they reported. By the early 1960s, old-physics hadron-beam experiments at the major PSS had isolated two broad classes of phenomena. At low energies, cross-sections were 'bumpy' – they varied rapidly with beam energy and momentum transfer. At high energies, cross-sections were 'soft' – they varied smoothly with beam energy and decreased very rapidly with momentum transfer. The bumpy and soft cross-section regimes were ascribed different theoretical meanings. The low-energy bumps were interpreted in terms of the production and decay of unstable hadrons. As more and more bumps were isolated, so the list of hadrons grew. This was the 'population explosion' of elementary particles, discussed in Section 1. The following sections review various theoretical attempts to get to grips with the explosion. Section 2 outlines the way in which conservation laws were used to sharpen the distinction between the strong, weak and electromagnetic interactions and to classify the hadrons into families. This approach resulted in the Eightfold Way classification scheme, of which the quark concept was a direct descendant. Section 3 discusses how HEP theorists attempted to extend the use of quantum field theory from the electromagnetic to the weak and strong interactions. This attempt bore fruit in the early 1970s with the elaboration of gauge theory. But in the 1950s and early 1960s the field-theory programme did not prosper. For the weak interaction, the field-theory approach was at least pragmatically successful, but for the strong interaction it failed miserably. Section 4 reviews the struggle to salvage something from the wreckage. What emerged was the 'S-matrix' approach to strong interaction physics. The S-matrix was founded in quantum field theory but achieved independence from it and was seen, in the 'bootstrap' formulation, as an explicitly anti-field-theory approach. In the late 1950s the principal referent of

S-matrix theorising was the growing list of hadrons. But in the early 1960s one strand of S-matrix development, 'Regge theory', came to dominate the analysis of the high-energy soft-scattering regime.

Thus, from the early 1960s onwards, the old physics had a two-pronged structure: at low energies, the emphasis was on bumps in cross-sections and their interpretation as unstable hadrons; at higher energies, the emphasis was on soft scattering, interpreted in terms of Regge theory. The low-energy work led directly to quarks, as we shall see in the next chapter. The Regge-oriented traditions of theory and experiment led nowhere – in the sense that they contributed little to the eventual emergence of the new physics of quarks and gauge theory. My account of the Regge traditions in Section 4 is correspondingly brief, and in the following chapters I will not discuss them in any detail. But it is important to bear their existence in mind. During the 1960s and into the 1970s, the Regge traditions constituted a major component of old-physics research.[2] They enshrined a world-view quite different from that of the new physics. And as long as they were actively supported by HEP theorists and experimenters, the new physics could not be said to be established. The Regge traditions constituted the resistance which quarks and gauge theories had to overcome.

3.1 The Population Explosion

Writing in 1951 on 'The Multiplicity of Particles', HEP theorist Robert Marshak from the University of Rochester looked back upon a golden age of elementary particles:

> In the year 1932, when James Chadwick discovered the neutron, physics had a sunlit moment during which nature seemed to take on a beautiful simplicity. It appeared that the physical universe could be explained in terms of just three elementary particles – the electron, the proton, and the neutron. All the multitude of substances of which the universe is composed could be reduced to these three basic building materials, variously combined in 92 kinds of atoms. An atom consisted of a tight nucleus, built of protons and neutrons, and a swarm of electrons revolving around the nucleus like planets around a sun.[3]

In December 1951, Marshak counted 15 elementary particles – the list was already beginning to look untidy. As the years went by, the list grew longer, and this section traces out the principal early developments.[4]

In the late 1940s, cosmic-ray experimenters decided that they were observing the production of two different types of particle where hitherto they had suspected the existence of only one. One particle,

47

the muon (μ), was recognised as a lepton – an electron-like particle immune to the strong interactions. Having a mass of 105 MeV, the muon was around 200 times heavier than the electron (e), but otherwise appeared identical to it. The second particle, the pi-meson or pion (π), displayed the large interaction cross-sections with nuclear matter characteristic of a strongly interacting particle or hadron. With a mass of around 140 MeV, the pion was much lighter than other hadrons (the proton, for example, had a mass of 940 MeV) and was identified as the carrier of the strong interactions predicted by the Japanese physicist Hideki Yukawa in 1935 (see Section 3 below).

In 1953 another lepton was discovered: the electrically neutral and massless neutrino (v) (see also Section 3). From then on, the list of leptons – e, μ, v – remained static for many years. The list of hadrons continued to grow. Between 1947 and 1954 a succession of hadrons were identified in cosmic-ray experiments and in accelerator experiments at the Brookhaven Cosmotron: the K-meson or kaon (K: 500 MeV in mass); the lambda (Λ: 1115 MeV); the sigma (Σ: 1190 MeV); and the cascade or xi (Ξ: 1320 MeV).[5] All of these particles were unstable and decayed to lower mass particles. They were observed as tracks a few centimetres long in visual detectors, and their lifetimes before decay were accordingly deduced to be around 10^{-8} to 10^{-10} seconds.[6] This was the time scale expected for processes mediated by the weak interaction, and observations on the decay of these particles constituted an important new source of information on the weak interaction in the early years of HEP.

In 1952, a particle known as the delta (Δ: 1230 MeV) was observed at a low-energy accelerator, the Chicago cyclotron. The delta then appeared to be unique in the shortness of its lifetime. Unlike the quasi-stable particles discussed above, which produced experimentally measurable tracks, the delta existed only for an infinitesimal time, around 10^{-23} seconds, between its production and decay, and was impossible to observe directly. Instead, its existence was inferred from a large bump in pion–nucleon cross-sections around the CM energy of 1230 MeV. Figure 3.1 shows a modern compilation of cross-sections for the interaction of negative pions with protons; the delta is the first and largest peak, situated at around 0.34 GeV/c beam momentum.[7] The energy-dependence of cross-sections in the vicinity of the delta peak was typical of 'resonance' phenomena already well known in nuclear physics, and the delta was often referred to as a 'resonance' – marking physicists' uncertainty as to whether such a short-lived entity should be regarded as a genuinely fundamental particle.[8] Whatever its nature, the delta's lifetime of 10^{-23} seconds was characteristic of strong-interaction processes, and thus it followed that the decay as well as the production of the delta was mediated by the strong interaction.

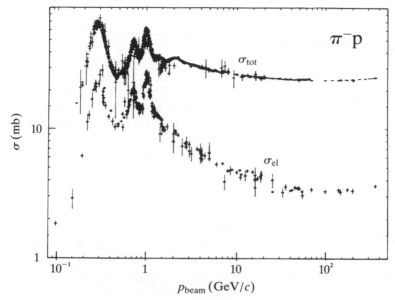

Figure 3.1. Cross-sections (σ) as a function of pion beam momentum (p_{beam}) for negative pion–proton interactions. The upper curve, labelled σ_{tot}, shows the total cross-section, corresponding to the probability of any kind of interaction taking place. The lower curve, σ_{el}, shows the cross-section for elastic scattering, i.e. the relative probability that the pion will scatter off the proton without production of any additional particles.

As one might guess from the several peaks evident in Figure 3.1, the uniqueness of the delta was, like the particle itself, short-lived. In the early 1960s, more and more entries were added to the list of hadronic resonances, all of which were entities decaying on a time-scale of 10^{-23} seconds. In 1961 four new resonances were found in experiments at the Brookhaven Cosmotron and the Berkeley Bevatron: the eta (η: 550 MeV), rho (ρ: 770 MeV), omega (ω: 780 MeV) and K-star (K*: 890 MeV). In 1962, the Brookhaven AGS added the phi (ϕ: 1020 MeV). It was at around this time that the population explosion really took off. Masses of cross-section data were emerging from the Bevatron, the AGS, the CERN PS and the other new machines of the 1960s, and 'bump-hunting' for resonance peaks became the central experimental preoccupation of the old physics in the low-energy regime. More and more resonances were found, and it would serve little purpose to list them individually. A review article published in 1964 noted that 'only five years ago it was possible to draw up a tidy list of 30 sub-atomic particles', but that 'since then another 60 or 70 sub-atomic objects have been discovered',[9] and this is sufficient to make the point.

One last group of particle discoveries needs to be mentioned here: that of antiparticles. Antiparticles are states having the same mass as the corresponding particles, but with opposite electric charge and other quantum numbers (see below). The first antiparticle to be discovered was the antielectron or positron (e^+), a particle apparently identical to the electron (e^-) except in carrying a positive rather than a negative charge. Observed in 1932 by Carl Anderson in a cloud-chamber experiment, the positron came to be regarded as confirmation of Paul Dirac's quantum theory of electromagnetism. According to Dirac's theory, all particles were expected to have associated antiparticles, but many physicists remained sceptical until the antiproton ($\bar{p}$), the antiparticle of the proton (p), was discovered at the Berkeley Bevatron in 1955.[10] From that point on, it became a routine assumption that all particles had antiparticles. Many species of antiparticle were experimentally observed, swelling the list of hadrons even further.

3.2 Conservation Laws and Quantum Numbers: From Spin to the Eightfold Way

As the experimenters generated increasing quantities of data on increasing numbers of particles, so theorists set out to create some order in what was being found. One strategy was to theorise in detail about the constitution of elementary particles and their interactions, as discussed in Sections 3 and 4 below. Here I will discuss a second strategy: the use of conservation laws.[11] Conservation laws embody the idea that some quantities are fixed come what may, and have an honoured place in the history of physics. Particle physicists followed a traditional path in using them to make sense of their subject matter. On the one hand, they applied already established laws as definitions, more clearly to establish what was new in the interactions of elementary particles; on the other, they constructed new, empirical, conservation laws in order to systematise experimental data and to characterise the fundamental interactions.

We can begin with the truly definitional conservation laws which were rooted in fundamental beliefs about space–time. These were the energy–momentum and angular momentum conservation laws. One important consequence of the former was that it implied constraints upon particle decays: particles could only decay into other particles whose combined mass was less than or equal to the mass of the parent.[12] Conservation of energy implied, for example, that the delta (mass 1230 MeV) could decay into a proton plus a pion (combined mass 1080 MeV) but not into a lambda plus a kaon (combined mass 1615 MeV). By virtue of its definitional status, conservation of energy–momentum could also be used to extract information about the interactions of electrically neutral particles. Neutral particles leave no direct record in particle detectors, but experimenters routinely inferred their presence by insisting that energy–momentum was conserved in all interactions: any apparent energy–momentum imbalance had to be due to undetected neutrals. It was through such reasoning that neutral particles like the lambda were discovered. Through its defining status, conservation of angular momentum served similar purposes. It also served as foundation for a system of labelling elementary particles, and a discussion of this will serve to introduce the important concept of a 'quantum number'.

Modern conceptions of the atom derive from the planetary model first developed by Niels Bohr in the early years of this century. According to Bohr's model, electrons orbit the nucleus as do the planets the sun. Associated with the orbital motion is angular momentum: the faster an electron or planet travels in an orbit of given radius, the more angular momentum it carries. The difference between the planetary and atomic systems is that the former is macroscopic and believed to be well described by the laws of Newtonian classical mechanics, while the latter is microscopic and believed to require a quantum-mechanical analysis. According to quantum mechanics, orbital angular momentum is not a continuous quantity free to assume arbitrary values. Instead, it is *quantised*; restricted to values which are integral multiples of a fundamental unit, denoted by $\hbar$.[13] The number of these units associated with a given orbit is known as the orbital angular momentum *quantum number*, denoted by L: to say that the quantum number of an electron orbit is L is to say that the orbital angular momentum associated with it is $L\hbar$.

As the result of a combination of experimental and theoretical work in the late 1920s, physicists concluded that it made sense to ascribe angular momentum or *spin* to elementary particles themselves as well as to their orbital motion.[14] It was as though particles rotated about an internal axis just as planets rotate about theirs. Like angular momentum, the spin of elementary particles was supposed to be quantised – this time in either integral or half-integral multiples of $\hbar$. Each species of particle was supposed to carry a fixed spin which could not be changed in any physical process, and thus spin was recognised as an identifying characteristic of a given species. In the pre-Second-World-War era, spin values were assigned to the known particles – the electron, proton and neutron were all identified as spin $\frac{1}{2}$ objects; the photon, the quantum of the electromagnetic interaction, was assigned spin 1. As more particles were discovered in post-war HEP they were routinely assigned spins on the basis of experimental measurements of their production and decay. Deltas, for example, were identified as spin $\frac{3}{2}$; pions and kaons, spin 0; and rho, omega and phi particles spin 1. Together with the spin concept came another layer of jargon. Particles with half-integral spin, like the electron and proton, became generically known as *fermions*; particles with integral spin, such as the pion and rho, as *bosons*.[15] All of the known leptons were fermions, but fermions and bosons were equally well represented amongst hadrons. Hadronic fermions were generically christened *baryons*, while hadronic bosons were named *mesons*.

One subtlety concerning spin remains to be discussed. Angular momentum, spin and orbital, is a vector quantity: it has both magnitude and direction. The spin angular momentum of a planet, for example, has a magnitude determined by how fast the planet rotates, and a direction determined by the orientation in space of the axis about which it rotates. Now, just as quantum mechanics requires the quantisation of the magnitude of angular momentum so, speaking loosely, it requires the quantisation of orientation. To be more precise, it requires that the component of the total angular momentum as measured relative to any axis fixed in space is itself quantised. This is best explained in concrete terms. Consider a spin 2 particle, and suppose that one sets out to measure how much of the spin is oriented along a chosen axis (conventionally referred to as the 3-axis). Clearly the answer must be between 2, for a parallel alignment of the spin with the chosen axis, and -2, for an antiparallel alignment. In classical mechanics, any answer in this range is conceivable, but in quantum mechanics only answers differing by integers are conceivable: $+2$, $+1$, 0, -1, -2. A particular measurement may result in any one of these values – for example $+1$ – and one would then say that the 3-component of the particle's spin was $+1$. Similarly, the 3-component of spin of a spin $\frac{1}{2}$ particle can only take on the values $+\frac{1}{2}$ or $\frac{1}{2}$; that of a spin $\frac{3}{2}$ particle $+\frac{3}{2}$, $+\frac{1}{2}$, $-\frac{1}{2}$ or $-\frac{3}{2}$; and so on. Such considerations are central to the experimental analysis of elementary particle interactions, but their particular relevance here will become evident below when we discuss 'isospin'.

Three further conservation laws have also enjoyed definitional status in the analysis of elementary particle interactions. These are the conservation of electric charge, the conservation of baryon number and the conservation of lepton number. The first of these states that electric charge can be neither created nor destroyed in any physical process: a Δ^+ (charge $+1$), for example, can decay to a π^0 (0) and a proton ($+1$), but not to a π^+ ($+1$) and a proton ($+1$). Charge conservation is grounded in the theory of electrodynamics, while the conservation laws of baryon and lepton numbers represent analogical extensions of the charge concept to explain purely empirical regularities. All baryons are assigned baryon number 1, all antibaryons -1, and all other particles 0. Baryon number is supposed to be an additive, charge-like quantity – so that, say, the baryon number of a system of five protons is 5 – and is supposed to be absolutely conserved. This has the experimentally confirmed consequence that antibaryons can only be produced in conjunction with baryons (zero net change of baryon number), and that the lightest baryon, the proton, is absolutely stable (since all the lighter particles to which it could decay are mesons or leptons with baryon number zero). Likewise, the observed regularities of lepton production and decay are systematised by assigning an additive lepton number $+1$ to leptons, -1 to antileptons, and 0 to other particles, and insisting that total lepton number is conserved. (In the late 1970s the conservation of baryon and lepton number came under theoretical challenge: we will discuss this in Chapter 13).

Restricted Conservation Laws
The conservation laws so far discussed were well established in pre-war physics, and were regarded by particle physicists as having unlimited and universal scope. We can now turn to a second class of laws, those which were believed to apply to some interactions and not to others. Let us begin with *parity* conservation. Parity is a concept with no classical analogue. It has meaning only within the framework of quantum mechanics, and is best thought of here as a book-keeping device.[16] The parity quantum number can take on only two values, $+1$ (positive parity) or -1 (negative parity), and can be associated with both individual particles and assemblages of particles. Every type of particle has its own intrinsic parity – all pions have negative parity, all protons positive, and so on. Parity is a multiplicative rather than an additive quantum number, so that, say, a system of two stationary pions has positive parity ($-1 \times -1 = +1$). Relative orbital angular momentum is also significant in assigning parities to multiparticle states: each unit of orbital angular momentum contributes a factor of -1 to the overall parity. Thus, a system of two pions orbiting one another with one unit of angular momentum has

negative parity $(-1 = -1 \times -1 \times -1)$. The parity and spin quantum numbers together form the basis for more jargon. Physical quantities having spin 0, positive parity – denoted 0^+ – are known as *scalar* quantities; 0^- quantities are *pseudoscalars*; 1^+, *axial vectors*; and 1^-, *vectors*. This nomenclature is used in many circumstances: for example, it is applied to particles themselves – pions and kaons are referred to as pseudoscalar mesons; rhos, omegas, phis and photons as vectors – and to the weak currents (see Section 3 below).

For many years, it was thought that parity conservation was absolute since it followed from rather basic considerations concerning the mirror-symmetry of physical processes. The world of HEP was therefore shaken in 1956 by the discovery that in certain processes parity-conservation was violated. Nuclear transitions and particle decays were observed in which the parity of the product state was not the same as that of the parent. These transitions and decays were all mediated by the weak interactions, and it became part of the conventional wisdom of HEP that, while the strong and electromagnetic interactions did conserve parity, the weak interaction did not. Thus the weak interaction was distinguished from the other forces not only by its characteristically weak effects but also by its irreverent attitude to a hitherto well-established conservation law. Parity was a 'good' quantum number for strong and electromagnetic processes but it was meaningless in the characterisation of processes mediated by the weak interaction.

A similar comment applies to two other quantum numbers invented by HEP theorists, *strangeness* and *isospin*. In the early 1950s, kaons and lambdas were known as 'strange' particles. Their strangeness lay in the fact that although they were copiously produced in particle interactions with typical strong-interaction cross-sections, they decayed (relatively) slowly, having lifetimes typical of weak decay processes. In 1952, US theorist Abraham Pais proposed his 'associated production' hypothesis.[17] According to this, kaons and lambdas interacted strongly only in pairs: a kaon and a lambda could be produced in association with one another via the strong interaction, but in isolation they could only decay via the weak interaction. This idea was refined by the American theorist Murray Gell-Mann, and independently by the Japanese theorist Kazuhiko Nishijima, who proposed that the strange particles carried a new additive quantum number called, appropriately enough, strangeness.[18] Kaons would carry strangeness $+1$, lambdas -1, and other particles, such as the pion and proton, strangeness zero. Strangeness was supposed to be exactly conserved by the strong interaction, and therefore every time a lambda particle, say, was produced it had to be accompanied by a kaon. Like parity, strangeness was not supposed to be conserved by the weak interactions, and thus kaons and lambdas could decay to nonstrange particles with the observed weak-interaction lifetimes.[19] As the population explosion got under way, associated production was observed to obtain for many hadrons, and appropriate values of strangeness were assigned to these: -1 for the sigma baryons, -2 for the xis, and so on.

Strangeness could be thought of as a conserved charge, like electric charge, baryon number and lepton number. Isospin, on the other hand was, as the name suggests, a vector quantity like spin. The isospin concept was an extension of the pre-war 'charge-independence' hypothesis,[20] and asserted that hadrons having similar masses, the same spin, parity and strangeness, but different electric charges, were identical as far as the strong interactions were concerned. Thus, for example, in their strong interactions the neutron and proton appeared to be indistinguishable, as did the three charge states of the pion, π^+, π^0 and π^- (the superscripts indicate electric charge). This was formalised by assigning each group of particles an isospin quantum number (I) analogous to the usual spin quantum number. Thus the neutron and proton, now collectively known as the *nucleon*, were assigned isospin $\frac{1}{2}$; the pion, in all its charge states, isospin 1; and so on. Because isospin was supposed to be a vector quantity, its 3-component (I_3) was supposed also to be quantised, and different I_3 values were taken to distinguish between different members of each isospin family or *multiplet*. For the isospin $\frac{1}{2}$ nucleon, possible I_3 values were $+\frac{1}{2}$, identified with the proton, and $-\frac{1}{2}$, identified with the neutron. Similarly for the isospin 1 pion, an I_3 value of $+1$ corresponded to the π^+, of 0 to the π^0, and of -1 to the π^-. As more and more particles were discovered they were routinely assigned to isospin multiplets – (K^+, K^0), ($\bar{K}^0$, K^-), (Δ^{++}, Δ^+, Δ^0, Δ^-), (Σ^+, Σ^0, Σ^-), and so on. Particles with no partners, like the neutral Λ, were assigned to one-member groups or 'singlets'.

Isospin, like parity and strangeness, was supposed to be exactly conserved by the strong interaction. But to explain the small observed mass differences within isospin multiplets (1 MeV between the proton and the neutron, for example) it was assumed that, unlike parity and strangeness, isospin conservation was violated by the electromagnetic as well as the weak interaction. The small observed violations of isospin conservation – the mass differences or *splittings* between members of a given multiplet – were ascribed to both weak and electromagnetic effects. These represented a minor perturbation in relation to the dominant strong interaction, and isospin conservation

The Prehistory of High-Energy Physics

thus remained a powerful tool in the analysis of strong-interaction processes.

SU(3): The Eightfold Way

Strangeness and isospin constituted new and empirically useful analytic tools with which to make sense of the properties of the many new particles discovered in the 1950s. Isospin, moreover, brought economy as well as order by grouping different particles into multiplets. As far as the strong interactions were concerned theorists had effectively fewer particles to deal with, since they had only to consider the overall properties of multiplets rather than the individual members of each. In the latter half of the 1950s many theorists attempted to build upon this success, looking for ways to achieve even greater economy by grouping particles into larger families. Their efforts culminated in 1961 with the 'Eightfold Way' or su(3) classification of hadrons. The content of the Eightfold Way scheme is perhaps most readily explained in terms of quarks, but this would be to invert the historical sequence. Here I will sketch out the process which led to the establishment of su(3); further clarification can await Chapter 4 wherein quarks themselves appear.

To explain what the Eightfold Way scheme entailed, it is useful to discuss isospin conservation in more formal terms. In theoretical physics there is a direct connection between conserved quantities and 'symmetries' of the underlying interaction. Thus the strong interaction, which conserves isospin, is said to possess an isospin symmetry. This corresponds to the fact that the choice of axis against which to measure the 3-component of isospin is arbitrary. A change of axis would entail different definitions of physical particles – if one reversed the direction of the 3-axis, for example, the π^+ would have $I_3 = -1$ instead of $+1$ – but this would be irrelevant as far as the strong interaction is concerned: the strong interaction is said to be 'invariant' under transformations of I_3. In mathematical terms, such changes of axes are known as 'symmetry operations', and form the basis of a branch of mathematics called 'group theory'. Different symmetries correspond to different symmetry groups, and the group associated with the arbitrariness of the orientation of the isospin axis is conventionally denoted as 'su(2)'.[21] Associated with each mathematical group is a set of 'representations' describing the families of objects which transform into one another under the relevant symmetry operation. The representations of su(2) can accommodate any integral number of objects, and each isospin multiplet can be identified with an appropriately sized representation of the group.

By making the connection with group theory, physicists translated the search for a classification wider than isospin into the search for a more comprehensive group structure – a higher symmetry of the strong interactions with representations suitable to accommodate particles of different isospin and strangeness. Many groups were tried and found wanting, and I will discuss only the one which was finally accepted. First proposed in 1961 by Murray Gell-Mann and the Israeli theorist Yuval Ne'eman, this was the group which mathematicians denote 'su(3)' and which Gell-Mann, in a reference to Buddhist philosophy, christened the Eightfold Way.[22] It is interesting to note that, in their original papers, neither Gell-Mann nor Ne'eman was aiming simply at the classification of hadrons. Both were also attempting to set up a detailed quantum field theory of the strong interaction. Indeed, as discussed further in Chapter 6, both formulated their proposals on the basis of gauge theory – the particular variant of quantum field theory destined to constitute the conceptual framework of the new physics.[23] However, the early 1960s saw the rapid development of a radical alternative to the field theory approach to strong-interaction physics (see Section 4 below) and field theory itself quickly went out of fashion.[24] Thus, very soon after its invention, su(3) was divorced from its roots in gauge theory and left to stand or fall on its merits as a classification system.

Those merits were the following. According to su(3), particles having the same spin and parity but different isospin and strangeness can be grouped into large families or multiplets. These families contain fixed numbers of particles – 1, 3, 6, 8, 10, 27, etc. – which are determined by the representation structure of su(3), and which fall into characteristic patterns when plotted against strangeness and the third component of isospin. Figure 3.2 shows the su(3) multiplet assignments for the lower mass mesons, and Figure 3.3 shows the lower mass baryon families. One point of confusion may possibly arise here. As indicated in Figure 3.3, the lower mass baryons can be assigned to 8 and 10 member families – an 'octet' and a 'decuplet' – corresponding to the 8 and 10 member representations of su(3). But the mesons are shown in Figure 3.2 as belonging to two 9 member families – 'nonets' – and from the list just quoted it is evident that there is no 9 member representation of su(3). The existence of meson nonets is explained by the quantum-mechanical concept of 'mixing'. Mixing comes about because one member of each su(3) meson octet has zero isospin and strangeness – the same quantum numbers as the su(3) singlet. Since su(3) is not an exact symmetry (see below) the singlet and octet mesons can mix: the physically observed isospin and strangeness zero mesons are not pure su(3) octets or singlets, but are instead mixtures of both. Thus, for mesons, octets and singlets lose their separate identities and are better represented as nonets.

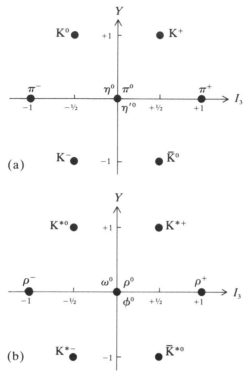

Figure 3.2. Su(3) classification of low-mass mesons: (a) spin 0 nonet; (b) spin 1 nonet. The hypercharge quantum number (Y) is given by the sum of the baryon and strangeness quantum numbers, and is the relevant group-theoretical variable.

Agreement that the su(3) multiplet structure was to be found in nature was only reached after several years' debate within the HEP community. One key problem was that early experiments indicated that the lambda and sigma baryons had opposite parity.[25] This made the su(3) assignment of the sigma and lambda to the same family (Figure 3.3(a)) impossible, and favoured alternative classificatory schemes. Only in 1963 did experimenters at the CERN PS report that the lambda and sigma had the same parity, making the su(3) assignment tenable.[26] A further obstacle to the acceptance of the Eightfold Way classification was that the mass differences within su(3) multiplets were of comparable size to the particle masses. This

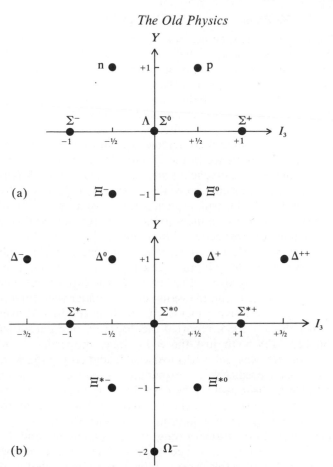

Figure 3.3. Su(3) classification of low-mass baryons: (a) spin $\frac{1}{2}$ octet; (b) spin $\frac{3}{2}$ decuplet.

was especially true of the lowest mass meson nonet, where the splitting between the pion and kaon masses was of the order of 360 MeV, around three times greater than the mass of the pion itself. Unlike the small splittings within isospin multiplets, such large mass differences could not be explained away as electromagnetic or weak perturbations: they were presumably intrinsic to the strong interaction. SU(3) was not thus an exact symmetry of the strong interaction like isospin; it was an approximate or 'broken' symmetry.

In order to deal with the novel problem of a broken symmetry, Gell-Mann and the Japanese theorist Susumu Okubo followed the strategy of assuming a simple but plausible form for the SU(3)-breaking component of the strong interaction, which they used to derive mass relations between members of su(3) multiplets.[27] The Gell-Mann–Okubo mass formula fitted the su(3) assignments for the nucleon octet and spin 1 meson nonet extremely well, and the spin 0 meson nonet at least fairly well. For the spin $\frac{3}{2}$ baryon decuplet shown in Figure 3.3(b), the formula reduced to an equal-spacing rule: the mass-differences between Δ and Σ^*, Σ^* and Ξ^*, and Ξ^* and Ω^- were all predicted to be equal. In 1961 only the Δ had been experimentally observed. At an international conference held at CERN in 1962 the first evidence for the existence of the Σ^* and Ξ^* was reported, with masses confirming the equal-spacing rule. Gell-Mann took this as an opportunity to make a prediction of the mass and properties of the hitherto unseen Ω^-.[28] The equal-spacing rule gave a mass of 1685 MeV for the Ω^-, and its assignment to the su(3) decuplet required it to carry three units of strangeness. This combination implied that it could not decay via the strong interactions, instead decaying only weakly and with a very long lifetime. Such a massive yet long-lived particle offered a highly distinctive signature for experiment, and attempts to observe the required events soon got under way. First to the post was a group of experimenters working at the Brookhaven AGS who, in February 1964, reported the discovery of a particle of mass 1686 ± 12 MeV with just the right decay properties.[29] The discovery of the Ω^- was only the most visible and straightforward manifestation of the predictive and explanatory successes of the su(3) scheme, and from 1964 onwards there was little argument within the HEP community against the conclusion that su(3) was the appropriate classification system for hadrons. 1964 was also the year in which quarks were born, transforming the su(3) approach and, for the time being, thrusting its gauge-theory origin into yet deeper obscurity. We will take up this thread of the story again in Chapter 4. First some more background on prior developments in theoretical HEP is needed.

3.3 Quantum Field Theory

The use of conservation laws, symmetry principles and group theory brought some order into the proliferation of particles. Su(3), for example, not only provided a classification scheme for hadrons but also predicted relations between cross-sections for the interaction of particles from different multiplets. But this was a broad approach and gave no clue to the detailed dynamics of the interactions which gave rise to those cross-sections. In pursuit of a detailed dynamical scheme, HEP theorists again picked up a set of tools inherited from their pre-war forebears and attempted to adapt them to contemporary needs.

Quantum Electrodynamics

Theoretical HEP's inheritance from the pre-war era was quantum field theory.[30] Quantum field theory is just what its name implies – the quantum-mechanical version of classical field theory. When the tools of quantum mechanics had been assembled in workable form in the 1920s, one of their first applications was to the field theory of electromagnetism developed by Maxwell. The quantised theory of the interactions of charged particles – quantum electrodynamics, or QED, as it became known – was successful in explaining an enormous range of topics in atomic physics, and this success contributed to the general acceptance of quantum mechanics itself. Even so, pre-war QED suffered from a theoretical disease which was only cured in the late 1940s. The cure transformed QED into the most powerful and accurate dynamical theory ever constructed, and was the single most influential theoretical achievement in the formative post-war years of HEP. The disease and its cure cast their shadow over all that followed and bear examination in some detail. We can begin with a general outline of calculational procedures in QED. This will provide essential background to the discussion of QED's disease, and will also serve to typify the traditional approach to the construction of field theories of the weak and strong interactions. For reasons which will quickly become evident, this traditional approach is often known as 'perturbative' or 'Lagrangian' field theory.

The standard way in which to formulate a quantum field theory is to construct a mathematical quantity known as the Lagrangian of that theory. In highly schematic notation, the Lagrangian for QED can be written as:[31]

$$\mathcal{L}(x) = \bar{\psi}(x)D\psi(x) + m\bar{\psi}(x)\psi(x) + (DA(x))^2$$
$$+ eA(x)\bar{\psi}(x)\psi(x).$$

Here $\mathcal{L}(x)$ is the so-called Lagrangian density at space–time point x, with $\psi(x)$ and $\bar{\psi}(x)$ representing the electron and positron fields at point x, and $A(x)$ being the electromagnetic field. D is a 'differential operator', so that $D\psi$ and DA represent field gradients in space–time. e and m represent the charge and mass of the electron respectively. As in classical electrodynamics, all of the properties of systems of charged particles are supposed to be derivable from the QED Lagrangian. I will not go into the mathematics of such derivations, but I will try to outline some of the salient features. The simplest way to do this is diagrammatically. Each term in the QED Lagrangian can be represented by a diagram, understood as a shorthand notation for a well-defined mathematical expression. The first term $\bar{\psi}D\psi$ gener-

61

ates the diagram of Figure 3.4(a). This is known as the electron 'propagator', and represents an electron (or positron) travelling freely through space in the direction indicated by the arrow. If only the first term were included in the Lagrangian then the mathematical expression associated with Figure 3.4(a) would be appropriate to a zero-mass electron. When the second term, $m\bar{\psi}\psi$, is included, the diagram is unchanged but the associated expression becomes that appropriate to an electron with mass m. Like the first term, the third term, $(DA)^2$, generates a diagram corresponding to a particle travelling freely through space. In this case the particle is the *photon*, the quantum of the electromagnetic field, as shown in Figure 3.4(b). Note that there is no mA^2 term in the Lagrangian analogous to the $m\bar{\psi}\psi$ term for the electron. This indicates that the photon is being represented as a zero-mass particle. The absence of a mass-term for the photon arises from the fact that electromagnetism is a long-range force which acts over macroscopic distances. According to the Uncertainty Principle the range over which forces act is limited by the mass of the particle responsible for them, and only zero-mass particles can give rise to forces appreciable over macroscopic distances.[32]

If only the first three terms are included in the Lagrangian, QED is an exactly soluble theory. The dynamics of any collection of electrons, positrons and photons can be exactly described – whatever particles are present simply propagate freely through space. Unfortunately, this 'free-field' solution is trivial. It is precisely the interactions of particles which HEP theorists seek to analyse, and these are not described by the first three terms of the QED Lagrangian. Enter the fourth term, $eA\bar{\psi}\psi$. Unlike the first three terms which are 'bilinear', containing only two fields, the fourth term is 'trilinear': it contains two electron fields and a photon field. It represents the fundamental *interaction* or *coupling* between electrons and photons, as indicated in Figure 3.4(c). Figure 3.4(c) represents the electron-photon *vertex*, and is associated with a mathematical expression essentially given by the magnitude of the electron charge, e.

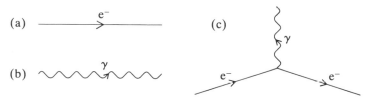

Figure 3.4. Basic diagrams in QED: (a) electron propagating through space; (b) photon propagating through space; (c) electron–photon vertex.

When diagrams of the type shown in Figure 3.4(c) are taken into consideration, QED becomes a physically interesting theory capable of describing interacting particles. Thus, for example, when the full Lagrangian is used, electron–electron scattering can be represented by Figure 3.5(a). Here two electrons (e⁻) travel in from the left as indicated by the arrows; they interact and scatter off one another as a photon (γ) carrying energy and momentum passes between them; and then move off to the right. A well-defined mathematical expression can be associated with Figure 3.5(a) and is found to reproduce quite well the measured cross-sections for electron–electron scattering. However, in QED there is no reason why Figure 3.5(a) should lead to exact predictions for the interactions of charged particles. In principle, processes of much greater complexity could intervene in the scattering process. For example, the exchanged photon could convert to an electron-positron pair which would subsequently recombine, as in Figure 3.5(b), or one of the incoming electrons might emit a photon and reabsorb it on the way out, as in Figure 3.5(c). And, in general, the exchange of arbitrarily large numbers of photons, electrons and positrons can contribute to electromagnetic interactions.

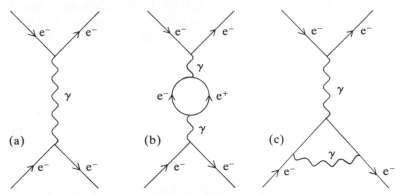

Figure 3.5. Electron–electron scattering in QED: (a) one-photon exchange; (b) and (c) more complex exchanges.

The inclusion of the interaction term in the QED Lagrangian makes the theory physically-interesting. But it also makes the theory extremely complex mathematically, since very complicated multiparticle exchanges have to be taken into account in the analysis of physical systems. Indeed, no exact solutions to the QED equations are known, nor have such solutions ever been shown rigorously to exist. There is, however, one consolation for the QED theorist. I noted above that the overall magnitude of the expression corresponding to the electron–photon vertex (Figure 3.4(c)) is given by e, the charge of the electron. The overall magnitude of Figure 3.5(a) is then given by e^2 – there are two electron–photon vertices, one for the emission of the photon the other for its absorption. The magnitude of the expressions for Figures 3.5(b) and (c), on the other hand, is e^4, since there are two photons and four vertices. Similarly, whenever three photons appear the magnitude of the appropriate expression is e^6 and, in general, for n photons the magnitude is e^{2n}. For every photon line added to a diagram, the magnitude of the associated expression changes by a factor of e^2. For purposes of practical computation this is extremely important. e^2 is a small number (roughly $1/137$), which means that QED diagrams can be arranged in order of decreasing size – a so-called 'perturbation expansion' – making possible the computation of sensible and systematic approximations to the unknown exact predictions of QED. In QED, the dominant contribution – the 'first-order' approximation – to electron–electron scattering is given just by the one-photon exchange diagram, Figure 3.5(a). The theorist can calculate this, and feel confident that diagrams involving more photons will contribute terms which are at least 137 times smaller and less important. The diagrams involving more photons correspond to small corrections or 'perturbations' relative to the first-order term. For greater precision, the theorist can calculate the second-order approximation corresponding to diagrams involving two photons, such as those of Figures 3.5(b) and (c), and still feel confident that the terms which he is omitting are at least 137 times smaller than those which he is including. The even more determined theorist can go to the third-order approximation by calculating and summing the three-photon diagrams, and so on. In this way more and more precise approximations to the exact but unknown predictions of QED can be systematically developed by evaluating successive terms in the perturbation expansion.

One problem with the perturbative approach to QED remains to be discussed, the theoretical disease mentioned above, but before we come to that it may be useful to clarify some basic features of the quantum field theory approach. One general feature which is implicit in the diagrammatic analysis but which should be made explicit is that in quantum field theory *all forces are mediated by particle exchange*: action-at-a-distance plays no part in quantum field theory. In QED the force-transmitting particle is the photon; the analogous particles for the weak and strong interactions will be discussed below. It is equally important to stress that the exchanged particles in the diagrams are *not observable*: the diagrams of Figure 3.5 should not be thought of as resembling, say, bubble-chamber pictures. In a bubble chamber, it would be possible to photograph the tracks of the incoming and outgoing electrons, but the exchanged particles – the photons and electron–positron pairs – would leave no record. To explain why this is so, it is necessary to make a distinction between 'real' and 'virtual' particles. According to the theory of special relativity, there is a fixed relation between a particle's energy (E), momentum (p) and mass (m): $E^2 = p^2 + m^2$.[33] All observable particles, like the incoming and outgoing electrons in Figure 3.5, obey this relation and are said to be 'real'. If one applies the law of energy–momentum conservation to, say, Figure 3.5(a), one finds that the photon is not real: $E^2 - p^2$ is not zero for this photon, as it should be for a real zero-mass particle. A similar calculation for the intermediate electron and positron of Figure 3.5(b) shows that they are not real either: their energies and momenta do not conform to the relation $E^2 - p^2 = m^2$. Such particles with unphysical values of energy and momentum are said to be 'virtual' or 'off mass-shell' particles. In classical physics they could not exist at all – the concept of a virtual particle is classically meaningless. In quantum physics, in consequence of the Uncertainty Principle, virtual particles can exist, but only for an infinitesimal and experimentally undetectable length of time. In fact, the lifetime of a virtual particle is inversely dependent upon how far its mass diverges from its physical value. Only when the energy and momentum of a particle conform to relation $E^2 - p^2 = m^2$ can the particle persist indefinitely through time – as one expects real particles to do.

Renormalisation

We now come to the disease of QED and its eventual cure. From the perspective of calculation there are two important differences between diagrams (b) and (c) and diagram (a) of Figure 3.5. On the one hand, diagrams (b) and (c) contain two photons and are thus a factor of 137 times smaller than diagram (a) (for the sake of brevity, I shall speak of diagrams as equivalent to the associated mathematical expressions from now on). On the other hand, diagrams (b) and (c) contain 'closed loops' of particles – the e^+e^- loop of diagram (b) and the $e^-e^-\gamma$ loop of diagram (c) – which create great computational difficulties. This is because although the total flow of energy–momentum through any loop is given by the law of energy–momentum conservation, the way in which the energy–momentum is shared between the particles of the loop is not so determined. For example, whatever energy–momentum is carried by the virtual photons of diagram (b) can be divided in an infinite number of ways between the virtual electron and positron which make up the intermediate loop.

Calculation of diagram (b) requires that one take this into account by summing over the infinite range of momenta of, say, the virtual positron. Now, in principle there is no problem in performing these sums as mathematical integrals over the infinite range of loop momenta. But when one performs them they 'diverge' – the integrals are formally infinite. Diagrams with closed loops – (b) and (c) of Figure 3.5, and the vastly more complicated diagrams which appear at higher orders of approximation in QED – are all infinite.[34] But these diagrams are supposed to correspond to experimentally measurable quantities, like the cross-section for electron–electron scattering, which are manifestly not infinite but finite. Thus much was recognised by pre-war theoretical physicists as the underlying disease of QED. The infinities caused no great practical problems – for most applications physicists could content themselves with the first-order approximation to QED (Figure 3.5(a), which contains no closed loops) and forget the rest – but, even so, they were taken by many to indicate a defect in the conceptual fabric.[35] As HEP theorist Steven Weinberg wrote: 'Throughout the 1930s, the accepted wisdom was that quantum field theory was in fact no good, that it might be useful here and there as a stopgap, but that something radically new would have to be added in order for it to make sense'.[36] Possible revisions included hypothesising a granular structure of space–time or abandoning field theory altogether, but these went by the board in the late 1940s.

In June 1947 a four day conference on the foundations of quantum mechanics was held at Shelter Island, New York. The centre of attention was a precision measurement of the hydrogen atom spectrum made by Willis Lamb and R.C. Retherford. Using microwave techniques developed during the war, Lamb and Retherford found that the energies of the first two excited states of hydrogen, predicted to be equal by the first-order QED calculation, differed by about 0.4 parts per million. The measurement of the 'Lamb shift', indisputably a second-order effect if QED were to make sense, brought matters to a head and resulted in the triumph of the 'renormalisation' programme. This programme ultimately accomplished the following. If one computes an electromagnetic process – such as the electron–electron scattering shown in Figure 3.5 – to an arbitrarily high order of approximation, including an indefinite number of closed loops, one finds that only a small number of distinct types of infinities occur. These can be understood as infinite contributions to the electron mass, charge and so on. If one puts up with this and, at the end of the calculation, sets the apparently infinite mass and charge of the electron equal to their measured values, one arrives at perfectly sensible results. This 'renormalisation' of the theory by absorption of infinities into physical constants appeared intuitively suspect. But it immediately gave a quantitative explanation of the Lamb shift and subsequently went from strength to strength. In renormalised QED, calculations of electromagnetic processes could be pushed to arbitrarily high orders of approximation and always seemed to come out right. As Weinberg remarked in 1977:

> For instance, right now the experimental value of the magnetic moment of the electron is larger than the Dirac value by 1.15965241 parts per thousand, whereas theory gives this anomalous magnetic moment as 1.15965234 parts per thousand, with uncertainties of about 0.00000020 and 0.00000031 parts per thousand, respectively. The precision of agreement between theory and experiment here can only be called spectacular.[37]

The renormalisation idea had been suggested before the war by Weisskopf and Kramers, and was carried through in the late 1940s by Sin-Itiro Tomonaga and his colleagues in Japan, and by Julian Schwinger and Richard Feynman in the USA. In 1949 the British theorist Freeman Dyson completed the proof of the renormalisability of QED to all orders of approximation and showed the equivalence of the different approaches of Tomonaga, Schwinger and Feynman; these three later shared the 1965 Nobel Prize for Physics for their work. The most intuitively transparent renormalisation procedure was that developed by Feynman. His approach utilised diagrams like those in Figures 3.4 and 3.5, which subsequently became known as 'Feynman diagrams'. He derived a set of rules for associating a mathematical expression with each diagram, and these became known as 'Feynman rules'. The divergent integrals became 'Feynman integrals'.[38] The diagrammatic approach characterised much of the subsequent development of quantum field theory – as Schwinger recalled in 1980: 'Like today's silicon chip, Feynman diagrams brought calculation to the masses'.[39]

The success of the renormalisation programme transformed the quantised version of that most venerable of field theories, Maxwell's electromagnetism, into a theoretical instrument capable of describing the electromagnetic interactions of charged particles to an apparently arbitrary degree of accuracy. Not surprisingly, therefore, in the early 1950s almost all HEP theorists attempted to build upon this success and to reach a fundamental understanding of the remaining forces – the weak and strong interactions – in terms of quantised theories of appropriate fields. Soon, however, this enterprise was looking even sicker than pre-war QED. To quote Weinberg again:

> For a few years after 1949, enthusiasm for quantum field theory

was at a high level. Many theorists expected that it would soon lead to an understanding of all microscopic phenomena, not only the dynamics of photons, electrons and positrons. However, it was not long before there was another collapse in confidence – shares in quantum field theory tumbled on the physics bourse, and there began a second depression, which was to last for almost twenty years.[40]

The depression lasted until 1971, when there was an explosion of interest in one particular class of quantum field theories – gauge theories. We can pick up this thread of the story in Chapter 6 below. For the moment, the question is why did the depression begin? The quantum field theory approaches to the weak interaction and to the strong interaction failed for different reasons. Let us examine them each in turn.

The Weak Interaction

Beta decay, the emission of electrons and positrons from unstable nuclei, was one of the principal themes of radioactivity research in the early decades of the twentieth century. It was an especially puzzling phenomenon in that electrons were emitted over a range of energies; their energy spectrum was continuous, rather than discrete as expected for transitions occurring in a quantised system.[41] Amongst the fathers of quantum mechanics, Bohr, Heisenberg and Pauli each proposed radical explanations for this observation – Bohr, that energy was not exactly conserved; Heisenberg, that space–time was not continuous – but it was Pauli's proposal that won the day. In 1930, Pauli suggested that in beta-decay two particles rather than one were emitted: the electron was accompanied by an electrically neutral massless lepton which carried off energy undetected. Because the latter particle, the *neutrino*, went unseen, a truly quantised total energy spectrum appeared experimentally as a continuous spectrum of electron energies. In 1934, Enrico Fermi formalised Pauli's suggestion by constructing a quantum field theory of the force responsible for beta-decay, the weak interaction.[42] Fermi modelled his theory as closely as possible upon the existing theory of QED. Figure 3.6 will help to explain how he did this.

Figure 3.6(a) shows a possible mechanism for the weak decay of the neutron, directly modelled upon QED. The neutron converts into a proton by emitting a particle, labelled W, which is the weak-interaction analogue of the photon. The W particle is a prototypical intermediate vector boson, of the type already mentioned in connection with the weak interactions in Chapter 2.3 (recall that particles which have spin 1, like the photon and the hypothetical W, are

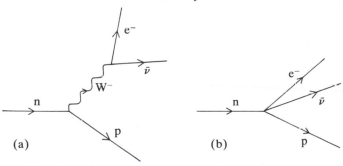

Figure 3.6. Neutron beta-decay.

generically known as vector particles). Because the neutron and proton differ in electric charge by one unit, the W itself must carry a charge (unlike the photon which is electrically neutral). The emitted W-particle then disintegrates into an electron–antineutrino (e–ν̄) pair, in the same way as a photon can convert into an electron–positron pair. However, Figure 3.6(a) does not correspond exactly to the weak-interaction theory enunciated by Fermi. The photon has zero mass which, according to the Uncertainty Principle, corresponds to a force of infinite range (as necessary to represent the macroscopic effects of electromagnetism). The weak force was known to be of short range, even on the typical length-scale of nuclear physics (10^{-13} cm). According to the Uncertainty Principle again, short-range forces correspond to the exchange of massive particles, and Fermi effectively assigned the W infinite mass by contracting the weak interactions to a point in space, as shown in Figure 3.6(b). The interaction shown there was known as a current–current interaction, since it was analogous to two electric currents (here, n–p and e–ν̄) interacting directly without exchanging a photon. In the following years, a great deal of evidence was amassed in favour of Fermi's theory, but one key element was lacking – evidence for the existence of the neutrino itself.[43] This was forthcoming in 1953, when events induced by the neutrino flux from a nuclear reactor were reported by US experimenters F.Reines and C.L.Cowan.[44]

In 1956, HEP theorists T.D.Lee and C.N.Yang proposed, in response to the so-called 'τ–θ' problem concerning the weak decays of kaons, that parity conservation might be violated by the weak interaction. This proposal was quickly confirmed in several nuclear physics and HEP experiments, and Lee and Yang shared the 1957 Nobel Prize for Physics for their part in this development.[45] The discovery of parity nonconservation required detailed amendments to weak interaction theory, but the general form of Fermi's current–current interaction emerged unscathed in the 'V minus A' (V – A) theory of the weak-interactions published in 1958 by Feynman and Gell-Mann, and independently by two other US theorists, R. Marshak and E. Sudarshan.[46] Following the discovery of parity nonconservation, it was argued that the weak interaction respected a less restricted conservation law known as CP invariance. Although parity was no longer supposed to be a good quantum number of the weak interaction, the combined operation of charge-conjugation (C: replacing particles by antiparticles) and space reflection (P: inverting spatial coordinates) was still supposed to yield a good quantum number, CP. However, in 1964 further experiments on kaon decays pointed to the conclusion that the weak interactions did not even conserve CP.[47] Theorists accommodated to this observation as they had to parity-violation, and once more the overall current–current picture of the weak interaction emerged unscathed.[48]

Throughout the 1950s and 1960s, then, the current–current quantum field theory was, in various guises, successful in organising a wide range of weak-interaction data. It suffered, though, from two major theoretical shortcomings. The first was that it was non-renormalisable. As in the case of QED, the diagram of Figure 3.6(b) represents not the entire content of the Fermi theory but only the first approximation to its solution. Higher-order approximations correspond to more complex diagrams in which closed loops and numerically infinite integrals appear. When they came to compute these higher-order diagrams, theorists concluded that current–current theories involved ever more types of infinity at succeeding levels of the perturbation expansion. This plethora of infinities could not be conjured away by redefining a few parameters (as was done in QED) and higher-order calculations therefore appeared meaningless. In this respect the current–current weak-interaction theory was sicker than QED had ever been, and this was the first roadblock encountered by the attempts to model the weak interactions on electromagnetism.

One response to the sickness of the Fermi theory was the same as that taken by many theorists to the pre-war sickness of QED: to use the well-defined first-order diagram for calculation of weak-interaction processes, and to disregard higher-order corrections. But here, too, clouds loomed on the horizon. According to the Feynman rules for two currents interacting at a point in space, weak cross-sections were expected to increase in direct proportion to beam energies. This was nice for experimenters, since it promised easier detection of weak processes as higher-energy beams became available. On the other hand, it was obvious that at sufficiently high energy, weak cross-sections must violate the 'unitarity limit'. Unitarity was simply a statement of the conservation of probabilities in HEP terms. As such it was strongly held, and violation of unitarity, even at energies only attainable with some future machine, was regarded as pointing to a failing of the first-order diagram.

One way in which to avert, or at least postpone, the 'unitarity catastrophe' was to work with theories of the type shown in Figure 3.6(a). In these, the weak interaction was mediated by a charged W-particle of finite mass, an intermediate vector boson, with the result that cross-sections were predicted to be better behaved at high energies. Theoretically, this line of thought was one strand leading eventually to the unified electroweak gauge theories discussed in Chapter 6. Experimentally, W-particle theories inspired a redirection of weak-interaction physics. In the 1940s and 1950s, experimental investigations had focused upon weak *decay* processes – the systematics of pion, kaon, lambda, sigma decay, and so on. In the 1960s experiments began which instead hoped to produce W-particles in high-energy *scattering* experiments using either neutrino or hadron beams. None of these experiments bore the expected fruit of the W (until, perhaps, 1983). But, as discussed in later chapters, they quite unintentionally laid the experimental basis of the new physics in PS experiments.

The Strong Interaction
The quantum field theory of the weak interaction, despite its manifest theoretical shortcomings, offered a coherent organising principle for weak-interaction research in the early decades of HEP. This could not be said of the attempts which were made at a field theory of the strong interactions. While the failings of the Fermi theory were problems of principle, those of strong-interaction field theories were immediate and pressing.

The prototypical quantum field theory of the strong interaction was that proposed by Yukawa in 1935.[49] Like Fermi's theory of the weak interaction, Yukawa's was explicitly modelled upon QED. Following the reasoning discussed above, Yukawa argued that since the strong interaction was a short-range force, it must be carried by a non-zero-mass particle. Typical nuclear dimensions were of the order of 10^{-13} cm, and Yukawa suggested that this would correspond to the existence of a particle having a mass of around 100 MeV. The properties of a suitable particle – the 140 MeV pion – were established in cosmic-ray and accelerator experiments in the late 1940s, and for a while the situation seemed rather straightforward. Theorists found no difficulty in writing down renormalisable field theories of the strong interaction in which, for instance, nucleons interact with one another via the exchange of pions, as shown in Figure 3.7(a). Neither, as more particles were discovered, did they find it difficult to arrange for the conservation of isospin and strangeness: Figure 3.7(b) shows the associated production of kaons and lambdas in the interaction of pions and protons.

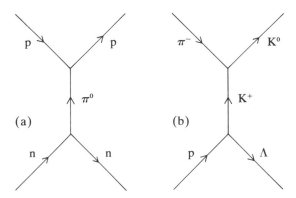

Figure 3.7. Strong interactions in Yukawa theory.

But despite their renormalisability and their respect for conservation laws, quantum field theories of the strong interaction were quickly perceived to be useless. The problem lay in a breakdown of the perturbation expansion. In QED, the contributions of complicated diagrams were suppressed by the appearance of factors of e^2. e^2 was a small number (1/137) and complex multi-loop diagrams made a similarly small contribution to the perturbation expansion. The strong-interaction analogue of e was the strong-interaction coupling constant (g) which fixed the strength of the hadron–hadron vertices (such as the meson–meson and meson–nucleon vertices of Figure 3.7). And the experimentally determined value of g^2 was around 15. Thus successively more complicated diagrams in strong-interaction field theories grew by a factor of 15, rather than diminishing by a factor of 137 as in QED.

In field theories of the strong interaction, then, it made little sense to rely upon a first-order term which was followed by an infinite series of 'perturbations' of increasingly *greater* magnitude. This was the conclusion of the HEP theorists who attempted to develop field theories of the strong interaction. They drew the moral that in order to make any kind of predictions in such theories, it was necessary first to sum the entire perturbation series in its infinite complexity. Unfortunately, they had no idea how to do this, and the field-theory approach to the strong interaction therefore appeared doomed to failure.[50] As Weinberg later wrote: 'It was not that there was any difficulty in thinking of renormalisable quantum field theories that *might* account for the strong interactions – it was just that having thought of such a theory, there was no way to use it to derive reliable quantitative predictions, and to test if it were true'.[51]

Thus, by the mid-1950s shares had tumbled in the field-theory approach to the weak and strong interactions. The problems besetting theorists were far greater in respect of the strong interaction, in terms of both the range of data to be explained and of immediate calculational impotence. In the following section we can see how they accommodated themselves to this uncomfortable situation.

3.4 The S-Matrix

The central problem with field theories of the strong interaction lay in their description of hadronic processes as a perturbation series of Feynman diagrams. Theorists did not know how to sum this series. However, they reasoned, the individual terms of the series have no necessary physical significance: one cannot directly observe nucleons, say, exchanging pions as they scatter in an HEP experiment. All that one can observe is that, in the course of an experiment, an initial collection of particles of specified quantum numbers and momenta is transformed into a different final state. All that can be known, therefore, is enshrined in the set of transition probabilities that from given initial states one will arrive at given final states. Thus, in the 1950s one reaction against the failure of field theory in strong-interaction physics was to turn away from intractable Feynman diagrams towards the transition probabilities themselves – perhaps at this level of analysis some sense could be made of the strong interaction.

For historical reasons, the entire array of probabilities covering transitions between all conceivable initial and final states was known as the Scattering- or S-matrix.[52] The basic framework of the S-matrix approach to HEP was laid down in the immediate post-war period, largely in response to the problem of infinities then plaguing QED. With the advent of renormalisation little further attention was paid to the S-matrix – until, that is, the problems of strong-interaction theory began to be acutely felt. From the mid-1950s onwards, an increasing number of theorists turned their attention to the S-matrix. Since they had no other theoretical tools available they inevitably – but ironically, in view of some later developments – used quantum field theory and diagrammatic techniques to explore the general properties of the S-matrix. As US HEP theorist Geoffrey Chew wrote in 1966: 'All results at this stage [1958] . . . were either motivated by, or derived from, field theory'.[53] Between 1956 and 1959, the work of Chew, Gell-Mann and others, established a theoretical result of primary significance: the S-matrix could be regarded as an 'analytic' function of the relevant variables.[54] The concept of an analytic function was one which was already well explored in the branch of mathematics concerned with functions of 'complex' variables.[55] This branch of mathematics therefore provided a pool of conceptual resources with which theorists could elaborate models of the S-matrix in its own right as an analytic function. Thus it became possible to proceed with S-matrix theory as an autonomous research programme, independently of its origins in quantum field theory. In the early 1960s, many HEP theorists took this opportunity to abandon the traditional field-theory approach to particle interactions. Some theorists persisted with diagrammatic, perturbative field-theory techniques, but now used them heuristically in sophisticated investigations of the analytic properties of the S-matrix.[56] Alternatively, and more adventurously, some theorists pronounced field theory dead – at least in respect of the strong interactions. The leader and spokesman of this group was Chew. From his base at the University of California at Berkeley Chew expounded the explicitly anti-field-theory 'bootstrap' philosophy.[57]

The essence of the bootstrap was this. Consideration of the analytic structure of the S-matrix led to an infinite set of coupled non-linear differential equations. No one knew how to solve this set of equations exactly – it would have amounted to a solution of strong-interaction physics – but Chew proposed that it did have a solution; that the solution was unique; and that, through the requirement of self-consistency, it determined all of the properties of all of the hadrons. If this were the case, the general properties of the S-matrix determined *everything* about hadrons, and reference to field theory was, at best, a waste of time. Chew described his approach as 'democratic' – since all of the hadrons, stable and unstable, were supposed to pull themselves up by their bootstraps as a self-consistent solution to the S-matrix equations – in contrast to the 'aristocratic' QFT approach, wherein each particle was assigned its own quantum field. The bootstrap was a statement of faith on Chew's part. Since the infinite set of equations could not be solved, it was hard to evaluate his assertions. Nevertheless, ways were found of implementing an approximate bootstrap programme. By truncating the infinite set of equations, for example, it was possible to calculate self-consistently the properties of rho mesons using as input only the properties of pions,[58] and to derive an SU(3) hadronic symmetry using as input only isospin.[59] Such approximate implementations of the bootstrap programme made it a viable alternative to the symmetry group approach to resonance physics in the early 1960s (until, that is, the advent of detailed quark-model analyses).

One final aspect of the S-matrix/bootstrap approach to strong-interaction physics remains to be discussed: Regge theory. The point of departure for Regge theory lay in work done in the late 1950s by the Italian theorist Tullio Regge. Working on formal problems connected with non-relativistic potential scattering, Regge pointed out the utility of examining analytic structure in terms of complex energy and angular momentum variables, rather than in terms of energy and momentum transfer.[60] Regge's work was seized upon by Chew and his collaborators and translated into relativistic, S-matrix language appropriate to the subject matter of HEP.[61] They argued that at high energies and small momentum transfers, the behaviour of the S-matrix could be understood in terms of the properties of a small number of 'Regge poles' – quasi-particles whose spin depended upon their energy. Whenever the energy of a Regge pole was such that its spin assumed a physical (integral or half-integral) value, it was supposed to manifest itself as an observable hadron. And, indeed, it was found experimentally that mesons and baryons appeared to lie upon 'Regge trajectories' with a linear relation between masses and spins, as shown in Figure 3.8.[62]

Thus the Regge picture led to a novel classification of hadrons according to the trajectory on which they lay. More importantly, though, Regge theorists argued that their analysis was applicable to the high-energy soft-scattering regime which lay above the resonance region. In quantum-mechanical terms, scattering cross-sections were essentially squares of quantities known as scattering amplitudes (A). In the Regge picture, high-energy amplitudes had the form

$$A(s,t) \sim \Gamma(s/s_0)^{\alpha(t)}$$

where s was the square of the centre-of-mass energy, t the square of momentum transfer, $\alpha(t)$ the spin of the appropriate Regge trajectory, and Γ and s_0 scale factors. Amplitudes of this form led immediately to two key predictions. First, high-energy cross-sections should depend smoothly upon s (for example, at zero momentum transfer, $A(s) \sim s^{\alpha(0)}$); secondly, cross-sections should be soft, falling-off exponentially fast with t, that is $A(t) \sim e^{\alpha' t \ln(s/s_0)}$, where a linear form, $\alpha(t) = \alpha(0) + \alpha' t$, has been assumed, to match the observed linear Regge trajectories. These two features were found to be characteristic of high-energy hadronic scattering. Figure 3.1 above shows the dependence upon beam momentum of the total and elastic

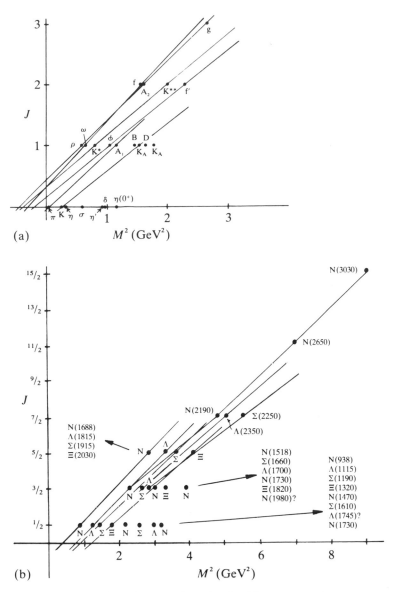

Figure 3.8. Various Regge trajectories for (a) mesons, and (b) baryons. Note the linear relation between spin (*J*) and mass-squared (*M*²).

cross-sections for pion–proton scattering. The bumpy resonance region extends up to beam momenta of around 2 GeV; beyond that cross-sections show the smooth variation predicted by Regge. Figure 3.9 shows the momentum-transfer dependence of high-energy pion–proton elastic cross-sections.[63] These decrease linearly with t when plotted on a logarithmic scale, corresponding again to the exponential fall-off expected in Regge theory.

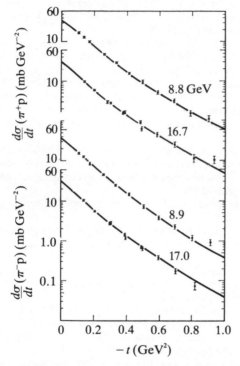

Figure 3.9. High-energy elastic cross-sections ($d\sigma/dt$) for π^+p and π^-p scattering as a function of the square of momentum transfer (t). The curves are labelled by the pion beam momenta at which they were measured.

Experiment at the major accelerator laboratories generated enormous quantities of high-energy soft-scattering data during the 1960s. Regge theory offered a well matched set of analytical resources. Together, soft-scattering experiment and Regge theory constituted the high-energy wing of the old physics.[64] Regge theory led in the late 1960s to the 'duality' conjecture and to the 'Veneziano' or 'dual resonance model' of hadrons.[65] But these developments were at most

The Prehistory of High-Energy Physics

tangential to the development of the new physics. To explore them here would lead us too far afield. Instead we must return to the low-energy branch of the old physics, where we can examine the origins of the quark concept.

NOTES AND REFERENCES

1   For the pre-war constitution of elementary particle physics as a distinct field of research emerging from the interplay of nuclear physics, cosmic ray physics, and quantum field theory, see Brown and Hoddeson (1982, 1983), Cassidy (1981) and Steuwer (1979).

2   Dr S. Wagner has made a keyword analysis of the titles and abstracts of HEP publications from 1947 to 1974. He finds that in the latter half of the 1960s, Regge papers outweighed those on quarks and the Eightfold Way. The Regge publication rate reached a peak of more than 700 papers per year in 1970. Combined publication rates on quarks and the Eightfold Way were around 600 per year during this period. I am grateful to Dr Wagner for permission to quote his unpublished data.

3   Marshak (1952, 23).

4   For popular accounts of the early years of the population explosion, see Gell-Mann and Rosenbaum (1957) and Chew, Gell-Mann and Rosenfeld (1964). For technical discussion and references to the literature, see Alvarez (1970) and Rosenfeld (1975).

5   The Greek and Latin letters used to name particles have only conventional significance.

6   This follows immediately from track length and the fact that particles travel at close to the speed of light (3×10^{10} cm/s) in HEP experiments.

7   Particle Data Group (1982, 48).

8   Nuclear resonances, for example, were regarded as energetically excited states of existing nuclei, and not as new species of nuclei in their own right. The calculation of the delta lifetime followed routine lines already laid down in nuclear physics: according to the Uncertainty Principle of quantum mechanics, the lifetime of resonances was assumed to be inversely proportional to the width of resonance peaks. The delta peak had a width of around 100 MeV, leading directly to the 10^{-23} seconds lifetime estimate. Similar peaks due to quasi-stable particles were unmeasurably narrow.

9   Chew, Gell-Mann and Rosenfeld (1964, 74).

10   Alvarez (1970, 3) recalls that following the Bevatron discovery, 'one of the most distinguished high energy physicists I know, who didn't believe that antiprotons could be produced, was obliged to settle a 500-dollar bet with a colleague who held the now universally accepted belief that all particles can exist in an antistate.'

11   For a popular account of the topics reviewed in this section, see Chew, Gell-Mann and Rosenfeld (1964).

12   Recall that mass and energy are interconvertible according to the theory of relativity. If the combined mass of the decay products is less than that of the parent, the excess mass-energy of the parent appears as kinetic energy of the products.

13 $\hbar = h/2\pi$, where h is Planck's constant.
14 Sec Goudsmit (1976) and Uhlenbeck (1976).
15 The significance of the general distinction between fermions and bosons lay in the different 'statistics' which each obeyed. This is discussed further in Section 7.2.
16 In quantum mechanics, physical systems are described by mathematical constructs known as 'wave-functions'. One can devise a 'parity operation', in which the coordinates of all of the particles within the system are reflected into their mirror-images. If the wave-function remains the same under this operation, then the system is said to be in a state of positive parity. If the wave-function changes sign, then the state is one of negative parity.
17 Pais (1952).
18 Gell-Mann (1953, 1956), Nakano and Nishijima (1953), Nishijima (1954).
19 In retrospect, the introduction of strangeness seems almost trivial. At the time, though, the situation was very complex, and resolution of the problem of associated production required a simultaneous reassignment of isospin values (see below) to the kaon and lambda: see Gell-Mann and Rosenbaum (1957).
20 For the history of the isospin concept, see Mukherji (1974) and Kemmer (1982).
21 Su(2) stands for special unitary group in 2 dimensions.
22 Gell-Mann (1961), Ne'eman (1961). These and other seminal papers on Su(3) are reprinted in Gell-Mann and Ne'eman (1964). For the history of the early years of Su(3) and a fascinating autobiographical sketch, see Ne'eman (1983). Gell-Mann's reference to the Eightfold Way derived from the fact that in group-theory terms Su(3) has eight 'generators'.
23 Ne'eman (1961) entitled his paper, 'Derivation of the Strong Interactions from a Gauge Invariance'. Gell-Mann remarked that 'The most attractive feature of the scheme is that it permits the description of eight vector mesons by a unified theory of the Yang-Mills [gauge theory] type' (Gell-Mann 1961, 1).
24 Ne'eman (1983, 15) recalls: 'Within a year [from 1961, field theory] had gone out of fashion. With the anti-field-theory drive . . . one had to apologize for using a Lagrangian [field theory]. Even the fact that *the gauge predictions worked* . . . did not matter. One even had to find an excuse for that fact . . . by the end of 1963 I was also refraining from using Lagrangians.'
25 Nambu and Sakurai (1961).
26 Courant *et al.* (1963).
27 Gell-Mann (1961), Okubo (1962).
28 Gell-Mann (1962a). Ne'eman (1983, 19) recalls that he was present at the same conference and intended to announce the same prediction. Gell-Mann got the first opportunity to speak, and took his chance to give the Ω^- its presently accepted name.
29 Barnes *et al.* (1964). For a popular account by the experimenters of the Ω^- discovery, see Fowler and Samios (1964). The discovery was the culmination of a race between experimenters at Brookhaven and CERN:

see Gaston (1973, 83–8). It is interesting to note that, in Gaston's interviews, Brookhaven physicists stressed the importance which they attached to obtaining good publicity for their discovery in order to strengthen their claims for future funding.

30 For outline histories of quantum field theory, see Weinberg (1977a), Redhead (1980) and Cushing (1982). For a popular account, see Weisskopf (1981).

31 Amongst other things, I have suppressed the space–time indices and the matrix structure of this equation, and I have been cavalier with phase-factors and differential operators. These simplifications are irrelevant to the subsequent discussion.

32 The Uncertainty Principle is a fundamental result of quantum mechanics. As first formulated by Werner Heisenberg in 1927, it asserts that the more closely one tries to measure the position of a particle, the less closely can one measure its momentum, and vice-versa. In mathematical terms $\Delta p \Delta x \gtrsim 1$ where Δp is the uncertainty in a particle's momentum and Δx is the uncertainty of its position (I have set Planck's constant equal to one). The relation between particle masses and the range of forces arises as follows. Creation of a zero-mass particle, like the photon, requires only an infinitesimal amount of energy and momentum. Therefore Δp can be very small, and Δx very large – corresponding to a long-range force, falling off according to an inverse square law (as $1/x^2$). Creation of a finite-mass particle requires a finite amount of energy and momentum (Δp) and thus implies a finite Δx. This corresponds to a short-range force, falling off exponentially fast (as e^{-mx}). The Uncertainty Principle can be reformulated in terms of energy (E) and time (t) variables as $\Delta E \Delta t \gtrsim 1$. This leads to the relation between resonance lifetimes and widths referred to in note 8 above.

33 If the particle has zero momentum, this equation reduces to the well-known relation $E^2 = m^2$, expressing the relativistic equivalence of energy and mass. The more usual form of this equation is $E^2 = m^2 c^4$ or $E = mc^2$ where c is the velocity of light, but I have written it in units such that $c = 1$ as is conventional in HEP.

34 Integrals arise of the form $\int_{-\Lambda}^{+\Lambda} d^4p/p^2$ and $\int_{-\Lambda}^{+\Lambda} d^4p/p^4$. Here p^2 is the square of the momentum running around the loop, and the integrals are over the four independent components of the momentum. These integrals diverge, in the sense that they have the values Λ^2 and $\log \Lambda$ respectively, and both become infinite when Λ is set to infinity. (Here, as elsewhere, it should be noted that, in relativistic computations, energy (E) and the three components of linear momentum (p_x, p_y, p_z) are combined in a four-component vector (E, p_x, p_y, p_z). The square of this vector, denoted p^2, is given by $E^2 - (p_x^2 + p_y^2 + p_z^2)$.)

35 For a more detailed discussion of the problem of infinities and references to the literature, see Weinberg (1977a). For an account of perceptions of the problems of QED in the 1930s, and of the often drastic conclusions which were held to follow from them, see Cassidy (1981).

36 Weinberg (1977a, 25).

37 Weinberg (1977a, 29).

38 For an account of Feynman's route to the renormalisation of QED, see his Nobel lecture: Feynman (1966). The seminal papers on a renormalised QED are collected in Schwinger (1958). For a biographical account of the renormalisation of QED, see Chapters 5 to 7 of Dyson (1979).

39 Remark at the International Symposium on the History of Particle Physics, Fermilab, May 1980.

40 Weinberg (1977a, 30).

41 For the early history of radioactivity research, see Trenn (1977).

42 For the early development of weak-interaction theory, see Brown (1978).

43 For developments concerning the status of neutrinos up to 1940, see Morton (1982).

44 Reines and Cowan (1953, 1956).

45 For the history of the discovery of parity nonconservation, see Franklin (1979). For the history of strange-particle physics up to and including the tau–theta problem, see Adair and Fowler (1963, 1–15).

46 Feynman and Gell-Mann (1958), Sudarshan and Marshak (1958). Fermi's original theory had postulated that the weak interaction had a vector (V) character, in direct analogy with the electromagnetic interaction (the photon is a spin 1, negative parity, vector particle). To explain parity-violation, theorists required the current–current interaction to be a mixture of vector and axial vector (A, spin one, positive parity) parts. Hence the notation 'V–A'. For the history of the establishment of the V–A theory, see Hafner and Presswood (1965).

47 The first observation of CP-violation was made in an experiment at the Brookhaven AGS led by US physicists Val Fitch and James Cronin, who shared the 1980 Nobel Prize in Physics for this work (Christenson *et al.* 1964). For extended discussions of this episode, see Fitch (1981), Cronin (1981), Kabir (1979) and Franklin (1983).

48 Quantitative sociological data and brief historical accounts of the discovery of parity nonconservation, of an apparent experimental falsification of the V–A theory in the early 1960s, and of CP non-conservation are given in White, Sullivan and Barboni (1979), White and Sullivan (1979) and Sullivan, Barboni and White (1981). These studies of citation and co-citation rates make it abundantly clear that although the basic current–current picture of the weak interactions persisted through all of these episodes, each in turn constituted a major focus for experimental and theoretical practice within HEP.

49 Yukawa originally intended his theory to apply to both the strong and weak interactions. I am ignoring the latter aspect for simplicity. For more details, see Brown (1981) and Mukherji (1974).

50 For the widely perceived failings of strong-interaction quantum held theories, see Bethe and de Hoffmann (1956, 20–9).

51 Weinberg (1977a, 31).

52 For outline histories of the S-matrix programme, see Redhead (1980) and Cushing (1982). For a popular account, see Chew, Gell-Mann and Rosenfeld (1964).

53 Chew (1964a, 4).

54 Chew (1964a, 4) cites the work of Chew, Gell-Mann, M.L.Goldberger,

L.Landau, F.E.Low, S.Mandelstam and I.Ya.Pomeranchuk as seminal to the S-matrix programme in HEP. The contributions of these authors is discussed in Chew (1964a) Redhead (1980) and Cushing (1982).

55 A complex variable is one which can be expressed in term of the form $x + iy$, where x and y are real numbers, and i is the 'imaginary' number $\sqrt{-1}$.

56 This was, for example, the approach of the Cambridge University HEP group: see Eden, Landshoff, Olive and Polkinghorne (1966).

57 See, for example, Chew and Frautschi (1961a). For popular accounts of the bootstrap philosophy, see Chew (1964b, 1968, 1970) and Capra (1979). I am grateful to Professor Chew for an interview and the provision of biographical materials.

58 Zacharaisen and Zemach (1962).

59 Abers, Zacharaisen and Zemach (1963).

60 Regge (1959).

61 Chew and Frautschi (1961b), Chew, Frautschi and Mandelstam (1962).

62 Collins (1971, 127, Fig. 5; 130, Fig. 8).

63 Collins and Squires (1968, 224, Fig. 8.13a).

64 For reviews of Regge theory and the Regge analysis of soft-scattering data, see Collins and Squires (1968) and Collins (1971). One major transformation of the Regge programme should be mentioned here. Through the 1960s high-energy experimenters focused upon *exclusive* measurements of soft-scattering cross-sections. They concentrated upon elastic scattering and processes in which a small number of particles were produced. In the late 1960s and early 1970s, they began to focus instead upon highly inelastic processes in which many particles were produced. Exclusive measurements were impractical when, say, 10 particles emerged simultaneously from each interaction, and experimenters contented themselves with *inclusive* measurements of soft-scattering processes such as $\pi p \rightarrow \pi X$. A great deal of data was thus generated, fostering new traditions of Regge analysis. In his key-word analysis (see note 2 above) Wagner finds that research on inclusive processes was the dominant concern of HEP in the early 1970s (more than 500 papers per year in 1974). Ironically, interest in inclusive processes was triggered by Richard Feynman's 'parton' analysis, which itself constituted a mainstay of the new physics (see Chapter 5, notes 24 and 25). For technical reviews of soft multiparticle production, see Van Hove (1971), Horn (1972), Slansky (1974), Aurenche and Paton (1976), Abarbanel (1976) and Ganguli and Roy (1980).

65 The basis of the dual resonance model was laid by Veneziano (1968). For technical reviews of subsequent developments, see Schwarz (1973), Frampton (1974), Jacob (ed.) (1974) and Veneziano (1974). For a popular account, see Schwarz (1975). The dual resonance model metamorphosed in the early 1970s into the 'string model' of hadrons (discussed in Chapter 9, note 22: for a popular account, see Nambu 1976).

PART II

CONSTRUCTING QUARKS AND FOUNDING THE NEW PHYSICS: HEP, 1964–74

4

THE QUARK MODEL

In the 1960s the old physics split into two branches. The high-energy branch focused upon soft scattering, analysed in terms of Regge models, while at low energies resonance physics was the order of the day. In the early 1960s, group theory provided the most popular framework for resonance analysis but in the mid-1960s quark models took over this role. We can examine here how and why they did so.

4.1 The Genesis of Quarks

In early 1964, Caltech theorist Murray Gell-Mann published an article entitled 'A Schematic Model of Baryons and Mesons'.[1] In it, he began by remarking that, 'If we assume that the strong interactions of baryons and mesons are correctly described in terms of the broken "eightfold way", we are tempted to look for some fundamental explanation of the situation', and he went on to propose that hadrons should be seen as composite particles, built up from more fundamental entities which themselves manifested an su(3) symmetry. He indicated two possible choices for the building blocks. The first made use of four elementary entities, each carrying an electric charge of either 0 or 1 (in units of the charge of the electron). By taking appropriate combinations of these objects the observed su(3) multiplet structure of hadrons could be reproduced. However, there was an awkward asymmetry in this choice of entities – one particle acting as a 'basic baryon', and thus standing somewhat apart from the other three. As Gell-Mann remarked: 'A simpler and more elegant scheme can be constructed if we allow non-integral values for the charges'.[2] The simpler and more elegant scheme was the quark model – proposed independently at around the same time by George Zweig, a Caltech graduate who was then visiting CERN as a postdoctoral fellow.[3] (Gell-Mann abstracted the name 'quark' from James Joyce's *Finnegan's Wake*; Zweig called his constituents 'aces' – as usual, Gell-Mann won.)

We shall see in the following two sections that Gell-Mann and Zweig formulated the quark concept in quite different ways, each of which was subsequently elaborated into a distinctive tradition of experimental practice. But before going into the differences between the two schemes, it will be useful to review their common features and to relate these to the common background from which they derived. Gell-Mann and Zweig agreed not only that hadrons should have constituents but also upon what the basic attributes of those constituents should be. Quarks were supposed to have spin $\frac{1}{2}$ and baryon number $\frac{1}{3}$. As far as isospin and strangeness were concerned there were supposed to be three distinct species of quark: the 'up' (u) and 'down' (d) quarks made up an isospin $\frac{1}{2}$ doublet of zero strangeness, and the 'strange' (s) quark an isospin 0 singlet of unit strangeness. Apart from the $\frac{1}{3}$-integral baryon number, the oddest thing about quarks was their electrical charges: $+\frac{2}{3}$ for the u quark, and $-\frac{1}{3}$ for the d and s quarks. From quark–antiquark pairs (q$\bar{q}$) mesons could be constructed, and three quarks (qqq) would make a baryon. Thus, for example, the (u$\bar{d}$) combination would have charge $+1$, strangeness 0 and baryon number 0 and could be identified with the spin 0 π^+ (if the quark spins were antiparallel) or the spin 1 ρ^+ (if the quark spins were parallel); similarly a spin $\frac{1}{2}$ (uud) combination, having charge $+1$, strangeness 0 and baryon number 1 could be identified with the proton; (udd) with the neutron; (uds) with the lambda; and so on.

That Gell-Mann and Zweig should have chosen to suggest that hadrons were composites of the same fundamental entities was no coincidence, as one can see from an examination of the historical background to their proposals. The idea of reducing the number of elementary particles by explaining some of them as composite was by no means new. As early as 1949, C.N.Yang and Enrico Fermi had written a paper entitled 'Are Mesons Elementary Particles?', in which they speculated that the pion was a nucleon–antinucleon composite.[4] When the strange particles began to be discovered, the Fermi–Yang approach was appropriately extended by several authors – most notably by the Japanese physicist S.Sakata in 1956.[5] The fundamental entities of the Sakata model were the proton and neutron (as in the Fermi–Yang model) and the lambda, required to provide the additional element of strangeness. In the late 1950s, colleagues of Sakata at the University of Nagoya formulated his model in group-theory terms[6] and, as Yuval Ne'eman later recalled, 'throughout 1961–64, the Sakata model . . . was indeed the most serious competition to my own'.[7] The competition was finally resolved in favour of the Eightfold Way by the 1964 discovery of the Ω^-: according to Gell-Mann and Ne'eman the Ω^- was the missing member of the spin $\frac{3}{2}$ baryon decuplet, while according to Sakata it contained at least three lambda particles – a most implausible hypothesis.

The Quark Model

The idea of composite hadronic systems was, then, quite familiar to HEP theorists when quarks arrived. Furthermore, the group-theory approach to hadronic symmetries positively invited such a viewpoint. As we noted earlier, according to su(3) hadrons were expected to fall into multiplets containing 1, 3, 6, 8, 10, 27, etc., members, characteristic of the representations of su(3). Amongst these representations, the triplet, containing 3 members, was known as the 'fundamental representation' of su(3) (Figure 4.1) because all other representations could be derived from it by appropriate mathematical manipulations. Thus, even working in the pure su(3) tradition, theorists almost inevitably found themselves performing mathematical operations on the fundamental representation of the group. From there it was but a small step to the identification of the fundamental triplet representation of su(3) with a triplet of fundamental entities: the quarks.

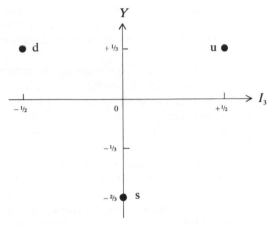

Figure 4.1. Quarks as the fundamental representation of su(3).

These two features – awareness of the possibility of composite models, and operations on fundamental representations – were conspicuously present in both Gell-Mann and Zweig's development of the quark concept[8] (as they were in the work of several other theorists working on symmetries of the strong interaction).[9] Beyond that, however, Gell-Mann and Zweig's formulations of the quark model had only one thing in common: they were open to the same empirical objection – a free quark had never been seen. Routinely associated in physicists' minds with the idea of compositeness was the idea that the primitive constituents should be experimentally detectable – the constituents of the atom, electrons and nuclei, could be routinely exhibited in the laboratory, as could the nucleonic constituents of nuclei. The properties attributed to quarks were quite distinctive and yet no such objects had ever been observed. The inventors of quarks, and the theorists who followed them, exhibited a certain squeamishness on this point. Gell-Mann, for instance, concluded his first quark paper thus:

> It is fun to speculate about the way quarks would behave if they were physical particles of finite mass (instead of purely mathematical entities as they would be in the limit of infinite mass). Since charge and baryon number are exactly conserved, one of the quarks (presumably $u^{2/3}$ or $d^{-1/3}$) would be absolutely stable . . . Ordinary matter near the earth's surface would be contaminated by stable quarks as a result of high energy cosmic ray events throughout the earth's history, but the contamination is estimated to be so small that it would never have been detected. A search for stable quarks of charge $-\frac{1}{3}$ or $+\frac{2}{3}$ and/or stable diquarks of charge $-\frac{2}{3}$ or $+\frac{1}{3}$ or $+\frac{4}{3}$ at the highest energy accelerators would help to reassure us of the non-existence of real quarks.[10]

This quotation, with its reference to reassurance about the non-existence of real quarks, neatly exemplifies the ambiguity felt about the status of quarks. Gell-Mann was particularly prone to suspicions concerning their status. Others were less so – amongst them, the hardy experimenter. From the experimental point of view, quarks had one key attribute: fractional electric charge. The belief had come down from the days of Robert Millikan that all matter carried charges in integral multiples of the charge on the electron. Now this belief had been challenged, and experimenters were not slow to respond. According to their differing expertise, experimenters followed up one of three general strategies. One was that advocated by Gell-Mann: to look for particles of fractional charge in accelerator experiments. All of the detectors used in such experiments react to electrically charged particles, and it appeared straightforward to detect quarks should they exist. The second strategy, deployed by cosmic-ray physicists, was to use similar techniques to look for quarks in the cosmic ray flux. The third option was to use up-dated versions of the 'oil-drop' apparatus – with which Millikan had established the primacy of the electronic charge half a century earlier – to look for fractional charges in the terrestrial environment. Experiments in each category got under way in 1964. If any of them had reported success the story of quarks would have been much simpler. As it was, no evidence for quarks was reported from the first wave of experimental searches. And indeed, although experimenters continued to look for quarks throughout the 1960s, the 1970s, and into the 1980s, an acknowledged quark was never found.[11]

The lack of direct evidence for quarks did much to undermine the credibility of the quark model in its early years. Against this had to be set the growing success of variants of the model in explaining a wide range of hadronic phenomena. There were two principal variants of the model, which can be traced back respectively to Gell-Mann and Zweig. As Zweig put it at the time, Gell-Mann's 'primary motivation' for introducing quarks 'differs from ours in many respects'.[12] Since Gell-Mann, inventor of strangeness and founder of the Eightfold Way, was a dominant figure in many developments of this period, it might seem reasonable to look at his style of work first. On the other hand, Zweig's approach, which I will call the constituent quark model (CQM), presents a much more readily comprehensible picture, so I will start there.

4.2 The Constituent Quark Model

George Zweig was born Moscow in 1937 but completed his higher education in the USA.[13] In 1959 he was awarded a BSC degree in mathematics by the University of Michigan, and then began postgraduate work at Caltech. There he worked for three years on an unproductive HEP experiment at the Berkeley Bevatron, before deciding in late 1962 to write a theoretical thesis under the supervision of Richard Feynman. Zweig began his theoretical work by looking at meson decay systematics. Following a circuitous route through the Sakata model (which he had already studied) and review articles on SU(3) symmetry, he arrived at the realisation noted above: the multiplet structure of the observed hadrons could be recovered if all hadrons were composites of two or three quarks carrying the appropriate quantum numbers. From there, Zweig proceeded in a simple and direct fashion. As one would expect for a neophyte to theoretical physics, he treated quarks as physical constituents of hadrons and thus, as discussed below, derived all of the predictions of SU(3) plus more besides.[14] In 1963 he was awarded a one-year fellowship to CERN, where he wrote up his findings, concluding: 'In view of the extremely crude manner in which we have approached the problem, the results we have obtained seem somewhat miraculous'.[15] Apparently the elders of the HEP community agreed with Zweig both as to the crudity of his approach and as to the miraculous nature of his results. He later recalled:

> The reaction of the theoretical physics community to the ace [quark] model was generally not benign. Getting the CERN report published in the form that I wanted was so difficult that I

finally gave up trying. When the physics department of a leading university was considering an appointment for me, their senior theorist, one of the most respected spokesmen for all of theoretical physics, blocked the appointment at a faculty meeting by passionately arguing that the ace model was the work of a 'charlatan'.[16]

In retrospect, it would be easy to dismiss the antagonism to Zweig's work as misguided, but it is important not to do this if one wants to understand the subsequent development of the CQM. We saw in the previous chapter that, in the early 1960s, there were two principal frameworks for theorising about the strong interaction: quantum field theory and the S-matrix. The CQM was distinguished by being unacceptable to protagonists of both. Consider quantum field theory first. In order to explain why free quarks were not experimentally observed, the obvious strategy was to assume that they were very massive (so that the accelerators of the day had insufficient energy to produce them). In the mid-1960s, this implied that quarks had masses of at least a few GeV. However, in combination with one another, quarks were supposed to make up much lighter hadrons – the 140 MeV pion being the extreme example. Theorists did not find it totally inconceivable that light particles should have heavy constituents. The idea that the mass of bound states should be less than the total mass of the constituents was familiar from nuclear physics, where the mass difference was known as the 'binding energy' of the nucleus. Unfortunately, the binding energy of quarks in hadrons was necessarily of the same order as the masses of quarks themselves – in contrast to the situation in nuclear physics, where binding energies were small fractions of nuclear masses. This implied very strong binding of the quarks, and strong binding, like strong coupling, was something which field theorists did not know how to calculate. Their techniques had been developed for weakly bound systems like atoms, where perturbative methods could be used. As discussed in the previous chapter, such methods failed for strong couplings, and thus the CQM left traditional field theorists cold. Of course, traditional field theory was in decline in the mid-1960s. The favoured approach was the S-matrix, and here opposition to quarks was even more clear-cut. At the heart of the S-matrix programme, especially in the bootstrap formulation, was the belief that there were no fundamental entities. The suggestion that all hadrons were built from three quarks was anathema to such thinking. As Zweig later commented, 'the idea that hadrons, citizens of a nuclear democracy, were made of elementary particles with fractional quantum numbers did seem a bit rich.'[17]

The CQM was hemmed in on all sides: experimentally, there was no sign of a free quark; theoretically, the model was disreputable in the extreme. These problems, and the quark modellers' response to them, were summed up by the Dutch theorist J.J.J.Kokkedee in his 1969 book on the CQM:

> Of course, the whole quark idea is ill-founded. So far quarks have escaped detection. This fact could simply mean that they are extremely massive and therefore difficult to produce, but it could also be an indication that quarks cannot exist as individual particles but, like phonons inside a crystal, can have meaning only inside the hadrons. In either case, nevertheless, the dynamical system of such quarks binding together to give the observed hadrons that has the properties demanded by the applications, is very difficult to understand in terms of conventional concepts. The quark model should, therefore, at least for the moment, not be taken for more than what it is, namely, the tentative and simplistic expression of an as yet obscure dynamics underlying the hadronic world. As such, however, the model is of great heuristic value.[18]

Enough has been said above about why the CQM was ill-founded; we can now examine the other side of the coin – its great heuristic value.

Phenomenology

Particle physicists make an imprecise distinction between doing 'theory' and doing 'phenomenology'. The former refers to such activities as the abstract elaboration of respectable theories (like quantum field theory or the analytic S-matrix); the latter to the application of less dignified models to the analysis of data and as a guide to further experiment. As theory, the CQM was a disaster; as phenomenology it was a great success. The great merit of the CQM as a phenomenological tool was that it brought resonance physics down to earth. Instead of manipulating the abstract structures of group theory, physicists could now work in terms of physical constituents. And instead of looking to the often unfamiliar realm of pure mathematics for inspiration, physicists could draw upon their own shared expertise in the analysis of composite systems. While it was unclear to many physicists what to do next in elaborating SU(3) as an abstract symmetry of hadrons, all physicists, by virtue of their training in composite systems, were well equipped to elaborate the CQM.

To put some flesh on these remarks, we can begin with three examples drawn from Zweig's 1964 work.[19] The first concerns the Eightfold Way classification of hadrons. The representations of SU(3) could accommodate a wide variety of hadron multiplets with 1, 3, 6, 8, 10, 27, etc., members. However, in 1964 it appeared that all of the hadrons were members of singlets, octets or decuplets – there were no evident candidates for the triplet, sextet or 27-plets of su(3). From the group-theoretic point of view, this was an empirical fact of no evident significance. The cqm, in contrast, promised at least the beginnings of an explanation. If one insisted that all mesons were built from a single quark–antiquark (qq̄) pair, and that all baryons were three quark (qqq) composites then, according to accepted rules for combining quantum numbers, one arrived at only singlets and octets of mesons (which could mix as nonets) and singlets, octets and decuplets of baryons.[20] Thus the apparent absence of certain possible hadron multiplets could be translated into an attribute of quarks: they combined only in qq̄ and qqq configurations. This did not constitute a complete solution to the problem of the missing multiplets, but it did transform the problem into a form which physicists had the resources to tackle, as we shall see in Chapter 7 in the discussion of 'colour'.

As a second example of the translation of a group-theory fact into a physical attribute of quarks, we can consider the mass splittings within su(3) multiplets. We noted in Chapter 3.2 that Gell-Mann and Okubo explained the splittings in terms of the group-theory properties of the su(3)-breaking component of the strong interaction. Zweig recovered the Gell-Mann–Okubo mass formula by assuming that the s quark was heavier than the u and d quarks. This rendered the calculation of su(3) mass splittings easily manageable, and became a standard assumption in the subsequent development of the cqm.

The third example to be discussed here concerns Zweig's analysis of hadron couplings and decay rates. This is best illustrated through consideration of what later became known as the 'Zweig rule'. In 1963 Brookhaven experimenters reported that the phi-meson decayed predominantly to a kaon–antikaon pair, rather than to a rho-meson plus a pion as expected from the usual systematics of hadron decays.[21] Okubo explained this in abstract group-theory terms, but admitted that his approach was quite *ad hoc*: 'Unfortunately', he wrote, 'the present author could not justify this "ansatz" on a more satisfactory mathematical ground.'[22] Zweig, in contrast, visualised phi-decay as shown in Figure 4.2. The phi-meson was believed to be a non-strange particle, but Zweig argued that it had 'hidden-strangeness': that it was composed of an s quark and an anti-s quark – an ss̄ pair – whose net strangeness cancelled out. He further argued that these constituent quarks persisted through the decay process, which entailed the materialisation of either a uū or dd̄ pair and hence led automatically to mesons containing strange quarks – kaons rather than non-strange rhos and pions.[23] This argument over the decay systematics of hidden-strangeness particles was recycled in the mid-1970s to hidden-charm particles with great effect, as we shall see in Chapter 9.

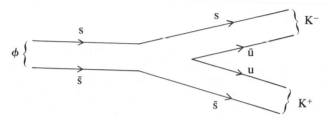

Figure 4.2. The decay $\phi \rightarrow K^+ K^-$ according to Zweig.

SU(6)

The advantage of reformulating the group-theoretical properties of hadrons in terms of quarks became increasingly evident as theorists began to load quarks with additional physical properties. One of the properties with which both Gell-Mann and Zweig had endowed quarks was spin. Quarks were assigned spin $\frac{1}{2}$, so that (forgetting about orbital angular momentum for the moment) $q\bar{q}$ composites would have total spin 0 or 1 and qqq composites $\frac{1}{2}$ or $\frac{3}{2}$. These results followed from the quantum-mechanical addition law for angular momenta, and reproduced the spins of the observed low-mass mesons and baryons (tabulated in Figures 3.2 and 3.3). In lectures given in August 1964 Zweig took this line of reasoning further. He suggested that the relative alignment of quark spins within hadrons was largely irrelevant to their binding: that for mesons whether the spins added to 0 or 1 was relatively unimportant, and similarly for baryons.[24] This amounted to the assertion of an additional if approximate symmetry of hadrons relating to their spins, and the combination of this new symmetry with the old su(3) was that of the group su(6).[25]

Two important consequences could be drawn from the assertion of an approximate su(6) symmetry. First, the representation structure of su(6) offered a higher classification system for hadrons. For example, the qqq representations of su(6) were multiplets containing 20, 56 and 70 particles, and the 56, in particular, was just the right size to embrace the low-lying octet and decuplet baryons.[26] Secondly, as discussed below, simple assumptions about the electromagnetic and strong interactions of quarks led, in the su(6) scheme, to predictions of hadronic properties in reasonable agreement with experiment. These were clear advances over the pure Eightfold Way approach which, for example, offered no explanation as to why the low-lying baryons should all have similar masses and either spin $\frac{1}{2}$ or $\frac{3}{2}$. But it is important to note that at this stage the CQM had not entirely outstripped abstract group theory. There was a second and independent route to SU(6) which avoided reference to quarks, and which pre-dated the latter by a couple of weeks. The pioneers of the alternative route were Turkish theorist F.Gürsey, Italian theorist L.Radicati and US theorist A.Pais, all working in the theory group at Brookhaven National Laboratory.[27] The three men shared a common background of work in group theory, and arrived at SU(6) not via quarks but via an analogical extension of a group-theory approach to *nuclear* physics. In 1936 Eugene Wigner had proposed that nuclei manifested an approximate SU(4) symmetry, corresponding to a combination of the SU(2) isospin independence of nuclear forces with a similar independence from the relative spin orientation of nucleons within nuclei.[28] Gürsey, Radicati and Pais proposed that hadrons should manifest a parallel symmetry, except that it should be SU(6) rather than SU(4) so as to accommodate the approximate SU(3) independence of the strong interaction. Thus they arrived at SU(6) via an abstract argument without any explicit reference to quarks.

In mid-1964 the pure group-theory approach to SU(6) was more appealing to many theorists than the CQM, which was open to all of the theoretical objections discussed above. However, to the theoretical purist, there was one immediate objection to SU(6) as a formal symmetry of hadrons: it was not in accord with the theory of special relativity; in technical terms it was not 'Lorentz invariant'. In essence this was because the existing SU(6) schemes made a distinction between spin and orbital angular momenta, and such a distinction was regarded as legitimate only in non-relativistic situations. Theorists therefore set out to repair this defect. As the Italian quark theorist (and quark searcher) Giacomo Morpurgo later noted: 'The whole end of 1964 and the winter of 1965 were spent by many people trying to construct theories simultaneously invariant under the SU(6) and the Lorentz group and realizing, finally, that this was impossible.'[29] This realisation derived from three sources. First, theorists proposed larger symmetry groups in which SU(6) could be embedded in a relativistically invariant fashion, but these led to predictions in manifest disagreement with experimental data. Secondly, mathematical inconsistencies were found in the formulation of these higher symmetries. And thirdly, general and mathematically rigorous 'nonexistence' theorems were proved, showing that no physically interesting marriage of the Lorentz group with internal symmetries of the Eightfold Way type was possible.[30]

The failure of the search for a relativistic extension of su(6) marked the demise of the CQM's only theoretically respectable rival, and from early 1965 onwards the CQM held the field of resonance phenomenology alone. Morpurgo summarised this phase of development succinctly:

> [T]hese efforts [towards a relativistic version of su(6)] were not entirely useless; indeed, as a sort of reaction they led to considering the brutal possibility of a nonrelativistic description of the internal dynamics of hadrons. The underlying line of reasoning was: It is not so tragic that it is impossible to have a theory which is simultaneously su(6) and Lorentz invariant; indeed we know from the start that su(6) is not an exact symmetry. Therefore all that is required is to have an approximate su(6) symmetry where the spin-dependent forces are small with respect to the spin-independent ones. This can easily be achieved in a nonrelativistic dynamics, where no problem arises in this respect. So why not explore the possibility that quarks are heavy (if they exist they should be heavy since they have not been seen), very strongly bound (the binding cancelling the largest fraction of their rest mass), and still, when bound, having nonrelativistic relative motion?[31]

Morpurgo himself showed that nonrelativistic motion was quite conceivable for heavy quarks bound in a deep but flat-bottomed potential well,[32] and in the latter half of the 1960s a growing number of HEP theorists indicated their willingness to join him in exploring the phenomenological possibilities of such a model. Similar images of composite systems were already familiar to physicists from their training in nuclear (and atomic) physics, and the order of the day was proclaimed by Morpurgo: 'proceed as you would do in nuclear physics'.[33] Different theorists followed this suggestion in different ways, and a discussion of two of the major lines of development – hadron spectroscopy and hadronic couplings – will serve to illustrate this.[34]

Hadron Spectroscopy
Spectroscopy is a venerable term in physics, relating to the study of the energy levels of composite systems. It originated in atomic physics before being taken over into nuclear physics. In elementary particle physics, the straightforward CQM conjecture was that the hadrons themselves constituted the spectrum of energy levels of composite

quark systems. One of the first and most influential theorists to take this idea seriously was Richard Dalitz. Dalitz was born in Australia in 1925 but completed his education in Britain, gaining a PhD in physics at Cambridge in 1950. He then held a variety of research positions before becoming Royal Society Research Professor at Oxford in 1963. Dalitz' thesis work was in nuclear physics, but during the late 1950s and early 1960s he moved into HEP, specialising in analyses of the strange hadronic resonances and of 'hypernuclei' – nuclei in which one of the usual nucleons had been replaced by a strange particle. He was thus an expert in both resonance physics and the analysis of composite systems, and in 1965 he set about elaborating the CQM accordingly.[35]

Dalitz's strategy, later elaborated by many authors, was straightforward and taken over directly from nuclear physics. It went as follows. The first problem was to explain the successes of existing SU(6) schemes in terms of quark dynamics. This was simple. Dalitz assumed that the dominant force binding quarks together was independent both of their spins and of the SU(3) quantum numbers of the resulting hadronic bound state. In this way he could explain the overall SU(6) symmetry of hadrons in just the same way as nuclear physicists explained the overall SU(4) symmetry of nuclear states. Dalitz then had to confront the observation that SU(6) was a broken rather than an exact hadronic symmetry: different hadrons assigned to the same SU(6) multiplets had different masses, not the same masses as expected for an exact symmetry. Here again, Dalitz followed the traditional route of nuclear physics. He assumed that, although the dominant quark binding forces were SU(6) invariant, there existed a hierarchy of lesser, perturbing, forces which acted upon quarks to break the SU(6) symmetry and to produce the observed mass-splittings within SU(6) multiplets. In this category he included forces depending upon the SU(3) quantum numbers of the hadron in question, and 'spin–spin' and 'spin–orbit' forces acting between pairs of quarks. Such forces were already familiar from nuclear physics, and the spin–spin forces, for example, served to create mass differences between hadrons depending on whether the spins of their constituent quarks lined up parallel or antiparallel to one another.

Such assumptions served to account for mass splittings *within* SU(6) multiplets. Dalitz then proceeded to consider the overall splittings *between* SU(6) multiplets. To generate these some further assumption was required. Once again nuclear physics provided a clue. The total spin of a nucleus is regarded as made up of the combined spins of its constituents plus their orbital angular momentum. The lowest mass nuclear state is typically that in which the orbital angular momentum is zero; higher energy states correspond to orbital (and vibrational) excitation of the constituents. Dalitz argued that the same was true of hadrons. The lowest mass su(6) multiplets corresponded to quark systems with zero orbital angular momentum; higher mass multiplets corresponded to systems with 1, 2, 3, etc., units of orbital angular momentum. This agreed with the observation that the lowest mass su(6) multiplets of mesons and baryons had spins consonant with the composition of quark spins alone, while higher mass resonances typically had larger spins which could only be accounted for by combining the spin of the quarks with one or more units of orbital angular momentum. Thus Dalitz suggested that there should exist a sequence of increasingly massive su(6) hadron multiplets, associated with increasing values of the orbital angular momentum of the constituent quarks.

Qualitatively, it was clear that the nuclear-physics inspired approach could make sense of the overall features of the hadron spectrum. To go further, Dalitz and many other theorists became quantitative, constructing mathematical models in which the strengths of the various su(6)-breaking and -splitting interactions were free parameters to be determined from experiment. They thus arrived, in effect, at highly complex mass formulae supposed to describe the entire hadronic spectrum. The phenomenologist's art was then to fit these formulae to contemporary data by appropriate adjustment of parameters (not an empty exercise, since there were far more resonances than parameters). Figure 4.3, taken from J.J.J.Kokkedee's 1967 CERN lectures on the CQM,[36] schematises how this was done, and may serve both to clarify the above discussion and to illustrate some of the phenomenological complexity of CQM spectroscopy.

The first thing to note about Figure 4.3 is that it is entirely typical of the spectroscopic diagrams drawn in nuclear and atomic physics. The horizontal bars represent energy levels – in this case hadron masses – and the higher the bar, the greater the energy or mass. The bar on the left is labelled [70, $L=1$], indicating that the hadrons in question are being assigned to a 70-dimensional representation of su(6) in which the quarks have one unit of relative orbital angular momentum (L). The Vs at the bottom of the diagram indicate the various forces included in the calculation. V and V' are su(6)-symmetric forces and, at this level of analysis, all of the hadrons within the su(6) multiplet have the same mass, as indicated by the single horizontal bar. As one moves towards the right of the diagram various su(6)-breaking forces are added into the calculation. V_σ

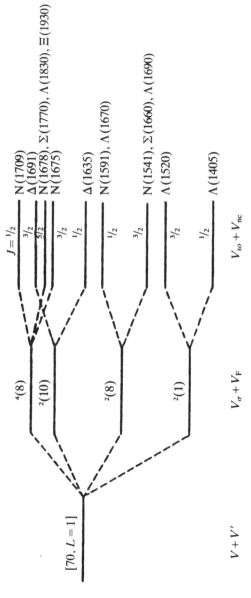

Figure 4.3. Hierarchy of mass splittings in the CQM.

represents spin–spin forces and V_F SU(3)-dependent forces, which together split the SU(6) multiplet into four SU(3) multiplets: two octets, a decuplet and a singlet. The superscripts on the SU(3) assignments of these levels indicate the total quark spin ('4' denotes total spin $\frac{3}{2}$, '2' total spin $\frac{1}{2}$). Finally, V_{so} and V_{nc} represent spin–orbit type forces, which induce further mass splittings through interaction of quark spins with their orbital motion. These separate the levels according to their total angular momentum (J) as shown on the right of the diagram. At this point the level structure has become sufficiently finely divided for assignment of candidate resonances to them, as indicated. Note that splittings *within* SU(3) multiplets have not been included here, so that baryons of different masses within the same SU(3) multiplet can appear at the same level, the N(1678), $\Sigma(1770)$, $\Lambda(1830)$, and $\Xi(1930)$ for example. Were such SU(3) splittings to be included, the diagram would be even more complicated.

During the latter half of the 1960s, the CQM analysis of hadron mass spectra grew into a complicated and sophisticated phenomenological tradition. Fits to the known resonances were never exact, but it proved possible to reproduce the overall sequence of levels in the spectra of mesons and baryons: most resonances had the appropriate quantum numbers and masses to be assigned to the expected SU(6) multiplet structure. There were few obvious misfits.[37] Neither was there any competing theoretical scheme capable of offering a comparably fine-grained alternative to the CQM. Thus, as Kokkedee summarised his 1967 lectures:

> We may conclude that the simplest quark model – qq̄ and qqq configurations for mesons and baryons with the possibility of rotational and eventually vibrational excitation – can reproduce qualitatively the gross features of the presently established resonance spectra of the hadrons. To be able to generate an SU(3) multiplet spectrum that agrees remarkably well with the observations is a challenging achievement.[38]

As the years went by, HEP experimenters explored the resonance spectrum in increasing detail, CQM modellers made increasingly sophisticated fits to the data, and the situation remained the same: with few exceptions, the resonances could be accommodated by the patterns expected in the CQM. Table 4.1 shows the accepted classification for the known baryons of masses up to around 2 GeV, as of 1981.[39]

Hadronic Couplings
Besides classifying hadrons and explaining their mass spectrum, CQM

Table 4.1. Assignments of the lower-mass baryons to su(6) multiplets.

N su(6),L	su(3)	J^P	States
0 56,0	8	$1/2^+$	N(939), Λ(1116), Σ(1193), Ξ(1318)
	10	$3/2^+$	Δ(1232), Σ(1385), Ξ(1533), Ω(1672)
1 70,1	1	$1/2^-$	Λ(1405)
		$3/2^-$	Λ(1520)
	8	$1/2^-$	N(1535), Λ(1670), Σ(1750)
		(two)	N(1700), Λ(1870)
		$3/2^-$	N(1520), Λ(1690), Σ(1670), Ξ(1820)?
		(two)	N(1700), Σ(1940)?
		$5/2^-$	N(1670), Λ(1830), Σ(1765)
	10	$1/2^-$	Δ(1650)
		$3/2^-$	Δ(1670)
2 56,2	8	$3/2^+$	N(1810), Λ(1860)
	8	$5/2^+$	N(1688), Λ(1815), Σ(1915), Ξ(2030)?
	10	$1/2^+$	Δ(1910)
		$3/2^+$	
		$5/2^+$	Δ(1890)
		$7/2^+$	Δ(1950), Σ(2030)
56,0	8	$1/2^+$	N(1470), Σ(1660)
	10	$3/2^+$	Δ(1690)

enthusiasts set about analysing hadronic production and decay mechanisms. Again, theorists drew their inspiration directly from the standard treatments of composite systems in atomic and nuclear physics. Again the procedure was one of loading the properties of hadrons onto their constituent quarks. As a protypical example, consider Becchi and Morpurgo's 1965 analysis of the electromagnetic decay of the positively-charged delta resonance: $\Delta^+ \rightarrow p\gamma$.[40] Just as in nuclear and atomic physics, Becchi and Morpurgo assumed that the photon was emitted not from the delta as a whole, but rather from a single quark within it, as indicated in Figure 4.4. In general, since the photon has spin 1, this could take place in one of two ways. Either the spin of the emitting quark could 'flip' from $+\frac{1}{2}$ to $-\frac{1}{2}$ (technically, a magnetic dipole or M_1 transition) or the relative orbital angular momentum of the three-quark system could change by one unit (an electric quadrupole, E_2, transition). However, Becchi and Morpurgo argued, both the delta and proton belonged to an su(6) multiplet in which the quarks had zero orbital angular momentum; therefore the latter possibility was ruled out. The electromagnetic decay of the delta had to proceed through a spin-flip M_1 transition: E_2 transitions were forbidden by a 'selection rule' of the CQM. Experimentally, M_1 and E_2 transitions corresponded to different angular distributions for the emitted photons, and it was indeed found that the E_2 transition rate in delta decay was less than four per cent of the M_1 transition rate.

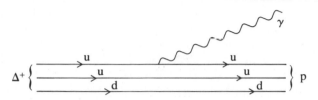

Figure 4.4. The electromagnetic decay $\Delta^+ \to p\gamma$ in the CQM.

This agreement between theory and data was one of the first successes of the application of the CQM to hadronic couplings,[41] and encouraged theorists to press onwards. Other selection rules were found relating to the electromagnetic production and decay of higher spin resonances, which were again supported by the data.[42] Finally, towards the end of the 1960s, theorists went beyond the search for selection rules, and mounted full-scale numerical calculations of electromagnetic resonance couplings using all of the details of resonance structure derived from fits to the hadron spectrum.[43] Here agreement with the data was not so impressive as the four-per-cent accuracy of Becchi and Morpurgo's selection rule, but the predictions in general matched the overall trend of the data, explaining why some resonances were observed to be strongly coupled to photons and others not.[44] Like the fits to hadron spectra, such coupling calculations were counted as a success of the CQM, but it is again relevant to note that this was in the absence of any significant competition from rival theories.

CQM modellers did not, of course, confine their attention to the electromagnetic interactions of hadrons. Models were also constructed of strong (and weak)[45] hadronic couplings. Here the analysis followed the same path as that of electromagnetic processes, with the obvious replacement of, say, pions for photons. This was somewhat paradoxical. The pion, although treated as an elementary particle in such calculations, was itself supposed to be a quark–antiquark composite. But quark modellers learned to live with such paradoxes. Early calculations focused upon the production and decays of the $\frac{3}{2}^+$ baryon decuplet and the spin 1 mesons, but were soon extended to the properties of higher spin resonances as well.[46] Like hadron spectroscopy, the CQM phenomenology of hadronic couplings also developed into a thriving and qualitatively successful tradition.[47]

The Symbiosis of Theory and Experiment

Thus far we have been looking at the CQM from the theorists' perspective, viewing it as an explanatory resource for the interpretation of data. But it is important to recognise that the traffic between experiment and theory was not one way: just as the theorists' practice was structured by the products of experimental research, so the experimenters' practice was structured by the products of the theorists' research. Through the medium of the CQM theorists and experimenters maintained a mutually supportive symbiotic relationship.

The roots of this symbiosis lay in the pre-CQM era. During the 1950s theory was largely parasitic upon experiment – experimenters reported the existence of new hadrons, and theorists did their best to explain them – but with the advent of the Eightfold Way the balance became more even. Experimenters began to take as well as to give. The clearest example of this centred on Gell-Mann's 1962 prediction of the existence of the Ω^-. Drawing upon the resonance data already provided by experimenters, Gell-Mann predicted the mass and decay characteristics of this unobserved particle, and thus pointed to a new topic for experimental practice. We saw in Chapter 3.2 that the experimenters took Gell-Mann's hint, invested a great deal of effort in the search, and in 1964 repaid him in kind with the particle itself. The existence of the Ω^- then formed an integral part of the next generation of Eightfold Way theorising which included the formulation of the quark concept.

With the advent of quarks and the CQM, the symbiosis between theory and experiment entered a new and more intimate phase. In essence this was because experimenters had begun to move on from the 'classic' hadrons – the lowest mass meson and baryon SU(6) multiplets – to the exploration of resonances of higher masses and spins. In this second generation of resonance experiments, 'bump-hunting' became a much more difficult and technical business than it had been in the first. Classic resonances like the delta could be associated with large and conspicuous peaks in experimental cross-sections (see Figure 3.1). But there were only a few such peaks and to find more resonances experimenters had to examine fine details of individual 'decay channels'.[48] In consequence, resonance identifications became less confident than they had been in the early days of the population explosion. George Zweig recalled that, even in 1963, 'Particle classification was difficult because many [resonance] peaks . . . were spurious', and that, of the 26 meson resonances then listed in an authoritative compilation, 19 subsequently disappeared.[49] Similarly, in his 1967 lectures on the CQM, Kokkedee prefaced his fits to the resonance spectrum with the comment: 'Because of the unstable experimental situation many detailed statements of the model are probably not guaranteed against the passage of time'.[50] The instability of the resonance data persisted through the 1960s and 1970s, as indicated in Table 4.2 taken from the 1982 edition of the *Review of Particle Properties* – the standard HEP catalogue of elementary particles.[51] Table 4.2(a) lists the mesons which had been reported from one or more high-quality experiments up to 1982. On the entries marked by an arrow, the reviewers comment: 'We do not regard these as established resonances'. Similarly, the entries in the baryon table (4.2(b)) are graded according to current status of the experimental evidence on a scale from four-star, 'good, clear, and unmistakable', to one-star, 'weak'. A couple of entries are marked 'Dead', indicating that they were once believed to exist but that their non-existence could now be regarded as proven.

Theory, in the guise of the CQM, intervened into this context of experimental uncertainty in two ways. First, CQM analyses offered experimenters specific targets to aim at in succeeding generations of experiment. Thus, for example, Zweig argued in 1964 that the strange resonance $\Lambda(1520)$ had been misidentified as a member of an SU(3) octet: according to his formulation of the CQM, the genuine octet member should have a mass of around 1635 MeV. Zweig noted that experiments had observed structure in cross-sections in this energy region, but that this had been ascribed to another nearby resonance, the $\Sigma(1660)$. The $\Lambda(1635)$ and $\Sigma(1660)$ resonances had isospin 0 and 1 respectively, so further experiment using different beams and targets could attempt to disentangle them.[52] Examples of this kind of heuristic feedback from theory to experiment could be multiplied almost indefinitely throughout the history of the CQM.[53]

The more general influence which the CQM exerted on experiment was that it made experimental resonance physics *interesting* – a field worthy of the expenditure of considerable amounts of time, money and effort.[54] It seems implausible that, without the sustaining interest of quark modellers, experimenters would have persisted in the painstaking exploration of the low-energy resonance regime. There would have been little point to such activity, and experimenters would have found other ways of exploiting the finite resources available for research.[55] As it was, the flourishing of the CQM tradition translated in experimental terms into determined programmes of precision measurements of cross-sections for a variety of choices of

Table 4.2(a). Mesons listed in Review of Particle Properties, 1982.

Non-strange (S=0; C,B=0)

Entry	$I^G(J^P)C_n$	Entry	$I^G(J^P)C_n$		Entry	$I^G(J^P)C_n$
π	$1^-(0^-)+$	f′ (1515)	$0^+(2^+)+$	→δ	(2450)	$1^-(6^+)+$
η	$0^+(0^-)+$	ρ′ (1600)	$1^+(1^-)-$	→e⁺e⁻	(1100–2200)	
ρ (770)	$1^+(1^-)-$	→θ (1640)	$0^+(2^+)+$	→ṄN	(1400–3600)	
ω (783)	$0^-(1^-)-$	ω (1670)	$0^-(3^-)-$	→X	(1900–3600)	
η′ (958)	$0^+(0^-)+$	A₃ (1680)	$1^-(2^-)+$	η_c	(2980)	$+$
S* (975)	$0^+(0^+)+$	φ′ (1680)	$0^-(1^-)-$	J/ψ	(3100)	$0^-(1^-)-$
δ (980)	$1^-(0^+)+$	g (1690)	$1^+(3^-)-$	χ	(3415)	$0^+(0^+)+$
φ (1020)	$0^-(1^-)-$	→φ (1850)	0	P_c or χ	(3510)	$0^+(1^+)+$
H (1190)	$0^-(1^+)-$	→X (1850)	(2^+)	χ	(3555)	$0^+(2^+)+$
B (1235)	$1^+(1^+)-$	→S (1935)		→η′_c	(3590)	$+$
→ρ′ (1250)	$1^+(1^-)-$	→δ (2030)	$1^-(4^+)+$	ψ	(3685)	$0^-(1^-)-$
f (1270)	$0^+(2^+)+$	h (2040)	$0^+(4^+)+$	ψ	(3770)	$(1^-)-$
A₁ (1270)	$1^-(1^+)+$	→π (2050)	$1^-(3^+)+$	ψ	(4030)	$(1^-)-$
→η (1275)	$0^+(0^-)+$	→π (2100)	$1^-(2^-)+$	ψ	(4160)	$(1^-)-$
D (1285)	$0^+(1^+)+$	→ρ (2150)	$1^+(1^-)-$	ψ	(4415)	$(1^-)-$
ε (1300)	$0^+(0^+)+$	→ε (2150)	$0^+(2^+)+$	ϒ	(9460)	$(1^-)-$
π (1300)	$1^-(0^-)+$	→ρ (2250)	$1^+(3^-)-$	ϒ	(10020)	$(1^-)-$
A₂ (1320)	$1^-(2^+)+$	→ε (2300)	$0^+(4^+)+$	ϒ	(10350)	$(1^-)-$
E (1420)	$0^+(1^+)+$	→ρ (2350)	$1^+(5^-)-$	ϒ	(10570)	$(1^-)-$

Strange (|S|=1; C,B=0)

Entry		$I(J^P)$
K		$1/2(0^-)$
K*	(892)	$1/2(1^-)$
Q₁	(1280)	$1/2(1^+)$
κ	(1350)	$1/2(0^+)$
Q₂	(1400)	$1/2(1^+)$
→K′	(1400)	$1/2(0^-)$
K*	(1430)	$1/2(2^+)$
→L	(1580)	$1/2(2^-)$
→K*	(1650)	$1/2(1^-)$
L	(1770)	$1/2(2^-)$
K*	(1780)	$1/2(3^-)$
→K*	(2060)	$1/2(4^+)$
→K*	(2200)	

Charmed (|C|=1)

Entry		$I(J^P)$
D	(1870)	$1/2(0^-)$
D*	(2010)	$1/2(1^-)$
F	(2020)	$0\ (0^-)$
→F*	(2140)	

Bottom (Beauty) (|B|=1)

→B	(5200)	
→Exotics		

N	Δ / Z	Λ	Σ	Ξ, Ω, etc.
N(939) P11 ****	Δ(1232) P33 ****	Λ(1115) P01 ****	Σ(1193) P11 ****	Ξ(1317) P11 ****
N(1440) P11 ****	Δ(1550) P31 **	Λ(1405) S01 ****	Σ(1385) P13 ****	Ξ(1530) P13 ****
N(1520) D13 ****	Δ(1600) P33 ***	Λ(1520) D03 ****	Σ(1480) *	Ξ(1630) **
N(1535) S11 ****	Δ(1620) S31 ****	Λ(1600) P01 ***	Σ(1560) **	Ξ(1680) S11 **
N(1540) P13 *	Δ(1700) D33 ****	Λ(1670) S01 ****	Σ(1580) D13 **	Ξ(1820) 13 ***
N(1650) S11 ****	Δ(1900) S31 ***	Λ(1690) D03 ****	Σ(1620) S11 **	Ξ(1940) **
N(1675) D15 ****	Δ(1905) F35 ****	Λ(1800) S01 ***	Σ(1660) P11 ***	Ξ(2030) 1 ***
N(1680) F15 ****	Δ(1910) P31 ****	Λ(1800) P01 ***	Σ(1670) D13 ****	Ξ(2120) *
N(1700) D13 ***	Δ(1920) P33 ***	Λ(1800) G09 Dead	Σ(1670) **	Ξ(2250) *
N(1710) P11 ****	Δ(1930) D35 ****	Λ(1800) *	Σ(1690) **	Ξ(2370) 1 **
N(1720) P13 ****	Δ(1940) D33 *	Λ(1820) F05 ****	Σ(1750) S11 ***	Ξ(2500) **
N(1990) F17 **	Δ(1950) F37 ****	Λ(1830) D05 ****	Σ(1770) P11 Dead	Ω(1672) P03 ****
N(2000) F15 *	Δ(2150) S31 *	Λ(1890) P03 ****	Σ(1775) D15 ****	Λ_c(2282) ****
N(2080) D13 ***	Δ(2160) *	Λ(2000) *	Σ(1840) P13 *	Σ_c(2450) **
N(2100) S11 *	Δ(2200) G37 **	Λ(2020) F07 *	Σ(1880) P11 **	Λ_b(5500) *
N(2100) P11 *	Δ(2300) H39 **	Λ(2100) G07 ****	Σ(1915) F15 ****	Dibaryons
N(2190) G17 ****	Δ(2350) D35 *	Λ(2110) F05 ***	Σ(1940) D13 ***	NN(2170) 1D2 ***
N(2200) D15 ***	Δ(2400) F37 *	Λ(2325) D03 *	Σ(2000) S11 *	NN(2250) 3F3 ***
N(2220) H19 ****	Δ(2400) G39 *	Λ(2350) ****	Σ(2030) F17 ****	NN(?) *
N(2250) G19 ****	Δ(2420) H311 ***	Λ(2585) ***	Σ(2070) F15 *	ΛN(2130) 3S1 ***
N(2600) I111 ***	Δ(2500) *		Σ(2080) P13 **	ΞN(?) *
N(2700) K113 *	Δ(2750) I313 *		Σ(2100) G17 *	
N(2800) G19 *	Δ(2850) ***		Σ(2250) ****	
N(3000) *	Δ(2950) K315 *		Σ(2455) ***	
N(3030) ***	Δ(3230) ***		Σ(2620) ***	
N(3245) *	Z0(1780) P01 *		Σ(3000) **	
N(3690) *	Z0(1865) D03 *		Σ(3170) *	
N(3755) *	Z1(1900) P13 *			
	Z1(2150) *			
	Z1(2500) *			

**** Good, clear, and unmistakable.
*** Good, but in need of clarification or not absolutely certain.
** Needs confirmation.
* Weak.

beam, target and decay channels, and into the development of increasingly sophisticated techniques of data analysis.[56] In connection with the former, it is significant to note that although hadron-beam experiments at PSS had led the way in resonance physics, the electromagnetic couplings calculated in the CQM could more readily be measured in 'photoproduction' experiments at electron accelerators.[57] As expected in the CQM, it also proved to be the case that certain resonances could only be detected in photoproduction experiments, and thus resonance physics became a principal preoccupation of research at electron accelerators as well as PSS in the latter half of the 1960s.

The upshot of the increasingly diversified and precise programme of experimental resonance physics was the steady flow of new resonances encapsulated in Figure 4.5.[58] This shows how the lists of resonances recorded in the Tables of *Review of Particle Properties* grew over two decades. Figure 4.5(a) refers to mesons, and includes both 'well understood' particles, for which all quantum numbers were known, and those which were simply 'listed' as having been sighted in one or more experiments. Figure 4.5(b) refers only to those baryons whose spin and parity had been determined, and reflects the intensity with which their properties had been investigated. It is important to note here that the 'properties' count is almost entirely made up of 'branching ratios'. These express how often a given resonance decays to a given decay channel and are precisely the quantities described by CQM calculations of resonance couplings.

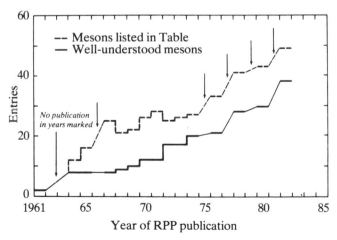

Figure 4.5(a). Growth of information on mesons.

106

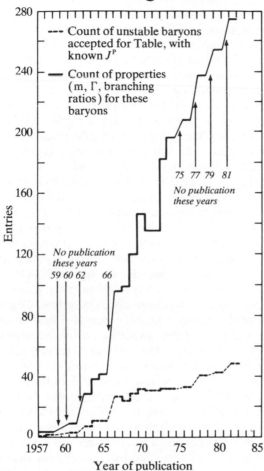

Figure 4.5(b). Growth of information on baryons.

Thus the activity of CQM modellers served to sustain and motivate a growing programme of experimental resonance physics. This programme generated more and more information on resonances and their properties, which, in turn, served to sustain and motivate more work on the CQM. The theoretical and experimental wings of resonance physics grew together in a symbiotic relationship through the 1960s and into the 1970s.[59]

The Death of Quantum Field Theory
One last point remains to be made before moving on from the CQM.

107

Two theoretical traditions dominated the old physics of the 1960s and early 1970s. At high energies there was the Regge analysis of soft scattering, discussed in Chapter 3.4; at low energies there was the CQM analysis of resonances. Neither of these traditions drew significantly upon the resources of quantum field theory. Indeed field theory was anathema to the bootstrappers amongst Regge theorists, while the CQM was itself repellent to the field-theory purists. Thus during the 1960s, the field approach to elementary-particle physics fell largely into disuse. Many of the older generation of field theorists ceased actively to practise their art, and many of the younger generation of HEP theorists – rapidly expanding in this period – received their training in a context from which field theory was effectively absent.[60]

From a sociological perspective, the decline of field theory in the 1960s creates an apparent problem in understanding its resurrection in the new physics of the 1970s: where did the required pool of theoretical expertise come from? Fortunately, the problem is only apparent. In the shadow of the CQM and Regge lay two lesser traditions oriented around the field-theory approach. One was devoted to the elaboration of the prototypical field theory, QED. The other was the 'current algebra' tradition discussed below. Together, these traditions offered a refuge wherein committed field theorists could make constructive use of their expertise and at the same time make contact with experiment. The QED and current algebra traditions kept the flame of field theory alive during the 1960s, and were the training ground for a new generation of field theorists. As discussed in Chapter 8.3, the advanced guard of the 1970s new physics resurgence of field theory was drawn from the ranks of QED theorists and, especially, current algebraists. I have already outlined the QED approach to electromagnetic phenomena and I will not go into further details here.[61] But some discussion of current algebra is needed. It was within this tradition that Gell-Mann grounded his earliest discussion of quarks and, as one of the leading figures in theoretical HEP, it was surely Gell-Mann rather than Zweig who first persuaded physicists to entertain the quark concept.[62]

4.3 Quarks and Current Algebra

There were two ways of formulating and elaborating the quark concept: the straightforward CQM way, and Gell-Mann's way. Gell-Mann's 'current algebra' approach was highly esoteric compared with the CQM, and had relatively little impact upon the overall practice of the HEP community.[63] I will therefore discuss it only briefly, and without going very far into technical details.[64] To begin the discussion it is illuminating to compare Gell-Mann's first quark paper with that of George Zweig.[65] Both authors listed the quark quantum numbers and stressed the importance of looking for free quarks, but there the resemblance ends. In his lengthy first draft (24 pages in preprint form) Zweig laid out the straightforward conse-quences of the composite analogy, writing down quark assignments for the spin 0 and spin 1 meson nonets, and the baryon octet and decuplet. Gell-Mann did no such thing. His paper covered less than two pages, and the only equations are 'commutators' for hadronic 'weak currents', evaluated for free quarks. I will come back to 'commutators' and 'weak currents' in a moment; initially I want just to emphasise that spectroscopic considerations, drawing directly on the composite analogy, were completely absent from this paper – Gell-Mann's interests were elsewhere. As he indicated in the text and footnotes, his aim was to plug in quarks as a new resource into an already evolving tradition which he had himself founded some years earlier. This was the tradition devoted to 'current algebra'. To understand the form it took it is useful to know something about Gell-Mann's approach to particle physics.[66]

Above all else, Gell-Mann was a field theorist. His first publication (written with F.E.Low in 1951) was portentously entitled 'Bound States in Quantum Field Theory'.[67] We noted in Chapter 3.2 how his enthusiasm for gauge field theory influenced his 1961 proposal of the Eightfold Way. We have also remarked on the obstacles theorists encountered during the 1950s in trying to extend the QED-type approach to any field theory of the weak and strong interactions. Like all of the theorists of this period, Gell-Mann was aware of these problems, but his response was a novel one. Characteristic of his work was the idea that one should take traditional field theory (the standard perturbative form, modelled on that of QED) and extract whatever might be useful in practice. He sought, that is, to use field theory heuristically, in order to arrive at new formulations or models which could then stand in their own right, independent of their origins. In Chapter 3.4 one way in which Gell-Mann attempted to do this was discussed: the exploration of the analytic properties of the S-matrix using perturbative quantum field theory as a guide. The other was the current-algebra approach, which led through SU(3) and quarks. The S-matrix approach related exclusively to the strong interactions of hadrons, while current algebra focused on their properties with respect to the weak and electromagnetic interactions.

In 1958 Feynman and Gell-Mann had written down the pheno-menologically successful V–A field theory of the weak interactions. This theory was directly modelled on QED, the weak interactions taking place between weak currents just as in QED electromagnetic interactions take place between electromagnetic currents. Questions arose, however, as to the nature of the weak currents. For the purely leptonic sector – the weak interactions of electrons, muons and neutrinos – the situation was clear: the currents were to be constructed from the relevant fields as in QED. But for hadrons the waters were murky. One problem concerned the population explosion: was every hadron to be ascribed its own personal field, contributing to an equation several pages long for the hadronic weak current? If so, was there any relation between the individual contributions to this current? These were some of the questions which Gell-Mann addressed in a 1962 publication – his first writings on the Eightfold Way to appear in a refereed journal.[68] The ideas in this paper were characteristic of his previous work and foreshadowed what was to follow. In effect, he proposed that in discussing both the weak and electromagnetic properties of hadrons one should forget about traditional quantum field theory, and instead regard weak and electromagnetic currents as the primary variables. At the same time, however, he used field theory as an aid to guessing the properties of the currents themselves. In particular, he hypothesised that the relationships between the various *hadronic* currents – technically speaking, the 'equal-time commutators' between the currents – could be modelled on those believed to hold in the V–A theory for the *leptonic* currents. If one considered a world containing only leptons one could show that the currents generated an $SU(2)$ algebra, and Gell-Mann proposed that the hadronic currents similarly generated an $SU(3)$ algebra – or, to be more precise, that the two orthogonal linear combinations of V (vector) and A (axial vector) currents generated two independent $SU(3)$ algebras. In technical terms, the hadronic currents were postulated to obey an $SU(3) \times SU(3)$ algebra – hence the name 'current algebra' for the work which grew up around this proposal.[69]

There are two characteristic points to note about Gell-Mann's strategy in arriving at the current-algebra formulation. First, current algebra had immediate phenomenological applications since currents were directly related to experimentally observable quantities via standard theoretical techniques (see below). Secondly, although Gell-Mann was clearly working within the field-theory idiom – his 1962 publication is littered with suggestive field-theoretic calculations – the algebraic structure relating the currents could not be said to be derived from any field theory. This was so because a derivation required a fully calculable theory of the *strong* interactions (recall that hadrons were at issue) and no such theory existed. Thus the invention of current algebra offers a perfect illustration of Gell-

Mann's strategy for arriving at self-sustaining theoretical structures from a shaky field-theoretic base, a strategy which he encapsulated in a culinary metaphor:

> In order to obtain such relations that we may conjecture to be true, we use the method of *abstraction* from a Lagrangian field theory model. In other words we construct a mathematical theory of the strongly interacting particles, which may or may not have anything to do with reality, find suitable algebraic relations that hold in the model, postulate their validity, and then throw away the model. We may compare this process to a method sometimes employed in French cuisine: a piece of pheasant meat is cooked between two slices of veal, which are then discarded.[70]

Having discussed the pre-existing current-algebra tradition, we are now in a position to understand the distinctive features of Gell-Mann's 1964 quark paper. In 1963, building upon Gell-Mann's 1962 current-algebra/Eightfold Way paper, the Italian theorist Nicola Cabibbo (working at CERN) had written down a general expression for the hadronic weak interactions, and in a detailed phenomenological analysis had shown this to be in agreement with the available experimental data on hadronic weak decays.[71] Gell-Mann, in turn, pointed out in 1964 that if one constructed suitable weak currents from free, non-interacting, quark fields, one could re-derive the current commutators he had postulated earlier in his Eightfold Way paper, and hence recover Cabibbo's successful phenomenological analysis of the weak interactions. Gell-Mann was careful to point out that he was not suggesting that hadrons were really made up of non-interacting quarks since, with no forces to bind them together, non-interacting quarks would simply fly apart from one another. In accordance with his standard strategy, he simply observed that the relationships between the weak currents required to explain the experimental data were the same as those between currents constructed from free quark fields.

Thus, working from his individual field-theoretic viewpoint, Gell-Mann used quarks to underpin su(3) results on hadronic weak interactions, just as Zweig and others were to use quarks to underpin the su(3) spectroscopy of hadronic resonances and their strong interactions. Also, as in the spectroscopic tradition, the introduction of the new quark degree of freedom enabled theorists to go beyond simple su(3) considerations in the current-algebra tradition, too. The point here was this. Gell-Mann had conjectured the su(3) × su(3) structure of current algebra on the basis of heuristic field-theoretic models. It turned out that certain current commutators were independent of the models used to arrive at them, in the sense that it appeared that whichever model one chose one arrived at the same result. It was thus reasonable to suppose that these commutators would remain unchanged in a full field-theoretic calculation (as distinct from the heuristic and approximate calculations actually done). On the other hand, certain other commutators depended crucially on the model field theory chosen – the free quark field theory, for instance, leading to different results from other theories. Thus, to the extent that theorists used the quark model commutators, and that these led to successful predictions, the quark model was centrally implicated within the current-algebra tradition.

Theorists tended initially to work with safe, model-independent, commutators and, in its early years at least, current algebra did not lend such persuasive support to quarks as did the CQM tradition.[72] The significance of current algebra lay rather in the fact that, although Gell-Mann advocated only a heuristic use of field theory, currents remained explicitly field-theoretic constructs and were manipulated and explored using characteristically field-theoretic techniques. And, since the problems entailed in extending the traditional field-theory approach beyond electromagnetism were manifest to all, current algebra became a haven in the 1960s for the majority of committed field theorists.

We can turn now from the conceptual roots of current algebra to the phenomenological uses which theorists found for it. Central to many applications was PCAC: the 'partially conserved axial current' hypothesis. PCAC was another instance of Gell-Mann prising a pheasant from the veal of perturbative field theory. In 1958, in their 'V–A' paper, Feynman and Gell-Mann had proposed that the weak vector current was exactly conserved, just like the electromagnetic current (another vector quantity). The conserved vector current (CVC) hypothesis explained the 'universality' of strength of weak decays: the fact that the muon and neutron beta-decay coupling appeared experimentally to be the same. In 1960, Gell-Mann and M. Lévy (of the University of Paris, which Gell-Mann was then visiting) went on to propose that the axial weak current was not exactly conserved, but partially. They supposed that the 'divergence' of the axial current (a measure of its non-conservation) came about through its coupling to the pion.[73] The PCAC hypothesis was, quite typically, argued heuristically to be true from a variety of field-theoretic models, and then postulated as a truth in its own right. To back this up, Gell-Mann and Lévy offered a PCAC re-derivation of the 'Goldberger–Treiman relation' – a quantitative relationship between the weak- and strong-interaction couplings of the pion which was known to hold experimentally to within around ten per cent.[74] Subsequently the PCAC hypothesis became an important adjunct to the current-algebra approach in the construction of low-energy theorems for processes involving pions. A range of phenomenological applications were devised in the mid-1960s, most of which seemed to agree quite well with the data.[75] (A notable exception was the calculation of the $\eta \rightarrow 3\pi$ decay rate, which came out quite wrong, and of which all that could be said was: 'The reason for this failure of the current algebra is not understood'.[76] This decay process remained an intransigent anomaly until the resurgence of gauge theory in the 1970s and, even then, the solution to the problem remained controversial.)

Another significant and popular line of development of the current-algebra approach was the construction of 'sum-rules'.[77] The first of these, dating from 1965, was the 'Adler–Weisberger relation'.[78] This was a relation between the axial vector weak-coupling strength (g_A) and an energy integral over pion–nucleon scattering cross-sections. One could attempt to evaluate this relation from experimental data (although, since the integral extended to infinite centre-of-mass energy, some theoretical assumptions were required) and it appeared to be reasonably well satisfied. The success of this sum-rule was regarded as a successful test of the underlying current algebra, and encouraged the formation of a sum-rule industry which was, in general, phenomenologically successful. On the one hand, sum-rules could be used to re-derive the low-energy theorems mentioned above, and hence take over their successes (and failures). On the other, the first new sum-rule which could be numerically evaluated from experimental data seemed to work. This was the Cabibbo–Radicati sum-rule, proposed in 1966,[79] and it may be useful to explain what I mean by 'seemed to work', in order to give a feeling for the kind of phenomenology at issue. The Cabibbo–Radicati sum-rule asserted that a certain mathematical integral over photon–nucleon cross-sections had the value zero. As usual, this integral extended over an infinite range of centre-of-mass energies and could only be evaluated approximately from the data. The four dominant contributions to the measurable, low-energy end of the integral were evaluated to be -0.8, $+1.6$, -2.8, and $+1.6$.[80] These added up to -0.4: a small number, but whether it was near enough to the predicted value of zero was clearly a matter of personal choice. Later sum-rules met with a similar degree of numerical success.[81]

The elaboration of sum-rules together with exploration of the meaning and consequences of PCAC constituted the principal lines of development of current algebra in the 1960s. In the early 1970s a third variant appeared, the 'light cone algebra'. This was again Gell-Mann's brainchild, and depended decisively upon the use of quark fields. Before we discuss it (in Chapter 7) we must review the phenomenon it was designed to explain: scaling. Scaling is the subject of the next chapter, which begins the analysis of the new physics. Here one topic remains to be explored . . .

4.4 The Reality of Quarks

In the latter half of the 1960s it was unarguable that the constituent quark model provided the best available tools for the analysis of hadronic resonances – the low-energy domain of the old strong-interaction physics. Likewise, there was no doubt that the current-algebra tradition, in which quarks were frequently called upon, was the cutting edge of theoretical research into the weak and electromagnetic interactions of hadrons. But particle physicists did not extrapolate from this considerable body of theoretical, phenomenological and experimental practice to the assertion that the existence of quarks was established. Throughout the 1960s and into the 1970s, papers, reviews and books on quarks were replete with caveats concerning their reality. As the British quark modeller Frank Close wrote in 1978 (when quarks had at last been certified as real):

> The quark model had some success in the 1960s; in particular it enabled sense to be made out of the multitude of meson and baryon resonances then being found. When interpreted as excited states of multi-quark and quark–antiquark systems, these resonances and their properties were understood. Even so, there was much argument at the time as to whether these quarks were really physical entities or just an artefact that was a useful mnemonic aid when calculating with unitary symmetry groups.[82]

The reasons for this state of affairs are not hard to find. Most obviously, quarks, with their fractional electric charges, were supposed to be highly conspicuous entities; yet repeated experimental searches had failed to turn up any communally acceptable candidates. This in itself was not immediately damning, since theorists could argue that quarks were too heavy to be produced in contemporary accelerator experiments. But even if this empirical problem were set on one side, it remained difficult to argue for the reality of quarks from their phenomenological utility. In the first instance the problem was that, as George Zweig put it, 'the quark model gives an excellent description of half the world'.[83] Whilst the quark model was central to the world of low-energy resonance physics, it was at most peripheral to the world of high-energy soft-scattering. In the former, hadrons indeed appeared to be quark composites, while in the latter they were Regge poles. Hadronic structure differed according to who was analysing it, making it very difficult to assert that hadrons actually were quark composites (or Regge poles). This manifest social and conceptual division of the theoretical HEP community encouraged an instrumentalist attitude towards hadronic structure. Quark and Regge modellers existed in a state of peaceful coexistence, each group focusing upon its own chosen domain of phenomena in its own terms, and generally claiming no special ontological priority for its description of hadrons.

Such social and conceptual divisions extended, moreover, into the quark theorists' camp itself. Even if one were willing to forget that the CQM was theoretically disreputable, and that the current-algebra tradition drew upon unrealistic quark models in a purely heuristic fashion, the fact remained that the image of hadrons and quarks differed between the two. This was a subtle point which caused some confusion in the early days of quarks, but in outline went as follows. In the CQM, theorists used 'constituent quarks' to represent hadrons of definite spin. The latter were the physical particles observed in experiment. In the current-algebra approach, theorists built hadrons out of 'current quarks', but these were *not* the individual hadrons observed in experiments: they were combinations of different hadrons having different spins. (Hence the succession of terms appearing in the current-algebra sum-rules, corresponding to an infinite sequence of hadronic resonances.) Current and constituent quarks were therefore in some sense different objects. Gell-Mann's view was that they were, nevertheless, related – that constituent quarks represented some complicated combination of current quarks, and vice-versa. A considerable amount of work was done in exploring the form of this relation and led to several new insights into the classification and properties of hadrons, but a full understanding was never achieved.[84] The CQM and current-algebra traditions remained separate, each with its own practitioners, its own quark concept and its own routines of phenomenological application.

The theoretical wing of the old physics in the mid-1960s was thus constituted from three principal traditions: Regge theory, the CQM and current algebra. In the first of these it was quite possible to discuss hadron dynamics without reference to quarks. In the second, quarks were central to any discussion. In the third, quarks were an optional heuristic extra but led to a different image of hadrons from that of the CQM. This social and conceptual diversity of theoretical practice fostered an instrumentalist rather than a realist view of quarks. In the 1970s a unification of practice around gauge field theory eventually made possible a consensual realist perspective on quarks. But the situation got worse before it got better. In the next chapter we will be discussing the arrival in the late 1960s of the quark–parton model: yet another variant on the quark theme, and distinct from both the CQM and current algebra.

NOTES AND REFERENCES

1 Gell-Mann (1964a). Here and throughout I use 'Caltech' as the recognised abbreviation for the California Institute of Technology, Pasadena.
2 Gell-Mann (1964a, 214).
3 Zweig (1964a, b).
4 Fermi and Yang (1949).
5 Goldhaber (1956), Sakata (1956). The Japanese theory group centred around S.Sakata and M.Taketani at the University of Nagoya ascribed their development of composite models of hadrons to adherence to a Marxist–Leninist philosophy of dialectical materialism (see Ne'eman 1974). As far as I am aware, this is one of the few instances in which there is a clear *a priori* connection between important developments in HEP and 'social interests' in the wider sense of the term.
6 Ikeda, Ogawa and Ohnuki (1959), Yamaguchi (1959). Like the Eightfold Way, the group-theory version of the Sakata model was based upon an SU(3) symmetry. The Japanese theorists, however, assigned particles to group representations differently from Gell-Mann and Ne'eman.
7 Ne'eman (1983, 8).
8 See Gell-Mann (1961, 1962b), Zweig (1964a, b).
9 See Ne'eman (1983). Ne'eman himself, for example, can possibly claim to have invented quarks in 1962 (Goldberg and Ne'eman 1963) although 'the physical picture in that paper still appears somewhat ambivalent' (Ne'eman 1983, 17).
10 Gell-Mann (1964a, 215).
11 For a review of quark search experiments, see Jones (1977). It is important to note that individual experimenters have from time to time reported quark sightings, but these reports have never been consensually accepted within the HEP community. For a review of positive quark sightings, see McCusker (1981, 1983). The most definitive and as yet unrefuted sightings have come from a Millikan-type experiment at Stanford University, California: La Rue, Fairbank and Hebard (1977), La Rue, Fairbank and Phillips (1979) and La Rue, Phillips and Fairbank (1981). For the history of the Stanford experiment and of a similar experiment performed at the University of Genoa with negative results, see Pickering (1981a).
12 Zweig (1964a, 18).
13 I am grateful to Professor Zweig for providing me with biographical information.
14 For a fascinating account of his route to quarks, see Zweig (1981).
15 Zweig (1964a, 33; 1964b, 39).
16 Zweig (1981, 458). Zweig also recalled (1981, note 13) that 'Murray

Gell-Mann once told me that he sent his first quark paper to Physics Letters for publication because he was certain that Physical Review Letters would not publish it.'

17   Zweig (1981, 458).
18   Kokkedee (1969, ix–x).
19   These are taken from Zweig (1964b).
20   In group-theory notation: $3 \times 3 = 1 + 8$; $3 \times 3 \times 3 = 1 + 8 + 8 + 10$. Here, '3' denotes the triplet representation of su(3), 3 the antitriplet (corresponding to antiquarks), and so on.
21   Connolly *et al.* (1968).
22   Okubo (1963, 165).
23   It is interesting to note that Zweig's personal route to quarks began with the anomalous decay of the phi. He first explained it in terms of the Sakata model as a lambda–antilambda composite. He then encountered Okubo's su(3) explanation of the same phenomenon, which led him to study the representation structure of su(3), to quarks as underlying those representations and, finally, to reformulation of the Zweig rule in terms of quark rather than baryonic constituents. See Zweig (1981).
24   Zweig (1964c). B.Sakita (1964) at the University of Wisconsin independently followed the same line of reasoning.
25   The 6 of su(6) comes from the 3 of su(3) multiplied by 2 for the number of possible spin orientations of spin $\frac{1}{2}$ quarks.
26   $56 = 8 \times 2 + 10 \times 4$: the multiplicative factors count the number of possible spin orientations of the spin-$\frac{1}{2}$ octet baryons and the spin-$\frac{3}{2}$ decuplet baryons. For an introduction to su(6), and reprints of the seminal papers, see Dyson (1966).
27   Gürsey and Radicati (1964), Pais (1964), Gürsey, Pais and Radicati (1964).
28   Wigner (1937).
29   Morpurgo (1970, 108).
30   This summary of the failings of relativistic su(6) is taken from Dyson (1966, vi). For the original literature, see reprints 13 to 25 of that volume. The non-existence theorems demonstrated that the achievement of a non-trivial synthesis of internal symmetries with relativity required theorists either to work in more than the accepted four dimensions of space-time, or to resort to representations containing an infinite number of particles.
31   Morpurgo (1970, 108).
32   Morpurgo (1965). Similar observations were also made by the us theorist Y.Nambu, and the Soviet theorist A.Tavkhelidze: see notes 27 and 28 of Morpurgo (1970). Morpurgo demonstrated his willingness to take the cqm seriously by performing a series of experimental searches for free quarks at the University of Genoa using the Millikan oil-drop approach. These were all unsuccessful (in the sense that no evidence for fractional charge was found) and in the 1970s were often held to contradict the positive sightings at Stanford (see note 11 above). For the history of Morpurgo's experiment, see Pickering (1981a).

Morpurgo's route to the cqm was similar to that of Dalitz (see below). Born in Florence, Italy in 1927, Morpurgo began his research career in theoretical physics at the University of Rome in 1948 and became Professor of Structure of Matter at the University of Genoa in 1961. In the 1950s and 1960s he worked in both nuclear and elementary particle physics. One aspect of his work in particle physics was the phenomenology of the strong interactions, especially concerning the strange particles (see Morpurgo 1961). This confronted him with the proliferation of data associated with the population explosion of hadrons, and

> Already in 1961 or 62 the growth in the number of 'resonance' states being discovered looked to me similar to the growth in the number of levels known in some nuclei, e.g. B^{10} [boron], that I had been following on the Nuclear Tables for some years in connection with my work [Morpurgo 1958] . . . Therefore when in the Autumn of 1964 the paper by Gürsey and Radicati appeared [on the $SU(6)$ classification scheme] . . . my attitude was different from that of the most part of my theoretical colleagues. They were, mostly, interested in trying to reconcile $SU(6)$ invariance with Lorentz invariance . . . instead I considered the results of Gürsey and Radicati as an indication that (in some approximation) the internal dynamics of quarks inside hadrons could be treated as non relativistic.

Thus, by redeploying his existing knowledge of nuclear physics to the $SU(6)$ symmetry of hadrons, Morpurgo arrived at the formulation of the CQM described in his first quark paper (Morpurgo 1965). This paper was devoted to showing that, in general terms, the CQM described a physically sensible system. Then followed more detailed calculations. 'A few days after I had submitted my paper . . . I realised that other tests of the model were possible (and in fact experimental data already available) by the electromagnetic decays of the hadrons'. This realisation bore fruit in two publications discussed below (Becchi and Morpurgo 1965a, b). It is significant to note that these concerned the so-called M_1 electromagnetic decays of hadrons – a topic which Morpurgo had already analysed in the realm of nuclear physics (Morpurgo 1958, 1959).

I am indebted to Professor Morpurgo for a long account of his involvement with the CQM, from which the above quotations are taken, and for the provision of biographical materials.

33 Morpurgo (1970, 111).
34 The literature on applications of the CQM is enormous. For a popular account, see Weisskopf (1968). For a selection of technical reviews see Kokkedee (1969), Morpurgo (1970), Lipkin (1973), Hendry and Lichtenberg (1978), Greenberg (1978, 1982), Close (1979), Gasiorowicz and Rosner (1981) and Dalitz (1982). The third main line of development of the CQM, which I will not discuss here, concerned its application to high-energy hadronic scattering. This led to the prediction of relations between cross-sections for the interactions of different hadrons, many of which agreed well with experiment. See Chapters 14 to 17 of Kokkedee (1969) for a review of this work.
35 Dalitz first laid out his version of the CQM approach in the summer of 1965, at the Oxford Conference (Dalitz 1966a) and at the Les Houches

Summer School (published as Dalitz 1966b). From 1965 onwards, he gave authoritative reviews of developments in CQM phenomenology at many major conferences (see, for example, Dalitz 1967). For instances of Dalitz' later contributions to the CQM see Horgan and Dalitz (1973) and Jones, Horgan and Dalitz (1977). I am grateful to Professor Dalitz for an interview and the provision of biographical materials.

36 Kokkedee (1969, 46, Figure 7).

37 A particularly recalcitrant resonance was the N(1470) (see Kokkedee 1969, 50). Despite its relatively small mass, the properties of this resonance made it hard to assign to any of the accepted low-lying SU(6) multiplets. Many solutions to this problem were proposed in the 1960s and 1970s. The general consensus was that it was one of very few examples of a vibrationally rather than rotationally excited state. This, of course, raised the problem of the apparent absence of all of the N(1470)'s expected partners in a vibrationally excited SU(6) multiplet. It must be stressed, though, that the N(1470) was an exception: most particles could be assigned to multiplets with many known members, and the more typical phenomenological problem lay rather in deciding between a variety of possible multiplet assignments.

38 Kokkedee (1969, 51).

39 Gasiorowicz and Rosner (1981, 971, Table 7).

40 Becchi and Morpurgo (1965a).

41 Other early applications were to the magnetic moments of the proton and neutron (Morpurgo 1965) and to the electromagnetic decays of the vector mesons (Becchi and Morpurgo 1965b). For the former, Morpurgo showed that the ratio between the proton and neutron magnetic moments was expected to be $-\frac{3}{2}$ in the CQM – very close to the measured value of -1.47. Against this, however, had to be set the fact that, when theorists attempted to compute the ratio of the vector and axial vector weak-coupling constants of the nucleon (g_A/g_V) under the same assumptions, they arrived at a value of 5/3, while the experimentally measured value was around 1.2 (see Lipkin 1973, 250–3 for a discussion of this point).

42 Moorhouse (1966).

43 Copley, Karl and Obryk (1969a, b), Faiman and Hendry (1969).

44 It is worth noting just how manifestly phenomenological such predictions were. They depended upon arriving at any appropriate list of factors to be included or excluded from the calculations, and of approved approximations. As Kokkedee (1969, 70) put it: '. . . the weak leptonic and electromagnetic decays considered here are consistently described by the nonrelativistic, independent quark model' provided that certain ' "rules of the game" are observed'.

45 For early work on weak decays in the CQM, see van Royen and Weisskopf (1967).

46 Mitra and Ross (1967), Lipkin, Rubinstein and Stern (1967), Faiman and Hendry (1968).

47 Note that although resonance spectroscopy and coupling calculations have been discussed sequentially here, they were interdependent exercises. Each drew upon the other for information upon the SU(6) and

su(3) assignments of individual resonances. Thus each generation of spectroscopic fits informed the succeeding generation of coupling calculations, and vice-versa.

48 The 'decay channels' of a resonance are the different combinations of particles into which it can decay. Thus, for example, if a resonance is observed to decay into πN, KΛ, KΣ and $\pi\pi$N combinations, experimenters would speak of measuring its probability of decay into the πN decay channel, the KΣ decay channel, and so on. The decay channel concept was taken over directly from nuclear spectroscopy as was the associated concept of a 'branching ratio'. The branching ratio of a given decay channel represents the fraction of the decays of a given resonance which lead to that decay channel.

49 Zweig (1981, 454–5).

50 Kokkedee (1969, 40).

51 Particle Data Group (1982, 14, 16). Based primarily in the University of California at Berkeley, the Particle Data Group came into existence in 1957. Its aim was to provide an authoritative assessment of particle data, as enshrined in *Review of Particle Properties* – a function which assumed increasingly greater importance within the HEP community as the population explosion accelerated during the 1960s. For the history of the Particle Data Group, see Rosenfeld (1975).

52 Zweig (1964b, 32–3).

53 To give a few relevant examples: (1) in their early work on the electromagnetic decays of vector mesons, Becchi and Morpurgo (1965b) made predictions of several unknown decay modes and suggested that these could be experimentally tested; (2) Dalitz in his major 1967 review noted, amongst many similar points, that the CQM required the existence of three unobserved spin 1 positive parity mesons and that 'More effort will be needed to search for these three missing nonet states' (1967, 219); and (3) in their analysis of resonance hadronic couplings, Mitra and Ross (1967, 1635–6) gave an extensive discussion of the resonances required by their model but missing from experimental reports, and of how they could be best detected.

Other less detailed but important examples of the heuristic value to experiment of the CQM were its role in motivating searches for quarks and 'exotics'. The latter were resonances which could not be described as q$\bar{\text{q}}$ or qqq resonances by virtue of their quantum numbers. Attention here focused upon the search for strangeness $+1$ baryons, collectively known as Z's. These could not be three-quark composites: the s quark had strangeness -1, the Z's must therefore contain a strange antiquark, and thus their composition must be at least as complicated as $\bar{\text{s}}$qqqq. The most direct way to search for Z's was to study the interaction of strangeness $+1$ mesons, kaons, with nuclear targets, and such measurements got under way in the mid-1960s. Sightings of five different Z's were reported from 1967 onwards, but became embroiled in debates over what exactly constituted a resonance. The emergent consensus was that the evidence for the existence of Z's was 'weak' (see Table 4.2(b)). For access to the experimental and theoretical literature on Z's, see Particle Data Group (1982, 232–4).

The Quark Model

54 Rosenfeld (1975, 564) notes that in 1967, 'about 2 million bubble chamber events were being measured annually, and about a thousand physicists were hunting through 10,000 to 20,000 mass histograms each year, in search of striking features, real or imagined'. Rosenfeld (1975, 567) also shows that the rate of production of experimental data on particle properties grew linearly from 100 results/year in 1964 to 400 results/year in 1970 before declining to less than 300 results/year in 1972 (these data production rates exclude non-strange baryon resonance and Z searches).

55 It is significant to note that in the context of the 1970s new physics, conventional resonance experiments were seen to be relatively uninformative and such experimentation ceased almost completely. The apparent conflict between this assertion and the content of Figure 4.5 below is resolved by noting that in the latter half of the 1970s a great deal of experimental effort went into the investigation of the unconventional 'new particles' (see Particle Data Group 1982, 291). These points will be discussed further in Part III of the account.

56 For a review of the various techniques of 'partial-wave' and 'phase-shift' analysis, see Particle Data Group (1982, vii–xi).

57 According to QED, electrons interact with other particles only via the exchange of photons (in most experimental situations, their weak interactions are irrelevant). The electron beams available at electron accelerators are thus effectively equivalent to photon beams. According to the experimental conditions the interacting photons may be either real, i.e. zero-mass physical states, or virtual, finite-mass particles, permitted by the Uncertainty Principle to exist only for a very short interval of time. Experiments using real photons are known as photoproduction experiments, those using virtual photons as electroproduction. Both photoproduction and electroproduction experiments were principally aimed at the exploration of resonances: the technically more complex electroproduction experiments tended to displace photoproduction during the 1970s.

58 Particle Data Group (1982, 291).

59 A further element of this symbiosis can be noted. Resonance experiments could be divided into two classes; 'formation' experiments and 'production' experiments. Formation experiments generated data on resonances alone, but production experiments also generated data of interest to Regge theorists. Thus, traditions of production experiments existed in symbiosis with traditions of both quark and Regge phenomenology. The distinction between formation and production experiments is the following. Formation experiments observe the direct production and decay of single resonances: $\pi N \to N^* \to \pi N$, for example, where N is a target nucleon, and N^* the resonance in question. Here the N^* appears as a bump in cross-sections at a centre-of-mass energy equal to the resonance mass. Since typical resonance masses are in the range 1 to 3 GeV, formation experiments need only low energy beams and do not explore the energy range of interest to Regge theorists. Production experiments, in contrast, use high-energy beams to observe the production of resonances in association with other particles: $\pi N \to \rho N \to \pi\pi N$,

121

for example, where the produced rho decays to a pair of pions. Here the rho appears as a bump in the exclusive ππN cross-section when the data are plotted as a function of the 'invariant mass' of the pion pair (the invariant mass is the mass of the parent particle decaying to the pion pair, computed from the measured energies and momenta of the pair). High-energy resonance production processes (e.g. πN→ρN) were found to show the same soft-scattering characteristics as elastic scattering (e.g. πN→πN) and thus provided a new domain for Regge analysis.

60 Cushing (1982, 83) notes that, 'The analytic S-matrix theory was seen as a sufficiently promising wave of the future that it was sometimes difficult to find a conventional field theory course taught in graduate physics departments in the United States in the early 1960s'. The growth of Regge phenomenology and the arrival of the CQM can only have exacerbated this situation.

61 In his key-word analysis of the HEP literature, S.Wagner finds that the publication rate of papers on QED rose smoothly during the latter half of the 1960s, from around 100 papers per year in 1965 to around 200 in 1970 (compare these rates with those quoted for the CQM and Regge in Chapter 3, note 2). The stimulus for this rise came largely from experiment. High-precision measurements on atomic systems and of the magnetic moments of electrons and muons became available for comparison with higher-order calculations in perturbative QED. And high-energy experiments at electron accelerators and colliders served to test the lowest-order predictions of QED in new regimes of energy and momentum transfer. As noted in Chapter 3.3, this work served to reinforce the conclusion that QED was indeed an accurate theory of the electromagnetic interactions. For access to the literature on QED, see Brodsky and Drell (1970), Lautrup, Peterman and de Rafael (1972), Calmet, Narison, Perrottet and de Rafael (1977), Farley and Picasso (1979) and Combley (1979).

62 Gell-Mann was born in New York in 1929, took his first degree in physics at Yale in 1948, and completed his PhD in physics at MIT in 1951. After holding several temporary posts, he accepted a permanent position at Caltech in 1955. In 1959 he was awarded the first Dannie Heineman Prize. Other honours followed for his work on strangeness, the V–A theory, the Eightfold Way, current algebra and quarks, culminating in the Nobel Prize for Physics in 1969.

63 For example, an analysis of theoretical papers published in the journal *Physics Letters* revealed that between 1967 and 1969 papers on current algebra made up around ten per cent of the total. In the period 1969 to 1970 they comprised less than half that. In both intervals papers on strong-interaction physics – principally concerning the CQM and variants of Regge theory – constituted around seventy-five per cent of the total. See Jacob (1971, Fig. 1). See also Jackson (1969, 63–4).

64 For more details and reprints of important publications on current algebra, see Adler and Dashen (1968). An alternative extensive review is to be found in Bjorken and Nauenberg (1968).

65 Gell-Mann (1964a), Zweig (1964a).

66 I am grateful to Professor Gell-Mann for an interview and the provision

of biographical materials.

67 Gell-Mann and Low (1951).

68 Gell-Mann (1962b).

69 To say that the currents generated an su(2) or su(3) algebra is equivalent to the assertion that they could be assigned to representations of the relevant symmetry group. The 'current algebra' terminology arose because in this instance the su(2) symmetry of leptonic weak interactions followed from detailed properties already assigned to the lepton fields, rather than being postulated in the first instance as a classification system.

70 Gell-Mann (1964b, 73).

71 Cabibbo (1963). Cabibbo's analysis required the introduction of a free parameter 'sin θ' (found to have the value 0.26) relating the strengths of strangeness-changing to strangeness-conserving weak decay processes. θ later became known as the 'Cabibbo angle'. It is interesting to note that Cabibbo excluded from his analysis a class of observed particle decays in which the change of strangeness (ΔS) was equal but of opposite sign to the change of the electrical charge (ΔQ) of the hadrons involved. $\Delta S = -\Delta Q$ events were in contradiction with the V–A theory, and were reported by several experimental groups between 1961 and 1963. From 1964 onwards, however, no more such reports were made, and the earlier experiments came to be regarded as somehow in error: see White and Sullivan (1979).

72 To be more explicit: the applications of current algebra discussed below were independent of quarks. The applications discussed in subsequent chapters on the new physics were not. See Bjorken and Nauenberg (1968, 254–5).

73 Gell-Mann and Lévy (1960).

74 Goldberger and Treiman (1958).

75 These applications concentrated on processes such as the low-energy limit of pion– (and kaon–) nucleon scattering, pion photoproduction and pion production by pion and photon beams, as well as the weak decays of kaons and strange baryons. For a review, and reprints of the original papers, see Chapters 2 and 3 of Adler and Dashen (1968).

76 Adler and Dashen (1968, 138).

77 As Gell-Mann (1962b, 1080–1) pointed out, the derivation of sum-rules from commutation relations was routine in non-relativistic quantum mechanics. Unfortunately, when theorists attempted to extend this approach to the relativistic domain appropriate to HEP, they found that, in general, various experimentally inaccessible processes made important contributions to the sum-rules. In early 1965, two Italian theorists showed that contributions from such processes could be suppressed by performing calculations in a peculiar kinematic frame in which the momentum of proton states was infinite, the so-called 'infinite momentum frame' (Fubini and Furlan 1965). The invention of this technical trick unleashed the sum-rule industry outlined below (see Bjorken and Nauenberg 1968, 242–4).

78 Adler (1965), Weisberger (1966).

79 Cabibbo and Radicati (1966).

80 Gilman and Schnitzer (1966).
81 For more on the theoretical and phenomenological elaboration of current algebra sum-rules, see Bjorken and Nauenberg (1968, 242–7), and the discussion and reprints of Chapters 4 to 6 of Adler and Dashen (1968).
82 Close (1979, v).
83 Quoted in Lipkin (1982, 1).
84 The most sophisticated attempt was made by one of Gell-Mann's students, H.J.Melosh, who explicitly constructed a possible transformation linking non-interacting current quarks and the constituent quarks of the CQM. Melosh (1974) gives details of this work and reviews the earlier history of the problem.

5

SCALING, HARD SCATTERING AND THE
QUARK-PARTON MODEL

The new physics began, appropriately enough, in California. In the late 1960s, experiments at the Stanford Linear Accelerator Center (SLAC) saw the discovery of the first new-physics phenomenon: scaling.[1] The circumstances of the discovery itself are the subject of Section 1 of this chapter. Section 2 deals with the explanation of scaling offered by Richard Feynman's 'parton' model, and aims to explore the central analogy of Feynman's approach. Sections 3 and 4 review the phenomenological applications of the parton model to electron scattering at SLAC, and to neutrino scattering at the CERN PS. These applications led to the identification of Feynman's partons with the quarks of Gell-Mann and Zweig. Section 5 outlines the applications of the parton model to other lepton–hadron processes and to unexpected hard-scattering phenomena seen in purely hadronic reactions in the first round of experiments at the CERN ISR. Experimental investigation of the phenomena to which the parton model was held to apply came, in the 1970s, to constitute the empirical backbone of the new physics. The parton model itself was subsequently subsumed into the gauge field theory of the strong interaction, QCD, as we shall see in Chapter 7.

5.1 Scaling at SLAC

Construction of a 22 GeV, two-mile long linear electron accelerator at the Stanford Linear Accelerator Center was first formally proposed in 1957 and began in 1962. By January 1967 the design energy had been achieved and the physics programme got under way.[2] At this time several lower-energy electron machines were in operation around the world, but had contributed little of major impact to the development of HEP. As discussed in the previous chapter, the principal focus of electron-beam experiments was the exploration of the detailed properties of hadronic resonances.

Amongst the 19 experimental proposals which had been approved at SLAC by July 1967 were three by a collaboration of physicists drawn from SLAC, Caltech and Massachusetts Institute of Technology (MIT).[3] These were typical first-round experiments at a new accelerator, conceived to 'act as a shakedown of the spectro-

125

meters' – major items of experimental apparatus, designed to measure the energies and momenta of scattered electrons – and to 'provide a general survey of the basic cross-sections which will be useful for future proposals'.[4] The experiments in question aimed at measurements of the elastic scattering of electrons and positrons from protons, and of inelastic electron–proton scattering. The discovery of scaling emerged from the last of these measurements, but a discussion of the first will provide some necessary conceptual background.[5]

Elastic Electron–Proton Scattering
An elastic scattering process is one in which beam and target particles retain their identity and no new particles are produced. Elastic electron–proton scattering is, in obvious notation, the process ep→ep. Electrons are believed to be immune to the strong interaction. Thus the routine assumption in analysing electron-scattering processes is that they are predominantly electromagnetic and well described in terms of the exchange of a single photon between the incoming electron and the target proton (Figure 5.1).

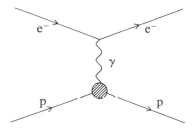

Figure 5.1.   Elastic electron–proton scattering
via single photon exchange.

The primary interest of all electron–proton scattering experiments derived from the belief that electron–photon interactions were fully understood in terms of QED. Experimental measurements could therefore be construed as exploring the details of photon–proton interactions. As discussed further in Section 5.2, according to QED (in first-order approximation) electrons interacted as structureless point particles, occupying an infinitesimal volume in space, while protons, by virtue of their strong interactions, were expected to have *structure* and to be extended over a finite volume of space. Measurements of electron–proton scattering could therefore be considered to probe the proton's structure and, in particular, to explore the distribution in space of the proton's electric charge.

The first electron–proton scattering measurements were made at

126

Stanford using the lower-energy electron accelerators which were the predecessors of the SLAC machine. These early experiments served to mark out a significant difference between elastic electron–electron and electron–proton scattering. It was found that cross-sections at large scattering angles for the latter were much smaller than those for the former (see Figure 5.4, below). Electrons frequently scattered from other electrons at large angles, while electron–proton cross-sections peaked at small angles, with very few electrons being scattered at large angles. In loose but suggestive language, electrons seemed to act as 'hard' objects, bouncing violently off one another, while protons seemed to be 'soft' and capable only of giving a gentle nudge to passing particles. This was the origin of the terminology of hard and soft scattering which serves to characterise many of the phenomena of the old and new physics. Speaking more formally, the observed angular dependence of the electron–electron cross-section was that predicted by QED for point-like electrons, while the angular dependence of the electron–proton cross-section resembled that familiar to physicists from optical diffraction phenomena. And, indeed, in the pre-SLAC era the accepted explanation of the elastic electron–proton scattering data was that the proton was a diffusely extended object which served to diffract the incoming electron beam. As in optical physics, a connection could be made between the dimensions of the diffracting object and the shape of the diffraction pattern it produced, and the electron–proton scattering data were taken to show that the proton had a finite diameter of around 10^{-13} cm.

In 1961, R. Hofstadter was awarded the Nobel Prize for Physics for his pioneering work at Stanford on electron scattering. The SLAC-MIT-Caltech experiment made measurements of elastic electron scattering at hitherto inaccessible energies, but no major discoveries resulted.[6] For our purposes, the point to note is that the observed electron–proton cross-sections retained their characteristic diffraction peak at high energies, and continued to be identifiable with a diffusely extended proton structure.

Inelastic Electron–Proton Scattering
When the elastic electron (and positron) measurements were complete, the Caltech group decided not to take part in the third experiment. The SLAC-MIT group led by Richard Taylor (SLAC) and Henry Kendall and Jerome Friedman (MIT) went on alone to measure inelastic electron–proton scattering.[7] The experimental runs took place in the second half of 1967 and the procedure was a simple one. The experimenters fired the electron beam at a proton (liquid

hydrogen) target, and counted how many electrons emerged with given energies at given angles relative to the beam axis. They thus measured inclusive cross-sections for the process ep→eX, where X represents whatever hadronic debris resulted from the electron–proton collision. Such inclusive measurements were considered to be less informative than exclusive measurements in which the particles comprising X were detected and identified. But they were also much simpler, since no instrumentation was needed to register the hadronic debris. The SLAC-MIT experiment was therefore regarded as a quick and crude survey, designed to investigate the new energy regime, in preparation for later exclusive measurements. As usual, inelastic electron–proton scattering was thought to be mediated by single-photon exchange (Figure 5.2(a)) and, in particular, it was expected that the most important process would be resonance production (Figure 5.2(b)).[8]

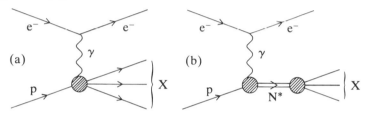

Figure 5.2.   (a) Inelastic electron–proton scattering; (b) Resonance (N*) production in inelastic electron–proton scattering.

Analysis of the inelastic data began in the spring of 1968. For the small-angle scattering of lower-energy electrons the experimenters found what they expected. Figure 5.3 shows a typical plot of cross-sections versus the mass of the hadronic system produced, and three bumps are clearly evident, corresponding to the $\Delta(1232)$, $N(1518)$ and $N(1688)$ resonances.[9]

However, as the experimenters examined the data for larger angle scattering and higher beam energies they encountered a surprise. Their expectation was that resonance production cross-sections would decrease rapidly at large angles like elastic electron–proton cross-sections, exhibiting a diffractive fall-off characteristic of the spatial dimensions of the resonances. What the experimenters found was that although the individual resonance peaks quickly disappeared, the measured cross-sections remained large at large angles and high energies (technically, at high momentum transfers, q^2). This behaviour is illustrated in Figure 5.4.[10] Here the inelastic cross-sections are shown divided by the corresponding electron-electron

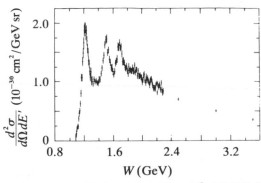

Figure 5.3.   Inelastic cross-sections ($d^2\sigma/d\Omega dE'$) for small-angle (6°) scattering of relatively low-energy (10 GeV) electrons at SLAC. W is the mass of the hadronic system X.

scattering cross-section (σ_{Mott}). It is evident that the q^2-dependence of the inelastic electron–proton and elastic electron–electron cross-sections is roughly the same. For comparison the elastic electron–proton cross-section is also shown, which falls off very quickly in proportion to σ_{Mott}.

From the preceding discussion it is perhaps clear that a straightforward inference from the parallelism between the high-q^2 inelastic cross-sections and σ_{Mott} was that the proton contained hard point-like scattering centres which, in some sense, resembled electrons. The significance of the large inelastic cross-sections was, however, far from immediately obvious to the SLAC-MIT experimenters. Their initial reaction was to suspect that their findings were illusory. The grounds for this suspicion were that electrons were well known to radiate photons rather freely, and measurements of electron energy and scattering angle could therefore not be directly translated into values of momentum transfer to the target proton. Allowance had to be made for the possibility that radiated photons had carried off energy and momentum undetected. The experimenters suspected that the large inelastic cross-sections would disappear when the appropriate radiative corrections were made to allow for the effects of processes like that of Figure 5.5.

Radiative corrections could only be estimated approximately, and the appropriate calculations were, in Richard Taylor's words, 'difficult and tedious'.[11] Nevertheless, such calculations went ahead at SLAC and MIT, and it appeared that the radiative corrections could not account for the large cross-sections observed at high momentum-transfers.[12] This was the experimental situation in early 1968, but it

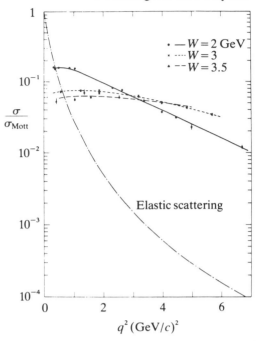

Figure 5.4. SLAC inelastic cross-sections (σ) as a function of momentum-transfer (q^2). The lower curve represents elastic cross-section measurements.

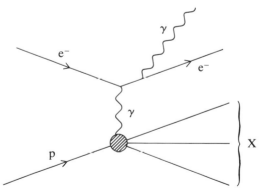

Figure 5.5. Electron energy and momentum loss via photon radiation.

could not be said that at this stage a discovery had been made. The SLAC-MIT experimenters felt fairly confident that their findings were genuine, but they had no idea what they meant: what had they found? At this point, the experimenters began to pay attention to a member of the SLAC IIEP theory group, James D.Bjorken.

Bjorken and Scaling

Bjorken was awarded a PhD in theoretical particle physics by Stanford University in 1959. After another four years there (as Research Associate and Assistant Professor) he joined the SLAC theory group at its foundation in 1963 (becoming a full Professor in 1967). At that time the SLAC accelerator was still under construction and Bjorken, like the other resident theorists, directed his efforts towards laying the groundwork for the interpretation of high-energy electron scattering. He was a field-theorist and a devotee of current algebra,[13] and one task with which he became particularly 'obsessed' was that of predicting the behaviour of inelastic electron scattering. From a variety of non-rigorous perspectives – principally from consideration of a modified version of a neutrino scattering current-algebra sum-rule – he concluded that 'the only credible guess' was that the inelastic cross-sections would be large – as they indeed proved to be.[14] He also derived a second and more quantitative result, which went as follows. In 1964, Stanford theorists S.D.Drell and J.D.Walecka had shown that in general the inelastic electron–proton scattering cross-section could be expressed in terms of two independent quantities, W_1 and W_2, known as 'structure functions'.[15] In principle, the two structure functions were each functions of two independent kinematic variables, usually chosen to be q^2 and v (where v was a measure of the energy lost by the electron in the collision). From his current-algebra calculations, Bjorken inferred that in an appropriate kinematic limit where q^2 and v became large but with a fixed ratio (the 'Bjorken limit', as it became known), W_1 and the product vW_2 should not depend upon v and q^2 independently. Rather, W_1 and vW_2 should only be functions of the ratio v/q^2: W_1 and vW_2 should fall upon unique curves when plotted against the variable $\omega = 2Mv/q^2$ (M is the mass of the proton).[16]

Perplexed by their observations, the SLAC-MIT experimenters were readily prevailed upon by Bjorken to plot their data against v/q^2, and the data were indeed seen to exhibit the required features, at least approximately, and for values of v and q^2 greater than around one or two GeV, the so called 'deep inelastic' region. For illustration, the universal curves for W_1 and vW_2 are shown in Figure 5.6.[17] These became known as 'scaling' curves – because the v-dependence of the data points could be compensated for by an appropriate scaling of q^2 – and the phenomenon they enshrined became known as 'scaling'. In passing, it is interesting to note that the variable ω' measured along the horizontal axis of Figure 5.6 is closely related but not identical to the scaling variable ω chosen by Bjorken as the relevant parameter.[18] When plotted against ω, scaling was clearly only an appropriate phenomenon; ω' was a so-called 'improved scaling variable', chosen to display scaling as an exact phenomenon. With the advent of QCD (Chapter 7) theorists became interested in measurements of *deviations* from exact scaling, and the use of improved scaling variables went out of fashion. For simplicity, I will therefore refer to ω as the scaling variable in what follows.

Once their puzzling observations had been transmuted into the discovery of the scaling phenomenon, the SLAC-MIT experimenters decided to make their findings public. The first presentations were made at the Vienna HEP Conference in September 1968. In a rapporteur's talk, SLAC's Director, Wolfgang Panofsky, mentioned the scaling property of the deep-inelastic data,[19] but this created little excitement at the time.[20] The problem was that, despite Bjorken's work, the significance of scaling remained unclear. Bjorken's calculations were not only heuristic: to experimenters and the many theorists unfamiliar with the outer reaches of current algebra they were esoteric to the point of incomprehensibility. When the deep-inelastic data appeared in a refereed journal in 1969 the response of the HEP community was more enthusiastic.[21] But by 1969 the theoretical context had been transformed by the intervention of Caltech theorist Richard Feynman.

5.2 The Parton Model

Late in 1968, shortly after the Vienna conference, Feynman paid a brief visit to SLAC. While there, he formulated the 'parton model' explanation of scaling, an achievement which will reverberate throughout the remaining pages of this account. Here I will trace out Feynman's route to the parton model, and explain how it was applied to deep-inelastic electron scattering.[22]

Feynman was a leading field theorist – we have already noted his work on the renormalisation of QED and on the V − A theory of the weak interactions – and during the 1960s he attempted to develop some useful field-theoretic insight into the strong interaction. The general obstacles encountered by field theorists in discussing the strong interaction have been outlined in Chapter 3, but it will be useful to reformulate them here in terms more appropriate to Feynman's approach.

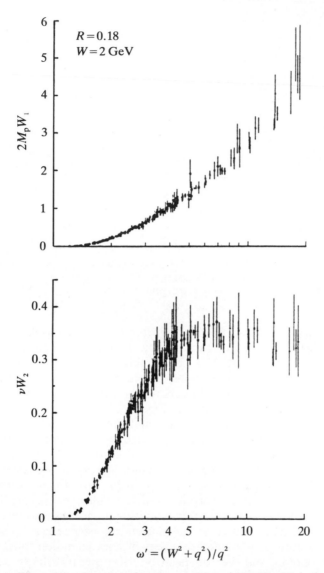

Figure 5.6. Structure functions (νW_2 and $2M_p W_1$, M_p is the proton mass) for deep-inelastic electron–proton scattering as measured at SLAC. Note that measurements at different values of q^2 all fall on the same curves when plotted against the scaling variable ω'.

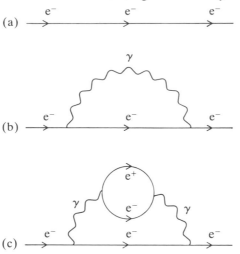

Figure 5.7. Feynman diagrams for (a) non-interacting electron; (b) emission and reabsorption of a photon; and (c) photon conversion to an electron–positron pair.

I have referred to the electron as structureless and point-like, and to the proton as having structure, and these terms have a particular meaning for field theorists. To explain this, let us begin with the electron and consider an electron moving freely through space. If the electromagnetic interaction did not exist, the Feynman diagram for this electron would be that shown in Figure 5.7(a) corresponding to the motion of an entity occupying only a mathematical point in space. When the electromagnetic interaction is taken into account the situation becomes more complicated: the electron can emit and reabsorb photons as it travels along, and the photons themselves can convert back and forth into electron–positron pairs. Figure 5.7(b) shows the emission and reabsorption of a single photon, as given by the first-order approximation to QED, and Figure 5.7(c) shows how, in second-order approximation, the photon can itself convert to an electron–positron pair. At higher orders of approximation to QED ever more photons and electron–positron pairs appear. Thus, when these higher order terms are taken into account, the electron appears not as a single point-like entity, but as an extended cloud of electrons, positrons and photons, which together carry the quantum numbers, energy and momentum of the physically observed electron. However, because of the weakness of the electromagnetic interaction (the smallness of the fine-structure constant) the higher-order QED

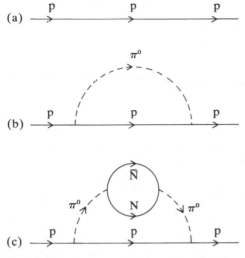

Figure 5.8. Feynman diagrams for (a) non-interacting proton; (b) emission and reabsorption of a pion; and (c) pion conversion into a nucleon–antinucleon pair.

corrections to Figure 5.7(a) represent only small perturbations. Physically, this corresponds to the fact that for most practical purposes it is sufficient to regard electrons as point-like particles. This is the field-theoretic significance of describing the electron as structureless.

A similar argument can be made for the proton (or any hadron). Figure 5.8(a) shows a non-interacting proton travelling through space. However, in considering the proton's structure, the strong as well as the electromagnetic interactions have to be taken into account. Thus, for example, in strong interaction field theories of the type discussed in Chapter 3.3, mesons and nucleon-antinucleon pairs appear as indicated in Figures 5.8(b) and (c). By virtue of its strong interaction, the proton appears in field theory as a spatially extended cloud of nucleons, antinucleons and mesons. So far the analogy between electrons and protons is exact. But it breaks down when one recognises that the strong coupling constant is large and that, as explained in Chapter 3.3, successive terms in the perturbation expansion are larger than their predecessors (rather than smaller, as is the case in QED). From the perspective of field theory it is therefore not sufficient to regard the proton as the structureless entity of Figure 5.8(a). All of the higher-order perturbations – those of Figures 5.8(b) and (c) and infinitely more complex ones, too – have to be taken into account. In field theory the proton has an intrinsically complex structure, constituted from a whole cloud of ephemeral particles.

This was the situation which Feynman confronted as a field theorist. In one way, the field-theory image of proton structure was welcome. The ascription of a finite size of around 10^{-13} cm served to explain why the elastic electron–proton scattering cross-sections decreased diffractively with momentum transfer. In another way, the image was most unwelcome. To have to deal with the proton as a particle-cloud rather than as a single particle made the field theorist's task that much more difficult, and constituted another reason for the abandonment of the traditional field theory approach to strong interaction physics. Feynman, however, was not so easily deterred. He took the view that protons (and all hadrons) were indeed clouds of an indefinite number of particles. In a field theory of mesons and nucleons, the clouds would contain mesons, nucleons and antinucleons; in a quark theory, they would contain quarks and antiquarks. Feynman was agnostic: he simply assumed that the swarms contained entities of unspecified quantum numbers, and christened them 'partons'.

Suppose, reasoned Feynman, one considers a high-energy collision between two protons. Because of their high relative velocity, each proton will see the other as relativistically contracted along its direction of motion to a flat disc or pancake. Furthermore, because the strong interactions are of short range, the two pancakes would only have a very short time to interact with one another, during which they would effectively see a frozen 'snapshot' of the partons within each pancake. Feynman therefore envisaged high-energy hadronic collisions as taking place between *individual partons* of each pancake and, because the available interaction time would be small, he regarded the interaction of partons *within* a given pancake as negligible: during a high-energy collision the partons of each proton would act as independent, quasi-free, entities. This image was the basis of the parton model; in technical language, the trick of visualising relativistic protons as frozen pancakes was known as working in the 'infinite-momentum frame' (a mathematical device which had been introduced into current-algebra calculations in 1965)[23] and the trick of treating partons as free particles was known as the 'impulse approximation' (familiar from nuclear physics).

Feynman developed this picture in the mid-1960s in an attempt to understand the interactions of hadrons and when, in 1969, he first published his work he addressed the problem of analysing multiparticle production in high-energy hadronic soft scattering.[24] However, this was the province of Regge-modellers,[25] and partons instead found a home in the analysis of deep-inelastic electron scattering. During his 1968 visit to SLAC Feynman realised that, while his analysis of hadronic soft-scattering was qualitative and intuitive,[26] he could give a very simple and direct parton-model explanation of scaling. He visualised deep-inelastic scattering as a process wherein the incoming electron emits a photon which then interacts with a *single free parton* (Figure 5.9). By construction, each parton was itself supposed to be a structureless particle, which would interact with photons just like an electron. Thus the similarity between the momentum-transfer dependence of electron–proton and electron–electron scattering shown in Figure 5.4 was immediately explicable in terms of partons. Furthermore, the mathematical definition of the structure functions W_1 and W_2 was such that each received a contribution at a given value of the scaling variable ω only when the struck parton carried a fraction $x = 1/\omega$ of the total momentum of the proton. Effectively, then, the structure functions measured the momentum distribution of partons within the proton, and this measurement was dependent only upon ω, the ratio of v and q^2, and not upon their individual values. Thus scaling emerged as an exact prediction of the parton model.

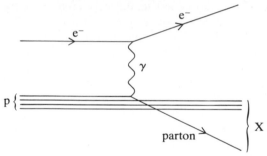

Figure 5.9. Deep-inelastic scattering in the parton model.

In the years which followed the discovery of scaling, the parton model became central to the practice of increasing numbers of HEP experimenters and theorists. We can discuss below some of the uses to which it was put, but first some general discussion of its appeal is appropriate. It is instructive to compare the explanations of scaling offered by Feynman's parton approach and Bjorken's current-algebra calculations. The first point to stress is that the attraction of the parton model did not lie in its success in explaining scaling *per se* – Bjorken could provide an alternative explanation, and other theorists had other explanations which worked just as well.[27] Neither can the popularity of the parton model be ascribed to the uniform validity of its predictions. As we shall see in the next section, straightforward extensions of the model often led to conflicts with observation. The difference between the parton and current-algebra approaches to deep-inelastic electron scattering lay rather in the different *meanings* with which each endowed scaling. Like Feynman, Bjorken related his prediction of scaling to the existence of point-like scattering centres within the proton,[28] but the two men interpreted the significance of this relation differently. Bjorken's calculations led first to the prediction of scaling, then to the observation that scaling could be said to correspond to point-like scattering (an observation which he traced back to the structure of the current-algebra commutators).[29] Feynman, instead, began with point partons, and thence arrived at scaling. Although, in the first instance, they led to the same predictions, the two approaches differed in the conceptual frameworks in which they were set and upon which physicists could draw for their elaboration. Only a few theorists counted themselves sufficiently expert to evaluate and exploit Bjorken's current-algebra approach. Feynman's parton model, in contrast, drew upon the common culture of all trained physicists, and was available for appreciation and exploitation by HEP experimenters and theorists irrespective of their particular expertise: physicists found the parton model both comprehensible and convenient.

The comprehensibility of the parton model derived from its relation to models already established in other areas of physics. The success of the parton model in explaining scaling indicated that in some circumstances, at least, one could profitably view hadrons as composites of effectively free particles. This kind of picture was far from unfamiliar to physicists, as they repeatedly pointed out, to each other in the technical literature, and to the wider public in popular scientific accounts. In a generalised sense the parton model established an analogical connection between the novel phenomenon of scaling and composite systems known to physicists from their professional childhood.

The analogy most often presented in the popular writing of scientists was that between deep-inelastic scattering and Rutherford's experiments in the early years of this century. Rutherford had bombarded specimens with α-particles (helium nuclei) and had observed that a surprisingly large number of α-particles were scattered at large angles. The SLAC-MIT group had observed similar behaviour with highly energetic electron beams. Just as Rutherford had interpreted his results in terms of a point-like scattering centre, the nucleus, within the atom, so Feynman's model showed how scaling could be interpreted in terms of point-like partons within the proton. Thus, for example, Wolfgang Panofsky and Henry Kendall wrote in *Scientific American* that 'the Stanford experiments are fundamentally the same as Rutherford's.'[30] In so doing, however, they were not quite telling the whole story. An atom, physicists believe, comprises a single central nucleus surrounded by an extended cloud of electrons. The parton model, on the other hand, asserted that the proton comprised a single amorphous cloud of partons. In this respect the proton was better to be conceptualised in terms of the electron cloud of the atom or the assemblage of nucleons within the nucleus, and this closer analogy with multi-constituent composite systems was brought out and exploited in many research papers and review articles in the technical literature.[31]

Even more significant than specific detailed analogies between protons and composite atoms or nuclei was the convenient way in which Feynman formulated the parton model. His approach served, in effect, to bracket off the still intransigent strong interactions in the application of the model: all of the effects of the strong interaction were contained in the parton momentum distributions. Knowing or guessing these distributions, calculations of electron–hadron scattering processes could be reduced to simple first-order QED calculations in which partons behaved like point-particles. The only difference between such calculations and those for electron–electron scattering was that the relevant quantum numbers of the partons were *a priori* unknown. Needless to say, various conjectures as to the nature of the partons and their quantum numbers were available in the late 1960s culture of HEP. One popular speculation was that the partons had the same quantum numbers as quarks and, as we shall see, this quickly came to be generally accepted.

In general terms, then, the appeal of the parton model was very much the same as that of the constituent quark model: it provided a simple and readily understood framework, ready for elaboration for the various purposes of theorists and experimenters in terms of the various resources available within the culture of HEP. We can examine in the following section the historical development of the model and the experimental traditions which sustained it, but two further points should be clarified in advance.

First, as a composite model, the parton approach to the structure of hadrons was very similar to that of the constituent quark model. The two models were, however, quite distinct; whilst the parton model came to inherit many resources from earlier quark models it remained a separately identifiable tradition. The reason for this was that in order to explain scaling one had to treat partons as *free particles*, whereas the existence of strong inter-quark *interactions* binding quarks into hadrons were at the heart of the constituent quark model. Besides this conceptual division, it is also important to emphasise that the parton model and the CQM addressed different experimental domains. The CQM found its greatest success in low-energy resonance physics, whereas the parton model was applied to the newly discovered high-energy, high-momentum-transfer scaling regime. Whilst attempts were made to explain scaling in terms of sums of resonances, it was clearly impossible to explain the details of hadron spectroscopy in terms of free partons.

Secondly, the idea that the parton model 'worked' as an explanation of scaling requires some clarification. As we shall see below, partons quickly became identified with quarks by most particle physicists, and this had one unfortunate consequence. The quark-parton explanation of scaling required that the quark struck in the electron–proton interaction behave as a free particle. If this quark continued to behave as a free particle one would expect it to shoot out of the proton and appear amongst the debris of the collision. However, quarks were no more observed in the final-states of electron scattering than in any other class of reactions. The debris of electron scattering was just a shower of normal hadrons, and some assumption had to be added to the parton model to explain this. This assumption was simply that although the quark behaved as a free particle in the initial hard interaction, it must subsequently undergo a series of soft, low momentum-transfer, strong interactions with its fellow partons. These interactions were supposed somehow to ensure that hadrons and not quarks would appear in the final state and, being soft, were assumed not to invalidate the explanation of scaling itself. Such assumptions were unavoidable and theoretically unjustifiable but, as the quark–parton model became central to the practice of increasing numbers of particle physicists, the HEP community learned to live with this unsatisfactory state of affairs.

5.3   Partons, Quarks and Electron Scattering

We can now turn from the conceptual foundations of the parton model to its historical evolution. Although the model was Feynman's creation, he did not publish on the analysis of scaling until 1972.[32] Instead, the parton model was taken up at SLAC, and first appeared in the HEP literature in 1969 (with ample acknowledgment to Feynman) in an article by Bjorken and E.A.Paschos.[33] A large and diverse body of theoretical and phenomenological work elaborating upon the basic form of the model quickly came into being.[34] Following Feynman's lead, field theorists began openly to practise their art in the realm of the strong interaction for the first time in more than a decade. This work led eventually to the formulation of QCD, as we shall see in Chapter 7. In the present chapter, though, I will focus on the phenomenological use of the parton model, for two reasons. First, because this was the sense in which the model was used to give structure and coherence to the experimental programme at SLAC and, later, elsewhere: as Kendall and Panofsky wrote in 1971, the model 'supplied the motivation for several experiments now in the planning stage'.[35] And, secondly, because the identification of quarks with partons was grounded in the phenomenological analysis of scaling.

Feynman's initial analysis showed that scaling should be observed in deep-inelastic scattering whatever the nature of the partons. Subsequent phenomenological work concentrated upon trying to make detailed fits to the available data with the aim of specifying more closely the attributes of partons. At issue here was the fact that besides their scaling property, the magnitudes and shapes of the structure functions could themselves be measured in a variety of processes. Each species of parton was supposed to contribute to the total structure functions in a calculable way, given by QED and dependent only upon the spin and electric charge of the parton in question. Thus, it was argued, measurements of different structure functions would yield information on the parton quantum numbers.

For example, one conclusion which followed directly from QED was that the relative magnitudes of the W_1 and W_2 structure functions in electron–proton scattering were determined just by the parton spins. A particular mathematical combination of W_1 and W_2 denoted by 'R' was predicted to be zero in the Bjorken limit if the partons had spin $\frac{1}{2}$ (as expected for quarks) and non-zero for spin 0 or 1 (as expected if the proton-cloud included elementary mesons).[36] The initial SLAC data were insufficient to determine R, but subsequent measurements covering a wider range of energies and momentum transfers yielded a value of $R = 0.18 \pm 0.10$.[37] This was 'certainly small', as SLAC theorist Fred Gilman put it in a 1972 review, and 'given possible systematic errors it is possible, although unlikely, that $R = 0$'.[38] In default of any better proposal than spin $\frac{1}{2}$ quarks for the proton's constituents, Gilman continued: 'since experiment suggests that R is small one thinks in terms of mostly spin-$\frac{1}{2}$ partons'.[39] Besides the measurement of electron-proton scattering, the SLAC-MIT physicists also measured electron–deuteron scattering, and the scattering of muons from both protons and deuterons. In all of these processes deep-inelastic scaling was observed to hold, consistently with the identification of partons as spin $\frac{1}{2}$ objects.[40] Theorists had little difficulty in concluding that the spin $\frac{1}{2}$ objects were very possibly quarks.

More information on the parton quantum numbers was sought from an examination of the individual structure functions. According to QED, each species of parton was supposed to contribute to the total structure function in proportion to the square of its electric charge. For example, the relative contributions from u, d and s quarks, having charges $+\frac{2}{3}$, $-\frac{1}{3}$ and $-\frac{1}{3}$ respectively, were expected to be in the proportions $\frac{4}{9}$, $\frac{1}{9}$ and $\frac{1}{9}$ (compared with a unit contribution expected from any unit-charge parton). Hence, by making assumptions about the parton composition of the proton, estimates of the proton structure functions could be made. Since the parton momentum distributions were *a priori* unknown, the estimates related to the overall size of the structure functions – mathematical integrals over the entire range of the scaling variable ω (i.e. the areas under the curves of Figure 5.6).

These estimates proved to be erroneous. The three-constituent (uud) quark-parton model overestimated the measured size of the proton structure functions by a factor of around two. Neutron structure functions could be extracted from measurements on deep-inelastic electron-deuteron scattering,[41] and the CQM assumption that the neutron was a three-quark udd composite again led to a factor of two overestimate. But in such adversity the parton model flourished. Feynman's fundamental image of the proton and neutron was that of a cloud containing an indefinitely large number of partons, and theorists argued that to approximate the cloud by just three quarks was not likely to be realistic. The model was elaborated accordingly, by dividing the cloud of quark-partons into two components. One component contained the minimum number of quarks required to construct the SU(3) quantum numbers of the hadron at issue – uud for the proton, for example. These were known as 'valence' quarks, in an explicit reference to the atomic analogy. The second component comprised a quark-antiquark 'sea': an SU(3) singlet cloud containing an indefinite number of $q\bar{q}$ pairs. The relative magnitudes of the valence and sea components could not be calculated in advance, but inclusion of the sea lent an additional degree of interpretative flexibility to the phenomenological analysis of the structure functions. However, even when they allowed for an arbitrary sea component, theorists found that they still overestimated the measured structure functions.[42]

Undeterred, theorists elaborated the quark-parton model still further, and introduced 'glue' into the constitution of the proton. The argument here was that if the nucleon was simply a composite of non-interacting quarks it would fall apart. Some interacting entity was required to 'glue' the quarks together: in field-theoretic terms, for example, some other particle must interact with the quarks to provide the requisite attractive interquark forces. This hypothetical particle, about which much more will be said later, was known as the 'gluon'.[43] Two theorists at MIT, Victor Weisskopf and Julius Kuti (a visitor from Eotvos University in Budapest) took the gluon idea seriously, and in 1971 produced a detailed model of the structure functions.[44] They assumed that the gluons were electrically neutral and would therefore not contribute directly to electron scattering. The gluons would, however, share in carrying the total momentum of protons and neutrons. Thus the gluons would act as an important but invisible component of protons and neutrons in electron-scattering experiments. The effect of allowing for a gluonic component was to reduce still further the estimates of the size of the structure functions. Like the qq̄ sea, the gluon component was a free parameter in the Kuti–Weisskopf model, and with this further degree of explanatory freedom it was found to be possible to achieve a 'fair quantitative fit'[45] to the SLAC data.

By this stage, though, the quark–parton model was in danger of becoming more elaborate than the data which it was intended to explain. A critic could easily assert that the sea and gluon components were simply *ad hoc* devices, designed to reconcile the expected properties of quarks with experimental findings. Field theorists could argue that the sea and glue would be required in any sensible field theory of quarks (although they had no actual candidate for such a theory); but most particle physicists in the late 1960s and early 1970s were not field theorists, and many were openly sceptical of such arguments. Something more was required to persuade the HEP community at large of the validity of the parton model, and of the equation of partons to quarks. Neutrino scattering experiments provided what was needed.

5.4 Neutrino Physics
One of the processes to which theorists quickly extended the parton model was neutrino (and antineutrino) scattering from nucleons.[46] Except for the fact that three independent structure functions were required to describe neutrino scattering (as against two for electron scattering), the weak neutrino–parton interaction could be seen as very similar to the electromagnetic electron–parton interaction.[47] This is indicated in Figure 5.10 where neutrino–parton scattering is depicted as mediated by exchange of a W-particle (see Chapter 3.3) in order to make the analogy with Figure 5.9 explicit. From a theoretical point of view the principal difference between electron and neutrino scattering lay in their different dependences upon the parton

quantum numbers: the photon coupled to the electric charges of the partons while the W coupled to the weak charges. Thus theorists argued that if one combined data on electron and neutrino scattering – assuming that scaling would hold for the latter – one could obtain more information than from electron scattering alone.

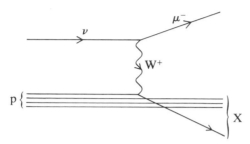

Figure 5.10. Inclusive neutrino–proton scattering in the parton model.

From an experimental point of view, the important difference between electron and neutrino scattering was one of reaction rates. Being mediated by the weak rather than the electromagnetic interactions, neutrino events were extremely infrequent and hard to observe. The first accelerator neutrino experiments were carried out in the early 1960s, using secondary neutrino beams produced at the newly operational Brookhaven AGS and CERN PS.[48] The objective of these experiments was three-fold: to demonstrate the technical feasibility of HEP neutrino physics; to investigate the 'two-neutrino hypothesis'; and to search for W-particles – the intermediate vector bosons (IVBS) of the weak interaction.[49] The first and second of these objectives were achieved in the earliest Brookhaven experiment and quickly confirmed at CERN: events induced by neutrinos were identified, and the two-neutrino hypothesis was confirmed. The latter asserted that there were two distinct species of neutrinos, v_e and v_μ, associated with electrons and muons respectively.[50] None of the first round experiments was successful in detecting the production of IVBS.[51] More neutrino experiments were performed during the 1960s, most of them at CERN where neutrino physics became a local specialty, but they were now directed towards more mundane topics: 'elastic' neutrino scattering (e.g. $vn \rightarrow \mu^- p$) and resonance production.[52] These exclusive processes, in which only a small number of particles were produced and all were experimentally identified, were those of principal theoretical interest at the time.[53]

This was the situation when the parton model entered the scene. In 1969 parton modellers began to argue, in analogy with electron scattering, that inelastic, inclusive neutrino scattering, in which any number of hadrons could be produced, was the truly interesting aspect of neutrino physics. The advocates of the parton model constituted a new audience for the neutrino experimenters, and the latter did their best to respond. The most extensive set of relevant neutrino data came from a series of bubble-chamber experiments carried out at CERN between 1963 and 1967. Analysis of these data had hitherto focused upon the exclusive channels mentioned above but, by virtue of the indiscriminate character of bubble-chamber photographs, inclusive cross-sections could also be extracted from them. A total of around 900 neutrino events had been recorded on film. This was insufficient to provide detailed information on the three neutrino structure functions at different values of v and q^2 and hence to test scaling directly, but it was sufficient to test a cruder prediction of the parton model. Bjorken and others had shown that if the neutrino structure functions did scale, then total neutrino cross-sections should be directly proportional to the energy of the neutrino beam.[54] The CERN experimenters analysed their data accordingly and in September 1969 announced that, within considerable experimental uncertainties, a linear relation between beam energies and cross-sections did obtain.[55]

Here matters rested until 1971, when the French-built bubble-chamber 'Gargamelle' was installed at CERN.[56] Gargamelle was almost ten times larger than the bubble chambers used in previous neutrino experiments at CERN, and promised to collect much more extensive data in a reasonable space of time.[57] In February 1964, when the construction of Gargamelle was first proposed, its virtues for the measurement of exclusive processes and searches for intermediate vector bosons had been stressed,[58] but as the time approached for its first operation priorities had begun to change. A quotation from Bernard Gregory, then Director-General of CERN, serves to indicate both the response of the HEP community to the SLAC data and the influence of the parton model on the planning of the forthcoming neutrino experiments: the SLAC scaling data, he wrote in 1970,

> appear to be produced by the recoil of point-like constituents inside the proton. Here . . . we have a mathematical picture which may or may not have a direct physical significance. The name 'partons' has been given to these postulated constituents. It will be possible to repeat such experiments at CERN, using the neutrino as an excellent probe. In a few months time, combining

a refined neutrino beam and the large Gargamelle bubble chamber, an experiment will be carried out on this important new subject.[59]

In 1971, Gargamelle produced around 500,000 pictures; in 1972, 400,000; in 1973, 700,000 (despite a serious breakdown); in 1974, 240,000; and in 1975, the final year of Gargamelle's operation at the PS, 250,000.[60] In 1972 a team of more than 50 physicists, drawn from CERN and six European universities,[61] began to analyse the film. In 1973, the Gargamelle group published their first results.[62] They had analysed 95,000 pictures taken in a neutrino beam and 174,000 pictures taken in an antineutrino beam, and had positively identified around 2,500 instances of neutrino interactions (neutrino events) and 1,000 antineutrino events. They were able to show that cross-sections increased linearly with neutrino energy, as expected from the parton model, albeit with very large uncertainties above 4 GeV where few events had been recorded. At this time, the Gargamelle group did not report a full analysis of the neutrino structure functions, but did report the overall size of W_2 (i.e. the integral of W_2 over all values of ω). This was given by the difference of the neutrino and antineutrino total cross-sections, and the observed value was quoted to be 0.49 ± 0.03. In the parton model, this number corresponded to the fraction of the proton momentum carried by non-interacting partons. Theorists assumed that the gluons were immune to the weak as well as the electromagnetic interaction, and therefore regarded the Gargamelle result as showing that around fifty per cent of the proton's momentum was carried by the gluon component – supporting the similar conclusion already reached in the analysis of deep-inelastic electron scattering.

In 1974 the Gargamelle group submitted for publication a more detailed study of their existing sample of neutrino and antineutrino events.[63] Here they announced that an analysis of the energy and momentum-transfer dependence of their data showed them to be consistent with scaling. They also announced two further results on the integrated magnitudes of the structure functions. First the ratio of the W_2 structure functions for neutrino- and electron-scattering was reported to be 3.6 ± 0.3. In the parton model this ratio was sensitive to the parton charges, and the experimental result supported the identification of partons with fractionally charged quarks. Indeed, the quark–parton model predicted a ratio of exactly 3.6 (18/5), while the assumption of integrally charged partons led to a ratio of 2.[64] Secondly, the Gargamelle experimenters reported that the integrated value of the W_3 structure function was 3.2 ± 0.6. The quark–parton prediction here was 3, corresponding to the three valence quarks of the model. Again, different identifications of partons led to different numerical predictions.[65]

The most detailed Gargamelle data appeared in a refereed journal in 1975. Preliminary announcements had, however, been made at conferences in the summer of 1973,[66] and by that time there was general agreement that the parton model, with the identification of quarks with partons, was a promising approach to the analysis of both electron and neutrino inelastic scattering. Theoretical arguments could still be brought against the quark–parton idea – why were quarks not produced as free particles in deep-inelastic scattering; what was the mysterious gluon component? – but Feynman felt justified in remarking that 'There is a great deal of evidence for, and no *experimental* evidence against, the idea that hadrons consist of quarks. . . . Let us assume it is true.'[67] Many physicists did just that, and in the early 1970s the quark–parton model became central to the planning and interpretation of lepton scattering experiments.

5.5 Lepton-Pair Production, Electron–Positron Annihilation and Hadronic Hard Scattering

The quark–parton model drew its early empirical support primarily from electron- and neutrino-scattering experiments. The model was, though, extended to certain other processes. In the late 1960s little experimental information was available on these processes but, in the fullness of time, they all came to constitute important strands of the new physics.

Inclusive Lepton-Pair Production

Inclusive lepton-pair production, the process $pp \rightarrow l^+l^- X$ (where 'l' represents either an electron or muon, and 'X' represents any collection of hadrons) was readily conceptualised in the quark–parton model. As indicated in Figure 5.11, it could be visualised as the annihilation of a quark and an antiquark (coming from the sea of either the beam or the target proton) to form a photon which then materialised as a lepton pair. Many authors, amongst whom SLAC theorists Sidney Drell and Tung-Mow Yan were most influential, gave estimates of the appropriate cross-sections, using QED to evaluate the probabilities for the production and decay of the photon, and the measured deep-inelastic structure functions to specify the parton momentum distribution.[68]

Experimentally, appropriate measurements were very hard to make. The major problem was to detect the few lepton pairs expected amongst the large numbers of hadrons produced in high-energy proton–proton collisions. In the late 1960s and early 1970s, the only

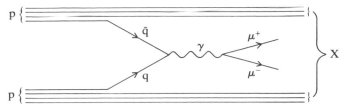

Figure 5.11. Lepton-pair production in the parton model.

relevant data came from an experiment performed at the Brookhaven AGS in 1968.[69] Carried out by Leon Lederman and his colleagues from Columbia University, the primary aim of this experiment, like the neutrino experiments of the 1960s, was to search for IVBS.[70] Like the neutrino experiments, the Columbia dilepton experiment failed to find any IVBS and was, instead, embraced by the parton modellers. Cross-sections were found to fall off roughly as the sixth power of the combined mass of the lepton pair, and the Drell–Yan model gave a 'qualitatively correct description'[71] of this behaviour. More important than the qualitative agreement of prediction and data, the Columbia experiment showed that high-energy lepton-pair measurements were technically possible, and the parton model made them interesting (even if no IVBS were to be found). Further experiments were planned along similar lines, and developments in this area will be reviewed in Part III of this account.

Electron–Positron Annihilation
In the quark–parton model, electron–positron annihilation to hadrons was envisaged as shown in Figure 5.12.[72] Here the electron and positron annihilate to form a photon which materialises as a quark–antiquark pair. This was the simplest of all processes to calculate in the parton model: since the momenta of the quark and antiquark were precisely determined by momentum conservation, no recourse to momentum distributions specified by measured structure functions was necessary. Of course, assumptions had to be made to explain how the produced quarks would rearrange themselves into hadrons, but theorists followed the usual route and asserted that such rearrangement would not affect the parton model predictions. In the late 1960s, the principal source of information upon electron–positron annihilation was the ADONE electron–positron collider which began operation at Frascati in Italy in 1967. Early data from ADONE confirmed the general expectations, at least, of the quark–parton model.[73] Quantitatively, however, the model failed to explain the

data. This merely encouraged theorists to endow quarks with yet another set of properties known as 'colour', but discussion of colour will be postponed to Chapter 7 when it can be related to several other developments connected with quarks and the parton model.

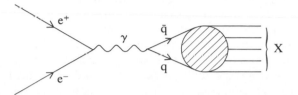

Figure 5.12. Electron–positron annihilation to hadrons in the parton model.

Hadronic Experiments

The earliest applications of the parton model were all to processes involving leptons in either the initial or final states. In the late 1960s and early 1970s there were no obvious analogues of scaling to be seen in the realm of purely hadronic experiment. Only with the advent in 1971 of the proton–proton collider, the Intersecting Storage Rings (ISR) at CERN, did this situation begin to change.

The ISR made available a centre-of-mass energy considerably greater than that available at any other facility, and in the first round of ISR experiments several groups set out to hunt for IVBs.[74] These hunts were extensions of the 1968 Brookhaven experiments discussed above. Any IVBS produced were to be identified by their decay to leptons, and the crucial technical problem at the ISR was to detect the leptons against the large background of hadrons associated with high-energy collisions. Lower-energy experiments had shown that inclusive hadron production (e.g. $pp \rightarrow \pi X$) showed similar diffractive characteristics to elastic electron–proton scattering. Most collisions were soft, and cross-sections decreased rapidly as a function of the transverse momentum (p_T) of the detected hadron. Typical inclusive hadron cross-sections fell-off exponentially fast, roughly as e^{-6p_T}. The experimenters therefore decided to look for high-transverse-momentum leptons, because at high transverse momenta the background of hadrons would be small and relatively easy to cope with.[75]

As IVB searches, these experiments failed. It was found that although inclusive hadronic cross-sections did fall fast with p_T at the ISR, they did not fall as fast as expected. Figure 5.13 shows an early compilation of ISR cross-sections for inclusive pion production, $pp \rightarrow \pi^0 X$.[76] The straight line is an extrapolation of the low-p_T data,

149

and the high-p_T measurements stand well above this. Although the high-p_T cross-sections found at the ISR were small in absolute terms, having fallen by six orders of magnitude in going from transverse momenta of 1 GeV to 4 GeV, they were still very large compared with prior expectations – by a factor of 10^2 to 10^3 at a transverse momentum of 4 GeV. The consequence of this was that hadronic backgrounds remained sufficient, even at high p_T, to make it impossible to detect leptons with existing apparatus, and hence to vitiate the IVB searches. However, just because the hadronic high-p_T cross-sections were unexpectedly large, they could be argued to represent a new phenomenon in their own right. One of the first-round experimenters at the ISR was Leon Lederman, pursuing the IVBs to higher energies after his earlier experience at the Brookhaven AGS and, as he later remarked: '[The hadronic] background became the signal for a new field of high transverse momentum hadrons. This subject literally exploded in the period 1972–1975.'[77]

The excess production of high-p_T hadrons was observed by three groups at the ISR in 1972 and the first results were published in 1973.[78] The observations were quite unexpected, at least by the experimenters, but in the theoretical context of the time were immediately, if tentatively, interpreted as the first evidence in purely hadronic reactions for the hard scattering of point-like constituents – partons – within the proton. As W. Jentschke, CERN's Director-General, wrote in 1972:

> An intriguing question over the past few years was why partons were noticed only in electromagnetic but not in strong interaction processes . . . . However, two recent ISR experiments have given results which might indicate point-like constituents also in strong interactions. . . . This is perhaps the most exciting, unexpected, result from the ISR.[79]

In ascribing the excess high-p_T production of hadrons to hard-scattering between point-like constituents, physicists were merely following the basic Rutherford-type analogy at the heart of the parton model. Indeed, the ISR observations had been qualitatively predicted in advance by parton modellers.[80] However, problems arose when attempts were made at more quantitative analyses of the ISR data. The high-p_T cross-sections were found to be several orders of magnitude larger than the earliest parton-model calculations. This was not very surprising. The calculations had assumed that hard hadronic scattering was mediated by photon-exchange between quark-partons, in direct analogy with the other processes discussed above. While such an assumption was routine for processes involving leptons (which

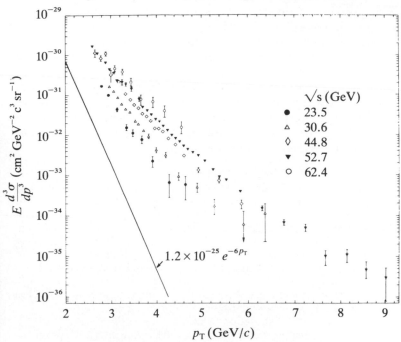

Figure 5.13. Inclusive pion-production cross-sections ($Ed^3\sigma/dp^3$) as a function of pion momentum transverse beam axis (p_T), measured at the ISR by the CERN–Columbia–Rockefeller (CCR) collaboration. The different data point symbols correspond to the different centre-of-mass energies given in the key in the upper right.

were believed to couple *only* to photons and IVBS) it was neither routine nor plausible for quarks. One could easily imagine quarks interacting with one another by exchanging gluons, or even conventional mesons. Furthermore, if one took seriously the gluonic component of hadrons, then one had also to consider the possibility of gluon–gluon hard scattering (one gluon from each proton) and gluon–quark scattering.

Thus much of the simplicity of the parton model was lost when it was applied to processes in which no leptons were present to guarantee the primacy of photon (or W) exchange. It was not clear how the parton model should be extended to hadronic hard scattering, nor was it clear what was to be learned from such extensions.[81] But, for the present, the significant point to note is that the interest of parton-modellers served to crystallise the ISR discovery of large high-p_T hadronic cross-sections as that of an important new phenomenon worthy of further exploration. From 1972 onwards a distinctive tradition of hard-scattering experiments emerged, first at the ISR and later at Fermilab and the CERN SPS, which aimed at detailed investigation of large transverse momentum hadronic processes. And, under the aegis of the quark-parton model, this tradition found a place in the experimental new physics alongside the investigation of deep-inelastic lepton scattering, lepton-pair production and electron–positron annihilation. We will encounter all of these traditions again in subsequent chapters.

NOTES AND REFERENCES

1 I am very grateful to the SLAC physicists for their hospitality and assistance with the research discussed here and in later chapters. For interviews, provision of biographical materials, and access to unpublished documents, I thank, in particular, J.D.Bjorken (now at Fermilab), E.D.Bloom, S.D.Drell, F.J.Gilman, G.Goldhaber (Lawrence Berkeley Laboratory), W.T.Kirk, M.L.Perl, C.Y.Prescott, B.Richter, R.Schwitters (now at Harvard), R.E.Taylor and W.Wenzel (LBL).

2 Neal (1968).

3 For details of the early experimental programme at SLAC, see Neal (1968, Table 2.3, p. 19) and Ballam (1967).

4 SLAC–MIT–CIT collaboration (1966, 2).

5 A brief comment is needed on the second experiment, elastic positron–proton scattering. The SLAC machine could accelerate positrons as well as electrons, and the object of the positron experiment was to look for differences between elastic electron and positron scattering. As discussed below, a routine assumption in lower-energy experiments was that the interactions between electrons and nucleons were mediated by the exchange of a single photon. When SLAC was built, physicists suspected that this assumption would no longer hold good, and that processes in which two photons were exchanged would play an important role (complicating the interpretation of data considerably). The comparison of electron and positron elastic scattering was designed to investigate this: two-photon exchange was expected to produce measurable differences between cross-sections for the two processes. In the event it was found that the cross-sections were the same, and physicists continued to presuppose the dominance of one-photon exchange in the analysis of SLAC data.

6 Coward *et al.* (1968), Mar *et al.* (1968).

7 According to one Caltech experimenter, the Caltech group dropped out of the inelastic scattering experiment because they 'thought it would be a bore'. Although there was 'no promising theory', like everyone else they expected that cross-sections would be small and that radiative corrections (see below) 'would be a big problem': the experiment would be long and arduous with little reward (interview with B.C.Barish).

8 SLAC–MIT–CIT Collaboration (1966, 33–42).

9 Bloom *et al.* (1969, 933, Fig. 1(b)).

10 Briedenbach *et al.* (1969, 935, Fig. 1).
11 Interview, R.E.Taylor, 12 May 1981. The potential difficulties entailed by the necessity to make radiative corrections (of resonance production data) were extensively discussed in the original experimental proposal: SLAC–MIT–CIT collaboration (1966, 44–58, App. V).
12 Since the radiative corrections corresponded to photon emission from electrons they could, in principle, be calculated exactly from QED. In practice, however, two problems arose. First, in order to make the computations manageable, the QED formulae were approximated on the basis of various physical assumptions. Secondly, the measured electron energies were not in a one-to-one correspondence with radiatively corrected energies: depending upon how much energy was radiated away, a variety of 'true' electron energies could correspond to a given measured value (and vice versa). This implied that correction of the measured cross-section at a given value of energy and momenta transfer required measurements of cross-sections over a continuous range of energies and momentum transfers. However, measurements were actually made at discrete values of these parameters, and interpolation procedures had to be devised to fill in the gaps. The first calculations of radiative corrections to the SLAC data were made at MIT, where Friedman and Kendall developed a computer programme extending routine correction techniques from low-energy electron–nucleus scattering to the SLAC regime. This programme relied upon particular assumptions, approximations and interpolating techniques, but indicated that the corrections to the high momentum transfer data were not large. An alternative model for radiative corrections was developed at SLAC by theorist Paul Tsai and post-doctoral experimenter L.Mo. This depended upon different assumptions from the MIT model but pointed again to the conclusion that the corrections were not large. In their first publication on the inelastic scattering data, the SLAC–MIT group estimated that the possible error in the corrections which they applied to the data was perhaps 10% (Bloom *et al.* 1969, 933). The radiative correction procedures are discussed in more detail, with references to the original literature, in Friedman and Kendall (1972, 207–9).
13 Bjorken's unpublished 1959 PHD thesis was on 'Experimental Tests of Quantum Electrodynamics and Spectral Representations of Green's Functions in Perturbation Theory'. He was junior co-author with another SLAC theorist, S.D.Drell, of two standard texts on relativistic quantum mechanics and quantum field theory (Bjorken and Drell 1964, 1965).
14 Adler (1966) derived current-algebra sum-rules involving integrals over inelastic neutrino–proton scattering cross-sections. Since the weak and electromagnetic currents were related in the current-algebra approach, Bjorken (1966a,b, 1967) argued that similar sum-rules could be constructed for electron–proton scattering. The phrases quoted above are Bjorken's (interview).
15 Drell and Walecka (1964). The general formula was

$$\frac{d^2\sigma}{d\Omega dE'} = \frac{4\alpha^2 E'^2}{q^4} \left[2W_1(\nu, q^2) \sin^2 \theta/2 + W_2(\nu, q^2) \cos^2 \theta/2 \right].$$

153

Here $d^2\sigma/(d\Omega dE')$ is the so-called 'double differential cross-section', representing the measured cross-sections for scattering of electrons with energy E' into a given element of solid angle Ω, and α is the fine structure constant. ν is the energy loss of the electron, $E - E'$, where E is the initial energy of the electron, and E' its final energy. θ is the angle through which the electron is scattered. q^2 is the square of the momentum transfer between the electron and proton: $q^2 = 4EE'\sin^2 \theta/2$. W_1 and W_2 are the physically interesting parts of the cross-section to be investigated by experiment; the other factors simply reflect the requirements of relativity and gauge invariance (discussed in Chapter 6).

16 This conclusion was first published in Bjorken (1969). However, it was already implicit in Bjorken (n.d.), an unpublished SLAC memorandum written before experiment began at SLAC (this is clear from the opening sentence: 'The purpose of this note is to celebrate the imminent inaugural of the SLAC 20-BeV spectrometer with some speculative calculations').

17 Gilman (1972, 133, Fig. 11).

18 ω' was defined to be equal to $\omega + M^2/q^2$. At large q^2, ω and ω' were effectively indistinguishable; at small values of q^2 their difference was large, and plotting the data against ω' made scaling more apparent in that region.

19 Panofsky (1968).

20 R.E.Taylor, interview.

21 Bloom et al. (1969), Briedenbach et al. (1969). In a survey of the HEP literature published between 1969 and 1972, Irvine and Martin (1982b, 18, Table 3) found that only nine experimental papers were cited fifty or more times within the space of one year. Two were the above papers from SLAC. The remaining seven came from the new PSs at Serpukhov (2) and Fermilab (3), and from the CERN ISR(2). These seven all reported measurements of quantities of central interest in the Regge tradition, performed at hitherto inaccessible energies.

22 I thank Professor Feynman for an interview and provision of biographical materials.

23 Fubini and Furlan (1965). See Chapter 4, note 77.

24 Feynman (1969a,b). These papers introduced the inclusive–exclusive terminology to HEP. Hitherto inclusive measurements had been seen only as crude surveys in preparation for subsequent exclusive studies (see the comments above on the SLAC inelastic measurements). Feynman used his parton concept to argue that inclusive measurements could yield important information in their own right, and that because of this they deserved a dignified name.

25 For a review of subsequent theoretical developments in the analysis of soft hadronic multiparticle production, see Landshoff and Polkinghorne (1972) and Aurenche and Paton (1976). The systematic features which Feynman derived from partons were generally explained by an extension of the Regge programme known as the Mueller-Regge approach.

26 As Feynman himself put it (1969a, 1415): 'I am more sure of the conclusions than of any single argument which suggested them to me for

they have an internal consistency which surprises me and exceeds the consistency of my deductive arguments which hinted at their existence'.

27 Bjorken's approach was, of course, taken up by current-algebraists and Gell-Mann's 'light-cone algebra' was particularly influential (see Chapter 7). These current-algebra results, unlike those discussed in the previous chapter, were dependent upon choices of fundamental fields. When quark fields (i.e. fields carrying quark quantum numbers) were used, many parton-model results could be recovered in current algebra. Those which could not, such as the explanation of purely hadronic hard-scattering discussed in Section 5, were regarded as particularly suspect by theoretical purists. However, as SLAC theorist Fred Gilman put it (1972, 141): 'even if the quark–parton model fails in this last regard [going beyond the premises of current algebra], it is certainly a very useful guide and mnemonic, provides intuition in many cases where other approaches are inapplicable or fail, and gives a unified way of thinking about deep inelastic phenomena'. With the advent of quantum chromodynamics, which both underwrote the parton model and gave a prescription for the current-algebra commutators, the conceptual distinction between the two approaches was more or less dissolved, although the two sets of techniques remained distinct. Besides current algebra and the parton model, many other theorists sought to explain scaling in terms of their own favoured schemes: for a review, see Friedman and Kendall (1972, 235–46).

28 See, for example, Bjorken (n.d., 1; 1966a, 305).

29 Bjorken (n.d., 8).

30 Kendall and Panofsky (1971, 61).

31 See, for example, the reviews given by Kogut and Susskind (1973), West (1975) and Wilson (1977). Wilson traces the ancestry of the SLAC inelastic electron scattering experiments back to the experiments of Jesse DuMond at Caltech in the 1920s. DuMond measured the scattering of X-rays from metals, and interpreted his data in terms of the momentum distribution of the electrons within the metal, in just the same way as the SLAC scaling data could be interpreted in terms of the parton momentum distribution within nucleons. Wilson points out that inelastic electron scattering from *nuclei* can also be interpreted as yielding information on the momentum distribution of nucleons within nuclei, and that data on this process, so interpreted, has been used to yield important information on the 'shell' structure of nuclei. Wilson (1977, 1144) directly attributes Feynman's invention of the parton model to its analogical significance: 'As a research assistant to John Wheeler 35 or so years ago he [Feynman] was assigned the task of verifying whether the theories of electron momenta used by DuMond were valid. When this new problem [SLAC scaling] came to his attention, Feynman was attuned to it because it represents the same physics on a different scale of size'. The central analogy relating inelastic electron scattering on atoms, nuclei and nucleons is spelled out in great technical detail in West (1975). The title of this article – 'Electron Scattering from Atoms, Nuclei and Nucleons' – almost speaks for itself. West traces the lineage of the SLAC experiments back to the work of Franck and Hertz in

1914, and seeks (1975, 284) to 'emphasise how intuition gleaned from nuclear physics can be used to discuss the quark–parton model' (and vice-versa). Kogut and Susskind (1973, 78–9) note that the purpose of their paper 'is to provoke thought and germinate ideas, not to provide a comprehensive traditional review. The central theme in this work is the intuitive explanation of particle physics phenomena ... the use of intuitive, non-relativistic pictures drawn from atomic and molecular physics will be a recurring theme throughout'.

32 Feynman (1972).

33 Bjorken and Paschos (1969).

34 For reviews of the main lines of development of the parton model and access to the original literature, see Llewellyn-Smith (1972), Gilman (1972), Landshoff and Polkinghorne (1972) and Yan (1976).

35 Kendall and Panofsky (1971, 77).

36 $R = (1 + v^2/q^2) \, W_2/W_1 - 1$.

37 Miller et al. (1972).

38 Gilman (1972, 133).

39 Gilman (1972, 139).

40 See Friedman and Kendall (1972, 225–35).

41 The deuteron – the nucleus of 'heavy hydrogen' or deuterium – contains one proton and one neutron. Since the proton structure functions were known from measurements on hydrogen targets, the neutron structure functions could be extracted from deuterium data using various routine assumptions about scattering from composite nuclei. See Friedman and Kendall (1972, 209–13).

42 See Friedman and Kendall (1972, 237).

43 The term 'gluon' was introduced by Gell-Mann in his first publication on the Eightfold Way (1962b, 1073).

44 Kuti and Weisskopf (1971).

45 Gilman (1972, 140).

46 Bjorken (1969), Gross and Llewellyn-Smith (1969), Bjorken and Paschos (1970).

47 Recall that, as electrically neutral leptons, neutrinos experience *only* the weak interaction.

48 For the Brookhaven experiment, see Danby et al. (1962); for CERN, see Block et al. (1964) and Bienlein et al. (1964).

49 Several authors argued for the feasibility and physics interest of neutrino experiments at the Brookhaven AGS and the CERN PS: see Pontecorvo (1960), Schwartz (1960) and Lee and Yang (1960a).

50 The two-neutrino hypothesis had long been advocated to explain the observed absence of electromagnetic decays of muons to electrons, $\mu \rightarrow e\gamma$. For references to the original literature, see Danby et al. (1962).

51 Searches for intermediate vector boson production in the process $v + Z \rightarrow W^+ + l^- + Z$ followed by $W^+ \rightarrow \mu^+ + v_\mu$ had been advocated by Lee and Yang (1960b). Here Z stands for an atomic nucleus of charge Z, and l^- stands for a negatively charged lepton.

52 For references to the original literature and for technical details and findings of the 18 accelerator neutrino experiments which had been conducted up to 1976, see Wachsmuth (1977).

53 Exclusive neutrino data were analysed in terms of the CQM and current-algebra approaches discussed in the previous chapter.

54 Bjorken (1969).

55 Budagov *et al.* (1969). It is interesting to note that rising inclusive neutrino cross-sections had been evident in all of the CERN neutrino experiments, but were treated as facts of no evident significance until the advent of the parton model. For example, D.H.Perkins, one of the leading neutrino experimenters at CERN, presented unpublished inclusive neutrino data at a 1965 HEP School, and commented: 'One of the earliest and most striking results of the CERN neutrino experiment was the observation of the very large cross-sections for the inelastic process at high neutrino energy' (Perkins 1965, 77 and Figure 12). The transformation of these data into a significant fact illustrate clearly one role of theoretical context in the process of experimental discovery.

56 For technical descriptions of Gargamelle, the history of its construction and its use in neutrino physics, see Cahier Technique No. 6 (1973) and Musset (1977).

57 Since neutrinos interacted so rarely, their interaction rate was assumed to be directly proportional to the mass of the experimental target. For this reason Gargamelle, like the earlier bubble chambers used in CERN neutrino experiments, was designed to contain a heavy liquid (freon) instead of liquid hydrogen.

58 Lagarrigue, Musset and Rousset (1964).

59 Gregory (1970, 20).

60 See the *CERN Annual Reports* for 1971 to 1975.

61 These were the Universities of Aachen, Brussels, Paris (Ecole Polytechnique and Orsay), Milan and London.

62 Eichten *et al.* (1973a).

63 Deden *et al.* (1975a).

64 Llewellyn Smith (1970, 1971).

65 Gross and Llewellyn Smith (1969).

66 See, for example, Franzinetti (1974). Franzinetti also discussed the high-energy data then beginning to emerge from two neutrino experiments at the newly operational Fermilab PS. The Fermilab experiments are discussed in the following chapter.

67 Feynman (1974, 608).

68 Drell and Yan (1970).

69 Christenson *et al.* (1970, 1973).

70 Lederman (1976, 152).

71 Friedman and Kendall (1972, 249).

72 Bjorken (1969).

73 Friedman and Kendall (1972, 248).

74 See Lederman (1976, 154) and Jacob (1974, 68).

75 For a discussion of the advantages of the ISR for measurements of high transverse momentum processes, see Chapter 2.2.

76 Büsser *et al.* (1973, 474, Fig. 3).

77 Lederman (1976, 154).

78 The observations were made by the Saclay–Strasbourg group (Banner *et al.* 1973), the CERN–Columbia–Rockefeller group (Büsser *et al.* 1973)

and the British–Scandinavian collaboration (Alper *et al.* 1973)

79 Jentschke (1972, 13).
80 Berman and Jacob (1970), Berman, Bjorken and Kogut (1971).
81 For reviews of applications of the parton model to hadronic hard-scattering phenomena, see Sivers, Brodsky and Blankenbecler (1976) and Michael (1979).

6

GAUGE THEORY,
ELECTROWEAK UNIFICATION AND
THE WEAK NEUTRAL CURRENT

The theoretical wing of the new physics incorporated two principal sets of resources. One set consisted of quark models of hadron structure, the other a class of quantum field theories known as gauge theories. In the previous two chapters we discussed the birth of quarks in the old physics of resonance phenomena, and their transplantation to the new-physics quark–parton model. In this chapter we can review the early development of gauge theory.

Looking back in 1979 on the history of gauge theory, one of the leading contributors spoke of 'threads in a tapestry... made by many artisans working together',[1] and the story to be told here is a long and complex one covering a great diversity of contributions over a period of two decades. The account begins in Section 1, with a discussion of the invention of gauge theory by C.N.Yang and R.L.Mills in 1954. Modelled closely upon QED, gauge theory enjoyed some popularity amongst theorists at the time and was one of the resources which fed into the Eightfold Way symmetry classification of hadrons. Soon, however, quantum field theory went into decline and gauge theory with it. Sections 2 and 3 deal with developments in quantum field theory and gauge theory during the lean years of the 1960s, and refer to the activities of those field theorists who continued to practise their art. Section 2 reviews various theoretical approaches to 'electroweak' unification – the representation of the electromagnetic and weak interactions in terms of a single gauge theory. Section 3 reviews the drawn out struggle to show that gauge theory was, like QED, renormalisable. In 1971, seventeen years after gauge theory was invented, the struggle was brought to a satisfactory conclusion by Gerard 't Hooft working at the University of Utrecht.

Once it had been shown to be renormalisable, gauge theory ceased to be a minority pursuit and became a major theoretical industry. Section 4 reviews one important branch of this industry, the construction of unified gauge-theory models of the electroweak interaction. Such models predicted the existence of new phenomena in neutrino experiments and one such phenomenon – the weak neutral current – was reported from the Gargamelle neutrino experiment at CERN in

159

1973 and confirmed at Fermilab the following year. Thus the gauge theory proposed by Yang and Mills in 1954 finally made contact with experiment nineteen years later. The repercussions of this confluence of theory and experiment were enormous, and the concluding section of this chapter therefore explores the experimental discovery of the neutral current in some detail, emphasising the changes in experimental procedure which were an integral part of the production of the new phenomenon.

6.1 Yang–Mills Gauge Theory

The story of gauge theory begins with HEP theorist Chen Ning Yang (the same C.N.Yang who shared the 1957 Nobel Prize for his work on parity violation in the weak interactions).[2] Yang was born in 1922 in Hofei, China. His early training in physics was completed in China, but in 1945 he moved to the USA. In 1948 he was awarded a PhD in HEP theory for his work on the weak interactions at the University of Chicago, and in 1949 he took up a position at the Institute for Advanced Study in Princeton. While at Chicago, Yang became interested in the strong interactions, as did many theorists of the day, and studied quantum field theory – principally referring to articles published in 1941 by Wolfgang Pauli in *Reviews of Modern Physics*.[3] Pauli emphasised the importance of 'gauge invariance' in QED, and, since this concept is important in much that follows, a short digression is appropriate here.

There is a certain arbitrariness in classical electromagnetism as embodied in Maxwell's equations. Maxwell's equations are formulated in terms of electric and magnetic fields, which can themselves be expressed as derivatives of vector and scalar potentials. The arbitrariness arises because one can modify the potentials in a space- and time-dependent way without changing the associated fields. Classical electromagnetism is said to exhibit a 'gauge invariance': one can make local transformations of the potentials – i.e. transformations which vary from one space–time point to another – without affecting the predictions of the theory.[4] This invariance carries over into the quantised version of electromagnetism, QED. The QED Lagrangian, written schematically in Chapter 3 as

$$\mathscr{L}(x) = \bar{\psi}(x)\,\mathrm{D}\psi(x) + m\psi(x)\bar{\psi}(x) + (\mathrm{D}A(x))^2 \\ + eA(x)\bar{\psi}(x)\psi(x),$$

is invariant under the transformations

$$\psi(x) \rightarrow \psi(x)\mathrm{e}^{ie\theta(x)}, \text{ and}$$
$$A(x) \rightarrow A(x) + \mathrm{D}\theta(x).$$

Gauge Theory

Here $\theta(x)$ is a quantity which varies from one space–time point (x) to another, and the transformations correspond to time- and space-dependent redefinitions of the electron and photon fields. When written out in full, it is straightforward to demonstrate that the QED Lagrangian is unchanged by such gauge transformations of the fields, and thus that the physical predictions of the theory, which follow from the Lagrangian, are themselves unchanged. An important remark here is that the part of the QED Lagrangian which refers only to electrons and positrons (the first two terms) is not in itself gauge-invariant: under local transformations of the electron fields it acquires an extra piece due to the presence of the differential operator D, in the first term. This extra piece is cancelled in the full Lagrangian by an equal and opposite contribution coming from the fourth term, which describes the interaction of electrons with photons. Thus the existence of photons, interacting with charged particles in a specified way, is a formal requirement of a gauge invariant theory of electromagnetism.

While at Chicago, it occurred to Yang that it might be worthwhile to model a field theory of the strong interactions upon QED in a very direct manner. In group-theoretical language, the local gauge transformations under which QED is invariant are those of the group U(1). Yang's idea was to construct an analogous theory which would be invariant under local transformations of the strong-interaction isospin symmetry group, SU(2).[5] It is interesting to note that group theory played little part in the training of most physicists working in HEP in those days, but that Yang was an exception to this rule: indeed, the title of his BSc thesis, completed in China, was 'Group Theory and Molecular Spectra'. Yang began by considering a Lagrangian consisting of the first two terms of the above equation, but where ψ (and $\bar{\psi}$) were taken to represent not electrons but isospin multiplets of hadrons. Thus, for example, ψ might be a two-component field (p, n) representing the nucleon isospin doublet. He then computed the change in the Lagrangian due to local redefinitions of isospin coordinates, and found that it was non-zero: as in the case of QED, the first-two terms of the Lagrangian were not gauge-invariant under SU(2) transformations. However, Yang found that gauge invariance could be repaired in his theory by introducing a set of 'gauge particles', which I shall denote by the symbol W. Analogues of the photon, the Ws had to be spin 1, vector particles, but had now to form an isospin triplet (W^+, W^0, W^-). Yang found that it was straightforward to write down a term describing the interaction of the W-particles with nucleons such that the overall Lagrangian was gauge-invariant. He was, however, unable to construct an SU(2)-

invariant analogue of the $(DA)^2$ term of the QED Lagrangian for the W-particles. This was a major problem since, as discussed in Section 3.3, such a term was necessary if the theory was to describe the propagation of the gauge particles through space. At this point, Yang put his ideas on gauge invariance to one side, and worked for the next few years on other topics in strong-interaction physics and statistical mechanics.[6]

There matters rested until 1954 when Yang spent a year on leave from Princeton at the Brookhaven National Laboratory. Besides devoting much of his time there to the analysis of data emerging from the newly operational Cosmotron, Yang attempted once more to construct a gauge-invariant field theory of the strong interaction. He shared an office with another HEP theorist, Robert Mills, and between them they quickly devised an SU(2)-invariant analogue of the $(DA)^2$ term of QED. The unusual feature of the term which Yang and Mills constructed was that it included not only bilinear combinations of W-fields but also trilinear and quadrilinear combinations. As in QED, the bilinear combination described the propagation of Ws through space, while the other combinations described the self-interactions of Ws, the couplings of three and four Ws respectively (Figure 6.1).

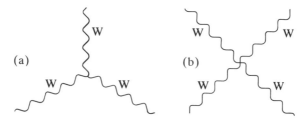

Figure 6.1. Self-interactions of gauge fields in Yang–Mills theory: (a) 3-W coupling; (b) 4-W coupling.

The self-interaction of the gauge fields in the Yang–Mills theory was necessary in order to ensure the desired SU(2) gauge invariance, and corresponded to the fact that the W particles coupled to all particles carrying isospin, including themselves. The self-interaction did, however, mark a significant departure from QED, in which the photons coupled to all electrically charged particles but not, being electrically neutral, to each other. Thus although the Yang–Mills theory was closely modelled upon QED there was an important difference between the two. The difference served to make the Yang–Mills theory interesting to theorists, but it also led to great computational problems. Having written down their Lagrangian,

Yang and Mills attempted to follow the standard route to obtain physical predictions from it via Feynman rules and diagrams, but failed. The mathematical complexities introduced by the self-interacting Ws were such that, as Yang later recalled, 'We completely bogged down – the non-linear [self-interaction] terms were so involved we never got out of the greater and greater complications that we were sinking into'. But, nevertheless, 'we felt that the whole idea was interesting enough, so without having pushed this point through, we published a paper'.[7] The paper appeared in 1954,[8] and constituted the starting point of the gauge-theory tradition in HEP.[9]

In the mid-1950s, field theory was still dominant in HEP, and many authors elaborated upon Yang and Mills' approach. Several lines of development emerged. One line was primarily formal: the attempt to construct a self-consistent set of Feynman rules for gauge theory, and to investigate its renormalisability. Such attempts are discussed in Section 3 below. Other lines had more direct phenomenological relevance. The particular theory constructed by Yang and Mills could be seen as the prototype of a whole class of gauge theories. Yang and Mills had written down an SU(2)-invariant theory of interacting nucleons, and several authors extended their work to other gauge groups, and to the interactions of different particles; and here it did not go unremarked that gravitation could also be derived from a gauge principle.[10] Besides their formal interest, particular realisations of gauge theory – particular choices of group and fundamental particles – were formulated according to their phenomenological appeal. Theorists aimed to bring the predictions of gauge theory into contact with the data – at least at the lowest order of perturbation theory, where closed-loop diagrams and the problems associated with renormalisation could be ignored. Two such early attempts at gauge theories of the strong interaction are discussed below, and the construction of electroweak gauge theories is the topic of the following section. Before examining particular applications of gauge theory, however, it will be useful to review a common obstacle to all of them: the 'zero-mass problem', which went as follows.

We saw in Chapter 3.3 that the QED Lagrangian contains no mass-term for the photon. Being modelled upon QED, mass-terms for the W-particles were similarly absent from the Yang–Mills Lagrangian. The straightforward inference was therefore that the Ws, like the photon, were massless. And, according to the Uncertainty Principle, any forces mediated by the exchange of massless Ws should be long-range. Unfortunately, the only long-range force of interest to particle physicists, electromagnetism, was already acknowledged to be well described by QED. The other forces – the weak and strong interactions – were short-range. Gauge theory thus appeared to be inapplicable to any recognised phenomena.

Various strategies were invoked by theorists in response to the zero-mass problem. One was, of course, to agree that gauge theory indeed had nothing to do with the world of elementary particles (although perhaps it represented a way forward in the construction of a quantum field theory of gravitation). Another was to argue that gauge theory was very complicated and imperfectly understood. Perhaps when theorists had investigated the theory further it would become clear that the Ws did not have zero mass. This was the perspective adopted by Yang, who recalls being reduced to silence at a Princeton seminar in 1956 when pressed by Pauli to give an opinion on the mass of the Ws.[11] A third strategy, and the only one with immediate phenomenological relevance, was to give the gauge particles masses 'by hand', by inserting an appropriate mass-term in the gauge theory Lagrangian. This disturbed the analogy with QED and destroyed the gauge invariance of the theory, but did make gauge theory into a realistic candidate for the description of the short-range weak and strong interactions.

One of the first theorists to follow the third route was the Japanese theorist J.J.Sakurai. Working at the University of Chicago he adapted the prototype Yang–Mills theory, assuming that beyond the original spin 1 isospin triplet of gauge particles there existed two other vector particles coupled respectively to the strangeness and baryon-number quantum numbers of hadrons. In an influential paper published in 1960, he gave masses to these particles by hand, and thus effectively predicted the existence of the ρ, ω and ϕ vector mesons which were experimentally observed shortly afterwards.[12] This predictive success was not unique to the gauge theory approach – as discussed in Chapter 3.4 the S-matrix bootstrap programme could also generate vector mesons – but was regarded by gauge theorists as justification for further work.

In 1961, Gell-Mann and Sheldon Glashow produced a catalogue of the various groups on which gauge-invariant field theories could be based.[13] Included in the catalogue was the group SU(3) and, as noted in Section 3.2, gauge theory was a central concern of both Gell-Mann and Ne'eman in their construction of the Eightfold Way of hadrons. However, the Eightfold Way contained the seeds of gauge theory's downfall as an approach to the strong interaction. In gauge theory, the vector mesons were special and fundamental entities, the quanta of the gauge fields. But as the Eightfold Way prospered and transmogrified into the quark model, the vector mesons came to be

164

seen as ordinary hadrons, quark–antiquark composites just like all the other mesons. Only in the 1970s did another set of candidate gauge particles emerge in the realm of the strong interactions – the gluons, which we glimpsed fleetingly in the last chapter and which will reappear in the next. Of more lasting significance were the early attempts to apply the Yang–Mills idea to the weak interactions, as outlined below.

6.2 Electroweak Unification and Spontaneous Symmetry Breaking

In Section 3.3 we discussed the Fermi theory of the weak interactions, and the V–A theory which succeeded it in 1958. Both the Fermi and the V–A theories envisaged the weak interaction as taking place between two currents at a point in space, and both were therefore open to two theoretical objections: they were non-renormalisable, and their first-order predictions violated the physically-reasonable 'unitarity limit' at ultra-high energies. Various theorists conjectured that it might be possible to overcome these objections if the weak force was represented not as a contact interaction between two currents but as mediated by particle exchange. In order to reproduce the space–time structure of the V–A theory, it was necessary that the exchanged particles – the carriers of the weak force – should be spin 1, vector, mesons, often referred to as intermediate vector bosons (IVBS). Two such IVBS – one positively charged, the other negative – were sufficient to recover the established V–A phenomenology. Once Yang and Mills had constructed their prototype gauge theory, a possible conjecture was that, in a suitable formulation, the vector gauge particles could be identified with the IVBS of the weak interaction. Furthermore, since gauge theory was closely related to QED, it was tempting to speculate that, in some sense, the weak and electromagnetic interactions were manifestations of a single underlying 'electroweak' force.[14] The twin possibilities of a gauge theory of the weak interactions and of the electroweak unification of forces were pursued by a small band of field theorists in the late 1950s and through the 1960s. In this section we can review how their programme developed.[15]

Electroweak Unification

One of the first authors to attempt an explicit unification of the weak and electromagnetic interactions was Julian Schwinger of Harvard University. Noted for his work on the renormalisation of QED, in 1957 Schwinger published 'A Theory of the Fundamental Interactions'.[16] As the title suggests, this was a not unambitious paper and in it, amongst other things, Schwinger suggested that the photon and the W^+ and W^- IVBs were members of a single family – just as, for instance, π^+, π^0 and π^- were considered to be members of a single isospin multiplet. By putting the photon and the IVBs in the same family Schwinger had, in a straightforward sense, unified the weak and electromagnetic interactions: they were manifestations of a single cause – exchange of members of a single family of particles. But there were problems with this. Empirically, the weak and electromagnetic interactions were quite distinct: the electromagnetic interactions were orders of magnitude stronger than the weak; they were of infinite range, while the weak interactions were of very short range ($\sim 10^{-15}$ cm); they conserved parity, while the weak interactions did not. The first two dissimilarities could be dissolved by supposing that the IVBs were very massive (the photon being massless): according to the tenets of quantum mechanics this would imply the short range of the weak interactions, and also that at energies low compared with the masses of the IVBs the weak interactions would be highly suppressed relative to the electromagnetic. But how could massive IVBs be seen as members of the same family as the massless photon? Schwinger conjectured the outlines of a positive answer to this question in his own idiosyncratic field-theory formalism (Feynman's diagrammatic techniques had been almost universally adopted by this time) but it was not taken up by other theorists.

The next significant step in the IVB theory of the weak interaction came from Sidney Bludman of the University of California at Berkeley. Ignoring the electromagnetic interactions, Bludman in 1958 proposed that the IVBs of the weak interaction were nothing more than the gauge particles of an SU(2) Yang–Mills gauge theory.[17] He suggested that there were three of these particles which formed an SU(2) triplet – W^+, W^0, W^- – in a 'weak isospin' space, and gave them a large mass by hand, inserting an appropriate term into the Lagrangian of the theory. Bludman did not attempt to unify the weak and electromagnetic interactions and he therefore had no problem in arranging for the former to violate, and the latter to respect, parity conservation.

Towards the end of the 1950s, the themes of electroweak unification and gauge theory were combined on both sides of the Atlantic. In the United States, the most important contribution came from Sheldon Lee Glashow. As a postgraduate student of Schwinger's at Harvard he worked from 1955 to 1958 on electroweak unification in gauge theory, and in 1961 he published a model setting out much of the theory which was to dominate the 1970s.[18] Glashow's model incorporated both a triplet and a singlet of IVBs; in technical terms, the gauge symmetry was SU(2) × U(1). The triplet comprised one positive, one negative, and one electrically neutral IVB, while the singlet IVB was also neutral. By judicious choice of the mass terms which he incorporated in the gauge theory Lagrangian, Glashow ensured that the singlet and the neutral member of the triplet should 'mix' in such a way as to produce one very massive particle (Z^0) and one massless particle which could be identified with the photon. The masses inserted into the theory also arranged that the two charged members of the triplet (W^+, W^-) became very massive. By a similarly judicious assignment of leptons to representations of $SU(2) \times U(1)$, Glashow ensured that the electromagnetic interactions (mediated by the photon) conserved parity, while the weak interactions (mediated by W^+, W^- and Z^0) did not.

Elsewhere, a similar path was trodden by the Pakistani theorist Abdus Salam (of Imperial College London and, later, the International Centre for Theoretical Physics, Trieste) and the British theorist J.C.Ward (of various US institutions). Both men were expert field theorists. Ward had made important contributions to the renormalisation programme for QED, and was well known for the 'Ward identities' – relations expressing the consequences of gauge invariance in perturbative QED.[19] Salam, too, had worked on the renormalisation of QED, and had been closely involved in the development of gauge theory from its inception.[20] At Cambridge in the mid-1950s, his student, Ronald Shaw, had independently formulated a Yang–Mills type Lagrangian.[21] At Imperial College, Salam was Yuval Ne'eman's supervisor from 1958 to 1961, and discussions between Salam and Ne'eman on symmetries and gauge theories were instrumental in the latter's formulation of the Eightfold Way.[22] From 1959 onwards, a series of joint publications by Salam and Ward expressed their common interest in gauge theory, culminating with a 1964 paper embodying a unified electroweak gauge theory similar to that of Glashow.[23]

A common feature of the electroweak models of Glashow and Salam and Ward was that the IVBs were given mass by hand. This made the models phenomenologically realistic but several authors, including Salam, argued that it also rendered them non-renormalisable.[24] In this respect the next important step came in 1967, with the formulation of the 'Weinberg–Salam model'. In the early 1970s this model was retrospectively recognised to be renormalisable. Its distinctive feature was that the IVBs acquired masses by 'sleight of hand', with no explicit IVB mass terms appearing in the Lagrangian. To understand how the trick was done, we must turn aside from electroweak unification to wider developments in field theory in the early 1960s.

Spontaneous Symmetry Breaking and the Higgs Mechanism

In the perturbative approach to quantum field theory developed in HEP during the late 1940s and 1950s, a direct correspondence was assumed between terms in the Lagrangian and physically observable particles. In QED, for example, the first two terms in the Lagrangian, $\bar{\psi}D\psi + m\bar{\psi}\psi$, were taken to refer to the propagation through space of real massive electrons, and the third term, $(DA)^2$, to the propagation of real massless photons. In the early 1960s such assumptions began to come under challenge. Various authors agued that when interaction terms, such as the $eA\bar{\psi}\psi$ term in QED, were taken into account, the direct correspondence between primitive fields and observable particles might be broken: the physical spectrum of observable particles might not be that which was naively read-off from the Lagrangian. The impetus behind this challenge came from solid-state physics, where all sorts of quasi-particles were used to explain experimental observations – particularly in superconductivity – and these particles did not directly map onto the fundamental fields of a field-theory approach. There was, however, a considerable conceptual gulf between the two areas of physics – field theorists in solid-state physics dealing with non-relativistic phenomena, while HEP theorists called for fully relativistic theories – and the transplanting of ideas from solid-state to high-energy physics was by no means straightforward.

One of the first physicists to attempt such a transplantation was HEP theorist Yoichiro Nambu of the University of Chicago. Born in Tokyo in 1921, Nambu moved to the USA after gaining a BSc degree in physics from Tokyo University in 1942 and a DSc from the same institution in 1952. While in Japan, he worked on quantum field theory and its applications to many-body phenomena in both solid-state and particle physics.[25] His early work in the US was mainly in HEP but he retained an interest in solid-state physics, especially in the phenomenon of superconductivity (the vanishing of electrical resistance in certain metals below a critical temperature). The late 1950s were an exciting time for research into superconductivity – the details of the Bardeen–Cooper–Schrieffer (BCS) theory of superconductivity were then being worked out[26] – and Nambu and his colleagues at Chicago joined in this work. Having a foot in both the superconductivity and HEP camps, Nambu (in collaboration with G.Jona-Lasinio, a post-doctoral visitor to Chicago from Rome) attempted to extend his insight from one to the other. This led to the publication in 1961 of two papers with the same self-explanatory title: 'A Dynamical Model of Elementary Particles Based Upon an Analogy with Superconductivity'.[27]

The work of Nambu and Jona-Lasinio was stimulating to particle physicists in two respects. First, it indicated the fruitfulness of solid-state physics as a source of inspiration in HEP, and in the following years resources were often taken over from one field to the other.[28] Secondly, and more specifically, in their 1961 papers Nambu and Jona-Lasinio introduced a new conceptual resource to relativistic field theory, making it available for elaboration by other theorists according to their expertise and context. The new concept was that of 'spontaneous symmetry breaking' (SSB). The thrust of SSB was that it is possible for a field-theory Lagrangian to possess a symmetry which is not manifest in the physical system which the theory describes. The symmetry is 'spontaneously broken' by the physical states of the theory, in the common parlance of HEP – although 'hidden' is a better description, since the symmetry of the Lagrangian remains exact. The active analogy in work on SSB was with superconductivity but physicists often explained the concept, to each other and to the wider public, in terms of an analogy with the more familiar phenomenon of ferromagnetism, and I will follow this practice here.[29]

Physicists picture ferromagnetic material (a bar magnet, say, or a compass needle) as an assemblage of spinning particles – these are the atoms which make up the material. Magnetism is produced by the mutual interaction of the atomic spins – each spin behaving like a little magnet, and alignment of spins being energetically favoured. Now, the Lagrangian for a system of interacting spins shows no preference for any particular direction in space – it is 'rotationally invariant'. None the less, when the spins of a ferromagnet line up to produce the macroscopically observed magnetism, they necessarily line up in a particular direction. By definition this direction is 'special', in the sense that the magnetism is directed there and nowhere else: the physical state of a ferromagnet lacks the symmetry of the Lagrangian which describes it. Thus ferromagnetism is a phenomenon which displays SSB, and superconductivity can be understood likewise.

SSB became a topic of great interest to HEP theorists in the early 1960s, principally in connection with the strong interactions. The SU(3), Eightfold Way, symmetry of the hadrons which was then becoming established was evidently only approximate, and it was an obvious conjecture that SSB had something to do with the breaking of exact SU(3) symmetry. Unfortunately, there was a problem with this conjecture, as was first pointed out by Cambridge HEP theorist Jeffrey Goldstone. Goldstone shared with Nambu a background of active research in the BCS theory of superconductivity and, working independently, Goldstone also concluded that the idea of SSB might be useful in particle physics.[30] His first publication on the subject appeared in 1961, and in this paper he concluded that SSB must be accompanied by the appearance of massless, spin-zero particles – 'Goldstone bosons' – as physical states of the theory.[31] This result was derived more rigorously the following year by Goldstone, Salam and Steven Weinberg (who was at Imperial College on leave from the University of California at Berkeley) and subsequently became the object of increasingly formal proofs.[32] On the face of it, the Goldstone theorem, as it was known, was extremely bad news. It implied that a theory of the spontaneous breaking of the SU(3) invariance of the strong interactions was out of the question: since the strong interactions were of short range, massless particles could play no part in them. Eventually, however, theorists made a virtue of necessity. The pion, whilst not massless, was much lighter than all other hadrons and could be regarded as a 'pseudo-Goldstone boson'. This line of thought was developed to great effect in the mid-1960s by theorists working in the current-algebra tradition, the pion being considered to arise from a spontaneous breaking not of the SU(3) symmetry of strong interactions, but of the SU(2) × SU(2) invariance of the weak currents of hadrons.[33] Current algebra provided the principal context for the elaboration of ideas on SSB in the 1960s, but to follow this line of development would take us too far from the concerns of this chapter. We must now turn to an apparent paradox of the Goldstone theorem, which will lead us back to electroweak unification.

The paradox was this. In 1962 Goldstone, Salam and Weinberg had shown, in a wide class of field theories, that SSB must be accompanied by the appearance of massless particles. But the original inspiration for work on SSB came from superconductivity, and in superconductors there are no massless particles. Even the photon acquires an effectively non-zero mass in a superconducting medium, as P.W.Anderson, a leading solid-state physicist, had argued in 1958.[34] In 1963, Anderson returned to the fray, pointing out this apparent exception to the Goldstone theorem, and arguing that it should hold true in any gauge-invariant theory.[35] Something of a controversy then surfaced in the theoretical HEP literature over whether phenomena seen in superconductivity persisted in relativistic contexts.[36] Several physicists participated in the resolution of the controversy, the upshot of which was that there existed a class of relativistic field theories wherein the Goldstone theorem could be evaded, namely those theories having a local gauge invariance: QED and Yang–Mills theories.[37]

The most accessible contribution to the debate came in papers published in 1964 and 1966 by the English theorist Peter Higgs. Higgs began his training as a physicist at King's College London in 1947 where he gained a PhD in theoretical physics in 1954.[38] His thesis work was on the application of quantum mechanics to molecular physics, but on leaving King's he followed his earlier inclinations and moved into theoretical HEP. He held a variety of research fellowships before becoming a lecturer at the University of Edinburgh in 1960, and his early work in HEP covered a variety of topics including quantum field theory, particle symmetries and the study of gravitation. He followed Nambu's work on superconductivity and SSB and the subsequent debate over the Goldstone theorem, making a decisive intervention with what became known as the 'Higgs mechanism'. Higgs' accomplishment was to exhibit and analyse the evasion of the Goldstone theorem in a very simple field-theoretic model. The model consisted of the standard QED Lagrangian augmented by a pair of scalar (spin 0) fields which were coupled to the photon and to one another in such a way as to preserve the gauge invariance of electromagnetism. Higgs found that, if he gave the scalar fields a negative mass term in the Lagrangian, the physical spectrum of the theory contained a massive photon and one massive scalar particle – a 'Higgs particle' – rather than a massless photon and two negative-mass scalar particles as one would naively have expected from reading-off particle properties from the Lagrangian.[39] The physical interpretation of the Higgs mechanism was that massless photons can only be polarised in two directions, while massive vector particles have three possible axes of polarisation: the massless photon could therefore be seen as 'eating up' one of the scalar particles which provided the third polarisation component and gave the photon a mass.

As mentioned above, theoretical interest in SSB centred originally upon possible approaches to SU(3) breaking in the strong interactions, and the Higgs mechanism generated considerable excitement amongst the small band of field theorists working in this area. Higgs and others pursued this line but produced no significant results; instead the mechanism found its first application in gauge theories of the electroweak interaction.

The Weinberg–Salam Model

In 1967, a crucial step in the development of electroweak theory was taken by Steven Weinberg (then at MIT) and Abdus Salam. Working independently, Weinberg and Salam took over the unified electroweak gauge-theory model proposed by Glashow in 1961 and by Salam and Ward in 1964, and replaced the IVB mass terms previously generated by hand with masses generated via the Higgs mechanism (straightforwardly generalised from electromagnetism to Yang–Mills theory). The first to publish was Weinberg,[40] who had rediscovered the Higgs mechanism in his own field-theoretic work on gauge theory, SSB and current algebra.[41] Weinberg was by then a leading field-theorist who, as we have seen, had played an important role in developing spontaneously broken field theories in HEP. He also had a long-standing interest in the weak interactions, dating back to his 1957 thesis from Princeton on the renormalisability of weak interaction field theories. Thus, having encountered the Higgs mechanism for himself, it was but a small step for Weinberg to redeploy it in the construction of a unified electroweak model. Weinberg's model was essentially that proposed six years earlier by Glashow, except that certain mass relations between the IVBs were determined in Weinberg's model in terms of a single free parameter – θ_w, the 'Weinberg angle' – which were not so determined in earlier formulations.[42] Salam's contribution can be similarly understood. We have already seen that Salam and Ward constructed a unified electroweak gauge theory in 1964, and that Salam was one of the principals of the SSB saga. As early as 1962, he had discussed the possibility of mass generation in gauge theories by means of SSB.[43] Salam learned of the Higgs mechanism from a colleague at Imperial College, T.W.Kibble, who had himself played an important role in the investigation of spontaneously broken gauge theories.[44] Salam then reformulated his 1964 electroweak model using the Higgs mechanism, lectured on the subject at Imperial College in 1967,[45] and presented his work at the Nobel Symposium held in Aspenäsgarden, Sweden, in May 1968.[46]

In view of its subsequent impact, the reception of the Weinberg–Salam model is very interesting. It has been said of the model, by HEP theorist Sidney Coleman, that 'rarely has so great an accomplishment been so widely ignored'. The citation record of Weinberg's paper from 1967–73 was, according to Coleman (and including self-citations): 1967, 0; 1968, 0; 1969, 0; 1970, 1; 1971, 4; 1972, 64; 1973, 162.[47] Several factors contributed to the lack of attention paid to Weinberg's paper in the period from 1967 to 1971. First, of course, this was a period when field theory was in decline: the potential audience for such a publication was small. Secondly, Salam and Weinberg both confined their attention to the weak interactions of *leptons*, for which very few relevant data existed. One reason why they did this was that the extension of the model to hadrons was straightforward and, when made, was wrong: as we shall see in Section 4 below, it led to predictions in conflict with established experimental results. The phenomenological utility of the Weinberg–

Salam model therefore appeared to be zero. Finally, except as an exercise in theoretical virtuosity, the model had no special appeal for theorists. Weinberg and Salam had simply conjoined the Higgs mechanism with earlier electroweak gauge models. The significance of this only became apparent in 1971 as a result of the developments discussed in the following section. At that point theorists recognised that Weinberg and Salam *had* done something special – they had constructed the first renormalisable theories of the weak interactions, with unification as an added bonus – and interest in unified gauge theories exploded, as the citation data show.[48]

6.3 The Renormalisation of Gauge Theory

The prototype for all other field theories was quantum electrodynamics. By virtue of its renormalisability, calculations in QED could be pursued to arbitrarily high orders of approximation. The gauge field theory first written down by Yang and Mills in 1954 was very closely modelled on QED, in the hope of extending its theoretical and phenomenological success to other interactions beyond electromagnetism. Yang and Mills, though, failed to construct a satisfactory set of Feynman Rules for their theory, and without these it was impossible to follow the conventional route to the demonstration of renormalisability. Thus it was quite unclear whether Yang–Mills gauge theory shared the desirable properties of QED, and, if it did, how to exploit them.

In this Section I want to discuss the work which proceeded very slowly throughout the 1960s, leading to the demonstration in 1971 that electroweak gauge theories were indeed renormalisable. The research programme which culminated in this demonstration was that of Dutch HEP theorist Martin Veltman who, looking back upon the 1960s, remarked that progress had been 'slow and painful' and that 'almost everybody made a mistake of one kind or another, or landed up in a blind alley.'[49] The proof of the renormalisability of QED had itself been a mathematical and conceptual *tour de force* and gauge theory, with its complex group structure, was even more difficult to handle. Reviewing recent developments in gauge theory in 1973 for an audience of particle physicists, Veltman quailed at the prospect of explaining the technical details of renormalisation. 'I have tried as much as possible to avoid the use of complicated mathematical or graphical arguments', he said, 'This type of consideration cannot be expected to be palatable to a general [sic] audience'.[50] In the present context, any attempt to go into the subtleties of the renormalisation programme would be futile. I will focus instead upon the history of Veltman's involvement with gauge theory, and its culmination in the work of his student, 't Hooft.[51]

Veltman and Massive Yang–Mills Theory

Veltman was born in Waalwijk, Netherlands in 1931. He studied physics at the University of Utrecht from 1948 to 1961 (with a break for military service from 1957–9). In 1957 he completed his 'doctoral' work (the equivalent of an American masters degree) on formal properties of QED, under the supervision of HEP theorist Leon Van Hove (who became Research Director-General of CERN in 1976). In 1959 he began research for a PhD, again supervised by Van Hove. He developed a Feynman diagram approach to the field theory of unstable particles, and made detailed calculations in IVB theories. In 1961 he moved to CERN, where he worked mainly on phenomenological topics of local experimental interest.

In 1966, Veltman visited the theory group at Brookhaven for four months before returning to Utrecht as Professor of Physics. At Brookhaven, Veltman finished 'a rather phenomenological work on electromagnetic decays of the mesons' and decided that he 'wanted to do something more fundamental.'[52] As he saw it: 'At that time the big issue was current commutators, introduced by Gell-Mann, and there seemed to be great confusion as to how and when they could be trusted.' The problem here lay with the so-called Schwinger terms. In 1959, Julian Schwinger had published a simple argument to the effect that, in certain circumstances, commutation relations derived from formal mathematical manipulations in quantum field theory were invalid, and this argument undermined many potential applications of current algebra.[53] Veltman decided to try to understand current algebra, PCAC and the anomalous Schwinger terms using techniques developed in his thesis work (which was published in 1963).[54] As he put it: 'on revient toujours à son premier amour'. His thesis work had been in Lagrangian field theory, and he attempted to recover current-algebra results from manipulation of quantum fields. He found that he could rederive all of the current-algebra sum-rules which were not bedevilled by Schwinger terms from two simple equations. These 'current equations' expressed the divergences of the vector and axial-vector hadronic currents (the quantities appearing in the CVC and PCAC relations) as products of the currents themselves with vector fields representing the photon and the IVBs of the weak interaction (plus a term representing the pion field, in order to recover PCAC).[55]

Veltman's two equations 'had an amazing structure' which, had he known it at the time, was characteristic of a Yang–Mills theory. This was not entirely surprising: in all of his work in the early 1960s on the Eightfold Way and current algebra, Gell-Mann had contrived to erect a structure which was consistent with gauge theory, before discarding the gauge-theory base and leaving the structure to stand alone. Veltman had, in effect, reversed Gell-Mann's strategy. In August 1966 Veltman met Richard Feynman at an HEP conference in Berkeley, California. Feynman had himself worked on gauge theory,[56] and explained to Veltman that the latter's equations were of the Yang–Mills type. Feynman, however, followed the line of reasoning on gauge theory advocated by Sakurai and later Gell-Mann and Ne'eman in their work on the Eightfold Way. He argued that the gauge structure which Veltman had found involved the vector mesons ρ, ω and ϕ, and related to the strong interactions rather than to the IVBS of the weak interaction. Confused by Feynman's explanation of his current equations in terms of strongly interacting vector mesons, Veltman returned to Utrecht in September 1976. Meanwhile, Veltman's friend and collaborator, CERN theorist John Bell, had become interested in Veltman's current equations. Bell attempted to find a Lagrangian which would correspond to the equations, and in 1967 published a paper arguing that what was needed was a Yang–Mills structure for the *weak* interactions.[57] Veltman later recalled that Bell's paper on the subject 'became a great mystery to me, for a while. It kept on going in my mind.'

In April 1968, Veltman was invited to spend a month at Rockefeller University in New York, and 'I decided to spend the whole month just to think about the state of affairs, and to choose the direction that I would go. It finally dawned upon me that the current equations were a consequence of a Yang–Mills type structure of the weak interactions.' This was precisely the import of Bell's 1967 paper and, seen in this light, Feynman's explanation of the current equations in terms of ρ, ω and ϕ mesons was a red herring: the vector fields appearing in the equations should instead be identified as the IVBS of the weak interaction. Veltman's realisation that the weak interaction might be described by a gauge theory was, of course, by no means unique amongst HEP theorists, but his route to it was idiosyncratic, as were the implications he drew from it. Impressed by his new understanding of his 1966 current equations, Veltman devoted the remainder of his time at Rockefeller to investigating the *renormalisability* of gauge theory, drawing further upon the resources he had acquired in his thesis work.[58] In particular he decided to investigate massive Yang–Mills theory, the version of the theory in which masses were inserted *by hand*, since the pure, gauge-invariant theory with massless gauge particles was manifestly unrealistic.[59]

This plunged Veltman even further into idiosyncracy, since the received opinion within HEP at that time was that all theories involving massive vector mesons were non-renormalisable.[60]

At Rockefeller, Veltman set about computing higher-order diagrams in the simplest non-trivial gauge theory – a theory of massive self-interacting IVBs, with no leptons or hadrons. He began with diagrams containing a single closed loop (the standard starting point for higher-order calculations in perturbative field theory). Here he encountered the first of the infinite mathematical integrals – divergences – which were commonly assumed to make the theory non-renormalisable. But he quickly convinced himself that many, if not all, of the divergences in the theory *cancelled*. Because of the high degree of symmetry of the gauge-theory Lagrangian (even when mass terms were included) the divergences from different diagrams were constrained to be equal but of opposite sign: in total, their contributions to physical processes was zero. Thus Veltman concluded that it was conceivable, at least, that massive gauge theory might be renormalisable – contrary to the conventional wisdom. Suitably encouraged, Veltman persevered with this line of thought. He had found the cancellations by brute force calculations and now looked for a more elegant line of approach. He reformulated the Feynman rules of the theory in such a way that the cancellation between divergences was manifest by inspection from the beginning. To achieve this, he found that he had to include in the set of diagrams the interactions of a 'ghost' particle – a particle which appeared only in closed loops and not as a physical incoming or outgoing particle.[61] In May 1968, Veltman returned to Utrecht where he wrote up this work as a lecture series for an HEP summer school held in Copenhagen in July of that year.[62]

In August 1968, Veltman joined the HEP theory group at Orsay, near Paris, on one year's leave from Utrecht. There he came into close contact with many of the other principals of the gauge-theory story,[63] and there he learned for the first time of the work done on *massless* gauge theories by a small band of theorists scattered around the USA, Europe and the USSR.[64] This work had resulted in a set of Feynman rules for the massless theory which included ghosts and which were similar to those which Veltman had derived for the massive theory. Furthermore, the analysis of the massless theory had been carried through for arbitrarily many loops, and the theory had been shown to be renormalisable.[65] Veltman studied the existing work on massless gauge theory and refined his analysis of the one-loop diagrams in the massive theory, publishing his reflections on the latter in 1968 and presenting them at an HEP Conference at CERN in January 1969.[66] In collaboration with a graduate student, J. Reiff, he then set about investigating two-loop diagrams in the massive theory, and here they established a puzzling result: when the mass of the vector particles was set to zero, the results from the massive theory did not reduce to those already established for the massless theory.[67] Veltman had no idea why this paradoxical result should hold. In the search for an answer, he checked through all of the calculations from the beginning, and mastered both the 'path integral' formulation of quantum field theory used by other exponents of gauge theory[68] and the methods of 'source theory' used by Julian Schwinger in his work on the renormalisation of QED.[69]

Armed with these new techniques, Veltman in 1970 proved the result he least desired: at the two-loop level, in massive gauge theory, there were non-renormalisable divergences.[70] At this point it appeared that Veltman's programme was doomed. The conventional wisdom was right: massive Yang–Mills theories were non-renormalisable. However, the paradoxical difference between the zero-mass limit of the massive theory and the zero-mass theory itself was still unclarified. And 'together with Van Dam [a visitor to Utrecht from the University of North Carolina] I unraveled the great mystery. Finally I learned to count to three!' The difference between massive and massless theories lay in the fact that while massive vector particles have three possible spin orientations in space, massless vector particles have only two. The extra degree of freedom associated with the spin of a massive vector particle was responsible for the non-renormalisability of the massive theory, and this degree of freedom did not vanish in the zero-mass limit.[71]

Having at last come to grips with the difference between massive and massless theories, Veltman pondered upon what to do about it. In early 1971, he began to entertain the possibility that perhaps appropriate scalar fields could be introduced into the massive Yang–Mills Lagrangian in such a way as to cancel the two-loop divergences. As events quickly proved, the introduction of scalar fields was the way to progress, but the architect of this part of the story was not Veltman but his student, 't Hooft.

't Hooft and Spontaneously Broken Gauge Theory

Gerard 't Hooft was born in Den Helder, Netherlands in 1946. He went to school in the Hague before moving in 1964 to the University of Utrecht to study physics and mathematics. At Veltman's suggestion, for his 'doctoral' examination 't Hooft submitted in 1969 a study of the renormalisability of the 'σ-model'.[72] The σ-model was a simple Lagrangian field theory often used in heuristic current-algebra calculations. It was not a gauge theory, but displayed spontaneous symmetry breaking, the consequent Goldstone bosons being interpreted as pions.[73] In 1969 't Hooft began full time research at Utrecht, studying for a PhD in theoretical HEP under Veltman's supervision. One of his first tasks was to write up the notes of Veltman's lectures on path integrals.[74] For his PhD, 't Hooft's inclination was to work on gauge theory. Veltman agreed, although he felt that it 'was so much out of line with the rest of the world that very likely one was producing specialists in a subject that nobody was interested in', and he suggested that 't Hooft work on the renormalisation of *massless* gauge theory. Although Veltman had believed since 1968 that the massless theory was renormalisable, problems remained with details of the 'regulator method' used to attribute finite and manageable values to formally infinite integrals in the course of actual computations. 't Hooft devised a suitable regulator method and, taking over many of Veltman's techniques, published an account of his work in 1971.[75] Although the conclusion was already accepted by a handful of cognoscenti, this paper constituted the first detailed argument that massless gauge theory was renormalisable to appear in the HEP literature.

't Hooft's work served to bring Veltman up to date on the massless theory, and to crystallise the latter's thinking on the difference between the massive and massless cases. Veltman recalls a conversation with 't Hooft early in 1971 in which he told 't Hooft that, while massless gauge theory was very elegant, what was needed was a realistic theory involving massive vector particles: translated from the Dutch, the discussion went:

M.V.: I do not care what and how, but what we must have is at least one renormalizable theory with massive charged vector bosons, and whether that looks like nature is of no concern, that are details that will be fixed later, by some model freak. In any case, all possible models have been published already.

G.'tH.: I can do that.

M.V.: What do you say?

G.'tH.: I can do that.

M.V.: Write it down and we will see.[76]

't Hooft proceeded to write his second paper of 1971.[77] Eleven pages long, this was the paper that ushered in the full panoply of the new physics (though, needless to say, it was not apparent at the time). 't Hooft used the technique of spontaneous symmetry breaking, familiar to him from his earlier work on the σ-model,[78] to give masses to the vector bosons of the pure gauge theory. By adding multiplets of scalar particles into the massless Yang–Mills Lagrangian, 't Hooft in effect re-invented the Higgs mechanism. From that point on, few problems remained. 't Hooft had the analytical tools already in hand to write down the Feynman rules of the spontaneously broken theory and to investigate its renormalisability. Like the pure massless theory, but unlike the massive-by-hand theories explored by Velt-man, 't Hooft found that gauge theories in which vectors acquired masses by SSB were renormalisable.

Courtesy of 't Hooft, Veltman now had just what he had always wanted: a physically realistic renormalisable theory of massive vector bosons. Having checked the details of 't Hooft's calculations for himself, Veltman arranged for 't Hooft to present his findings at the Amsterdam HEP conference in the summer of 1971, and 't Hooft's paper appeared shortly afterwards.[79] Veltman now regarded matters as settled, but the battle to demonstrate renormalisability was not yet quite won. The tools which Veltman had developed for analysing gauge theory were unfamiliar to many physicists, and the path-integral formalism which 't Hooft had inherited from him was widely regarded as being mathematically dubious. Furthermore, to the HEP community at large 't Hooft was an unknown Dutch graduate student: not the most likely person to have solved a problem which had defeated some of the world's leading field theorists over a period of nearly two decades.[80]

For all of these reasons, 't Hooft's work stood in need of support. Enter the influential US theorist Benjamin Lee. Lee was born in Korea in 1935, but moved to the US in 1956, becoming a US citizen in 1968. He studied for a PhD at the University of Pennsylvania under the supervision of Abraham Klein, and worked at several US institutions during the 1960s before joining the newly formed theory group at Fermilab in 1971 (on leave from the State University of New York at Stony Brook).[81] It was a joint paper by Lee and Klein on SSB in QED which in 1964 precipitated the debate leading to the construction of the Higgs mechanism,[82] and in subsequent years Lee made many important contributions to the analysis of SSB and gauge theories.[83] His 1964 work in SSB led him to the current-algebra tradition wherein the pion was regarded as a pseudo-Goldstone boson. In turn, this led to his 1969 analysis of the renormalisability of the σ-model.[84] During a visit to Caltech in the spring of 1971, Lee attempted to investigate the renormalisability of a theory of spontaneously broken QED. He discovered that many divergences cancelled in one-loop diagrams, but had got no further when he attended the 1971 Amsterdam Conference. There, he recalled, 'as soon as we exchanged greetings, Tini Veltman handed me two preprints by his student with the statement that the student solved the massive Yang–Mills theory'.[85]

After the Amsterdam conference, Lee returned to the US and his study of spontaneously broken QED. He soon proved to his own satisfaction that the theory was renormalisable and, in collaboration with Jean Zinn-Justin (a visitor to Stony Brook from Saclay, France) extended the proof to spontaneously broken gauge theories.[86] Lee's work was circulated during 1971 and published in 1972. His authority, coupled with a more conventional mathematical approach than 't Hooft, sufficed to guarantee general acceptance that the problem was solved: gauge theories in which the IVBs gained masses by the Higgs mechanism were renormalisable.[87]

Thus, in the period around late 1971 and early 1972, the status of Yang–Mills gauge theory was transformed. In a very short space of time, gauge theory ceased to be regarded as a mathematical curiosity, and was seen instead as a respectable, even profound, field theory 'every bit as good as quantum electrodynamics' (in the words of Martin Veltman). The question which HEP theorists now faced was one of what to do with the new theory. In essence, the answer they gave was: find out if it works. In his investigation of massive-by-hand gauge theories, Veltman had always had the weak interaction in mind, and the spontaneously broken theories investigated by 't Hooft and Lee had precisely the form of the Weinberg–Salam unified electroweak model. This model had now been renormalised – but how did it fare in comparison with the data?

6.4 Electroweak Models and the
     Discovery of the Weak Neutral Current
In 1974, Weinberg recalled that, following 't Hooft's work:
> it immediately became apparent to a great many physicists that 't Hooft's methods could be used to test the renormalizability of unified theories of weak and electromagnetic interactions. . . .
> As soon as it was realised that spontaneously broken gauge theories are renormalizable, there was a great explosion of theoretical effort devoted to detailed calculation of higher-order weak and electromagnetic 'radiative' corrections and to the construction of alternative models.[88]

Here we can concentrate upon the latter effort, electroweak model-building.[89] This was the theoretical line which led to the involvement of the experimental HEP community via the search for, and discovery of, the weak neutral current. As we shall see below, the prototypical Weinberg–Salam (WS) model, first presented by Weinberg in 1967, was by no means perfect, and many theorists developed variants on the electroweak theme.

Weinberg had based his model upon the simplest choice of gauge group, $su(2) \times u(1)$; he had used the minimal set of scalar particles to give masses to the gauge vectors by the Higgs mechanism; and he had made a particular choice concerning the multiplet structure of the known leptons (recall that at this stage the model did not include hadrons). The route to the construction of alternative models was to choose a larger unifying group, a more complex pattern of symmetry breaking and different multiplet assignments. It quickly became apparent that such work was essentially trivial, as was nicely illustrated when several authors gave explicit general recipes for the construction of unified, renormalisable models of the electroweak interaction.[90] By following such a prescription any theorist could choose a gauge group, the manner of its symmetry breaking, and a set of fundamental particles, and come out with a well defined, renormalisable theory. Thus it became clear that constructing models at random was not a very interesting or edifying exercise, and the question arose as to why one particular choice of model should be preferred over another. There were two mutually compatible approaches to this question. One was the search for 'naturalness', emphasised by Weinberg in his 1974 review:

> In my view, the most important criticism of the $su(2) \times u(1)$ model and also of all other existing gauge models, is that none of these theories is sufficiently *natural*. That is, the parameters in these theories have to be carefully rigged so as to achieve even a qualitative agreement with experiment. In particular, these models all contain small parameters, such as m_e/m_μ or $(m_p - m_n)/m_p$ which we feel ought to be calculable in any fundamental theory, but which in the existing theories have to be put in by hand. . . . Various kinds of natural gauge theory are described [later in this review], but so far none of them is very realistic. We need a theory that is both natural and realistic, but so far this has eluded us.[91]

The search for a 'natural' theory continued through the 1970s but with inconclusive results, and I will focus here upon the second approach to model-building: the search for a theory that was realistic. The ws model and later alternatives differed significantly in their predictions from the phenomenological V–A model, and the question arose as to whether support for any of the new unified models could be found in the data.

Phenomenology of Electroweak Unification

The principal novelty of the unified models of the 1960s was, of course, their dependence on massive iVBs (augmented by a single massive Higgs particle in the ws model) as intermediaries of the weak interaction. No such particles were experimentally established, but since their predicted masses in the ws model were of the order of 50 GeV this was not surprising – the energy required to produce such heavy particles was unattainable at existing accelerators. However, the Weinberg–Salam model did make one prediction which conflicted with the established V–A theory, which appeared to be accessible to experiment, and which went as follows.

The standard V–A theory could be regarded, in its phenomenological predictions, as equivalent to a theory in which the weak interactions were mediated by two *electrically charged* ivbs, conventionally denoted as W$^+$ and W$^-$ (the superscripts indicate charges). Thus, for example, the V–A prediction for neutrino-electron scattering cross-sections could be derived from the diagram of Figure 6.2(a). Note that in this figure, the leptons change their identity and charge at each vertex. At the upper vertex, the incoming neutral neutrino becomes an outgoing negatively charged electron; to maintain charge conservation, the ivb carries off a positive charge to the lower vertex, and there the incoming electron converts to a neutrino. Events such as these, involving changes in charge at the ivb vertex between incoming and outgoing particles were known, in the jargon of hep, as 'charged-current' events. And, as discussed in more detail below, it was routinely accepted in the 1960s that all weak interactions were of the charged-current type: only processes which could be mediated by charged ivbs were believed to exist. However, in the ws model there existed not only the charged W$^+$ and W$^-$ ivbs but also the massive partner of the photon, the Z^0. The Z^0, being electrically neutral, could mediate 'neutral-current' processes: weak interactions in which no change in charge occurred between incoming and outgoing particles, such as that shown in Figure 6.2(b).[92]

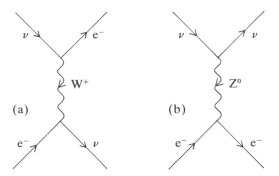

Figure 6.2. Neutrino–electron scattering: (a) charged current; (b) neutral current.

Gauge Theory

Experimental data on neutral-current processes like that of Figure 6.2(b) thus promised to distinguish between the WS unified model and standard V–A theory. As it happened, though, neither charged-current nor neutral-current neutrino-electron scattering had been experimentally observed by the early 1970s,[93] and no such empirical discrimination could be made on the basis of contemporary data – at least as long as the WS model were applied only to leptons. When theorists extended the WS model to include hadrons as well as leptons, however, a different situation obtained.[94] In the theoretical context of the early 1970s the obvious extension to hadrons was through quarks: the idea that quarks were the vehicles of the hadronic weak and electromagnetic currents was central to the current-algebra and constituent quark model traditions and, furthermore, the success of the parton-model analysis of deep-inelastic electron (and later neutrino) scattering, suggested that quarks were point-like entities – just like the leptons in terms of which Weinberg and Salam had originally formulated their model. Theorists therefore incorporated hadrons into the WS model by assuming that the IVBS coupled to u, d and s quarks in just the same way as they coupled to leptons. In consequence, neutral currents mediated by the Z^0 spilled over into the weak interactions of hadrons, and here a conflict with accepted data arose.

The conflict was sharpest with reference to the decays of the K-meson (kaon). Z^0s could mediate various neutral-current kaon decay modes, such as the decay $K^0 \rightarrow \mu^+\mu^-$ shown in Figure 6.3(a). HEP experimenters had sought to observe such decays throughout the 1960s but without success. All that had been established was that such processes could at most have very small cross-sections – much smaller than those expected from the WS model. In fact, the upper limits set by experiment on the neutral-current kaon decay modes were so low as to constitute a recognised anomaly even for the V–A theory, and this was to prove the WS model's salvation. The V–A theory was a theory with no Z^0, so no neutral currents were expected in it from first-order diagrams like Figure 6.3(a). But, in higher orders of perturbation theory, neutral currents were expected to arise in the V–A theory: the exchange, say, of both a W^+ and a W^- would mimic that of a single Z^0 as in Figure 6.3(b). The V–A theory was non-renormalisable and therefore, strictly speaking, theorists could say nothing about such higher-order diagrams. None the less, theorists did perform heuristic and approximate calculations of what might be expected if some satisfactory renormalisation procedure could be found, and these all agreed on one point: the relevant kaon decay rates were experimentally several orders of magnitude smaller than reasonable theoretical expectations. As a higher-order correction to the V–A theory, such effects could be expected to be suppressed, but no argument could be found to explain the extent of the observed suppression.[95] This anomaly was perceived against the accepted backdrop of V–A theory during the 1960s, and was only exacerbated when construed in terms of the ws model. The ws model, by virtue of the existence of the Z^0, predicted the existence of neutral currents of the same order of magnitude as charged currents, and was thus in even greater conflict with the K-decay data than was the V–A theory.

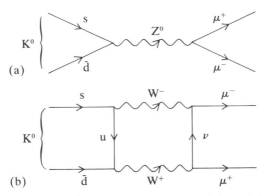

Figure 6.3. $K^0 \rightarrow \mu^+ \mu^-$ decay in: (a) Weinberg–Salam model; (b) V–A theory.

The GIM Mechanism and Alternative Models

Several theorists tried their hand at the kaon-decay anomaly, and the solution which eventually triumphed came from Harvard, in a 1970 paper from Sheldon Glashow, J.Iliopoulos and L.Maiani (then visiting Harvard from Rome).[96] The key to the GIM mechanism, as it became known, was an idea proposed in 1964 by several theorists: the existence of a new conserved quantum number of the strong interactions.[97] This amounted to an extension from su(3) to su(4) of the classification scheme for hadrons, and required the existence of a fourth 'charmed' quark (c) in addition to the usual u, d and s quarks. Amongst the promulgators of this idea (though not the first) was Glashow – the inventor of the su(2) × u(1) structure for electroweak models – and it was Glashow who gave the new quark its exotic name. The 1964 papers on charm were part of the first wave of enthusiasm for the quark model, and the extension to su(4) was motivated primarily by the symmetry between the four quarks and the four leptons which resulted, rather than by any requirements of the data. Indeed, from the existence of a charmed quark the existence of a whole new family of charmed hadrons could be inferred, and no evidence for such particles was known – either in 1964 when charm was invented, or in 1970 when it was resurrected by GIM. Undeterred by this empirical shortcoming, GIM pointed out that if the fourth quark existed, then a quantum-mechanical cancellation – the GIM mechanism – could be arranged in neutral-current calculations. Provided that the charmed quark carried suitable quantum numbers – it had to have charge $+\frac{2}{3}$, like the u quark – and was appropriately coupled to the IVBS, neutral currents could be suppressed in agreement with the K-decay data. GIM argued this conclusion in V–A-type theories, and the GIM mechanism at first received attention only from the small group of theorists then interested in the K-decay anomaly. However, as unified models became a growth industry following 't Hooft's work, the question of neutral currents acquired prime importance, and the GIM mechanism was identified as a suitable candidate for the defence of the Weinberg–Salam model against the data. GIM's 1970 publication became one of the central pillars of an ever-growing programme of work on unified electroweak models.[98]

The neutral currents cancelled by the GIM mechanism were known as 'strangeness-changing' neutral currents, since in the relevant kaon decays one unit of strangeness vanished. The mechanism did not, however, cancel 'strangeness-conserving' neutral currents – processes in which no change of strangeness occurred. Even when augmented with a charmed quark, the WS model thus predicted the existence of strangeness-conserving neutral-current effects of the same order of magnitude as the conventional charged-current effects. Here, a conflict remained between the WS model and accepted data. Data on strangeness-conserving neutral currents came from neutrino scattering experiments.[99] Figure 6.4(a) depicts a typical strangeness-conserving charged-current neutrino event, $\nu n \rightarrow \mu^- p$, and Figure 6.4(b) an analogous neutral-current neutrino event, $\nu n \rightarrow \nu n$. During the 1960s, the latter class of events were regarded as hypothetical. In 1963 an upper limit on the neutral-current rate was reported from the first CERN neutrino experiment,[100] with the comment: 'clearly neutral lepton currents cannot be admitted on a symmetrical basis with the charged.'[1] In 1970, the upper limit on neutral currents in neutrino experiments was revised upwards to around ten per cent of the charged-current rate,[2] but the new limit was still below the prediction of the WS model. It remained the case that the weak neutral current was officially non-existent, in neutrino scattering as in kaon decays.

One response from gauge theorists to this empirical problem was

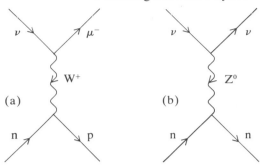

Figure 6.4. Neutrino–nucleon scattering: (a) charged current; (b) neutral current.

the construction of alternative electroweak models in which neutral currents were either absent or heavily suppressed. The prototypical models in this class were those devised at Fermilab by Benjamin Lee, at CERN by J.Prentki and B.Zumino, and at Harvard by Sheldon Glashow and Howard Georgi (where the latter was then a post-doctoral Research Fellow).[3] These models differed in their details but all followed the same strategy, introducing hypothetical and unobserved particles in order to cancel or suppress neutral-current effects in neutrino scattering (and kaon decay) experiments. Lee, Prentki and Zumino made do with new heavy leptons (more massive analogues of the electron and muon) while Georgi and Glashow resorted to new quarks as well.

The upshot of this stream of model-building was, then, that the prediction of neutral currents in neutrino experiments could only be avoided at the expense of introducing hypothetical particles, which should themselves be observable in neutrino experiments (and elsewhere). Thus in the early 1970s a mismatch developed between the growing band of electroweak gauge theorists and their experimental colleagues. Theorists found themselves articulating models referring to phenomena which seemed to have no experimental counterparts. They consoled themselves, however, with the thought that neutrino experiments were very difficult and still in their infancy relative to other fields of experimental practice. In the early 1970s, preparations were in hand at Fermilab for two high-energy neutrino experiments using electronic detectors, and at CERN the Gargamelle bubble-chamber neutrino experiment was already under way (see Section 5.4: the data discussed there concerned charged-current neutrino interactions). Perhaps these experiments would discover phenomena which had gone unobserved in earlier generations of research.

So reasoned the gauge theorists, and they communicated their feelings to the experimenters concerned in the most direct fashion.[4] In the US in late 1971, Weinberg contacted the leaders of the Harvard–Pennsylvania–Wisconsin–Fermilab (HPWF) neutrino collaboration. He persuaded them of the importance of an active search for neutral currents at Fermilab, and the HPWF group reconceptualised and redesigned their equipment accordingly.[5] At the same time in Europe, CERN theorists arranged a presentation for the Gargamelle experimenters, again stressing the importance of searches for neutral currents (or other new phenomena). And, as discussed further in the following section, the theorists' pleas did not go unrewarded. When, in early 1972, the CERN experimenters began to examine the photographs from Gargamelle, they found that quite frequently the incoming neutrino appeared to emerge unscathed (rather than converting to a muon as expected in a charged-current event). And, in July 1973, the Gargamelle experimenters submitted for publication a paper announcing their discovery of the weak neutral current. From an examination of 290,000 photographs taken in a neutrino beam they had identified around 100 examples of the neutral-current process $vN \rightarrow vX$ (compared with around 400 examples of the charged-current process $vN \rightarrow \mu^- X$).[6]

In 1974 the Fermilab neutrino experimenters followed suit, with reported sightings of the neutral current from the HPWF collaboration and, later, from a Caltech–Fermilab collaboration.[7] From this point onwards the existence of the weak neutral current was taken to be well established. Experimental estimates of the neutral-current event rate were crude, but were of the order of magnitude suggested by the WS model (augmented by GIM) and the alternative models were thus rendered redundant.[8] Neutrino data now constituted an argument in favour of, rather than against, the WS model and, twenty years after its invention by Yang and Mills, gauge theory had at last found a solid empirical base. This was a crucial event in the history of the new physics, and in succeeding chapters we will be able to examine some of its many repercussions. To close the present chapter it is appropriate to explore in some detail the experimental practices which lay behind the epoch-making neutral-current discovery, and their relation to the theoretical trend which they served to foster.

6.5 Neutral Currents and Neutron Background
How should the relationship between the experimental discovery of the weak neutral current and the development of unified electroweak

gauge theory be conceptualised? In the archetypal 'scientist's account', the former would be seen as an unproblematic observation of the state of nature and an independent verification of the latter. But can this view withstand historical scrutiny? I want to suggest that it cannot, through a comparative examination of the Gargamelle discovery experiment and its predecessors at CERN.[9] Two connected arguments will be involved. First I will argue that the observation reports on neutral currents which emanated from CERN in the 1960s and 1970s were all grounded in interpretative procedures which were pragmatic and, in principle, questionable. The procedures, and the observation reports which derived from them, could quite legitimately be challenged. Physicists had therefore to *choose* whether to accept or reject the procedures and the reports which went with them. As indicated in Chapter 1, this element of choice is the key element omitted from the 'scientist's account'. I will then show that there were significant differences between the neutrino experimenters' interpretative procedures in the 1960s and those of the Gargamelle collaboration, and that the differences were central to the existence or non-existence of the neutral current. Finally, I will argue that the communal decision to accept one set of interpretative procedures in the 1960s and another in the 1970s can best be understood in terms of the symbiosis of theoretical and experimental practice.

Neutron Background

Let us begin with the data. Figure 6.5(a) is a photograph of a neutral-current event in Gargamelle, and Figure 6.5(b) is a schematic representation of the recorded tracks. Needless to say, the equivalence between Figures 6.5(a) and (b) is not transparent to the untrained eye, but nor is it to the trained eye. For example, the dashed lines in diagram (b) representing the incoming and outgoing neutrinos characteristic of a neutral-current event cannot be read-off from the photograph. Being electrically neutral, neutrinos leave no tracks, and their presence can only be inferred from experimental observations. The necessity to make such inferences creates significant problems in the interpretation of bubble-chamber photographs in neutrino experiments, and I will focus upon just one of those problems: that of the 'neutron background'.

Neutron background is a well known source of interpretative difficulties in neutrino experiments and originates as follows. The 'active volume' of a bubble chamber – the volume of liquid in which tracks are formed and photographed – represents only a small fraction of the total mass of the detector. In Gargamelle, for example, an active volume of around 18 tonnes of freon was surrounded by

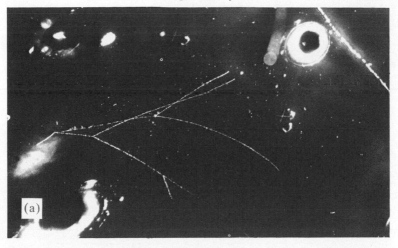

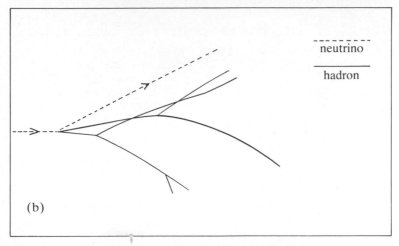

Figure 6.5. (a) Neutral-current event in Gargamelle
(reproduced by permission of CERN), and (b) Schematic
representation of the same event.

1000 tonnes of ancillary equipment (the chamber body, equipment
for controlling the temperature and pressure of the freon, a large
electromagnet, etc.). Neutrinos interact indiscriminately throughout
the detector, both within the active volume and outside it, in the
ancillary equipment. Often neutrons are amongst the products of
such interactions, and neutrons, being electrically neutral, leave the
same marks on bubble-chamber film as do neutrinos: they are

similarly invisible. Thus, there is no simple way for experimenters to distinguish between photographs of (a) genuine neutrino events within the active volume, and (b) neutron-induced events, where the neutron originates in an unseen neutrino interaction in the ancillary equipment and subsequently enters the active volume (Figure 6.6).

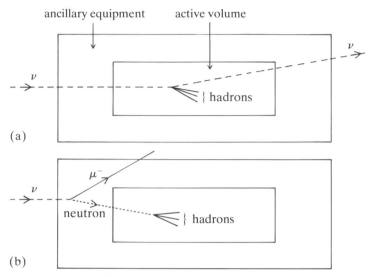

Figure 6.6. (a) Neutrino-induced neutral-current event; (b) Neutron-background event.

Events of the type shown in Figure 6.6(b) are known to neutrino experimenters as 'neutron background', and the identification and exclusion of such events is an integral part of the analysis of any neutrino experiment. This is a relatively minor task in the analysis of charged-current events, where the presence of an outgoing muon can be taken as signalling an incoming neutrino (rather than an incoming neutron). But neutral-current events are distinguished by the *absence* of an outgoing muon, and their overall characteristics are just those expected for neutron background. The problem of eliminating neutron background is correspondingly more difficult in neutral-current analysis. In what follows, I will compare and contrast how this problem was handled in the CERN neutrino bubble-chamber experiments of the 1960s and 1970s.

Neutrino Experiments in the 1960s
As noted in Section 6.4, a very tight upper limit (three per cent) for the neutral-current to charged-current ratio was quoted in 1963 from an

analysis of the first CERN neutrino experiment. How did the experimenters arrive at this figure? The first point to note is that it is not correct to assume that they failed to observe any neutral-current type events. They observed many, but suspected that they were due to neutron background. They then reasoned as follows. Neutron background events are secondary interactions, consequent upon primary neutrino events in the ancillary equipment. Thus the total energy of neutron background events will on average be less than that of genuine neutrino events. The experimenters therefore imposed an energy 'cut' upon their data (a standard HEP procedure), disregarding all events having a total energy of less than 250 MeV. Having done so, they were left with no elastic neutral-current events ($\nu p \rightarrow \nu p$) and with 31 'elastic' charged-current events ($\nu n \rightarrow \mu^- p$). These statistics translated directly into the three-per-cent upper limit.[10] It is important to emphasise that the imposition of an energy cut was a pragmatic procedure, and that the particular energy at which the cut was made represented a compromise between conflicting objectives. The higher the energy of the cut, the more confident could the experimenters feel that they had excluded the neutron background, but the poorer were the experimental statistics – more and more events of all kinds were excluded from the analysis.

One of the few explicit attempts to investigate the reliability of the cutting procedure was made by CERN experimenter E.C.M.Young. In his 1967 PhD thesis, Young attempted to *calculate* the expected flux of background neutrons in the series of neutrino experiments performed at CERN between 1963 and 1965.[11] He found that he could account for the majority, but not all, of the observed neutral-current type events. Left unaccounted for were around 150 neutral-current type events, to be compared with around 570 positively identified charged-current events: a ratio of roughly 1 'neutral-current' event to every 4 charged-current events – the same as that subsequently reported from Gargamelle.[12]

In 1967, however, the apparent neutral-current signal was ascribed to some shortcoming of Young's calculations rather than to the existence of a new phenomenon, and it remained standard practice to impose a pragmatic energy cut on the neutrino data before attempting to estimate the neutral-current signal.[13] In 1970, for example, upper limits were quoted on neutral-current rates from the CERN 1967 bubble-chamber experiments. The experimenters imposed a cut on the energy of the incoming 'neutrino', this time at 1 GeV, and found that 4 events of the elastic neutral-current type remained, and 4 which resembled single-pion production by the neutral current ($\nu p \rightarrow \nu \pi^+ n$). These they translated into upper limits of twelve per cent and eight per cent for the ratio between neutral-current and charged-current cross-sections for the corresponding processes.[14]

The Gargamelle Experiment

The interpretative procedures by which the Gargamelle experimenters deduced the existence of the weak neutral current differed significantly from those which had supported the opposite conclusion in earlier analyses. Like their predecessors, the Gargamelle experimenters imposed a cut at 1 GeV upon their data, but now a different set of pragmatic considerations had come to the fore. Because Gargamelle was much bigger than earlier CERN bubble chambers, and because it was operating in an improved neutrino beam, a relatively large number of events remained after the 1 GeV cut. The Gargamelle experimenters further enhanced their sample of candidate neutral-current events by analysing the inclusive processes ($vp \rightarrow vX$, where X is any hadron or combination of hadrons) rather than just the exclusive elastic and single-pion production channels on which earlier experiments had focused.[15]

The experimenters then proceeded in various ways to estimate the neutron background contribution to the apparent neutral-current signal, and concluded that the former could account for only around ten per cent of the latter. The remaining ninety per cent of the neutral-current events were therefore presumably genuine, and this was the basis of the Gargamelle discovery report. It is important to stress, though, that the Gargamelle estimates of the neutron background above the energy cut were essentially only refinements of those made earlier by E.C.M.Young and were similarly questionable in principle. The most detailed estimate (reported in 1975) relied upon a complex computer programme to simulate neutron production by neutrinos in the ancillary equipment of Gargamelle, the subsequent propagation of the neutrons into the active volume of Gargamelle, and their interactions therein.[16] Built into the programme were various assumptions concerning unmeasured physical parameters, and the determined critic could legitimately take issue with the assumptions (as well as with other aspects of the simulation) thus calling into question the discovery itself.[17] And, indeed, the Gargamelle discovery was regarded as controversial within the HEP community for several months after its announcement. Only in 1974, following the reported observation of the neutral current by the HPWF collaboration at Fermilab, did the discovery come to be regarded as established. And only then did the Gargamelle neutron-background simulations come to be generally accepted as an adequate representation of the experimental state of affairs. Here it might seem that, with an independent confirmation of the CERN result, we have returned to the 'scientist's account'. But, once again, matters were not so simple, and the 'scientist's account' is historically inadequate. Although I will not go into details here, the HPWF experiment was only the latest in a series of electronic neutrino experiments, and the HPWF neutral-current report was itself grounded in new and potentially questionable interpretative procedures.[18] Pragmatic interpretative shifts were intrinsic to the Gargamelle discovery and to its replication by HPWF at Fermilab. It is hard to see the work of either group as a direct and unproblematic observation of the state of nature.

Thus, in both branches of neutrino physics – bubble chamber and electronic experiments – the pattern was the same. The 1960s order, in which a particular set of interpretative procedures pointed to the non-existence of the neutral current, was displaced in the 1970s by a new order, in which a new set of interpretative procedures made the neutral current manifest. Each set of procedures was in principle questionable, and yet the HEP community chose to accept first one and then the other. Why did this transformation come about? The answer is obvious when one takes into account the dynamics of scientific practice, but it will be useful to spell it out.

The Symbiosis of Theory and Experiment

The history of the neutral current can be divided into two periods of stability separated by a brief interval of turbulence. The first stable period extended through the 1960s up to 1971, and was characterised by communal agreement (deriving principally from the kaon-decay experiments) that the neutral current did not exist. The second stable period extended from 1974 onwards, and was characterised by communal agreement that the neutral current did exist. And, within the two periods, the different interpretative procedures of the neutrino experimenters were each sustained in a mutually rewarding symbiosis of experiment and theory. In each period, experimental practice generated both justification and subject matter for theoretical practice, and vice versa. In the first period of stability, the neutrino experimenters reported data on topics of interest to theorists (principally elastic charged-current scattering) and theorists responded by elaborating their treatment of these topics, giving point to further generations of experiment. On both sides, interest in neutral currents was dormant. Experimentally, neutral-current processes were much harder to identify and interpret than charged-current processes (due to the neutron-background problem) and, theoretically, physicists working in the V–A tradition had no need for them.

There was therefore no incentive for anyone to challenge the apparently innocuous interpretative procedures which made the neutral current non-existent.

In the explosion of work on electroweak theories which followed 't Hooft's work a rift developed between gauge theorists and neutrino experimenters. Their symbiosis was temporarily broken. The theorists found themselves articulating models encompassing phenomena not known to the experimenter, and this prompted a reappraisal of the interpretative procedures current in neutrino physics. As a leading neutrino experimenter observed in 1979:

> In retrospect, it is likely that events due to neutral currents had been seen as early as 1967. Data from the CERN heavy-liquid bubble chamber . . . showed a surprisingly large number of events with hadrons in the final state, but with no visible muon. The calculations of neutron and pion-initiated background were so uncertain as to render the observed number of events inconclusive. In 1967 there was little pressure to rectify these uncertainties. Five years later the theoretical climate had changed dramatically, so that there were persistent but cautious efforts to conclusively resolve whether such events were actually anomalous.[19]

A leading theorist expressed the same sentiment more forcefully, when he noted in 1978 that neutral currents could well have been reported from the neutrino experiments of the 1960s, but that the experimenters had only begun to reconsider their interpretative procedures when 'theorists all over the world really started screaming for neutral currents.'[20]

The neutrino experimenters did reappraise their interpretative procedure in the early 1970s and, as we have seen, they succeeded in finding a new set of practices which made the neutral current manifest. The new procedures remained pragmatic and were, in principle, as open to question as the earlier ones. But, like the earlier ones, the new procedures were sustained in the 1970s within a symbiosis of theory and experiment. In adopting the new procedures, the neutrino experimenters effectively got something for nothing. They had only to take seriously the kind of neutron-background calculations performed (but not taken seriously) in the 1960s in order to bring into being a whole new phenomenon: the neutral current. They already had the tools in hand to investigate this phenomenon further, and the work of gauge theorists guaranteed that such investigations would be seen to be of the highest importance. Murray Gell-Mann noted in 1972 that 'the proposed [electroweak] models are a bonanza for experimentalists',[21] and so it proved. In the wake of the

Gargamelle discovery, programmes of neutrino experiment grew rapidly at the major PS laboratories where they constituted the vanguard of the new physics (recall that neutrino experiments were also intrinsic to the new physics by virtue of their relation to the quark–parton model: here the emphasis was on charged-current rather than neutral-current data). The discovery of the weak neutral current was, in turn, a bonanza for theorists. By accepting the discovery reports and, implicitly, the neutrino experimenters' new interpretative procedures, gauge theorists armed themselves with justification for their contemporary practice and with subject matter for future work.[22] Thus, in the years which followed the Gargamelle discovery, gauge theory and neutrino experiment traditions grew and prospered together. Theorists had the tools to elaborate the former, experimenters the latter, and the two traditions generated mutually supporting contexts for one another. The neutral current was both the medium and the product of this symbiosis, and acceptance of novel interpretative procedures was the price of its existence.

In Part III of this account we can follow the subsequent development of electroweak gauge theory and neutrino experiment along with that of the other traditions which made up the new physics. But, for the present, it is appropriate to close on a cautionary note. The discovery of the neutral current, welcome though it was, left theorists with one cloud on the horizon. The GIM mechanism created a significant difference between kaon decay and neutrino scattering experiments, making it conceivable that neutral currents should be observed in the latter but not in the former. It did this, but only at the expense of the introduction of new and unobserved charmed particles. The neutral current thus only partially completed the gauge theorists' shopping list. What became of their unfulfilled wants we shall see in Chapter 9. It is now time to return to the quark–parton model, and to see how it, too, was embraced by gauge theorists.

NOTES AND REFERENCES

1 Glashow (1980, 539).
2 I am grateful to Professor Yang for an interview and provision of biographical materials.
3 Pauli (1941).
4 The 'gauge' terminology was introduced by Hermann Weyl (see Weyl 1931) and refers to the possibility of redefining hypothetical measuring rods or 'gauges' at different points of space–time.
5 The strong interaction quantum field theories discussed in Chapter 3.3 were designed to be invariant under 'global' SU(2) transformations. That is, Lagrangians were so defined as to be unchanged if isospin coordinate

axes were transformed in exactly the same way at every point of space–time (i.e. if the function $\theta(x)$ introduced above were just a constant, independent of x). Yang's intention in modelling his theory on QED was to go beyond a global SU(2)-invariance to a local, x-dependent, SU(2)-invariance.

6 Like his interest in symmetry principles and group theory, Yang's interest in statistical mechanics was grounded in his early training in China. His MSc thesis, completed in 1944, was entitled 'Contributions to the Statistical Theory of Order–Disorder Transformations'.

7 Yang, interview.

8 The first public presentation of Yang–Mills theory was at the Washington American Physical Society Meeting on 30 April 1954 (Yang and Mills 1954a). A formal paper on the subject was submitted for publication in June and published in October 1954 (Yang and Mills 1954b).

9 Although QED is itself a gauge theory, I will use the term below to denote only theories of the Yang–Mills type in which the gauge particles are self-interacting. Technically, the difference between QED and Yang–Mills theory is that the former is invariant under Abelian gauge transformations while the latter is invariant under non-Abelian transformations. An Abelian group, like the U(1) of QED, is one in which repeated field transformations of the form $\psi \rightarrow \psi e^{ie\theta(x)}$, yield a result independent of the order in which they are performed. A non-Abelian group is one in which the result depends upon the sequence. SU(2) and SU(3) are non-Abelian groups: the functions $\theta(x)$ are respectively 2×2 and 3×3 matrices which do not commute with one another; products like $e^{ie\theta_1} e^{ie\theta_2}$ depend upon the ordering of terms.

10 Utiyama (1956).

11 Yang, interview. In 1957 Pauli similarly pestered Abdus Salam over the latter's attitude to W masses: 'There are dark points in your paper regarding the vector field . . . Every reader will realize that you deliberately conceal here something and will ask you the same question' (quoted in Salam 1980, 28).

12 Sakurai (1960). Sakurai subsequently developed his approach into the 'vector dominance' model, which played an important role in the phenomenology of the electromagnetic interactions of hadrons in the 1960s: for a review and access to the literature, see Williams (1979, 1706–14).

13 Glashow and Gell-Mann (1961).

14 My usage of the term 'electroweak' in this chapter is anachronistic. The term only became accepted in HEP in the late 1970s, following the developments covered in Chapter 10 below.

15 The development of unified electroweak gauge theories had a negligible effect on the course of HEP in the 1960s but was retrospectively ascribed great significance in the 1970s. Many good historical and conceptual reviews are now available. For a popular account, see Weinberg (1974a). For technical accounts, see Veltman (1974), Taylor (1976), Lee (1978), Sakurai (1978), Coleman (1979), Dombey (1979) and the 1979 Nobel Prize Lectures: Weinberg (1980), Salam (1980), and Glashow (1980).

16 Schwinger (1957).
17 Bludman (1958).
18 Glashow (1961).
19 Ward (1950).
20 I am grateful to Professor Salam for an interview and the provision of biographical materials. For his early contributions to the renormalisation programme, see Salam (1951a,b).
21 Shaw (1955). Learning that Yang and Mills had arrived at similar conclusions, Shaw did not publish his work (Salam, interview).
22 See Ne'eman (1983).
23 Salam and Ward (1959, 1961, 1964).
24 Komar and Salam (1960), Salam (1962).
25 I am grateful to Professor Nambu for an interview and provision of biographical materials.
26 Bardeen, Cooper and Schrieffer (1957). The BCS theory was given a particular elegant formulation by the eminent Russian theorist N.N.Bogoliubov (see Bogoliubov, Tolmachev and Shirkov 1958) and Bogoliubov's approach was most influential in HEP. For a quantitative analysis of the impact of BCS theory on the superconductivity literature, see Nadel (1981).
27 Nambu and Jona-Lasinio (1961a,b).
28 Further examples of the theoretical interplay between HEP and solid-state physics discussed below include work on the renormalisation group and the development of lattice theories (Chapter 7). The extent of the symbiosis was marked by the organisation of joint HEP–solid-state conferences (see, for example, Brézin, Gervais and Toulouse 1980).
29 The analogy with ferromagnetism did play a limited role in the development of HEP in this period. Even before the work of Nambu and Goldstone (see below), Werner Heisenberg had been attempting to develop a comprehensive theory of elementary particles drawing upon resources from this area. However, Heisenberg and his collaborators were pursuing an idiosyncratic 'nonlinear' theory and, as Nambu put it (1961a, 347), 'It is not easy to gain a clear physical insight into their results obtained by means of highly complicated mathematical techniques'. For the origins of Heisenberg's approach, see Cassidy (1981).
30 I am grateful to Professor Goldstone for an interview.
31 Goldstone (1961).
32 Goldstone, Salam and Weinberg (1962).
33 The original idea here came from Nambu and Jona-Lasinio (1961a,b). It was elaborated into the 'effective Lagrangian' approach to current algebra, in which the results of current algebra and the PCAC hypothesis were rederived and extended from a perturbative analysis of an appropriately constructed Lagrangian. For a review of this work, see Bjorken and Nauenberg (1968, 257–62).
34 Anderson (1958).
35 Anderson (1963). See also Schwinger (1962a).
36 Klein and Lee (1964), Gilbert (1964), Higgs (1964a).
37 Higgs (1964a,b, 1966), Englert and Brout (1964), Englert, Brout and Thiry (1966), Guralnik, Hagen and Kibble (1964), Kibble (1967).

38 I am grateful to Professor Higgs for several interviews, the provision of biographical materials and continuing helpful advice.

39 Strictly speaking, the Higgs mechanism involves the insertion of a negative mass-*squared* term for the scalar particles in the gauge-theory Lagrangian, from which an *imaginary* mass would be straightforwardly inferred. Also required is that the scalars be given a non-zero 'vacuum expectation value'. Physically, this corresponds to the non-zero magnetisation of a ferromagnet.

40 Weinberg (1967b). I am grateful to Professor Weinberg for an interview and for the provision of biographical materials.

41 See Weinberg (1974b, 256; 1980, 516) and Lee (1978, 150). Weinberg's electroweak model was submitted for publication in October 1967. In a paper submitted in January of that year Weinberg (1967a, 509, note 7) mentions developing an effective gauge-theory Lagrangian (see note 33 above) in which the mass-difference between the ρ and A_1 mesons was apparently generated by a kind of SSB (although their common mass was inserted by hand). It should be noted that Higgs spent the year 1965–6 on leave from Edinburgh at the University of North Carolina. He gave seminars on his work around the US, and it is clear that the Higgs mechanism was part of the theoretical culture in which Weinberg moved, even though Weinberg's route to appreciation of it depended upon his own researches.

42 The association between Weinberg and Glashow dated back to their schooldays at the Bronx High School, New York (interviews with Weinberg and Glashow). In 1967, Weinberg and Glashow were co-authors of two papers: Glashow, Weinberg and Schnitzer (1967a,b).

43 Salam (1962).

44 Guralnik, Hagen and Kibble (1964), Kibble (1967). In the latter paper, Kibble discussed the formal extension of the Higgs mechanism from QED to Yang–Mills gauge theory.

45 Salam, interview.

46 Salam (1968).

47 Coleman (1979, 1291). Equivalent figures are not available for Salam's 1968 paper, but since its place of publication was rather obscure, its citation pattern can safely be assumed to parallel that of Weinberg's paper. Certainly, when Gell-Mann summarised the meeting at which Salam presented his model he made no mention of electroweak unification (Gell-Mann 1968).

48 In their original papers, both Weinberg and Salam conjectured that spontaneously broken gauge theories might be renormalisable, but neither claimed to be able to demonstrate it. See also Salam (1962).

49 Veltman (1974, 439).

50 Veltman (1974, 429).

51 I am extremely grateful to Professor Veltman for a detailed account of his work on gauge theory and to Professor 't Hooft for an interview and provision of biographical materials. For a popular account of the renormalisation of gauge theory, see 't Hooft (1980).

52 These and the otherwise unattributed quotations below are from Veltman, private communication.

53 Schwinger (1959). For a discussion of Schwinger's argument in the context of current algebra, see Adler and Dashen (1968, 216–24).

54 Veltman (1963a,b).

55 Veltman (1966). See also Veltman (1968a).

56 Feynman (1963). See note 68 below.

57 Bell (1967).

58 Veltman (1963a) is a calculation of higher-order corrections in the electromagnetic interactions of IVBs (not necessarily gauge fields). Its particular relevance was to estimating IVB production rates in the first CERN neutrino experiments. Veltman and Bell were the resident theoretical experts at CERN while these experiments were in progress: see Bell and Veltman (1963a,b) and their contributions to Franzinetti (1963). During a visit to SLAC (Sept. 1963–April 1964) Veltman also wrote a computer programme (SCHOONSHIP) to handle the algebraic relations between the large number of diagrams appearing in higher orders of the perturbation expansion of IVB theories. SCHOONSHIP figured repeatedly in his subsequent investigations of gauge theory. Veltman (1963b) contains detailed considerations of the renormalisation of scalar field theory (specifically, in the presence of unstable particles). Veltman carried over to his work on gauge theory many of the conceptual and diagrammatic techniques developed in that paper.

59 Note that much of Veltman's work had been on the phenomenology of IVB theories of the weak interaction (see the previous note). Veltman's decision to work on the massive-by-hand theory points to the conclusion discussed in Section 6.2: mass-generation by SSB only came to be seen as fundamentally significant *after* its relation to renormalisability was established. In May 1968, Veltman heard about Salam's use of spontaneous symmetry breaking in electroweak unification from Rockefeller theorist Abraham Pais. Pais had attended the Nobel symposium where Salam presented his model, but had no details of Salam's work and Veltman forgot about it. A year or two later Glashow suggested that Veltman investigate spontaneously broken theories, to which Veltman replied 'I am not ready yet: I first have to understand this massive-by-hand case'.

60 Glashow (1959) claimed that massive vector meson theories were renormalisable, but his argument was quickly refuted by Komar and Salam (1960) (an experience which apparently deterred Glashow from further forays into renormalisation theory). Ne'eman (1983, 2, note 2) later observed: '"Current opinion" rejected quantum field theory, regarding it as hopelessly wrong . . . Hence, the quantization of Yang–Mills dynamics was finally achieved by workers who happened to be immune to that consensus' views. It required geographical remoteness (Fadeev-Popov, Veltman-'t Hooft [in the USSR and Utrecht, respectively]), professional remoteness (De Witt, working in gravitation theory) or strength of character (Feynman).' The work of the other authors mentioned by Ne'eman will be discussed below, but in the case of Veltman it is worth emphasising that he was both geographically isolated from mainstream HEP at Utrecht and professionally remote – his route to entanglement with gauge theory and the resources he

brought to bear were unique and distinctive.

61 The ghost particle was directly analogous in this respect to the unstable particles analysed in Veltman (1963b). The possibility of reformulating the Feynman rules of the theory was related to its gauge invariance. In making perturbative calculations it was necessary first to 'fix' the gauge, by adding a non-gauge invariant term to the Lagrangian. The test of a successful calculation was then to show that its results were independent of the chosen gauge. Veltman had first worked in a 'unitary gauge', in which it was clear that the Feynman rules corresponded to a unitary theory – one which conserved probability. In such a gauge, renormalisability was problematic. The rules in which the ghost particle appeared were those of a 'renormalisable gauge'. In this gauge, renormalisability was manifest, but unitarity was problematic and remained to be demonstrated. In what follows I will not distinguish between attempts to prove unitarity in renormalisable gauges and attempts to prove renormalisability in unitary gauges.

62 Veltman (1968a).

63 Veltman (private communication) mentions that amongst the theory group at Orsay were Bouchiat, Meyer and later Iliopoulos. Also visiting Orsay while Veltman was there were Mandelstam, Boulware and B.W.Lee. From 1968 onwards, the Orsay group organised a summer institute every August. Veltman regularly attended this and thus came into contact with, amongst others, Glashow, Coleman, Gross and Callan. Contributions from several of these theorists have already been discussed – others will appear later.

64 Veltman learnt of the work on massless gauge theory from two of the experts, D.Boulware and S.Mandelstam. The object of investigating massless theories was, as Veltman put it, twofold: 'either purely the mathematical amusement (Fadeev, Boulware, Mandelstam) or the understanding of gravitation (Feynman, De Witt, Fradkin)'. Veltman (1974, 439–47) discusses the work of these authors; see Feynman (1963), De Witt (1964; 1967a,b), Fadeev and Popov (1967), Fadeev (1969), Fradkin and Tyutin (1970) and Mandelstam (1968a,b). The connection between Yang–Mills theory and gravitation was that both could be derived from a principle of local gauge invariance. Theories of gravitation, however, involved massless *spin 2* fields, making them even more complex than Yang–Mills theory. Investigation of massless Yang–Mills theory was therefore regarded by the theorists mentioned above as a preliminary to the construction of a full quantum theory of gravitation.

It is important to stress that, apart from his interaction with Boulware and Mandelstam, Veltman received little encouragement for his work at Orsay. He recalls that 'Coleman once told me, literally: "Tini, you are just sweeping an odd corner of weak interactions" . . . I was generally considered as someone who was off in some odd and irrelevant direction. Some people told me so. Most people did not, but just switched off during seminars.'

65 According to Veltman (private communication) this result was accepted by the experts, but was not explicitly stated in the literature before the

work of 't Hooft discussed below.
66 Veltman (1968b, 1969).
67 Reiff and Veltman (1969). See also Slavnov and Fadeev (1970).
68 The use of mathematical quantities known as 'path integrals' to formulate quantum mechanics was pioneered by Richard Feynman in his post-war work on the renormalisation of QED. Feynman's path integral formalism was in many ways more transparent and comprehensible than conventional Lagrangian formalisms, but also involved rather dubious mathematical arguments. Feynman therefore translated his results on the renormalisation of QED into conventional terms before publishing them (see Feynman 1966). The first extended treatment of path integrals was Feynman and Hibbs (1965).

Feynman (1963) used path integrals in his work on gauge theory, and the theorists who followed him increasingly found his methods more appropriate than conventional formalisms. In the 1970s the use of path integrals in gauge theory became standard. As part of the process of mastering path integrals, Veltman offered to lecture on them at Orsay. He thus introduced many of the staff and visitors to Orsay to path integrals for the first time. He gave the same lectures (in collaboration with N.van Kampen) at Utrecht on his return there in 1969.
69 Veltman did not study Schwinger's original papers, but he did attend lectures on source theory given by H.Van Dam at Utrecht.
70 Veltman (1970). See also Boulware (1970).
71 Van Dam and Veltman (1970).
72 't Hooft (1969).
73 In their seminal paper on the PCAC hypothesis, Gell-Mann and Lévy (1960) introduced three heuristic field-theory models in which the divergence of the axial vector current was proportional to the pion field; one of these models they called the σ-model. See Adler and Dashen (1968, 24–7) for a discussion of models of this type.
74 See note 68 above.
75 't Hooft (1971a).
76 The account of the conversation is from Veltman (private communication) who remarks 'since that moment is burned in my memory I repeat here more or less verbatim how this went'.
77 't Hooft (1971b).
78 't Hooft (interview) notes that he had also learnt much from B.W.Lee's lectures on SSB and renormalisation in the σ-model at the 1970 Cargèse Summer School. Veltman (private communication) adds that 'there was a lot of σ-model going around at the Institute in Utrecht in that period'.
79 't Hooft's contribution does not appear in the *Proceedings* of the Amsterdam Conference since his paper was in press at the time.
80 't Hooft was awarded a PHD by the University of Utrecht in March 1972.
81 See *CERN Courier* (1977a). Lee died in a Chicago traffic accident, 16 June 1977.
82 See note 36 above.
83 The following account of Lee's career is based upon the detailed autobiographical narrative, Lee (1978).

Founding the New Physics

84 Lee (1969). 't Hooft attended Lee's 1970 lectures on this topic: see note 78 above.
85 Lee (1978, 152).
86 Lee (1972a), Lee and Zinn-Justin (1972a,b,c). Lee's involvement with gauge theories began when he tried to master Yang and Mills (1954) while spending the year 1960–1 at Princeton. Later, he refereed Weinberg's 1967 paper on electroweak unification, and discussed the renormalisability of gauge theory with Yang himself. Yang was not convinced of the renormalisability of gauge theory, but alerted Lee to the relevant literature on the massless theory. Lee spent the year 1968–9 at Orsay, where Veltman introduced him to the less accessible contributions to the literature.
87 Steven Weinberg (1980, 518) recalled:
   I have to admit that when I first saw 't Hooft's paper in 1971, I was not convinced that he had found the way to prove renormalizability. The trouble was not with 't Hooft, but with me: I was simply not familiar enough with the path integral formalism on which 't Hooft's work was based, and I wanted to see a derivation of the Feynman rules in 't Hooft's gauge from canonical quantization. That was soon supplied (for a limited class of gauge theories) by a paper of Ben Lee [1972a], and after Lee's paper I was ready to regard the renormalizability of the unified theory as proved.
   In 1967, Weinberg had conjectured the renormalisability of his unified electroweak model (see note 48 above) and, having read the work of 't Hooft and Lee, he published his own proof of the result (Weinberg 1971a). At Utrecht, 't Hooft and Veltman worked to improve the details of 't Hooft's proof of renormalisability. 't Hooft's new regulator method used the trick of working in 5 dimensions of space–time, and applied only to one loop diagrams. 't Hooft and Veltman (1972a) refined the method by working in a continuous (i.e. non-integral) number of dimensions, and showed that 'dimensional regularisation' could be applied to diagrams containing an arbitrary number of loops. 't Hooft and Veltman (1972b) redid the whole renormalisation procedure without relying on path integrals, which Veltman 'had learned to distrust'.
88 Weinberg (1974, 258).
89 Calculation of higher-order radiative corrections (diagrams involving closed loops) only became possible once the renormalisation procedure was well defined. This work was highly technical and did not lead to any immediate experimental repercussions: for a review, see Lee (1972b).
90 See, for example, Lee (1972b, 262–3).
91 Weinberg (1974, 258).
92 Note that first-order calculations, involving no closed loops, were sufficient to delineate the testable predictions of electroweak gauge theory. Thus while the renormalisability of the theory fuelled a great deal of purely theoretical work, it was irrelevant to the phenomenological work discussed here (except as a constraint in model-building: see note 98 below).
93 As discussed in Section 5.4, neutrino cross-sections were very small and

neutrino experiments were correspondingly technically difficult. Neutrino–electron cross-sections were expected to be especially small: theory (V–A and ws) suggested that cross-sections were proportional to CM energy, and relativistic kinematics dictated that, for a given neutrino beam energy, the CM energy for neutrino–electron scattering was much smaller than that for neutrino–nucleon scattering.

94 Weinberg (1972a).

95 For a historical account of the development of the kaon decay anomaly, references to the original literature, and a quantitative assessment of its impact in weak interaction physics, see Koester, Sullivan and White (1982).

96 Glashow, Iliopoulos and Maiani (1970).

97 Hara (1964), Maki and Ohnuki (1964), Amati, Bacry, Nuyts and Prentki (1964), Bjorken and Glashow (1964).

98 Crane (1980a, 42, Fig. 5) counts the combined citations per year in the HEP literature to Weinberg (1967b), Glashow, Iliopoulos and Maiani (1970), 't Hooft (1971b) and Lee and Zinn-Justin (1972a,b). Until 1970 there were no citations; in 1970, 3; in 1971, 6; and in 1972, 80. Thereafter the citation rate rose linearly, reaching more than 360/year in 1976 (the last year of Crane's study).

One rather esoteric point concerning the GIM mechanism should be noted here. Prior to the rise of gauge theory, Adler (1969) and Bell and Jackiw (1969) noted a problem in quantum field theories of fermions interacting with axial vector currents (for example, leptons or quarks interacting via the axial weak current). The details of the so-called 'Adler–Bell–Jackiw (ABJ) anomaly' need not concern us, except to note that it threatened to spoil the renormalisability of particular electroweak models. Members of the Orsay theory group, Bouchiat, Iliopoulos and Meyer (1972), showed that when u, d and s quarks were included, the ws model did become non-renormalisable due to the ABJ anomaly. They also showed, however, that with the addition of appropriate extra quark or lepton fields the anomaly could be cancelled, and that one appropriate choice was that of GIM (provided that the quarks were also 'coloured' – see Chapter 7). Since renormalisability was the key to the promise of electroweak gauge theories, this did much to enhance the status of GIM, and all subsequent models were so constructed as to avoid the ABJ anomaly.

99 No information on strangeness-conserving neutral-current interactions was available from observation of hadron decays, because strangeness-conserving decays were dominated by the *strong* interaction.

100 Bingham *et al.* (1963, 555, 569).

1 Bell, Løvseth and Veltman (1963, 586).

2 Cundy *et al.* (1970).

3 Lee (1972c), Prentki and Zumino (1972), Georgi and Glashow (1972). The particular interest of theorists at Fermilab and CERN related to programmes of neutrino experiment then under way: see below. B.Lee (1972d) also noted that the ws model was in conflict with neutral-current upper limits derived from a reanalysis of the second BNL neutrino experiment (which ran between 1963 and 1965). Confusingly enough,

the reanalysis was performed by HEP experimenter W.Lee (1972).

4 For extended historical accounts and detailed documentation of the events discussed below and in the following section, see Galison (1983) and Pickering (1984). For illuminating quantitative analyses of the relevant HEP literature, see Sullivan, Koester, White and Kern (1980) and Koester, Sullivan and White (1982).

5 The HPWF experiment was an electronic experiment, originally designed to trigger (i.e. record events) only on the muons produced in charged-current events of the type $vp \rightarrow \mu X$. Since the distinguishing mark of a neutral-current event is that no muon is produced – the neutrino emerges unscathed – neutral currents would have been invisible to this apparatus. On Weinberg's prompting, the apparatus was redesigned to trigger on hadrons – the shower of particles (X) resulting from any neutrino event, whether of the charged- or neutral-current type.

6 Hasert et al. (1973b). A more detailed account of the discovery was submitted for publication in January of the following year: Hasert et al. (1974). In the interest of completeness, it should be noted that besides these observations of hadronic neutral currents (i.e. neutral-current interactions between neutrinos and hadrons) the Gargamelle group also reported the discovery of purely leptonic neutral currents (neutral-current interactions between neutrinos and electrons, of the type shown in Figure 6.2(b)). Hasert et al. (1973a) reported one event of the latter type found in an examination of more than 700,000 pictures. In the text I will focus upon the discovery of the hadronic neutral current since this had the more immediate impact upon the HEP community (not surprisingly, since a single event would not be routinely acceptable as evidence for the existence of a major new phenomenon): see Figures 3, 4 and 5 of the quantitative cocitation analysis of the weak interaction literature performed by Sullivan, Koester, White and Kern (1980).

7 Early reports from the HPWF group were confused. When the Garga-melle discovery was first announced, HPWF reported that they could confirm it. Later in 1973, they came to the conclusion that they had evidence against the existence of the neutral current. Finally, in early 1974, HPWF submitted positive findings for publication: Benvenuti et al. (1974), Aubert et al. (1974a,b) (for more details of this sequence, see Galison 1983). For the Caltech–Fermilab observations, see Barish et al. (1975). The latter group also reported upon a negative search for heavy leptons of the type required by alternatives to the WS model (Barish et al. 1974), as did the Gargamelle group (Eichten et al. 1973b). For a popular account of the CERN and Fermilab observations, see Cline, Mann and Rubbia (1974).

8 For the rise of the WS model and the demise of its alternatives after the Gargamelle discovery, see the time-series co-citation analyses of the weak interaction literature given in Sullivan et al. (1980) and Koester et al. (1982).

9 I single out the CERN programme of neutrino bubble-chamber experiments for discussion because it was this programme which culminated in the neutral-current discovery. As noted below, the same analysis could be given of bubble-chamber experiments elsewhere, and of the develop-

ment of electronic neutrino experiments.

10 Bingham *et al.* (1963, 569).

11 Young (1967).

12 Young did not state this conclusion explicitly, but it follows directly from his background estimate (Young 1967, 58) and the counts of different types of events (Young 1967, 39, Table 2.3). In an unpublished 1972 memorandum one of the leading CERN neutrino experimenters, D. H. Perkins, reported a similar result: see Llewellyn Smith (1974, 460).

13 Having noted that neutron-production by neutrino interactions in the ancillary equipment could not account for the observed number of neutral-current candidates, Young (1967, 58) identified the remaining candidates as 'presumably due to leakage effects from the primary proton beam'.

14 Cundy *et al.* (1970).

15 This change in focus from exclusive to inclusive neutral-current events paralleled that earlier fostered in the analysis of charged-current events by the development of the quark–parton model (see Chapter 5.4). Advance predictions of inclusive neutral-current rates using the quark–parton model were given by Pais and Treiman (1972) and Paschos and Wolfenstein (1973). Before the Gargamelle discovery had been officially announced, one of the experimenters involved, Robert Palmer (a visitor from Brookhaven) attempted to precipitate events by submitting a paper on the parton-model interpretation of the neutral-current data (Palmer 1973). Publication was delayed, though, until after the official report had been published. I am grateful to Professor Palmer for an interview and provision of biographical materials.

A further difference between the cutting procedures of the Gargamelle group and their predecessors can be noted. In the 1960s, the experimenters made the cut on the 'visible' energy of each event: the total energy assigned to *all* of the tracks on film. Thus the visible energy of a neutral-current candidate corresponded to the sum of the energies of all of the visible hadron tracks, while the visible energy of a charged-current event corresponded to the sum of hadron energies plus the energy of the associated muon. This procedure made sense if one assumed that most of the neutral-current candidate events were in fact neutron background. But if they were genuine, then it was unfair to the neutral current. Part of the true energy of neutral-current events was carried off unseen by the emergent neutrino, implying that the visible energy of a neutral-current candidate was less than its true energy. In turn, since the intensity of neutrino beams fell-off sharply with increasing energies, this implied that comparisons of neutral- and charged-current rates based on a visible energy cut would badly underestimate the proportional importance of the neutral current. The Gargamelle experimenters abandoned the visible energy cut, choosing instead to make a cut on hadron energy. Here they calculated only the total energy of hadron tracks, and did not include the energy of the outgoing muon in charged-current events. This procedure made sense if the majority of neutral-current candidates were not neutron-background events but genuine.

16 Fry and Haidt (1975).

17 For a more detailed discussion of this point, see Pickering (1984). I should make it clear that in pointing to the potential for argument against the Gargamelle interpretative procedures I have no intention of suggesting that the experimenters acted in bad faith or that their achievement was non-trivial. Certainly, members of the Gargamelle group invested a great deal of effort in making their neutron background estimates as convincing as possible, drawing extensively upon the limited stocks of established knowledge concerning the relevant physical processes. My argument is simply that even after the investment of an unprecedented interpretative effort, the neutron-background simulation remained questionable (as the subsequent debate showed). Thus one can see that in a different overall cultural context (for example, that which prevailed in HEP in the 1960s) the Gargamelle photographs could have been ascribed a different phenomenal significance from that which they historically acquired.

18 See Pickering (1984).

19 Sciulli (1979, 46).

20 Sakurai (1978, 45). Sakurai argued that neutral currents could have been reported from 1960s neutrino experiments at both CERN and Brookhaven, and also from a 1970 'beam dump' experiment at SLAC. Neutral-current type events were observed at SLAC but were as usual dismissed as presumably due to neutron background.

21 Gell-Mann (1972, 336).

22 Co-citation analyses of the weak-interaction literature for the period from early 1972 to mid-1975 are given in Sullivan et al. (1980) and Koester et al. (1982). The bulk of this literature consists of theoretical rather than experimental articles (a ratio of around 4 to 1 in favour of the former is indicated in Sullivan, White and Barboni 1977: compare their Figure 1, p. 177, with Figure 2, p. 178) and the central position occupied in these co-citation plots by the Gargamelle and HPWF neutral-current reports is therefore a direct reflection of the use made of them by theorists.

7

QUANTUM CHROMODYNAMICS:
A GAUGE THEORY OF
THE STRONG INTERACTIONS

In the previous chapter we discussed the renormalisation of gauge theory and the consequent development of unified models of the electroweak interactions. In this chapter we will be looking at the construction of a gauge theory of the strong interactions – quantum chromodynamics. In Section 1, we return to the discussion of deep-inelastic electron scattering begun in Chapter 5. Here we can examine the specific response of field theorists to scaling. Their work culminated in the discovery in 1973 that gauge theory was 'asymptotically free'. The implication of this discovery appeared to be that gauge theories were the only field theories capable of underwriting the phenomenological success of the quark–parton model and hence of giving an explanation for scaling. In Section 2, we will examine the particular choice of gauge group made by field theorists in the construction of a realistic theory of the strong interactions. The chosen group was an su(3) of quark 'colours', invented by Greenberg, Han and Nambu in the mid-1960s. We will see that this choice was advocated by Gell-Mann and others even before asymptotic freedom was discovered. The gauge theory of coloured quarks and gluons was named quantum chromodynamics (QCD) in recognition of its explicit analogy with quantum electrodynamics. Although the construction of QCD was to have far-reaching implications for the development of HEP, in 1973 and 1974 it remained of interest only to field theorists. In Section 3 we can discuss why this was. We will see that a central obstacle to further development was the lack of any constructive approach to the problem of quark 'confinement' in QCD. We will also see that although QCD predicted deviations from the exact scaling behaviour predicted by the parton model, there was no prospect of immediate experimental investigation of those deviations, and thus little incentive for theorists to explore them in depth.

7.1 From Scale Invariance to Asymptotic Freedom
We saw in Chapter 5 that scaling was quickly recognised to be a novel phenomenon in HEP and as such provided a congenial topic for theoretical improvisation. Whilst the quark–parton model consti-

tuted a straightforward and intuitively appealing explanation of scaling, it was, in the late 1960s and early 1970s, by no means the only theoretical approach to the phenomenon. Considerable numbers of theorists of all persuasions attempted to bring their expertise to bear, generating a variety of different models of scaling.[1] In this section we will be looking at the field-theoretic approach which was eventually to prove most fruitful, an approach based on the concept of 'scale' or 'dilatation' invariance.[2]

In essence, scale invariance was a very simple concept, based upon the technique of 'dimensional analysis' common throughout physics. The idea of dimensional analysis is that one should only compare like with like. Thus, for instance, if one side of an equation has the dimensions of a mass (e.g. a weight measured in kilograms) then so should the other side. It would make no sense to equate a quantity measured in kilograms, say, to another measured in degrees Fahrenheit. Using reasoning like this it is often possible in physics to decide how particular quantities can depend upon the relevant variables. In deep-inelastic scattering the analysis went as follows. The deep-inelastic structure functions (W_1 and vW_2) have no dimensions – they are just dimensionless numbers. Therefore, inasmuch as they depend upon dimensional variables, the structure functions can only depend upon dimensionless ratios of those variables. Now, the two kinematic variables conventionally chosen to characterise deep-inelastic scattering are q^2 and v (the square of the momentum transfer between the incoming electron and target proton, and the energy loss of the electron respectively). q^2 is measured in units of mass-squared, while v is measured in units of mass. If one makes use of another quantity having dimensions of mass, for example the mass of the proton M, one can form a dimensionless ratio Mv/q^2. If this were the only dimensionless ratio which could be formed from the variables characterising deep-inelastic scattering, then it would be the only quantity upon which the structure functions could depend. And, since Mv/q^2 is the scaling variable ω, dependence upon this quantity is equivalent to exact scaling.

Theorists could, then, explain scaling on the assumption that ω was the only relevant dimensionless variable in deep-inelastic scattering. The problem remained of how to justify this assumption. Other dimensionless variables could be constructed from q^2, v and $M - q^2/M^2$ and v/M – which depended upon q^2 and v independently of one another. If the structure functions depended upon the latter quantities, they would not scale. To avoid this conclusion, theorists made the plausible assumption of 'scale invariance': when q^2 and Mv each became asymptotically large, the ratio Mv/q^2 remaining finite (the Bjorken limit) then M itself would be effectively zero in comparison and could be ignored. No significant mass-scale would remain in calculations, and ω would be the only important variable. Scale invariance could therefore be understood as approximately valid at high-energy and for large momentum transfer – that is, as an asymptotic phenomenon.

Remarkably, when in the late 1960s field theorists attempted to carry this logic through in detailed calculations, they found that it failed. The only theories which reproduced the required asymptotic behaviour were free-field theories – field theories of non-interacting particles. That free-field theories displayed such behaviour was no surprise – the parton model could itself be regarded as a free-field theory, since the partons did not interact with one another – but that no alternatives could be found was a great disappointment. The entire problem with the parton model was its unrealistic insistence on treating the quark-partons as independent entities, and its consequent inability to offer any clue as to how quarks were bound together. Field theorists had hoped to do better.

However, the way in which scale invariance broke down in field theory was instructive. Theorists confined their attention to renormalisable theories and it was with the renormalisation procedure that problems arose. As discussed earlier, renormalisation was a procedure designed to circumvent the numerical infinities which arose in field theory from mathematical integrations over infinite ranges of momenta. The infinities were handled in practical calculations by restricting integrations to a finite range, introducing a cut-off momentum (Λ) beyond which the integrand was taken to be zero. In renormalisable theories this tactic made sense because at the end of the calculation one could take Λ to be infinite and still arrange to obtain finite results. But Λ was a dimensional parameter (typically, a large mass) which was of necessity larger than v or q^2 in any meaningful calculation, and therefore non-negligible even in the asymptotic regime (in contrast to fixed masses like M). Thus the presence of Λ in field-theory calculations led to the prediction of large asymptotic violations of scale invariance. And hence, in the late 1960s and early 1970s it appeared that the field-theory approach to scaling was sterile, at least as regards the understanding of the phenomena observed at SLAC.

Scale invariance and its breakdown had, though, become something of a specialty in its own right amongst field theorists, following the publication in 1969 of a seminal paper by Kenneth G. Wilson. Since Wilson's work on field theory was both important and distinctive a few words on his background will be useful here.[3] Wilson majored in mathematics at Harvard in 1956 before embarking on a career in theoretical HEP as a postgraduate student at Caltech. His supervisor there, Murray Gell-Mann, suggested several thesis topics, but Wilson found them unsatisfying; he wanted to work on something 'very long term . . . something where it would take a long time to get results'. He studied quantum field theory and the desired very long-term problem began to emerge. Wilson decided to attempt to develop some field-theoretic understanding of the strong interactions – at just the time when most established field theorists were giving up and defecting to the S-matrix.

In 1959 Wilson returned to Harvard, and in 1963 he accepted a tenured position at Cornell University. While at Harvard, Wilson recalls that S-matrix theory was 'all the rage' but that he himself turned to work on 'strongly coupled' field theory.[4] The original papers on this topic had been written in the early 1940's by such authors as Pauli, Wentzel and Dancoff. Strongly coupled field theory was so called because the coupling, or interaction strength, between the basic fields was large, as appropriate to the strong interactions. The significant point was that the perturbative Feynman-diagram approach was useless for such theories (perturbative approximations require a small coupling parameter in which to expand, such as the charge of the electron in QED) and Wilson set out to devise novel, non-perturbative, calculational methods. Thus, by the time he moved to Cornell, Wilson was isolated from all of the contemporary trends of theoretical HEP: not only was he actively working on field theories of the strong interactions when most of his colleagues had given up, but he had abandoned the perturbative techniques which were the conventional mainstay of field theory itself. Not surprisingly, his research proceeded slowly. By 1968, nine years after completing his PhD, Wilson had published five papers, none of which had received a great deal of attention. His sixth paper, though, came out in 1969 and was quickly acknowledged as an important contribution to theoretical physics.[5] It was not an easy paper to read. I will not attempt to describe its arguments, but I will describe what other HEP theorists took from it: the 'operator product expansion' and the concept of 'anomalous dimensions'.

First the operator product expansion, or OPE. In field theory, quantities such as the electromagnetic current are represented by mathematical constructs known as operators. In 1969 it was pointed out that deep-inelastic scattering could be represented field-theoretically in terms of the product of two such operators acting at points separated by a short distance.[6] According to Wilson's OPE, this product of currents could be replaced by a sum of single operators, and the OPE thus provided a conceptual framework for the field-theoretic analysis of scaling. The idea of anomalous dimensions came into the analysis in the following way. The 'canonical' dimensions of the operators appearing in the OPE were those given by straightforward dimensional analysis. This procedure led immediately to the prediction that scaling should hold asymptotically. Against this, Wilson argued that the dimensions of operators would inevitably be modified in an *interacting* field theory. Operators should have 'anomalous' dimensions, differing from those given by dimensional analysis, and thus scaling should not hold.

Wilson arrived at the ideas of the OPE and anomalous dimensions via an idiosyncratic route and he formulated his arguments within a framework he had himself constructed. Few theorists were inclined to attempt to adopt Wilson's novel methods. Instead, they took over his results and attempted to recast them in the conventional language of perturbative field theory. In the late 1960s and early 1970s, this attempt took the form of a variety of field-theoretic studies of the OPE and analyses of scaling in simple field theories. Such work both corroborated Wilson's conclusions – no realistic interacting field theory which predicted scaling could be found – and contributed significantly to the resurgence of an active field-theory tradition in HEP.[7]

The Renormalisation Group
In 1969, too, certain other ideas of Wilson were coming to fruition. Throughout the 1960's he had been exploring, in his own characteristic way, a 'renormalisation group' (RG) analysis of strongly coupled field theory. The RG was a way of relating the predictions of field theory at different energy scales, so that if one knew the predictions of a theory in one range of energies one could calculate its predictions for other ranges. Typically, and especially for strongly coupled theories, one would not have a clear idea of the predictions of the theory in any energy range, but that did not imply that the RG was completely useless. The possibility remained that the RG transformation between different energy scales would itself constrain the predictions of the theory in particular limits – either asymptotically high or low energies, say.

The RG idea had been developed in investigations of QED in the 1950s by Gell-Mann (Wilson's supervisor at Caltech) and others,[8] but there it had languished until it was taken up again by Wilson. In the 1960s his interests primarily concerned the strong interactions of elementary particles but, through his contacts with solid-state physicists at Cornell and elsewhere, he came to believe that the problems he was engaging had direct analogues in solid-state physics.

211

In particular, he found that the RG techniques he had developed could be applied to the analysis of phase transitions in, for example, ferromagnets.[9] Furthermore, since solid-state phenomena were non-relativistic they were more readily amenable to concrete computation than those encompassed by the theories of HEP. Wilson was invited by his colleagues to give a solid-state seminar at Cornell in the fall of 1969. At that time ferromagnetic phase transitions had already been recognised as a scaling phenomenon subject to dimensional analysis, analogous to the scaling known in HEP. Also phase transitions were characterised by experimentally measured numbers known as critical exponents, which were known not to agree with the results of straightforward dimensional analysis. Wilson realised that the critical exponents of solid-state physics could be regarded as analogues of the anomalous dimensions of HEP, and that he could explicitly calculate them. Working with a very crude field-theory model of a ferromagnetic system, he made the calculation in preparation for the seminar, and found that his results were in good agreement with existing data. This was a very important development, since it constituted a quite new and original way of understanding critical exponents and the characteristics of phase transitions in general.

The first detailed exposition of Wilson's approach was published in 1971, although his seminal contribution to solid-state physics was not widely appreciated until Wilson and Fisher (a solid-state colleague of Wilson's at Cornell) published their work on the 'ε-expansion' in 1972.[10] The ε-expansion was a calculational technique based upon the idea of treating the number of dimensions of space and time as a continuous variable rather than as an integer (i.e. 4). The point of this strategem was that it allowed one to do numerical calculations using a perturbative Feynman-diagram approach. As Wilson later commented, this was 'more in line with what people were trained to think about', and greatly increased the accessibility of his work. Wilson's RG analysis of phase transitions blossomed into a major tradition of work in solid-state physics which was itself sufficient to attract the attention of HEP theorists in the early 1970s.[11] Wilson also published papers illustrating how the RG approach could illuminate the properties of simple, if unrealistic models of the strong interaction.[12] Characteristically, the HEP community showed no inclination to adopt Wilson's methods; once again, theorists set about rewriting his analysis in a format more congenial to conventional perturbative methods.

Asymptotic Freedom

The basic tool in the perturbative approach to RG analyses of quantum field theory was the so-called 'renormalisation group equation', first written down in 1970 by Curtis Callan (of the Institute of Advanced Study in Princeton) and Kurt Symanzik (from the DESY theory group in Hamburg).[13] Callan and Symanzik were both working within the tradition of field-theory investigations of scale invariance. Both also acknowledged the contributions to their work of Sidney Coleman of Harvard, and it was Coleman who did much to explicate and popularise the RG equation, particularly with his lectures at the Erice Summer School in 1973.[14] With the renormalisation group equation we have arrived at the crux of this chapter, and I will describe the conceptual use to which the equation was put before returning to the historical narrative.

The RG equation was a differential equation describing how the predictions of field theories changed in going from one momentum scale to another. The RG equation enabled such scale transformations to be understood in terms of changes in an effective coupling constant, 'g', of the underlying field theory, the behaviour of g being determined by a mathematical function $\beta(g)$. In general the precise form of this function was unknown, but for small values of g it could be calculated using conventional perturbative techniques. Now, the RG equation was appropriate to the analysis of any renormalisable field theory and, as we saw in the previous chapter, it became clear in the early 1970s that Yang–Mills gauge theories were members of that class. When the RG equation was applied to gauge theories a remarkable and unique property emerged which became known as 'asymptotic freedom'. Theorists found that when $\beta(g)$ was computed perturbatively in gauge theory, its form was such that at higher and higher momenta the effective coupling constant g became smaller and smaller, tending asymptotically to zero. In other words, at high momenta gauge theory behaved more and more like a free, non-interacting, field theory: gauge theory was, as theorists said, asymptotically free. And this, of course, was just the behaviour required to explain the success of the parton model and the scaling phenomena in deep-inelastic scattering – experimental observation of which had provoked this entire field-theoretic quest. Field theorists had, with the discovery of the asymptotic freedom of gauge theory, arrived at their goal: they had in their possession a genuine, realistic, and in certain respects calculable field theory of the strong interactions. Retrospectively speaking, the discovery that gauge theory was asymptotically free was one of the key conceptual advances in the history of the new physics, although, as we shall see below, it was not immediately perceived as such by the HEP community at large.

That the gauge theory β-function had the required form for asymptotic freedom was first remarked by Gerard 't Hooft at the Marseille conference on Yang–Mills Fields in June 1972. However, 't Hooft's remark went unpublished, and he himself did not appreciate the significance of the form of the β-function since he was accustomed at that time to working in the newly-flourishing tradition oriented towards the weak rather than the strong interactions.[15] The following year asymptotic freedom was rediscovered by two American graduate students, Frank Wilcek (at Princeton University) and David Politzer (at Harvard). Wilcek's supervisor at Princeton was David Gross.[16] Gross had already made important contributions to the understanding of deep-inelastic scaling and, coming from a field-theoretic milieu (his first post-doctoral position, like many other theorists in this story, was as a Junior Fellow at Harvard) he was interested in looking for field-theoretic explanations of the phenomenon. In 1972 Gross had begun to familiarise himself with renormalisation group techniques, learning from Curtis Callan who also had a position in Princeton and from Wilson who was spending a year on leave there. With the RG techniques in hand, Gross set out to investigate whether any renormalisable field theory could possibly explain scaling. With some assistance from Sidney Coleman who was a frequent visitor to Princeton, he concluded that all such theories, with the possible exception of gauge theories, were excluded. Gross suggested the problem of investigating the behaviour of the latter theories to Frank Wilcek. Wilcek at first concluded that gauge theories were, like other theories, not asymptotically free, but he soon discovered a sign error in his calculations. Thus was asymptotic freedom discovered at Princeton.

The discovery at Harvard was somewhat less coherent. Politzer was a graduate student of Coleman.[17] In early 1973 he was in his fourth year of graduate study and was feeling 'rather depressed', believing that he had accumulated insufficient material for a PhD thesis. He recalls that at the time there was great excitement around Harvard over the unified electroweak gauge theories, and that he had also just learnt to appreciate the power of renormalisation group arguments from a paper on spontaneous symmetry breaking by Coleman and Eric Weinberg (another of Coleman's students).[18] Putting two and two together, Politzer decided to do a renormalisation group analysis of gauge theory. Politzer then discovered that the β-function had the appropriate form for asymptotic freedom. However, he was not at that time thinking in terms of the strong interactions or scaling and he was at first dismayed by this result. It suggested that the *low energy* behaviour of gauge theories was not computable (see below). He had in mind here electroweak gauge theories, not surprisingly since little work on the strong interactions had been done in Harvard for several years. But prompted by his recollection of a paper by Symanzik on the asymptotic freedom of a simple model field theory,[19] Politzer soon realised that the form of the β-function meant that gauge theory was 'something to do with the strong interactions'. At this point Politzer became highly elated and announced to his supervisor 'Hey Sidney, this is stupendous'. Sidney leant a temporary check to his enthusiasm by relaying the information that David Gross believed that gauge theory was not asymptotically free (Wilcek had yet to discover his mistake). Politzer left Harvard for the spring vacation, rechecked his calculations, found them accurate and returned to Harvard to tell Coleman so. Coleman already knew – Wilcek had found his mistake – and papers by Politzer, Gross and Wilcek on the asymptotic freedom of Yang–Mills gauge theory appeared consecutively in the pages of *Physical Review Letters* in June 1973.[20]

The final part of this corner of the jigsaw was contributed by Coleman and Gross who, developing their earlier investigations of the asymptotic properties of renormalisable field theories, showed by straightforward enumeration that no theory without gauge fields was asymptotically free.[21] Thus, if one wanted a field theory of the strong interactions, it appeared that it had better be a gauge theory.

7.2 Quantum Chromodynamics

Once asymptotic freedom had been discovered, two questions remained to be decided in the construction of a realistic candidate gauge theory of the strong interaction: what were the fundamental fields which appeared in the Lagrangian of the theory, and under what symmetry group was the Langrangian invariant? An obvious conjecture, given that gauge theory reproduced the predictions of the quark–parton model, was that the fundamental fields were quark fields. In a gauge theory, the quark fields would interact with one another via the exchange of gauge vector fields, and an equally obvious conjecture was that the gauge vectors were the enigmatic gluons already put to use in parton-model phenomenology.

The choice of gauge group was not so obvious. A superficially tempting conjecture was that the theory should be invariant under local transformations of the su(3), Eightfold Way, symmetry group. But the consequences of this choice were disastrous. The weak interactions were already regarded as acting upon the su(3) quantum numbers of quarks: the charged weak current changed u quarks into d quarks, for example. It was therefore apparent that if the strong interaction acted upon the same set of quark quantum numbers the invariance properties of the weak interaction would spill over and mar those of the strong interaction. Since the invariance properties of the two interactions were quite different, theorists did not want this to happen.

To escape the intermingling of the weak and strong interactions, gauge theorists had resort to a stratagem. They argued that quarks carried not one but two sets of quantum numbers, which they referred to as 'flavours' and 'colours' (the choice of names had no physical significance). The flavours were the standard quantum numbers of the Eightfold Way which distinguished between u, d and s quarks. The colours were a second set of su(3) quantum numbers, say red, yellow and blue. Each quark colour was supposed to come in three flavours and vice versa, making nine quarks in all: there were red u quarks, blue u quarks, yellow u quarks, and so on. The two sets of quantum numbers were supposed to be quite independent. The electroweak gauge theories discussed in the previous chapter corresponded to a local gauge invariance of quark flavours, and the gauge particles required for such invariance were the IVBs of the weak and electromagnetic interactions. The strong interactions, in contrast, were assumed to arise from a separate local invariance under transformations of quark colours, and the theoretically required gauge particles were in this case an octet of coloured gluons. The introduction of colour, then, served to decouple the weak and strong interactions in gauge theory, and enabled theorists to respect their different conservation laws. But why should theorists have found it reasonable to assert that quarks came in three colours? The choice was not as arbitrary as it might seem.

Colour

To understand the origins of colour we must go back to the early days of the constituent quark model. We have already seen that several objections of theoretical principle could be raised against the CQM, and we now come to another. This concerned an apparent conflict between the CQM and the so-called 'spin-statistics' theorem, itself an extension of the Exclusion Principle invented by Wolfgang Pauli. The spin-statistics theorem referred to the quantum-mechanical wave functions used to describe multiparticle systems. It asserted that if the particles in question were fermions (half-integral spin), then the wave-function must be antisymmetric: it must change sign when pairs of particles are interchanged (this was the Exclusion Principle). If the particles were bosons (integral spin), then the wave-function must by symmetric: it must remain the same under particle interchange. The Exclusion Principle was experimentally well established in the domains of atomic and nuclear physics, and the spin-statistics theorem could be proven on very general grounds in quantum field theory. The theorem was therefore expected to apply to quarks as well. Quarks were fermions – they had spin $\frac{1}{2}$ – and thus the wave-functions of quark composites were supposed to be antisymmetric under constituent interchange. But it was found difficult to respect this condition in straightforward articulations of the CQM.

A discussion of the $\frac{3}{2}^+$ baryon decuplet (shown in Figure 3.3) will serve to illustrate the problems which arose. CQM wave-functions could be represented as products of three independent components: a component referring to the SU(3) quantum numbers of the constituent quarks; a component referring to the quark spins; and a component referring to the spatial distribution of the quarks relative to one another. In the $\frac{3}{2}^+$ decuplet, the first two components of the CQM wave-functions were of necessity symmetric.[22] Now, the $\frac{3}{2}^+$ decuplet was supposed to be the ground-state – the state of minimum energy – for the relevant quark configuration, and in all known composite systems the space-dependent part of the ground-state wave-function was symmetric.[23] Thus, if the $\frac{3}{2}^+$ decuplet followed the pattern of other composite systems, all three components of the wave-function, and hence the overall wave-function itself, had to be symmetric. Despite the fact that quarks were supposed to have spin $\frac{1}{2}$ the obvious CQM wave-function for the $\frac{3}{2}^+$ decuplet was symmetric, rather than antisymmetric as required by the Exclusion Principle. And, in general, it seemed that CQM wave-functions should be those appropriate to bosonic rather than fermionic quarks if a satisfactory phenomenological description of the hadrons were to be attained.

That spin $\frac{1}{2}$ quarks should be treated as bosons in the CQM made little sense to theorists, and three main avenues of escape from this conclusion were pursued. One was to assume that the space-dependence of the interquark force was somehow pathological, and that because of this the state of lowest energy, the ground-state, was antisymmetric under quark interchange. This was tenable but not popular; no clear-cut choice of such a pathological force existed and such forces inevitably made concrete calculations difficult. A second option was to forget about the Exclusion Principle. This was in accordance with the rough-and-ready philosophy of the constituent-quark model and was, in effect, what CQM phenomenologists did. Wave-functions were computed assuming a simple form for the interquark force (the so-called simple harmonic oscillator potential), the ground-state was symmetric, the model worked, and the Exclusion Principle went by the board.

There was, though, a third option, favoured by physicists of more refined theoretical taste. This option was first laid out in October 1964 by O.W.Greenberg from the University of Maryland.[24] In the early 1960s Greenberg had been possessed of a solution in search of a problem. From his days as a post-graduate student at Princeton University in the late 1950s he had worked on axiomatic field theory, and the tenor of his research was to 'stretch' field theory by the invention of new mathematical constructs. One such construct he called 'parastatistics'. Parastatistics, as formulated by Greenberg, formed a classification system of fields which went beyond the established fermion – boson dichotomy. As early as 1962, he had presented an invited conference paper on parastatistics in axiomatic field theory,[25] and in 1964 he began to publish a series of articles on the subject, written in collaboration with A.M.L.Messiah.[26] The only drawback with this sophisticated theoretical work was that it seemed quite unnecessary: at the time it appeared that all of the known particles were either fermions or bosons. Greenberg and Messiah admitted as much, referring in the title of one of their articles to 'the absence of paraparticles in nature'.

In the autumn of 1964, Greenberg moved to the Institute of Advanced Study at Princeton, on leave of absence from Maryland. While at Princeton he intended to work, like so many theorists of the time, on SU(6) (which he had first heard of in a telephone conversation with Gürsey). Constructing quark-model wave-functions, he encountered for himself the problem with the Exclusion Principle just discussed, and realised that parastatistics could solve it: quarks could be the paraparticles which had hitherto been absent in nature. There followed three weeks of work 'night and day', culminating in the publication of a 'paraquark' model of hadrons.[27]

Greenberg's solution to the problem of quark statistics was elegant but, emanating as it did from the depths of formal field theory, obscure to many particle physicists. This obscurity was overcome in early 1965 in a more transparent but essentially equivalent formulation given by two Japanese theorists, Yoichiro Nambu of the University of Chicago and M.Y.Han, then a graduate student at Syracuse University.[28] The prime mover of this collaboration was Nambu,[29] whose seminal work in the early 1960s on spontaneous symmetry breaking was discussed in the previous chapter. In his work on spontaneous symmetry breaking, Nambu had represented hadrons as composites of fundamental fermions, and when the constituent quark model was proposed in 1964 he took an active part in its elaboration. He addressed himself to the quark-statistics problem, and although he knew of Greenberg's work he felt that it was 'very formal'. He also knew of the 1964 publication of CERN theorists Bacry, Nuyts and van Hove which advocated the existence of not one triplet of quarks but two, and had actually constructed a similar model himself.[30] The CERN theorists had doubled the number of quarks in order to avoid the fractional charge assignment – with two triplets all quarks could be given electric charge zero or one. Nambu realised that one could preserve these integral charges and solve the statistics problem if one only went a little further. He proposed that there was not one triplet of quarks, not two, but three. These nine quarks should manifest a double SU(3) symmetry, the observed flavour hadronic SU(3) being joined by a colour SU(3) as discussed above.[31] The quark colours provided an extra degree of freedom which one could use to antisymmetrise an otherwise symmetric quark wave-function, and thus made it possible to reconcile the symmetry of the spatial wave-functions of the low-lying baryons with the overall antisymmetry required by the Exclusion Principle.

Beyond conformity with the Exclusion Principle, Nambu argued that coloured quarks offered two further advantages over their colourless relatives. First, as mentioned above, coloured quarks could be assigned integral electric charges. This seemed sensible in a world in which fractional electric charge had never been seen. But it should be noted that the attribution of colour to quarks did not *require* that quarks had integral charges. No problems arose in describing coloured quarks carrying the fractional charges with which Gell-Mann and Zweig had originally endowed them. And, as noted below, when coloured quarks came into vogue in the early 1970s, it was as fractionally rather than integrally charged objects.

The second advantage of colour, according to Nambu, was that it facilitated an attack on the 'saturation' problem in the CQM, which went as follows. All of the observed hadrons could be assigned to qq̄ or qqq multiplets. There seemed to be no need for, say, qqq̄ or qqqq structures. Interquark forces thus seemed to have a saturation property: bonding of a quark to an antiquark, or of three quarks, appeared to 'use up' all of the interquark attraction leaving nothing over for further bonding. In the phenomenological applications reviewed in Chapter 4 saturation was imposed by *fiat*: theorists simply agreed only to make calculations for qq̄ and qqq states. The interquark forces assumed in the calculations also produced quite sensible qqq̄ states and so on, but these were just ignored. Using the colour degree of freedom, Han and Nambu gave a simple analysis of quark dynamics based upon an analogy with atomic physics in justification of this neglect. They began from the observation that the simplest colour-neutral or colourless combinations which could be formed from quarks carrying an su(3) of colour were just the q$\bar{\text{q}}$ and qqq combinations required in CQM phenomenology. Thus colourless mesons could be formed by combining red quarks (of any flavour) with anti-red antiquarks, and so on for the other colours; and, according to the group theory rules for combining colours, colourless baryons could be formed by combining three quarks, one red, one yellow and one blue. Furthermore, according to the same rules, combinations like qq$\bar{\text{q}}$ and qqqq would inevitably be coloured. Han and Nambu then suggested that colour should be seen as a charge, analogous to electric charge. In atomic physics, electrically neutral atoms were the most stable and therefore, by analogy, in hadronic physics colour-neutral or colourless hadrons should be the lowest-lying in mass. Coloured hadrons (and coloured quarks) should only be found in some higher mass range. All of the known hadrons could be regarded as colourless states, while at higher masses coloured hadrons should also be found.

Having introduced colour and then pushed its most obvious manifestations into some unspecified higher mass range, Han and Nambu went on to ask themselves whether their model had any implications for contemporary CQM phenomenology. The answer was negative:

> How can we distinguish this and other different models . . .? Certainly different models predict considerably different structure of massive states . . . If we restrict ourselves to the low-lying states, only, however, it seems difficult to distinguish them without making more detailed dynamical assumptions.[32]

Thus, in the mid-1960s colour (and parastatistics) appeared to have little phenomenological relevance and received correspondingly little attention. It was in general regarded as a theorist's trick which served no purpose, except perhaps to legitimate the symmetric CQM wave-functions which would have been used anyway. Dalitz, for example, found no use for colour in his work on the CQM, and referred to parastatistics in a 1966 review as 'an unattractive possibility, since it represents a very drastic and far-reaching hypothesis which may raise more difficulties than it solves'.[33]

As the 1960s came to an end, though, two sources of empirical support for colour came into view. The first of these concerned the electromagnetic decay rate of the neutral pion, $\pi^0 \rightarrow \gamma\gamma$. The PCAC assumption, straightforwardly applied, predicted that the decay rate should be zero, in contradiction with experiment. However, in 1969 field-theoretic investigations led to the discovery of the Adler–Bell–

Jackiw anomaly which resolved the conflict.[34] Furthermore, the resolution came about in such a way that the theoretical prediction for the π^0 decay rate was sensitive to the number of quark colours: the predicted rate for tri-coloured quarks was 9 times greater than for quarks which did not carry colour. Experimentally the $\pi^0 \to \gamma\gamma$ rate was known fairly accurately and the colourless prediction was indeed too small by a factor of around 9.[35] The second source of support for colour also concerned a marked difference in predicted event rates depending on the existence or non-existence or colour. This was the data on R, the ratio of cross-sections for hadron production and muon production in electron-positron annihilation ($R = \sigma(e^+e^- \to$ hadrons$)/\sigma(e^+e^- \to \mu^+\mu^-)$). We will shortly be hearing much more about R, but briefly the situation in the early 1970s was as follows.

The first data on R at energies above 1 GeV were published in 1970.[36] They came from the newly operational electron-positron collider ADONE, built in Frascati, Italy. The data showed that, in the words of ADONE experimenter V.Silvestrini, 'there was a garden where a desert was expected', and the data 'were therefore greeted with surprise and scepticism'.[37] A straightforward extrapolation (using Sakurai's vector dominance model)[38] of the lower energy data on $e^+e^- \to$ hadrons led to the expectation of negligible hadron production – a 'desert' – in the energy range opened up by ADONE, whereas experiment showed it to be plentiful. Two years later the data had become more solid, following further work at ADONE and the inception of the CEA 'Bypass' project at Cambridge, Mass. The new data on R were reported by Silvestrini at the international HEP conference held at Fermilab in September 1972. Although the data still showed large error bars, it had by then become clear that in the range from 1 to 3 GeV (centre-of-mass energy) R was quite sizeable, being roughly in the vicinity of 2.[39] Now, although the direct extrapolation from lower energies led one to expect R to be very small at Frascati, this large value did not spread consternation in the theoretical camp. As will be discussed further in Chapter 9, both the parton model and field-theoretic approaches led one to expect such behaviour if one extrapolated from deep-inelastic electron scattering to electron–positron annihilation. Although possible objections to the latter extrapolation could be made it was generally agreed, amongst theorists working in the area, that R should tend to some constant value, and that that constant should be equal to the sum of the squares of the quark charges. For three colourless, fractionally-charged quarks, R was predicted to be $\frac{2}{3}[=(\frac{2}{3})^2+(\frac{1}{3})^2+(\frac{1}{3})^2]$; for the tricolour version and fractionally-charged quarks the prediction was three times larger – namely the observed value of 2; and for the

Han–Nambu model, with three colours and integrally-charged quarks, R was expected to be 4. Thus in 1972 it was apparent that the data on R, like that on π^0 decay, favoured the existence of colour and, moreover, favoured the existence of fractionally-charged coloured quarks over their integrally charged counterparts.

Gell-Mann and Light-Cone Algebra

Neither electron–positron experiments nor the observations of π^0 decay could be regarded as compelling evidence for colour: the data on R remained uncertain; $\pi^0 \rightarrow \gamma\gamma$ decay was of little intrinsic contemporary interest; and the bases of both calculations could be challenged. At this point colour needed some weight behind it, if it were to be noticed. Murray Gell-Mann provided the required intellectual bulk. In papers and review talks between 1971 and 1973 he preached the virtues of colour as a genuine dynamical variable.[40] These virtues he was well placed to appreciate from his long-standing interest in the problem of achieving a field theory of the strong interactions. Gell-Mann, it will be recalled, had in 1961 proposed the Eightfold Way symmetry of hadrons working from gauge-theoretic premises, and in 1964 he had gone on from this to propose the existence of quarks. In the early 1970's he was working towards an explicit reunification of fractionally charged quarks with the gauge-theoretic context from which he had conjured them. This work was set within a tradition of theoretical work known as the 'light-cone algebra'.

First proposed by Gell-Mann and Harald Fritzsch in 1971, the light-cone algebra was a field-theoretic approach to the understanding of scaling, and a direct extension of the current-algebra tradition which Gell-Mann had founded in the early 1960s.[41] In field theory, the deep-inelastic scattering structure functions were determined by the commutator of two currents, one current being situated at a point on the 'light cone' of the other (the 'light cone' of a given point in space–time is the region of space–time occupied by a light wave emitted from that point). Fritzsch and Gell-Mann proposed that the light-cone commutators were those expected in a *free* quark field theory, and showed how this assumption could reproduce the observed phenomenon of scaling. This was a repetition of Gell-Mann's veal and pheasant approach through which he had arrived at current algebra – the only change between the early 1960s and 1971 was the replacement of 'equal-time' with 'light-cone' commutators. Gell-Mann and Fritzsch also considered how the light-cone algebra would be modified in an interacting field theory of quarks. They began with Gell-Mann's 1962 gluon model: a QED-like theory, in which a single vector particle or gluon (the analogue of the photon) coupled to the colour charges of the quarks (instead of to the electrical charges, as in QED).[42] In this model, they were able to argue that, in all orders or perturbation theory, certain elements of the light-cone algebra were independent of the interaction, while others were sensitive to its form and strength.

However, having begun to consider the QED-like theory, Gell-Mann and Fritzsch noted an awkward asymmetry in it. The quarks, being coloured, could not decay directly to ordinary coloured hadrons; but the colourless gluons could − being able, for example, to convert directly to a rho-meson, in just the same way as could the photon. This made arguments over colour 'confinement' (see Section 3 below) difficult. In a major review in September 1972, Gell-Mann and Fritzsch proposed that a more realistic theory from which to extract the light-cone algebra would be a Yang–Mills gauge theory, in which a triplet of coloured quarks interacted with one another via an SU(3) octet of coloured gluons.[43] In such a theory the asymmetry between quarks and gluons would be removed, neither type of particle being directly coupled to colourless hadrons. Although in their 1972 review Gell-Mann and Fritzsch persevered with the QED-like model 'for convenience' in concrete calculations, they returned in collaboration with H. Leutwyler to the explicit advocacy of a gauge theory of colour in the following year (shortly after the discovery of asymptotic freedom).[44]

Even before the discovery that gauge theory was asymptotically free, then, several theorists led by the authoritative Gell-Mann were arguing the virtues of a specific gauge group: a colour SU(3), in which a triplet of coloured quarks interacted with an octet of coloured gluons. Thus when asymptotic freedom was discovered the colour SU(3) was already on the shelf, waiting to be picked up as the appropriate group structure for a gauge theory of the strong interaction. Gross and Wilcek noted that colour was 'particularly appealing' in their first paper on asymptotic freedom.[45] Politzer in his original paper on asymptotic freedom discussed gauge theory only in general terms, but in a long review the following year chose to discuss the strong interactions solely within the specific framework of the colour SU(3) model.[46] Although theorists often chose to work with a smaller gauge group − an SU(2) theory, for example, is also asymptotically free but technically simpler than SU(3) for detailed calculations − SU(3) was the obvious choice for a realistic theory of the strong interactions, and so it remained. For two or three years after the discovery of asymptotic freedom physicists spoke simply about Yang–Mills gauge theory (by implication, of the strong interaction) but as the theory became central to an ever-increasing body of practice some more compact and distinctive locution was required. Eventually 'quantum chromodynamics' – for distinction – and its acronym QCD – for compactness – became common usage (following, it appears, Gell-Mann, yet again).[47]

7.3 The Failings of QCD

We will see in Part III of the account that QCD was to become the backbone of the new physics. By around 1978 almost all experimental as well as theoretical HEP research concerning the strong interaction had become structured around QCD. But in 1973 and 1974, the period immediately following the discovery of asymptotic freedom, QCD received relatively little attention.[48] To close the present chapter I will discuss why this was so.

First, let me briefly review the virtues of QCD. QCD constituted the remarriage of two concepts which had been divorced in the early 1960's, gauge theory and the quark model. Furthermore, gauge theories had been shown in the interim to be renormalisable. Thus, on the face of it, calculations in QCD could be pursued to any desired degree of approximation, just as in QED. A close analogy had been manufactured between a field theory of the strong interactions, QCD, and that doyen of field theories, QED: the name QCD was coined in explicit recognition of this. Why, then, did not QCD emulate its electroweak sibling and rapidly come to dominate practice in strong interaction physics? Why did not theorists plunder the field-theory and quark-model traditions, transplanting resources into QCD-related activity? To answer these questions we must look more closely at the exemplary achievements of Politzer, Gross and Wilcek. Here we will see the severe problems facing theorists in both the theoretical and phenomenological exploitation of QCD.

Confinement

The strengths and weaknesses of QCD derived from the same source. As a theory of quarks and gluons, QCD promised to underpin and enrich the quark traditions of the 1960s. However, because the quarks and gluons appeared in the QCD Langrangian, a naive 'reading-off' of the physical particle states from the Lagrangian led directly to a world populated by real coloured quarks and gluons. Furthermore, the gauge theories which had been shown to be asymptotically free were *pure* gauge theories in which the gauge vector particles, the gluons, were massless. Many years of experimental effort had been expended without success in searches for quarks (coloured or not) and strongly interacting massless gluons were equally conspicuous by their absence. This unrealistic particle spectrum was a major failing of QCD and, as we saw in the previous chapter, had troubled theorists since Yang and Mills had constructed the prototype gauge theory in 1954.

In the unified electroweak models the problem of the unrealistic spectrum of gauge theories had been overcome in two ways. The massless gauge particles were given a realistic and credible mass via the Higgs mechanism, while questions concerning the existence of free quarks were set on one side: whether quarks were bound together or not was a question which strong rather than electroweak theories were expected to answer. As a theory of the strong interaction QCD could not avail itself of the latter excuse concerning free quarks – there was nowhere to pass the buck. Also, QCD theorists found themselves unable to appropriate the Higgs trick to give masses to the gluons, although they did try. The fruits of such attempts were reviewed by Politzer in 1974:

> Thus far, no model has been found which is both asymptotically free and, via the Higgs mechanism, has mass terms for all, or all but one, of the gauge fields. The search has been strenuous and fairly systematic but by no means exhaustive and, furthermore, has provided no insight as to why the two phenomena are apparently incompatible.[49]

Asymptotic freedom was the sole phenomenologically desirable feature of QCD and theorists were reluctant to abandon it for the sake of massive gluons (none of which had ever been observed). By 1974 gauge theorists were characteristically beginning to make a virtue of necessity, despite the lack of rigorous proof that the Higgs route was untenable. Their strategy was to argue that reading-off particle properties from the Lagrangian and thus imputing to QCD an unrealistic particle spectrum was illegitimate. They reasoned as follows. The asymptotic freedom of QCD was a reflection of the fact that the effective QCD coupling constant became smaller and smaller at larger and larger momentum scales. Thus at sufficiently high momenta it was legitimate to use perturbation theory to draw conclusions about the properties of the theory. However, by the same token, as one considered smaller and smaller momentum scales the effective coupling would become larger and larger. Thus at sufficiently low momenta the coupling must become so large that perturbative reasoning must break down. Now, according to the Uncertainty Principle, high momenta correspond to short distances and low momenta to (relatively) large distances, and the question of quark binding, or lack of it, was clearly related to the long distance properties of the theory. But at long distances perturbation theory was invalid, and thus reading-off particle states from the QCD field equations – an essentially perturbative activity – made no sense.

This argument effectively undermined naive assertions that the physical spectrum of QCD was unphysical, but conspicuously failed to erect anything in their place. Gauge theorists had talked themselves into a corner: perturbative arguments could not be applied to the long-distance properties of QCD, but unfortunately perturbative arguments were precisely the time-honoured stock-in-trade of field theory. Field theorists were mechanics without tools when it came to computing the low-energy properties of QCD, and action was replaced by words in the shape of the doctrine of 'confinement'. Gauge theorists admitted that once the effective coupling constant became comparable with unity they could no longer compute its evolution as one considered larger and larger distance scales. Instead, they simply asserted that the coupling continued to grow (or at least to remain large). Furthermore, they stated their faith that, as a consequence of this, colour was 'confined': all coloured particles – quarks and gluons – would interact so strongly with one another that only colourless combinations – hadrons – could exist in isolation. Quarks and gluons, that is, would be forever confined within hadrons, the so-called strong interactions between the colourless hadrons being a mere shadow of the forces between their coloured constituents.

In the years which followed, as QCD prospered in its phenomenological applications, the proof of confinement came to be seen as one of the most important problems in theoretical physics. But work on this problem was slow in getting under way, since it depended upon the importation of new techniques into field theory. In the years between 1974 and 1976 several new theoretical traditions – the '1/N expansion', 'monopoles', 'instantons' and 'lattice gauge theories' – sprang up, deriving from novel approaches to QCD.[50] Each of these traditions was fascinating in its own right for the hitherto unexpected properties of field theory which it elaborated, but suffice it to say here that during the 1970s none of them yielded an agreed solution to the confinement problem.

The failure of theorists to solve the confinement effectively blocked any immediate phenomenological application of QCD to the routine phenomena of strong interaction physics: hadron spectroscopy at low energy and small momentum-transfer scattering at high energies were both seen to be reflections of soft, large distance-scale, hadronic properties about which QCD had nothing constructive to say. In these areas, the constituent quark model and Regge models, respectively, remained the dominant traditions.

Scaling Violation

On the hard-scattering processes, described by the parton model, QCD did have something new to say, but even here its initial phenomenological impact was limited. I have so far implied that a consequence of asymptotic freedom was that at short distance scales QCD behaved as a free field theory, and hence that it exactly reproduced the results of the parton model. But this is not quite correct. Certainly, the early studies showed that the effective coupling constant of QCD went asymptotically to zero, but nonetheless the quantum mechanical operators governing the behaviour of the theory retained their anomalous dimensions (see Section 7.1) and these could be perturbatively computed to be non-zero. The non-zero anomalous dimensions in turn implied that scaling was not an exact phenomenon, and that *scaling violations* should be seen in processes such as deep-inelastic electron scattering. However, because the effective coupling constant tended asymptotically to zero, these violations were predicted in QCD to be rather mild, taking the form of an additional dependence upon the logarithm of q^2 over and above the free-field expectations, in contrast to the dependence upon powers of q^2 predicted from non-asymptotically free field theories. Power-law violations of scaling were clearly ruled out by the SLAC scaling data (except for small fractional powers of q^2) but the SLAC measurements had insufficient precision and energy range to rule out the logarithmic scaling violations predicted by QCD. Thus QCD remained tenable in the light of the SLAC data, and the predicted scaling violations promised an experimental acid-test of whether QCD was 'the reddest of herrings' or a serious candidate for a field theory of the strong interactions.[51]

Scaling violation, then, was the obvious area for phenomenological exploitation of QCD but theorists were slow to move into this area, too, for the simple reason that in 1973 there was no immediate prospect of reliable measurements over a sufficient energy range to differentiate between the predictions of QCD and the parton model. As we shall see in Chapter 11, in the latter half of the 1970s such data did begin to become available from Fermilab and, especially, the CERN SPS. At that time there was a considerable increase in phenomenological activity concerning scaling violations in QCD, but until then theorists perceived little point in carrying out analyses of scaling violation in any detail.

So, to sum up this chapter: the discovery of asymptotic freedom revealed QCD – the gauge theory of coloured quarks and gluons – to be a highly desirable theory, at least in the eyes of field theorists,

227

providing as it did a field-theoretic understanding of the successes of the parton model. However, the quark confinement problem was posed in a particularly acute form for believers in QCD, and in default of any solution to this problem it appeared that QCD had little to contribute to the phenomenology of the soft processes traditionally in the domains of the constituent quark model and Regge models. QCD did predict deviations from the exact scaling expected from the parton model in hard-scattering processes, but there was little immediate prospect of experimental illumination of this point. Thus, despite its manifest theoretical attractions, in 1973 it was quite unclear to theorists exactly what to do next with QCD, except to remark to one another, as they often did, that it was the only field theory of the strong interactions which was not obviously wrong! However, as we shall see in Chapter 9, fruitful lines of phenomenological application were shortly to open up.

NOTES AND REFERENCES

1 See Chapter 5, note 27.
2 For semi-popular accounts of scale invariance and field theory, see Jackiw (1972) and T.D.Lee (1972). For a technical review, see Carruthers (1970).
3 I am grateful to Professor Wilson for an interview and provision of biographical materials. Otherwise unattributed quotations below are from the interview. Wilson was awarded the 1982 Nobel Prize in Physics for the work discussed here.
4 For a brief review of the development of strong coupling theory, see Moravcsik and Noyes (1961, 130–1). These authors noted that: 'strong coupling theories were particularly fashionable in the decade of the forties when weak coupling theories appeared to be doomed because of the infinities', but that '[r]ecent work on strong coupling theory is scarce ... [it] so far has not yielded any practical nontrivial result for nucleon–nucleon scattering'.
5 Wilson (1969).
6 Brandt (1969), Ioffe (1969).
7 For a review of early field theory studies of scaling, see Frishman (1974).
8 Stueckelberg and Petermann (1953), Gell-Mann and Low (1954), Bogoliubov and Shirkov (1959).
9 Ferromagnets are said to exist in two phases. Below what is known as the critical temperature they act as permanent magnets, but above that temperature the magnetism disappears. A phase transition is said to occur at the critical temperature. Other examples of phase transitions are the boiling and freezing of liquids.
10 Wilson (1971a,b, 1972a), Wilson and Fisher (1972).
11 For access to the early literature on the RG and the ε-expansion, see Wilson and Kogut (1974). For a brief report on subsequent applications to solid-state physics, see Lubkin (1980). For a popular account of the renormalisation group, see Wilson (1979).

12 Wilson (1971c, 1972b).
13 Callan (1970), Symanzik (1970).
14 Coleman (1973).
15 't Hooft, interview. 't Hooft's contribution to the Marseille conference is well attested in the HEP literature: see, for example, Politzer (1974, 132).
16 I am grateful to David Gross for an interview and provision of biographical materials.
17 I am grateful to David Politzer for an interview and provision of biographical materials. Unattributed quotations below are from this interview.
18 Coleman and Weinberg (1973).
19 Symanzik (1973). Symanzik's model suffered from a theoretical 'disease' which was generally taken to disqualify it from consideration as a realistic theory of the strong interaction: see Politzer (1974, 132) and references therein.
20 Gross and Wilcek (1973), Politzer (1973).
21 Coleman and Gross (1973).
22 The $\frac{3}{2}^+$ decuplet included the Ω^-: this was built from three strange quarks, and the identity of the quarks implied a symmetric SU(3) wave-function. Similarly, it was assumed that the spins of the constituent quarks were all oriented in the same direction, implying a symmetric spin wave-function.
23 This corresponded to zero orbital angular momentum for the ground state.
24 I am grateful to Professor Greenberg for an interview and provision of biographical materials.
25 Greenberg, Dell'Antonio and Sudarshan (1964).
26 Greenberg and Messiah (1964, 1965a,b).
27 Greenberg (1964).
28 Han and Nambu (1965).
29 Nambu, interview. Nambu developed the ideas outlined below working alone and circulated them as a preprint. At the time, Nambu knew little group theory, and this element entered the paper when Han independently reworked and developed Nambu's preprint in group-theoretical terms. Han and Nambu then jointly developed the published paper.
30 Bacry, Nuyts and Van Hove (1964), Nambu (1965).
31 Nambu actually referred to the new set of quantum numbers as 'charm', but the meaning of this term later became identified with the new quark flavour operative in the GIM mechanism.
32 Han and Nambu (1965, B1010).
33 Dalitz (1967, 232). Similarly, Morpurgo recalls (private communication):

> When the idea of the non-relativistic [quark] model occurred to me I certainly did not think of accompanying an already heretical proposal (to treat the internal dynamics of a hadron non relativistically) with additional drastic and unjustified assumptions (such as it might have been the introduction of a new hidden [colour] degree of freedom) ... [T]here was really no logical need for this assumption, nor there was any physical fact that imposed it.

34 See Chapter 6, note 98.

35 It is interesting to note that Steinberger (1949) had shown that good agreement with the data was possible if the pion was viewed as a nucleon–antinucleon composite (as in the Sakata model). Sakata-type models were no longer popular in the late 1960s.

36 Bartoli *et al.* (1970).

37 Silvestrini (1972, 10).

38 See Chapter 6, note 12.

39 Silvestrini (1972).

40 See, for example, Fritzsch and Gell-Mann (1972), Bardeen, Fritzsch and Gell-Mann (1973) and Fritzsch, Gell-Mann and Leutwyler (1973).

41 Fritzsch and Gell-Mann (1971a,b). For a review of work in the light-cone tradition, see Frishman (1974, 24–33).

42 Gell-Mann (1962b).

43 Fritzsch and Gell-Mann (1972, 139–141).

44 Fritzsch, Gell-Mann and Leutwyler (1973).

45 Gross and Wilcek (1973, 1345).

46 Politzer (1974).

47 See Marciano and Pagels (1978, 139, note 1.1).

48 The papers reporting the asymptotic freedom of gauge theory – Gross and Wilcek (1973) and Politzer (1973) – appeared in the 25 June 1973 edition of *Physical Review Letters*; in the first year after their publication each received 76 citations in the HEP literature.

49 Politzer (1974, 149).

50 The $1/N$ expansion was a novel perturbative technique, the expansion parameter being the reciprocal of N, the number of quark colours. It was proposed by 't Hooft (1974a,b), and a popular account has been given by Witten (1980). The existence of magnetic monopoles and instantons (both manifestations of novel and essentially non-perturbative properties of gauge theories) were respectively proposed by 't Hooft (1974c) and Polyakov (1974), and by the Russian theorists Belavin, Polyakov, Schwartz and Tyupkin (1975) (see also 't Hooft 1976). Lattice gauge theories were first proposed by Wilson (1974). These theories were formulated on a discrete lattice of space-time points rather than in a space-time continuum, facilitating a perturbative expansion in *inverse* powers of the coupling constant, as appropriate to the long-distance properties of QCD.

51 Politzer (1974, 178).

8

HEP IN 1974: THE STATE OF PLAY

In the preceding chapters we examined the founding and growth of several important HEP traditions in the years between 1964 and 1974. We looked at the blossoming of quarks in the old physics CQM tradition, and at the early development of the field-theory oriented new physics traditions: current algebra, the parton model, unified electroweak theory and QCD. I now want to pause to take stock of the situation in HEP in mid-1974, since the character of events is about to change rapidly: 1974 was the year of the 'November Revolution'. Until November 1974 the new physics traditions of theory and experiment constituted only a minor aspect of HEP practice; the old physics remained dominant. In November 1974, the discovery of the first of a series of highly unusual elementary particles was announced. This announcement set in train the developments discussed in Part III of the account, which led within the space of five years to the eclipse of the old physics by the new. Thus the events of November 1974 constituted a watershed in the history of particle physics, and before discussing them I want to clarify some aspects of the social and conceptual context in which they took place. The first section of this chapter is conceptual, and reviews the theoretical structure of electroweak gauge theory and QCD, the theories at the heart of the new physics. The emphasis in the two following sections is social. Section 2 provides an overview of communal practice in HEP in mid-1974, taking as its text remarks made by James Bjorken in 1979. Finally, Section 3 focuses upon three of the leading actors in the November Revolution, and explains their revolutionary activities in terms of their earlier careers.

8.1 Gauge Theory – Reprise
By 1974 it was possible (though not compulsory) to represent each of the three fundamental forces of HEP – the strong, electromagnetic and weak interactions – in terms of gauge theory. The electromagnetic and weak interactions were gathered together in unified electroweak gauge theories, typified by the Weinberg–Salam (WS) model, while the strong interaction was described by QCD. The WS model and QCD will be at the heart of the discussions of Part III of this

account, and I want here to clarify their conceptual basis and structure.

First some generalities. Gauge theories are a class of quantum field theories. In gauge theories, as in quantum field theory in general, all forces are ascribed to particle exchange: there is no place for action-at-a-distance. For present purposes, gauge theories can be characterised as those quantum field theories in which the forces between fermions (quarks and leptons) are mediated by the exchange of *vector* (spin 1) particles. Furthermore, the vector, or gauge, particles are *massless* unless the theory is suitably modified. At the heart of any gauge theory is a mathematical group structure which encapsulates the relationships between the particles described by the theory: the particles are assigned to families which are representations of the group (and relationships between coupling strengths are similarly determined). Thus group structure and gauge particle masses (if any) serve to distinguish between individual gauge theories.

Now to particulars. It is convenient to begin by considering an imaginary world in which only the electromagnetic force is operative and which contains only a single species of fermion: the electron. The properties of this world are described by the gauge theory of electromagnetism, quantum electrodynamics. The electron is the only member of a one-membered family which can be denoted (e$^-$). There is a single massless gauge particle, the photon (this corresponds to the U(1) group structure of QED). The photon couples to the electric charge of the electron and thus mediates long range electromagnetic forces. The photon does not itself carry electric charge, and thus emission or absorption of a photon serves only to transmit energy and momentum and does not change the quantum numbers of the electron.

Now consider a more complicated world which includes both neutral and charged leptons, neutrinos as well as electrons, and in which both the weak and electromagnetic interactions are operative. According to the WS model, all of the forces acting between the leptons can be ascribed to exchanges of four species of gauge particles: the photon, and three intermediate vector bosons (IVBS) two of which are electrically charged, W$^+$ and W$^-$, and one of which is electrically neutral, Z^0 (this pattern of gauge particles arises from the SU(2) × U(1) group structure of the WS model). The photon continues to interact only with charged particles, electrons, but the IVBS mediate the weak interactions of electrons and neutrinos alike. The charged IVBS mediate charged-current weak interactions, which transform electrons into neutrinos and vice versa. Thus electrons and neutrinos can be regarded as members of a single family: the doublet

(ν_e, e^-). The neutral IVB, Z^0, mediates neutral-current weak interactions in which neutrinos and electrons maintain their identity (as in electromagnetic interactions). In the WS model the photon remains massless but the IVBS are given large masses through the Higgs mechanism. This guarantees that the weak interaction gives rise to small effects relative to electromagnetism and that it is operative only over short distances, as observed. The price to be paid is the existence of a scalar (spin 0) Higgs particle.

So far we have only a theory of the electroweak interactions of leptons. Suppose we now include quarks. It is convenient to begin with just two flavours, up (u) and down (d). In the WS model, the u and d quarks are assigned to a single family (u,d) analogous to that of the leptons. The photon and IVBS are coupled to the quarks as if they were leptons: the photon couples to the u and d quarks in proportion to their electric charges ($+\frac{2}{3}$ and $-\frac{1}{3}$ respectively); the charged IVBS convert u into d quarks and vice versa; and the Z^0 mediates the neutral-current weak interactions of the quarks, leaving their flavours unchanged.

How can we introduce the strong interaction into this toy world? According to QCD the answer is simple. Suppose that each flavour of quark comes in three colours: red, yellow and blue. We have hitherto regarded the u and d quarks as single species, but now regard them as triplets – (u_R, u_Y, u_B) and (d_R, d_Y, d_B) in obvious notation (these triplets correspond to the fundamental representation of the colour SU(3) group). The gauge particles of QCD are an octet of coloured vectors which are known as gluons. These mediate the strong interaction, and change quark colours – u_R to u_Y and so on – in the same way as the charged IVBS change quark flavours. Because leptons have no colours they cannot interact with the colour-changing gluons, and are therefore immune to the strong interactions. Thus the strong and electroweak interactions are separate in gauge theory: the coloured gluons which mediate the strong interactions couple only to quark colours and are 'flavour-blind', while the photon and IVBS which mediate the electroweak interactions couple only to quark and lepton flavours and are 'colour-blind'. One subtlety concerning QCD remains to be mentioned. The gluons are formally massless particles, like the photon but unlike the IVBS. Against this, though, the doctrine of confinement discussed in the previous chapter asserted that the properties of QCD were such that no coloured particles could exist in isolation. Only colourless combinations of quarks and gluons should be manifest in the laboratory, and these were to be identified with the observed hadrons.

Our imaginary world now approximates to the real world as

Table 8.1. Gauge theories of elementary particles.

Interaction	Gauge Theory	Gauge Group	Fundamental Fields	Basic Families*	Gauge Vector Particles:
Electroweak	Weinberg-Salam model (quantum flavour dynamics)	$SU(2) \times U(1)$: flavours	Quarks +leptons	$\begin{pmatrix} u \\ d' \end{pmatrix}_L, \begin{pmatrix} c \\ s' \end{pmatrix}_L$, u_R, d_R, s_R, c_R $\begin{pmatrix} \nu_e \\ e^- \end{pmatrix}_L, \begin{pmatrix} \nu_\mu \\ \mu^- \end{pmatrix}_L$, e^-_R, μ^-_R	3 IVBs: $\begin{pmatrix} W^+ \\ W^- \end{pmatrix}$, Z^0: very heavy Photon: γ: massless
Strong	Quantum chromodynamics	$SU(3)$: colours	Quarks	$\begin{pmatrix} q_R \\ q_Y \\ q_B \end{pmatrix}$	8 coloured gluons: all 'massless' but confined

* For electroweak interaction families:
L = left-handed, R = right-handed; $d' = d \cos\theta_C + s \sin\theta_C$, $s' = -d \sin\theta_C + s \cos\theta_C$, where θ_C = Cabibbo angle.[3]
For strong interaction families; R = red, Y = yellow, B = blue.

perceived by gauge theorists in mid-1974. The strong and electroweak interactions are fully described, but the complement of fundamental particles is not yet complete. Missing are two leptons, the muon and its associated neutrino,[1] and two quarks, the strange (s) quark and the charmed (c) quark (I include the c quark here because of its relevance to later developments). The gap is easily filled. Include a new family of leptons (v_μ, μ^-) alongside the (v_e, e^-) family and a new family of quarks (c, s) alongside the (u, d) family. Couple these new families to the photon, IVBs and gluons as before, and there one has the complete gauge-theory picture of elementary particle interactions.[2] Table 8.1 summarises the picture (with some additional technical details concerning electroweak theory).

8.2 The End of an Era

In July 1979 SLAC theorist James Bjorken addressed the 'Neutrino '79' HEP Conference at Bergen in Norway. As the title of his talk he chose 'The New Orthodoxy: How Can It Fail?'. The new orthodoxy which he perceived in HEP was the belief that the combination of the Weinberg–Salam electroweak model and QCD could explain all of the phenomena of elementary particle physics. 'During this last decade,' Bjorken began, 'neutrino-physics has evolved from its crude beginnings into a highly quantitative branch of particle physics'.

> And during this decade, particle physics itself has developed extremely rapidly and fruitfully into a state which is far-reaching, surprisingly rigid, and broadly accepted. In fact it is only about the last five years that the concepts central to the present orthodoxy really took hold and began to dominate our picture of particle interactions. All the concepts were actually in place in 1974, even before the November Revolution. Provided ones copy of the 1974 London Conference has not completely disintegrated, one need only refer back to John Iliopoulos' talk there to find a remarkably complete exposition of the contemporary orthodoxy.

We will shortly examine the poorly bound *Proceedings* of the London Conference in order to see what else interested physicists in July 1974 besides the contents of Iliopoulos' talk. However, the continuation of Bjorken's 1979 remarks bears repeating, just to emphasise the change which had come over HEP since the November Revolution:

> We have now reached a quiet time. Other than some residual confusion among the atomic parity-violation experiments, the smallest feasible gauge-theory-structure ($SU(2) \times U(1)$ for electroweak, $SU(3)$ for strong, and $SU(5)$ for grand unification)

accounts very well for the observations. The situation is remarkably satisfactory. It is no wonder that, after the false alarms and relatively extravagant theoretical responses of the past few years, there is at present such a mood of minimalism and complacency . . .

But whatever the pros and cons for accepting the orthodoxy, there are the attendant dangers common to any orthodoxy. With the risk of being banal, I feel compelled to express what I see as the biggest danger, which is that experiments become too sharply focused. While searches for what is predicted by the orthodoxy will proceed, searches for phenomena outside the orthodoxy will suffer. Even more important, marginally significant data which support the orthodoxy will tend to be presented to – and accepted by – the community, while data of comparable or even superior quality which disagrees with the orthodoxy will tend to be suppressed within an experimental group – and even if presented, will not be taken as seriously.[4]

The transformation to the 'new orthodoxy', and in particular the penetration of gauge theoretic concepts into the planning, execution and interpretation of experiments (as well as more arcane matters, such as the 'atomic parity-violation experiments' and 'grand unification') is the subject matter of Part III. Here I want to sketch in the backdrop against which that transformation was seen – the 'old orthodoxy', one might say, were it not for the social and conceptual fragmentation of HEP in the early 1970s. To provide such an overview I will briefly discuss the contents of the *Proceedings* of the London Conference, held in July 1974, since this was the last major HEP Conference to be held before the November Revolution.[5] Although many HEP conferences take place every year, the biennial 'Rochester' Conferences (named after their original venue) are regarded by the community as the forum for an in-depth review of the current state of the entire field, and the London Conference was the seventeenth in this series. Thus it is reasonable to take the *Proceedings* of this Conference as an accurate representation of the concerns of the HEP community in mid-1974.[6]

The *Proceedings* followed the format of the meeting and were divided into 5 sections, the titles and lengths of which were:

  I. Strong Interactions at High Energy (pp. 280)
  II. Resonance Physics (199)
  III. Weak Interactions and Unified Theories (115)
  IV. Lepton–Lepton and Lepton–Hadron Interactions (173)
  V. Large Transverse Momentum Reactions (80)

Each section comprised reports on recent developments, experimental and theoretical, in the appropriate sub-field of HEP. As can be seen from the space devoted to each topic, the staple diet of HEP in 1974 – 'strong interaction physics at high energy' and 'resonance physics' – had remained unchanged since the mid-1960s. The longest section, Section I, was devoted to the presentation and theoretical analysis of data on hadronic soft scattering, both exclusive and inclusive. The predominant theoretical orientations here were outgrowths of the Regge tradition.[7] Section II, the second-longest section, was devoted to low-energy resonance physics. The experimental identification of new resonances had become a fine art by 1974, and the bulk of the reports in this section were devoted to the presentation and analysis of data, comparisons between different experiments, and so on. The theoretical context here was that of the constituent quark model.[8] Taken together, Sections I and II represented the old physics of the strong interactions, Regge and resonance, and the space devoted to them in the *Proceedings* easily exceeded that given to the following three sections. But it was in these later sections that the shape of the post-revolutionary new physics was to be found.

Section III, 'weak interactions and unified theories', could be divided into two disconnected parts. The first part was concerned with new experimental measurements of standard weak decay parameters.[9] This was the old physics of the weak interactions, connected to the new physics only via reports on the non-existence of strangeness-changing neutral currents. The second and longer part of Section III was devoted to theoretical work on electroweak unification and gauge theory in general.[10] The unequal division of this section between old and new reflected the outburst of theoretical activity on unification which we examined in Chapter 6. Section IV was devoted to weak neutral currents, to deep-inelastic electron, muon, and neutrino scattering and to electron–positron annihilation. Neutral currents were dealt with here from an experimental perspective, though their relevance to theoretical speculation concerning electroweak unification was noted. The bulk of the reports in this section, however, were devoted to deep-inelastic lepton–hadron scattering and electron–positron annihilation. Most notable here was the *lack* of impact of QCD on the interpretative context within which new data were presented and analysed. This context remained that provided by the parton model and the quark light-cone algebra.[11] I have already indicated that this lack of impact could be ascribed to the difficulties experienced by theorists in finding something constructive to do with QCD, but it should also be noted that in 1974 QCD, like all of the models developed to explain scaling, appeared to be

237

refuted by the data on electron–positron annihilation – a topic which we shall explore in more detail in Chapter 9. Finally, the shortest section of the *Proceedings* was devoted to 'Large Transverse Momentum Reactions'. Data on purely hadronic reactions were reported in Section v principally from Fermilab and the CERN ISR. No consensus existed on their theoretical interpretation, although an explanatory framework was provided by crude quark–parton models suitably adapted from the standard model of scaling.[12]

Thus, while all the elements of Bjorken's 'new orthodoxy' – unified electroweak theory and QCD – were indeed in place in mid-1974, one can see from the *Proceedings* of the London Conference that their impact within the HEP community was limited. The dominant experimental traditions still related to soft processes: high-energy hadronic scattering and low-energy resonance physics. The dominant theoretical traditions were associated with these: the Regge and constituent quark models, respectively. Perspectives within weak-interaction physics were, however, undergoing re-organisation: neutral currents had become a principal focus of neutrino experiment and unification was *the* theoretical pursuit. Clouds remained on the theoretical horizon, none the less, since the new hadrons or heavy leptons required to suppress strangeness-changing neutral currents had not been observed. As far as the strong interactions were concerned, experimental perspectives were also in a process of transition. New traditions of experiment on hard-scattering proces-ses – deep-inelastic lepton-initiated reactions and large transverse momentum phenomena in purely hadronic reactions – had sprung up in parallel with the still-dominant soft-scattering traditions. Driven by experiment, new theoretical traditions had evolved – prin-cipally the pragmatic quark–parton model. But QCD, the gauge theory which promised to underwrite and amend the parton model, had had negligible impact by mid-1974, and all of the scaling models appeared to be contradicted by data from a new source: electron–positron annihilation.

8.3   Three Transitional Biographies

To conclude this chapter I want to look ahead to the subject matter of Part III, in order to expose and address a problem whose solution more properly belongs with this part of the account.

In November 1974, the discovery of the first of the 'new particles' was reported and theoretical speculation concerning their nature was soon rampant. Prominent in this debate were theorists who proposed that the new particles were manifestations of charm (the new quark flavour central to the GIM mechanism) and who used ideas drawn from both unified electroweak gauge theories and QCD to bolster their case. By 1976 the existence of charm had become generally accepted in the HEP community, and the Weinberg–Salam model was well on the way to becoming the accepted theory – the 'standard model' as it was then called – of the electroweak interactions. Furthermore, novel QCD-inspired traditions of strong-interaction phenomenology had been founded. These developments permeated both theory and experiment, and the momentum of the new physics became unstoppable. Their perceived significance was such that almost all new entrants to HEP received their training in a new-physics context and went on to reproduce this context in their research careers.

Thus the key phase in the communal establishment of the new physics centred on the debate over the new particles: the seeds of revolution were planted by those physicists advocating charm and gauge theory. Now, the advocacy of charm was sustained by two allied but distinct groups. On the one hand there was the old guard of gauge theorists, the principals of the preceding chapters. In this group, for instance, one would set Gell-Mann, Weinberg and Glashow. It should come as no surprise that such theorists were advocating the existence of colour and charm from a gauge-theory perspective even before the discovery of the new particles. On the other hand, we will see that a second group of younger physicists were equally, if not more, active than the old guard in advocating charm and gauge theory in the debate over the new particles. These were theorists who had entered HEP in the 1960s and whose training had not encompassed gauge theory. It will become clear in Part III that the existence of this group was crucial to the form taken by the subsequent development of HEP, and the question I want to address here is one of why these theorists should have aligned themselves at a crucial time with the gauge-theory camp.

My answer will have the same general form as that adopted throughout this account in analysing the cultural dynamics of scientific practice. It will depend upon the notion of 'opportunism in context' sketched out in the Introduction: the idea that individuals deploy the resources at their disposal (principally the intangible resource of expertise in the case of theorists) in contexts defined by the practice of their fellows, and that, in the process, new resources are acquired. I will examine the careers of three key theorists who entered HEP in the 1960s – a time when, to quote one of them, 'gauge theories didn't exist' – in order to show how the idea of opportunism in context can illuminate their determined advocacy of charm and gauge theory during the November Revolution.

The three theorists whose careers we will follow are Mary Gaillard,

239

Alvaro De Rújula and John Ellis. When the new particles were discovered Gaillard and Ellis were working at CERN, and they quickly became two of charm's most ardent protagonists in Europe; De Rújula was at Harvard, where the theory group was the spearhead of the US movement in favour of charm. Coming from the USA, Spain and England, respectively, Gaillard, De Rújula and Ellis followed quite distinct career paths up to the time of the November Revolution, yet they at once found themselves fighting on the same side of the barricade. I will try to show that the expertise acquired by these theorists in their careers – moving from one tradition to another, exploiting their existing expertise and at the same time mastering new resources – made it inevitable that they should have fought on the side of charm and gauge theory.[13]

Before turning to individual biographies, one preliminary remark is in order. The early careers of Gaillard, De Rújula and Ellis differed in many respects, but it is significant that they had a common starting point. All three began their research within the current-algebra tradition. We have seen that, unlike the dominant CQM and Regge traditions, current algebra never lost touch with its field-theory ancestry, and we should not therefore be surprised to learn that the new physics of gauge theory held particular attractions for theorists trained in current algebra. In examining the careers of Gaillard, Ellis and De Rújula we will, though, be able to go beyond such general considerations and to look at more local determinants of practice. It is possible to situate their careers in particular social contexts and to examine the bearing of those contexts upon individual practice. Repeatedly we will see that career development and the parallel acquisition of new expertise were structured by the local context of day-to-day practice rather than by the overall public culture of HEP (the professional literature and so on). In this respect the processes of acquisition and deployment of expertise have a contagious quality, and the biographies which follow exemplify a critical stage in an incipient epidemic of gauge theory.[14] In the context of the November Revolution the initially small band of committed gauge theorists infected a substantial fraction of the HEP community. The contagion soon spread until almost the entire community had succumbed and the new physics was established.

Alvaro De Rújula[15]

Alvaro De Rújula was born in Madrid on 29 January 1944. He studied physics at the University of Madrid, and was awarded a PhD in theoretical HEP in November 1968. From 1969 to 1971 he worked in the theory group at CERN. The year 1971–2 he spent in France, and in October 1972 he went to Harvard as a Research Associate, where he later held a tenured position.

De Rújula's route to gauge theory was a simple one. His postgraduate research was in the current-algebra tradition.[16] Leaving Spain for CERN he began to collaborate with other young theorists there. He contributed to a detailed and complex calculation in QED, gaining valuable experience in the use of perturbative methods in renormalisable field theories;[17] he worked on the analysis of neutrino scattering, a topic of considerable interest at CERN where the Gargamelle neutrino experiment was getting under way;[18] and he worked on deep-inelastic scattering and the parton model, again a matter of local concern (for the neutrino experiment) as well as more generally.[19]

Thus by the time he left CERN, De Rújula had some expertise in current algebra, detailed perturbative techniques in field theory, the phenomenology of weak-interaction experiments and the parton model. When he arrived in Harvard in October 1972 the electroweak unification industry was in full swing. At Harvard there were Sheldon Glashow – the inventor of charm and the GIM mechanism, one of the leading exponents of the model-building art and the most outspoken advocate of electroweak unification – and Sidney Coleman – one of the world's leading field theorists. Nearby at MIT was Steven Weinberg (who moved to Harvard in 1973) – the inventor of the prototype unified gauge model. De Rújula recalls that at Harvard he experienced 'peer pressure to understand field theory' and that, amongst field theories, gauge theories were 'the most interesting'. This was the kind of pressure to which he was ideally equipped by his previous experience to succumb. He first set about learning the principles of the unified electroweak gauge theories. To this end he agreed to lecture on the subject at the Formigal Winter School in Spain and then began a collaboration with Glashow and others at Harvard, exploring the phenomenological implications of unification.[20]

In 1973 at Harvard, Politzer discovered the asymptotic freedom of unbroken gauge theories. De Rújula had gained by that time some experience in gauge-theory calculations, and from his CERN days he was familiar both with sophisticated perturbative techniques and with the parton model. Once again, expertise and context combined neatly together towards the acquisition of new techniques, and in 1974 De Rújula published two papers on the phenomenological consequences of asymptotic freedom.[21] By July 1974, he was presenting a review of 'Lepton Physics and Gauge Theories' at the London Conference.[22] When the revolution came, De Rújula, along with the entire HEP theory group at Harvard, was on the side of the angels: gauge theory and charm.[23] To have adopted any other position, in the light of his expertise and local context, would have been little short of perverse.

John Ellis

John Ellis was born in London on 1 July 1946. Attending Cambridge University from 1964 onwards, he studied mathematics as an undergraduate and then transferred to theoretical HEP as a postgraduate. He was awarded a PhD in 1971. His final year as a postgraduate was spent at CERN. From 1971 to 1972 Ellis was a Research Associate at SLAC and from 1972 to 1973 he was a Research Fellow at Caltech. In September 1973 he returned to CERN, where he became a staff member in September 1974.

Ellis recalls that theoretical research at Cambridge in the late 1960's was Regge-oriented: 'mainly on the high-energy multi-Regge behaviour of scattering amplitudes and looking at properties of multi-Reggeon interactions, Regge cuts – stuff like that'. However, these problems 'struck me as being boring', and Ellis arranged for his postgraduate research to be supervised by Bruno Renner, an expert in field theory and group theory who had worked on current algebra at Caltech with Gell-Mann. Ellis' work at Cambridge typified contemporary developments in the field-theory tradition, moving from current algebra and its extension to 'chiral symmetry' to broken scale invariance.[74] He was untypical in his productivity, and in early 1971 he was invited to review the theoretical status of scale and chiral invariance at the prestigious Coral Gables Conference held at the University of Miami.[25]

Renner left Cambridge in 1970, and Ellis moved to CERN for his final year of postgraduate research. While there, he began to work in two fast-developing areas in HEP with which he had become acquainted at the Coral Gables Conference. One of these was a new development of the Regge tradition as applied to multiparticle production in purely hadronic reactions. Such processes were of great experimental interest at CERN, and Ellis collaborated with other CERN theorists in developing the novel approach.[26] The other new area of research pursued by Ellis at CERN was the use of the operator product expansion in the theoretical analysis of scaling. As he later put it:

> There was a lot of discussion of these things [OPEs] at the [Coral Gables] Conference, and I worked out an application for those ideas. . . .
>
> These ideas I'd sort of had in mind when I started to work on

broken scale invariance, but I thought I'd play myself in gently by starting with something which was a mix of old stuff and new stuff rather than jumping immediately into something totally new.

In 1971, Ellis left CERN for SLAC, choosing to go there because he 'wanted to learn properly about partons and scaling'. At SLAC, and at Caltech the following year, Ellis continued to collaborate with other theorists on the two new lines begun at CERN – Regge theory and the operator product expansion and scale invariance.[27] In 1973, Ellis lectured on scaling at the Santa Cruz Summer School. He and MIT theorist Bob Jaffe wrote up a two-hundred-page version of these lectures which constituted 'at that time the most complete view of classic partons, operator products and so on'.[28]

1973, too, was the year in which asymptotic freedom was discovered. This was, as we have seen, a very significant development for field theorists interested in scaling – such as John Ellis. Back at CERN in 1974, Ellis set to work on gauge theory 'since by that time it was obvious that QCD should be the theory of the strong interactions': it was 'the only sensible theory'. He first began to investigate the properties of QCD in a two-dimensional universe (one space and one time dimension). Field-theory calculations are much simpler in two dimensions than four, and well known work by Julian Schwinger in the early 1960s had shown that QED in two dimensions exhibited a phenomenon analogous to confinement. Ellis (like other theorists) thus hoped to gain useful experience in QCD by investigating the two-dimensional version. However, some months' work resulted in 'total failure' and no publications – although 't Hooft was shortly to point the way forward, and thus to found a new tradition of theoretical research.[29] Ellis was undeterred.[30] On a visit to the University of California at Santa Cruz, he had become acquainted with an approach to multiparticle production known as 'Reggeon field theory'. This approach had been founded by the eminent Russian theorist Gribov, and was enjoying a considerable resurgence on the West Coast of the USA. Its popularity derived from the fact that as a field-theoretic reformulation of the Regge approach it was amenable to renormalisation group techniques (including Wilson and Fisher's ε-expansion). Here Ellis found the opportunity to exploit his expertise in both field theory and multi-particle production and, at the same time, to acquire familiarity with the renormalisation group and asymptotic freedom 'while working on a problem which was new but not infinitely complicated'. A 1974 paper entitled 'An Asymptotically Free Reggeon Field theory' marked the beginning of an extensive collaboration between Ellis and West Coast theorists.[31]

In 1973 and 1974 a major topic for discussion in HEP was the data on electron–positron annihilation then emerging from the new colliding beam machine SPEAR at SLAC. These data were in manifest conflict with ideas of scale invariance, and Ellis recalls that in many quarters they were seen as 'total nemesis' for field theory. Because of his expertise in the area, Ellis was asked to give several major reviews on theoretical ideas concerning electron–positron annihilation in 1973 and 1974.[32] In his talk at the London Conference in July 1974 he reviewed the problems posed by the data for a variety of theoretical models, but he subsequently recalled that 'it was quite obvious to me that there was only one solution' – namely the existence of a new hadronic degree of freedom such as charm – and that he had expressed this sentiment in seminars he gave at CERN, DESY and the Niels Bohr Institute in Copenhagen. The difference between John Ellis' reaction and that of other theorists to the electron–positron data illustrates nicely the role of prior experience in determining responses in particular contexts. Ellis recalls that many of his colleagues took an attitude of 'who believed in that parton light-cone crap anyway?' He, on the other hand, reasoned that:

> The ideas of partons and operator products worked in so many different cases that they must be in some sense correct. So if there was a violation of those predictions for e^+e^- annihilation that could not be because the ideas of light-cone expansion and so on were breaking down. It had to be that there was some new parton coming in or some other new degree of hadronic freedom . . . By the autumn of that year [1974] it was clear that something like charm must exist.

Thus, when the first of the new particles was discovered (in electron–positron annihilation) Ellis had an explanation to hand – charm – and the tools with which to construct arguments in its favour – asymptotically free gauge theory. Unlike De Rújula and, especially, Gaillard, Ellis' preparation for the November Revolution was primarily through field-theoretic experience of the strong rather than the electroweak interactions, but the end result was the same: he decided upon the charm explanation 'about five minutes after I'd heard about the psi [the first new particle] . . . With the discovery of the psi it just became – for me at any rate – transparent that that [gauge theory] was the way the world was.' From that point on, Ellis and Gaillard were two of Europe's staunchest and most active advocates of charm and of the gauge-theory framework in which it was embedded.[33]

Mary Gaillard

Mary K.Gaillard was born in New Jersey on 1 April 1939. She was awarded a BA in Physics by Hollins College, Virginia in 1960, and an MA from Columbia University, New York in 1961. While at Columbia she married the French physicist J-M.Gaillard, and her postgraduate research in theoretical HEP was done at the University of Paris at Orsay. After completing her doctorate in 1968, Gaillard's institutional base was the theory group at CERN, although she spent the year 1973–4 visiting Fermilab.

Gaillard's postgraduate research in the mid-1960s was on the phenomenology of weak interactions.[34] 'In those days', she recalled, 'gauge theories didn't exist'. Her early work included analysis of the phenomenon of CP-violation in kaon decay, which had caused great excitement in HEP when first discovered in 1964.[35] Moving to CERN she continued between 1969 and 1972 to focus on the phenomenology of the weak interactions, particularly on various K-decay processes, working within the chiral symmetry tradition. Amongst other things, she worked on the anomalous neutral-current kaon decay modes which, as we saw in Chapter 6, were then recognised as posing a problem for all existing theories of the weak interactions.[36] By 1972 she had attained the status of an expert in weak-interaction phenomenology.[37]

In 1973 Mary Gaillard left CERN to spend a year at Fermilab and this marked the crucial phase of her involvement with gauge theory. The leader of the theory group at Fermilab was Benjamin Lee who, as we saw in Chapter 6, played an important part in the renormalisation of gauge theory and the elaboration of unified electroweak theories. Gaillard arrived at Fermilab at the time when the existence of neutral currents had just been reported from CERN, and this was a topic of great interest amongst the Fermilab neutrino experimenters. Thus at Fermilab Gaillard found herself in a local context where the phenomenological implications of electroweak theories were of pressing importance, and she was in daily contact with Lee, a leading gauge theorist. She brought with her expertise in weak-interaction phenomenology in general and in strangeness-changing neutral currents (the K-decay anomaly) in particular, and her theoretical approach was securely grounded in the field-theoretic current-algebra tradition. Given her expertise and local context, it would have been surprising if she had acted other than she did – which was to begin the investigation of the detailed phenomenology of electroweak theories, learning gauge-theory techniques in a collaboration with Lee.[38]

The pressing problem for electroweak gauge theories in 1973 was the absence of strangeness-changing neutral currents – in particular, the K-decay anomaly with which Gaillard had close acquaintance. Together, Lee and Gaillard set out to investigate, in more detail than hitherto, the phenomenological consequences of the existence of GIM charm. From gauge-theory calculations, they concluded that if the charmed quark had a mass of around 1 GeV then all of the data concerning strangeness-changing neutral currents could be understood, and that charm should therefore be taken seriously. The upshot of this work was a major 1974 review article 'Search for Charm' co-authored by Gaillard, Lee and J.L. Rosner (an expert on hadron spectroscopy).[39] This paper reviewed the attempts of Gaillard, Lee and Rosner to estimate the masses and widths of the as yet unobserved charmed hadrons, if the GIM mechanism were to be effective in suppressing the strangeness-changing neutral currents. In mid-1974 Gaillard reviewed prospective experimental searches for charm at an international conference in Philadelphia, and at the London Conference she reviewed gauge theories and the weak interactions.[40] She then returned to CERN in late 1974 armed with new expertise in both the phenomenology of charm and electroweak gauge theory in general. Her return to Europe was closely followed by the discovery of the new particles, which immediately became the central talking point of HEP. Gaillard at once sided with the charm camp.[41]

One further point can be noted. Arguments in favour of the charm interpretation of the new particles encompassed ideas drawn from QCD as well as electroweak gauge theory. Here again Gaillard had acquired expertise from her collaboration with Lee at Fermilab. Together they had attempted to elucidate certain long-standing problems in weak-interaction physics (CP-violation and the so-called '$\Delta I = \frac{1}{2}$ Rule') using the newly developed concept of asymptotic freedom.[42] They did not succeed in solving the problems, but the indirect result of this work was that in the debate over charm Gaillard was already equipped with a working familiarity with the unusual properties of QCD. Thus in the case of Mary Gaillard, one can see rather clearly how the idiosyncracies of personal biography, linking particular local contexts, could lead a theorist trained in an era when gauge theories 'didn't exist' to be at the forefront of the November Revolution. To emphasise the importance of the interaction of prior expertise with particular contexts in structuring future action – both in respect of Mary Gaillard and physicists in general – I will close this chapter with an extract from my 1978 interview with her:

MG: John [Ellis] and I .. we had sort of a long battle trying to convince people that charm was correct after I came back here [CERN, 1974] and then more recently we had this big battle in connection with future machines. We were convinced that an ep machine [electron–proton collider] would be a good thing because you could test asymptotic freedom. We met a lot of resistance amongst older people who just didn't take asymptotic freedom seriously at all. I don't know quite what it is, but I think that people who grew up doing chiral symmetry and things like that, sort of the general properties of chiral symmetry – and the parton model – you almost instinctively realise that there are all sorts of things that fall out of a theory of quarks and gluons that are flavour-blind, and to us it seemed incredible that people shouldn't take it seriously as the best candidate for strong interactions.

AP: Have you got any explanation of why they didn't?

MG: I don't know. In some cases it might be a kind of general scepticism. I mean, physics has been dragging along sort of in the dark for a number of years and some people of a certain generation I think have just got used to the idea that it's just going to keep on dragging in the dark. I think that well, there are also people who are just not used to thinking that field theory is going to be relevant to the strong interactions at all. They are not people who are so familiar with the details of the weak interactions and their implications for the strong-interaction symmetry that you see all these things falling out at once. I guess its largely a matter of kind of what you did before – what your perception of the world is [laughter].

NOTES AND REFERENCES

1   We saw in Section 5.4 that the first HEP neutrino experiment was taken to establish that there were two distinct species of neutrino, the electron-neutrino (v_e) and the muon-neutrino (v_μ). In charged-current weak interactions, the former species invariably converted to electrons, the latter to muons.

2   It is interesting to note that there was no gauge theory based upon the approximate spectroscopic SU(3) classification scheme of hadrons. That hadrons could be allotted to SU(3) multiplets was, from the point of view of electroweak theory and QCD, an accident deriving from the similar masses of the u, d and s quarks and the flavour-blindness of the strong interactions (QCD).

3   It is evident here that I have suppressed some intricacies of electroweak theory in the text. First, only left-handed particles – particles whose spin was aligned parallel to their direction of motion – were assigned to doublet families. Right-handed particles, with the opposite spin orientation, were assigned to singlets. This separation of left- and right-

handed particles made it possible to construct theories in which electromagnetic forces conserved parity while weak forces did not. The photon coupled equally to left- and right-handed particles, and therefore conserved parity (i.e. it expressed no preference for either handedness). The charged IVBs, in contrast, could only couple to members of doublets, the left-handed particles. This absolute preference for a particular handedness corresponded to the established 'maximal' parity-violation of charged-current weak interactions. The neutral IVB could couple to both doublets and singlets, and neutral-current weak interactions were therefore expected to violate parity conservation sub-maximally, to an extent determined by the Weinberg angle (Section 6.2).

Note also that the d' and s' members of the electroweak families were combinations of d and s quarks in proportions determined by the Cabibbo angle (Section 4.3). This corresponded to the experimental observation that in charged-current weak interactions u quarks converted most often to d quarks but occasionally to s quarks (and, according to GIM, that c quarks converted most often to s quarks but occasionally to d quarks).

4   Bjorken (1979, 9–10).
5   Smith (1974). References 7 to 12 below are all to Plenary Reports from the London Conference.
6   It could be argued that conference proceedings give a somewhat distorted view of the balance of communal *practice* because they tend to emphasise the novel over the routine. However, the aim of this section is to demonstrate that the old physics continued to outweigh the new in mid-1974, and this is already evident in the *Proceedings* of the London Conference. Since the new-physics traditions were the principal novelties discussed there, the above argument serves only to reinforce the demonstration of old-physics dominance.
7   Barger (1974), Halliday (1974).
8   Rosner (1974a). Resonance spectroscopy was construed in terms of the CQM as outlined in Section 4.2. Hadronic decays were interpreted in terms of 'current quarks', related to the 'constituent quarks' of the CQM via a Melosh-type transformation (see Section 4.4).
9   Kleinchnect (1974).
10 Iliopoulos (1974).
11 Gilman (1974).
12 Landshoff (1974).
13 This assertion deserves elaboration. I am not arguing in favour of any transcendental determination of thought processes. Undoubtedly Gaillard, De Rújula and Ellis could, in principle, have responded differently to the discovery of the new particles: they could have ignored them, or they could have sided with the theoretical opposition. However, in consequence of their expertise, they were amongst the theorists best-equipped to lay out and develop the gauge-theory arguments: it was a routine response for them, and it was what their colleagues expected from them. (The latter is meant in a constructive sense: experimenters sought guidelines as to what to investigate next, and turned to the

experts for them). Thus any other response would have seemed perverse (to Gaillard, De Rújula and Ellis themselves, and to their colleagues). This is the sense of inevitability I wish to convey both here and throughout the text.

14 The similarly contagious quality of experimental expertise is documented in Collins (1974, 1975b).

15 I am grateful to De Rújula, Ellis and Gaillard for interviews and the provision of biographical materials. The remainder of this chapter is based on those materials, and otherwise unattributed quotations are from the interviews. The publications cited below serve to typify the research careers of the three theorists, and are not an exhaustive list.

16 De Rújula (1968).

17 De Rújula, Lautrup and Peterman (1970).

18 De Rújula and De Rafael (1970).

19 De Rújula (1971).

20 See De Rújula, Georgi, Glashow and Quinn (1974).

21 De Rújula (1974a), De Rújula, Georgi and Politzer (1974).

22 De Rújula (1974b).

23 De Rújula and Glashow (1975). See also the works cited in later chapters.

24 Ellis and Renner (1969), Ellis (1970), Ellis, Weisz and Zumino (1971).

25 Ellis (1971).

26 Ellis, Finkelstein, Frampton and Jacob (1971).

27 Brower and Ellis (1972), Chanowitz and Ellis (1972).

28 Ellis and Jaffe (1973).

29 't Hooft (1974b). See Chapter 7, note 50, and Schwinger (1962b).

30 Neither was the experience wasted. In 1976 and 1977 he contributed to several papers and reviews on two-dimensional QCD: see Brower, Ellis, Schmidt and Weis (1977a,b).

31 Brower and Ellis (1974). The subsequent evolution of this collaboration is revealing of the relationship between expertise, context and the acquisition of new resources. An important element in the later papers of this series was the idea of quantising the theory on a discrete lattice of space–time points. Ellis and Brower were unfamiliar with this idea until they were invited to attend a conference on statistical physics at the International Centre for Theoretical Physics at Trieste in the summer of 1974. They were invited because the renormalisation group and field-theory techniques they used in HEP were, following Kenneth Wilson's work, of great interest to physicists studying phase transitions. At the conference, Ellis and Brower became familiar with the practice of solid-state physicists of working on a discrete lattice. They realised, as had Wilson before them (Chapter 7, note 50), that to adopt such a procedure in HEP offered considerable advantages in practical calculations, and they made it a central element of their later work (see Brower, Ellis, Savit and Zakrzewski 1975).

32 Ellis (1974).

33 See Ellis, Gaillard and Nanopoulos (1975) and the works cited in later chapters.

34 Gaillard (1968).

35   Gaillard (1965).
36   See Stern and Gaillard (1973).
37   Gaillard (1972).
38   Gaillard and Lee (1974a).
39   Gaillard, Lee and Rosner (1975). Although not published until 1975, this article was widely circulated in preprint form in 1974, before the discovery of the new particles.
40   Gaillard (1974).
41   See Ellis, Gaillard and Nanopoulos (1975) and the works cited in later chapters.
42   Gaillard and Lee (1974b).

PART III

ESTABLISHING THE NEW PHYSICS: THE NOVEMBER REVOLUTION AND BEYOND

9

CHARM: THE LEVER
THAT TURNED THE WORLD

9.1 The November Revolution
The November Revolution began in November 1974 and was essentially complete less than two years later. Its consequence was a dramatic rise to prominence of the new-physics traditions of theory and experiment. In this section I will outline the developments which made up the revolution; in the following sections I will discuss them in more detail. My aim is to set the revolution in the context of the traditions which have been described in Part II, and to show how, in turn, the revolutionary developments defined the context for the subsequent evolution of HEP.[1]

On Monday, 11th November 1974, the discovery was announced of the 'J-psi' – an extremely unusual elementary particle. The discovery was made simultaneously at the SPEAR electron–positron collider at Stanford, and at the Brookhaven proton synchrotron. In subsequent months, evidence for the existence of further, similarly anomalous, new particles came principally from electron–positron machines, and thus a primary component of the November Revolution was the rise to prominence of electron–positron physics. In parallel, though, novel programmes of experiment designed to investigate the production and properties of the new particles using photon, neutrino and hadron beams got under way at PSs around the world. Theorists advanced many speculations concerning the new particles, and by mid-1976 one of these was generally accepted within the HEP community: the new particles were manifestations of a new quark carrying the charm quantum number required by the GIM mechanism in unified electroweak gauge theories. The implications of this were manifold. First, the existence of a new hadronic quantum number implied the existence of a whole new family of hadrons, which in turn had implications for the new particle searches just mentioned. Secondly, GIM charm was, as we noted in Section 6.5, the missing piece of the electroweak unification jigsaw. Once the existence of charm had become established, the unified electroweak gauge theory proposed by Weinberg and Salam, augmented by the GIM mechanism, became the 'standard model' around which the subsequent programme of weak-interaction experiment was organ-

253

ised. Neutrino experiments were at the heart of this programme and were thus assured of a central place in HEP. Furthermore, in seeking to explain the new particles theorists drew not only on the resources of electroweak gauge theories; they also used the asymptotic freedom property of QCD. Theoretical models of the new particles were thus the first novel application of QCD. These applications constituted the exemplary achievements from which sprang new QCD-inspired traditions of hadron spectroscopy, around which subsequent experimental programmes were structured. Finally, the charm explanation of the new particles drew so directly upon the resources of atomic physics that the existence of quarks was put beyond debate: from 1976 onwards, the properties of the new particles were held to demonstrate the reality of quarks. In its composite aspect, at least, the new physics was established – although its gauge-theory component, especially QCD, was still open to dispute.

The November Revolution, then, comprised a constellation of related developments: experimentally, electron–positron physics achieved the highest status, neutrino physics was confirmed in its centrality, and novel programmes of charm-related research got under way at all laboratories; theoretically, that hadrons were quark composites was put beyond dispute, and gauge theory received a tremendous boost – the Weinberg-Salam plus GIM model became the standard electroweak model, while QCD became the basis of a new hadron spectroscopy. At the heart of these developments was charm. The protagonists of charm drew upon the new gauge-theory traditions of the early 1970s, and put them to work in the dramatic context of the discovery of the new particles. The triumph of charm was simultaneously a triumph for gauge theory, and it marked a watershed in the history of HEP. As John Ellis remarked later, 'charm was the lever that turned the world'.[2]

9.2   The R-Crisis

The story of charm began disastrously, with the failure of the data on electron–positron annihilation to conform to theoretical expectations. These expectations have been touched on in Sections 5.5 and 7.2 above, and I will briefly recapitulate their relevant features before turning to the empirical scene of the early 1970s. According to QED, electron–positron (e^+e^-) annihilation is a very simple process. Since the positron is the anti-particle of the electron, an electron and positron can cancel one another out to form a virtual photon (carrying all of the energy and momentum of the original leptons). The virtual photon can then decay into physical particles which share the available energy and momentum amongst them. One decay possibility is to a pair of muons, as shown in Figure 9.1 (a). An alternative possibility is that the photon converts into hadrons. In the quark–parton model this was assumed to be a two-stage process. The photon first converts into a quark–antiquark (q̄q) pair, which then rearranges itself into a shower of hadrons, as in Figure 9.1 (b). In accordance with the standard parton-model philosophy, theorists assumed that the production of a q̄q pair was, like that of a muon pair, exactly described by QED, and that the rearrangement of the quarks into hadrons did not materially alter the QED prediction. In the quark–parton model, then, the only difference between muon and hadron production in e^+e^- annihilation lay in the electric charges of muons and quarks. Total cross-sections for the respective processes were therefore expected to have the same energy dependence, given by QED, and the ratio R of total cross-sections for muon and hadron production was expected to be an energy-independent constant. The magnitude of this constant was expected directly to reflect the number of quark species and their electric charges (Section 7.2).

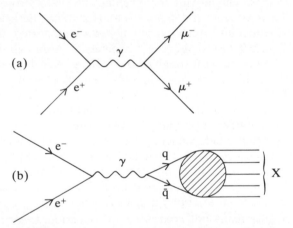

Figure 9.1. Electron–positron annihilation to form (a) a pair of muons, $e^+e^-\rightarrow\gamma\rightarrow\mu^+\mu^-$; (b) hadrons, $e^+e^-\rightarrow\gamma\rightarrow q\bar{q}\rightarrow X$.

The earliest data on e^+e^- annihilation at centre-of-mass energies above 1 GeV came from the Italian electron–positron collider, ADONE. As we noted in Chapter 7.2, by 1972 the ADONE data appeared to be in satisfactory agreement with parton-model predictions. Although reported with sizeable error bars, R seemed to be constant between 1 and 3 GeV (ADONE's maximum energy) at a value of around 2. This was the value expected if quarks came in three colours and three flavours (u, d and s), and parton modellers were thus encouraged to extend their calculations to more detailed aspects of e^+e^- hadronic production. However, problems arose in 1973 when data on e^+e^- annihilation were reported from the CEA 'Bypass' project. The CEA experimenters had made measurements at 4 and 5 GeV (above the region accessible to ADONE) and they reported values for R of around 5 or 6.[3] These numbers were markedly in excess of theorists' expectations and produced a feeling of unease amongst the theoretical community. The CEA data, though, consisted of only two points, with much larger error bars than those from ADONE, and the typical response was to prevaricate: the new purpose-built e^+e^- storage ring SPEAR was just coming into operation at SLAC; a similar machine, DORIS, was under construction at DESY; and these promised much more precise data in the near future.

Data-taking began at SPEAR in early 1973 (the year in which CEA closed down, and more than a year before physics began at DORIS). It quickly became apparent that e^+e^- annihilation refused to conform to theoretical expectations. The conflict came to a head in mid-1974, when Burton Richter – the moving force behind the SPEAR project, and leader of the SLAC team in the SLAC-LBL (Lawrence Berkeley Laboratory) experimental collaboration at SPEAR – reported the data shown in Figure 9.2 to the London Conference.[4] Although the SPEAR data were reported with sizeable errors, the straightforward inference from this plot was that R was linearly rising as a function of s (the square of the centre-of-mass energy) from a value of 2 at the highest ADONE energy, to around 6 at 5 GeV. This behaviour precipitated a crisis of confidence amongst theorists. They could accommodate almost any energy-independent value for R by suitable choices of quark flavours and colours, but they had no resources ready to hand for the explanation of a linear rise. At the London Conference, it fell to John Ellis to emphasise this point in his review of 'Theoretical Ideas about e^+e^- Hadrons at High Energies'. As he put it:

> The subject does really not need much introduction: theorists were almost unanimous in predicting that R . . . would tend to a constant . . . but these predictions have not been borne out . . . by experiments at CEA and SPEAR. Because of the gravity of this theoretical debacle, and the intensity of interest mirrored by the number of recent theoretical papers [on $e^+e^- \rightarrow$ hadrons] in this talk I allow myself to be deflected into mainly discussing this process in energy ranges presently (CEA and SPEAR) and soon to be (SPEAR II, DORIS) available.

There followed a discussion of various lines of theorising, some more outlandish than others, designed to account for the aberrant behaviour of R, leading Ellis to conclude:

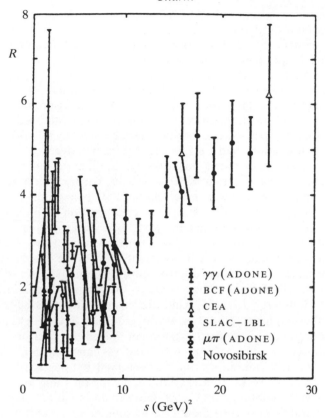

Figure 9.2. Electron–positron annihilation: *R* vs *s*, July 1974.

It is clear that there is no consensus among theoreticians working on e^+e^- annihilation – not even about such basic questions as whether or not it proceeds predominantly by one photon exchange, or whether or not to use parton ideas. At the moment developments in two directions seem necessary – the experimental exclusion of some of the theoretical alternatives, and finding good theoretical reasons for believing one model rather than another.[5]

To drive his remarks home, Ellis closed his talk by tabulating the predicted values for *R* from no less than 23 models: the predictions ranged from 0.36 to infinity.

9.3   The New Particles

In the summer of 1974 electrons and positrons threatened to annihilate the quark–parton model (and scale invariance and QCD) as well as one another. But theoretical disaster was averted when the R-crisis came to an abrupt close on Monday, 11 November 1974. This was the day on which the discovery of the J-psi – the first of the 'new particles' – was simultaneously announced from Stanford and Brookhaven. The news swept through the particle physics world, and when the implications had sunk in the R-crisis was over.

The J-psi was an extremely massive (3.1 GeV), extremely narrow particle. It was discovered at the Brookhaven AGS by an MIT group led by Samuel Ting, who were investigating e^+e^- pair-production by hadrons;[6] and at SPEAR by the SLAC-LBL collaboration led by Burton Richter and Gerson Goldhaber, who were investigating e^+e^- annihilation to hadrons.[7] Since it was SPEAR which set the pace of subsequent developments, I will first briefly discuss the Brookhaven experiment and then return to electron–positron annihilation and the resolution of the R-crisis. Ting's route to Brookhaven began in the 1960s in Hamburg.[8] In 1965, Ting and his collaborators began a series of experiments at the newly operational 7 GeV electron accelerator, DESY. The first experiment aimed to measure the production cross-section of e^+e^- pairs, and was conceived as a test of QED. An apparent violation of QED was observed, but was ascribed to production of the vector meson ρ which, having the same quantum numbers as the photon, could itself decay to e^+e^- pairs. Attention now switched from QED to the vector mesons, the ρ, ω and ϕ, and later experiments in the DESY series focused upon elucidation of their properties by detection of e^+e^- pairs in the appropriate mass-region (around 1 GeV). By the early 1970s, Ting and his group had unparalleled expertise in the detection of e^+e^- pairs, and Ting was looking for a new context in which to use it. He conceived the idea of a search for higher-mass vector mesons. To perform such a search a higher-energy beam was needed, and in 1972 Ting submitted a proposal for an experiment at the 31 GeV Brookhaven AGS. The first data were taken using a proton beam and a beryllium target in early summer 1974. The separation of e^+e^- pairs from a background of hadrons posed a much more severe problem in the Brookhaven experiment than it had in the electron–beam experiments at DESY but, after performing detailed checks, Ting's group were confident that they were observing a large number of e^+e^- events at a mass of around 3.1 GeV. They ascribed these events to the production and decay of a new vector meson which they named J, and this was the

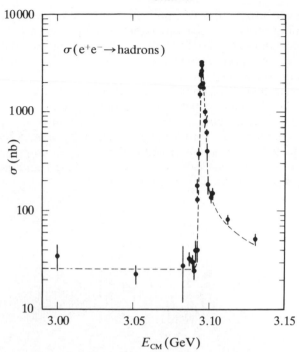

Figure 9.3. Cross-sections for electron–positron annihilation to hadrons as a function of centre-of-mass energy. The peak situated just below 3.10 GeV is the J-psi signal. At the peak the cross-section (and R) increases by a factor of around 2000.

discovery they announced on 11 November 1974.

On the same day, the discovery of a 3.1 GeV vector meson was reported from SPEAR.[9] The SPEAR experimenters named their particle psi (ψ) – hence the double-barrelled J-psi for the first of the new particles. (Later members of the new family emanated unambiguously from SPEAR, and were generically known as 'psis'). The discovery at SPEAR took the form of an enormous but narrow peak at 3.1 GeV in the total cross-section for e^+e^- annihilation to hadrons (Figure 9.3).[10] Peaks in cross-sections were routinely ascribed to particle production, and the SPEAR experimenters accordingly interpreted theirs as evidence of the production of a narrow vector particle with a mass of 3.1 GeV: $e^+e^- \to \gamma \to \psi \to$ hadrons.[11]

Having found the J-psi peak, the SPEAR experimenters set out to look for more. Changing the beam energy by small steps, they were rewarded two weeks later by a second narrow peak at 3.7 GeV. They ascribed this to production of another vector meson which they named psi-prime (ψ').[12] In the raw data on the psi and psi-prime peaks, cross-sections increased rapidly on the low-energy side but tailed-off more slowly on the high-energy side. The high-energy tails could be routinely ascribed to radiative corrections (see Chapter 5.1) which were calculable from QED. When such corrections were made and the effects of the ψ and ψ' peaks were subtracted, R was seen to be roughly constant at around 2 up to 4 GeV, whereupon complex structure became apparent until R stabilised again at around 5 above 5 GeV (Figure 9.4).[13] The apparent linear rise of R had disappeared; the R-crisis was over, and once more parton-modellers and gauge theorists felt free to ply their trade in this area.

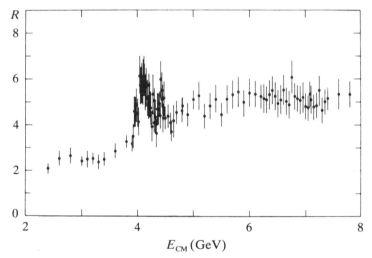

Figure 9.4. R as a function of centre-of-mass energy in electron–positron annihilation. The J-psi and psi-prime peaks have been removed from this plot: on this scale they would appear as vertical spikes, disappearing from the top of the frame at 3.1 and 3.7 GeV.

With the discovery of the ψ and ψ', attention switched from R to the new particles themselves. The developments which followed were fascinating but complex. I will summarise them here and go into more detail in the following sections. First, beginning in mid-1975, another family of massive long-lived particles began to emerge from observations on ψ and ψ' decays at SPEAR and the new e^+e^- collider DORIS at DESY. By mid-1976 the existence of five such particles was considered to have been established. And, in the spring of 1976, the first of yet another new family of particles – the D meson – was reported from SPEAR. As soon as the existence of the J-psi had become known, an army of theoreticians had advanced a whole range of models to account for it. In the wake of the accumulating data on the psis and their decay products, theoretical consensus rapidly converged upon the idea that these particles were manifestations of the existence of a *new quark flavour*. From the start, the favourite candidate for the new flavour was *charm* and the discovery of the D meson, which had just the properties predicted by charm models, set the final seal on this interpretation. Thus, by the time of the 18th Rochester Conference, held in Tbilisi in the Soviet Union in July 1976, charm was the accepted interpretation of the new particles. Later that year, Ting and Richter, the leaders of the two experimental teams which had set the whole process in motion, shared the Nobel Prize for Physics.

9.4 Charm

To particle physicists the properties of the J-psi and its relatives were extraordinary. The peculiarity of the new particles lay in their combination of large masses with long lifetimes. All things being equal it was a commonplace in HEP that the heavier a hadron, the shorter its lifetime. At 3.1 GeV, the J-psi was the heaviest particle then known, but it was also one of the *longest* lived: the lifetime of a particle is inversely proportional to its width, and at around 70 keV the width of the J-psi was around 2000 times smaller than straightforward expectations. This anomaly initially provoked speculation that the J-psi was not a hadron – perhaps it was the Z^0 intermediate vector boson of the weak interactions. But as data accumulated hadronic explanations quickly came to the fore, and it is with one of these, charm, that we are concerned here.[14]

As we noted at the end of the last chapter, gauge theorists had begun to speculate upon the properties of charmed hadrons even before the new particles were discovered. In the summer of 1974, for example, the detailed review 'Search for Charm' by Mary Gaillard, Benjamin Lee and Jonathan Rosner was widely circulated within the HEP community.[15] One of the many novel phenomena considered in this paper was the production of 'hidden-charm' states in e^+e^- annihilation: mesons composed of a charmed quark plus a charmed antiquark ($c\bar{c}$) in which the net charm cancelled out. As Gaillard, Lee and Rosner remarked, hidden-charm states would be expected to be relatively long-lived on the basis of the established phenomenology of the conventional hadrons. This expectation derived from the empirically successful 'Zweig rule' (discussed in Section 4.2). The Zweig rule stated that hadron decays involving quark–antiquark annihilation

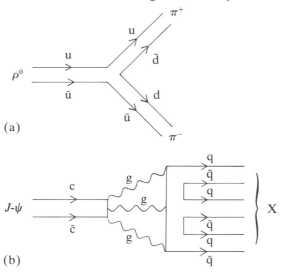

Figure 9.5. (a) Zweig-allowed decay, $\rho^0 \to \pi^+\pi^-$;
(b) Zweig-forbidden decay, J-$\psi \to$ conventional hadrons, X.

were suppressed relative to decays in which the constituent quarks maintained their identity (such as the $\rho \to \pi\pi$ decay shown in Figure 9.5 (a)). Because of the need to conserve energy and momentum, it appeared that this rule should apply to the lower mass hidden-charm states. If the lowest mass 'naked-charm' particles (c$\bar{q}$ or $\bar{c}$q, where q is a u, d or s quark) had masses greater than one-half of that of the psis, it would be energetically impossible for the psis to decay into a pair of them. Thus the only decay channels open to the psis would be Zweig-rule violating ones in which the c and $\bar{c}$ quarks annihilated one another. In QCD, for example, psi decays would proceed via c$\bar{c}$ annihilation to form gluons, which would then materialise as u, d and s quarks (Figure 9.5 (b)). The Zweig rule thus offered a qualitative explanation of the longevity of the psis. Quantitatively, however, the rule failed. Extrapolation of the suppression of conventional decays was insufficient to fully account for the anomalous properties of the psis: even when the Zweig rule was taken into account, their measured lifetimes were around 40 times longer than expected. This remaining discrepancy was, as Gaillard, Lee and Rosner put it, 'either a serious problem or an important result, depending on one's point of view'.[16]

The Charmonium Model

There were, of course, a number of theorists predisposed to adopt the
hidden-charm interpretation of the psi and thus to see its discrepant
lifetime as an important result rather than a serious problem. These
were the gauge theorists, and in justification of their position they
fastened upon the so-called 'charmonium model'. The inventors of
the charmonium model were two young Harvard theorists, David
Politzer – the discoverer of asymptotic freedom – and Thomas
Appelquist. In mid 1974, Appelquist and Politzer were looking for
novel consequences of QCD, and they hit upon the idea of considering
the production and decay of systems of new heavy quarks. Here they
believed that they could argue that because of asymptotic freedom – the decrease of the effective QCD coupling strength with
energy – a super-Zweig rule would obtain. A super-Zweig rule was
just what was needed to explain the longevity of the new particles, and
when these were discovered Appelquist and Politzer were not slow to
say so in print.[17] In late 1974 and early 1975 their work was taken up
and elaborated into the charmonium model by theorists at Harvard
and elsewhere, and became the central plank of the charm explanation of the new particles.[18]

The charmonium model explained the properties of the hidden-
charm states in terms of a direct and explicit analogy with one of the
simplest systems known in atomic physics: 'positronium'. First
observed by M. Deutsch in 1951, positronium was the name given to
the atomic bound states of an electron–positron pair.[19] The positronium atom was effectively a hydrogen atom – the simplest atomic
bound state, and the prototypical instance for the application of
non-relativistic quantum mechanics – in which the proton was
replaced by a positron. The standard quantum-mechanical analysis
of positronium therefore exactly paralleled that of the hydrogen
atom, except in one respect. Since the electron was the antiparticle of
the positron, it was possible for positronium, unlike the hydrogen
atom, to go up in a puff of smoke: the electron and positron
constituents of positronium could annihilate one another to form
photons.

This was one of the properties of positronium which was central to
the charmonium model. Just as positronium atoms decayed via e^+e^-
annihilation to photons so, according to perturbative QCD, the decay
of hidden-charm hadrons could be supposed to proceed via $c\bar{c}$
annihilation to two or three gluons (the gluons then rematerialising as
conventional quarks, as in Figure 9.5 (b)).[20] Because the hidden-
charm states were very massive, the energy and momentum carried by

these gluons had to be large in comparison with that involved in the Zweig-rule-violating decays of lower-mass hadrons. In QCD large values of energy–momentum implied small values of the effective coupling constant (asymptotic freedom). And this in turn implied that the decays of the hidden-charm states would be further suppressed relative to the Zweig-rule-violating decays of conventional hadrons. Thus the asymptotic freedom property of QCD guaranteed the operation of a super-Zweig rule in the decays of hidden-charm states – just what was needed to explain the most striking property of the psis.

The charmonium model worked, in the sense that it offered an explanation of the longevity of the psis, and it was transparent to any trained physicist by virtue of its analogical basis in atomic physics. The primitive formulation of Appelquist and Politzer thus served as a signpost, indicating how all of the resources of atomic physics could be brought to bear at the research front of HEP. One immediate and important development was that theorists began drawing atomic-physics style energy-level diagrams for the hidden-charm system, like that of Figure 9.6.[21]

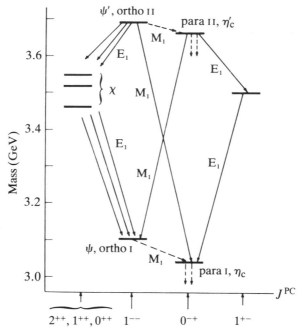

Figure 9.6. Masses and radiative transitions of charmonium states.

To construct such diagrams, theorists assumed that charmed quarks were very massive (around 1.5 GeV) so that their motion within hidden-charm states would be non-relativistic. This implied that it was legitimate to use the methods of non-relativistic quantum mechanics to compute the energy spectrum of the system, as was routinely done for positronium and the hydrogen atom. When these diagrams were first drawn, only the spin 1, negative parity J-psi and psi-prime (labelled ORTHO I and II, in analogy with the established nomenclature of the hydrogen atom) were known to exist. The existence of other hidden-charm particles with different spins and parities, corresponding to other states of the positronium spectrum, was a prediction of the model. The predicted masses of the additional particles, on a scale set by the masses of the J-psi and psi-prime, were dependent upon the interquark force, and it is important to stress that this force could not be deduced from QCD. To calculate the force from QCD would have amounted to a solution of the quark confinement problem, and no such solution was available. Instead, in their early work Appelquist and Politzer simply chose to use a Coulomb force – the electromagnetic force which acts between the electron and positron in positronium. The virtue of this choice was that quantum mechanical wave-functions for charmonium could then be extracted from textbooks on atomic physics, leading directly to numerical predictions. The disadvantage of the Coulomb force was that it no more confined quarks than it did the electrons and positrons seen every day in the laboratory. Theorists therefore soon switched to a force independent of quark separation, a 'linear potential' as it was known. The advantage of this choice was that it implied quark confinement: whatever the separation of the $q\bar{q}$ pair in space, there would be a constant force attracting the quarks back together again.[22]

The linear potential led to predictions of further hidden-charm particles beyond the ψ and ψ'. According to the relative spin orientation and angular momentum of the constituent quarks, these additional states would have the masses and J, P and C (spin, parity and charge conjugation) quantum numbers indicated in Figure 9.6. Unlike the psi and psi-prime, the new hidden-charm states did not have the same quantum numbers as the photon (1^{--}) and they were not therefore expected to appear as large peaks in the e^+e^- annihilation cross-section. Instead, as indicated by the arrows in Figure 9.6, the psi and psi-prime were predicted to decay via emission of a single photon into five of these particles (labelled χ, η_c and η_c').[23] Experimenters quickly set out to look for the photons which would signal the existence of such particles, and during 1975 and 1976 evidence for the existence of five additional hidden-charm particles was forthcoming from SPEAR and DORIS. The masses and decay rates reported were not exactly as predicted – the spin zero η_c and η_c' were especially deficient in this respect – but the very existence of the predicted number of states was taken to be a triumph of the charmonium model.[24] Discrepancies between prediction and data were taken as important results rather than serious problems: topics for further work rather than objections to the model. And, in confronting these discrepancies, theorists had at their disposal all of the resources of atomic physics: modified forces, spin–orbit couplings (now based upon perturbative QCD), relativistic corrections, and so on – the same conceptual tool-kit which had been put to use in constructing the CQM a decade earlier.

As a result of this fruitful stream of phenomenology, the existence of some new species of heavy quark had by 1976 become widely accepted as the explanation of the new particles. We will shortly see why it was also accepted that the new species had the properties required for charm. But before that two comments are appropriate on the establishment of the charmonium model. First, it is important to note that the transparency of the model (deriving from its connection to basic principles of atomic physics) made it attractive to experimenters as well as to theorists. The model provided a simple and compelling framework within which to develop an experimental strategy. It is clear that, at least after the discovery of the first χ states in mid-1975, experimenters at SPEAR and elsewhere oriented their research strategy around the investigation of the phenomena expected in the charmonium model. A similar comment applies to the search for naked charm described below. If, on the contrary, experimenters had oriented their strategy around some other model the pattern of experiments around the world would have been quite different, and the outcome of this whole episode might well have been changed.[25]

Despite this centrality of a particular model to experimental practice, it is still tempting to imagine that the discovery of the five intermediate hidden-charm particles constituted a direct verification of the charmonium model, along the lines implicit in the 'scientist's account': the empirical facts proved the theory was right. But if we look ahead to the subsequent fate of the 'facts' we will see that such an analysis is untenable. As noted above, two of the new particles, η_c and η_c' had reported masses and widths which were in conflict with the predictions of the charmonium model. In 1976 this conflict was seen as demanding further elaboration of the model. But, by 1979, a highly sophisticated experiment at SPEAR had convinced most physicists that the η_c and η'_c did not exist at all – at least, not with the masses originally reported. Thus the empirical basis of the charmonium model, a key element in the November Revolution, was retrospectively destroyed.[26] The original sightings of the η_c and η'_c had been made in a single experiment at DORIS and it was assumed that they were an instrumental or statistical artefact. There is nothing mysterious about such reasoning, but it does point to the role of theory in sustaining the perceived existence of natural phenomena. The choice of the HEP community in 1976 to accept the reported existence of the η_c and η'_c was structured by the centrality of the charmonium model to the contemporary symbiosis of experimental and theoretical practice. Data and theory did not stand in the adversarial relation of the 'scientist's account'.

Naked Charm

The successes of the charmonium model discussed above were seen as sufficient to justify a general belief that the new particles were manifestations of a new species of heavy quark. But the phenomenological analysis of the new particles as hidden-charm states was insufficient to justify an identification of the new quark with that required by the GIM mechanism. The GIM mechanism had to do with the suppression of strangeness-changing weak neutral-current decays, and the distinctive attributes of charmed quarks thus concerned their weak interactions. Observations confined to hidden-charm states could shed no light on the weak interactions of the constituent quarks: the decays of the ψ and ψ' to the χs, η_c, η'_c and to normal hadrons were all dominated by the strong interaction.[27] From 1974 onwards evidence had accumulated from neutrino experiments that a new species of quark having the weak interactions postulated for charm was being produced.[28] But all eyes were fixed on SPEAR: this was where the new particles had been discovered, and this was where the critical evidence for charm was expected to emerge. At SPEAR the experimenters searched for the production of hadrons carrying naked charm: $c\bar{q}$ and $\bar{c}q$ mesons, in which q was a u, d or s quark. Simple considerations of quark masses indicated that the lowest-lying naked-charm mesons should have masses somewhere in the vicinity of 2 GeV. The charm quantum number was supposed to be conserved by the strong interaction, so the only way in which such mesons could lose their charm would be through weak decays. In the GIM scheme, the charmed quark was a member of the same doublet family as the s quark, and the weak decays of charmed hadrons were therefore expected to lead to the production of strange particles, kaons.

Based on these considerations, the search for direct or indirect evidence for the production of naked-charm particles got under way at SPEAR very soon after the psis had been discovered. It proved unexpectedly long and arduous. In mid-1975 the SLAC-LBL group published a report on an unsuccessful analysis of extended measurements at 4.8 GeV centre-of-mass energy.[29] Further unsuccessful searches followed at 4 GeV and 7 GeV, leading one SPEAR experimenter to remark in February 1976: 'Nobody understands it – just nobody understands it'.[30] Perhaps the easiest way to understand it would have been to recognise that the new particles were not manifestations of charm, but in the theoretical context of the time the experimenters did not find this an attractive option.

In April 1976 Gerson Goldhaber, the leader of the LBL group at SPEAR, attended an HEP meeting at Madison, Wisconsin, where Glashow and others impressed on him the importance of finding naked charm. Returning to Berkeley, Goldhaber began a detailed examination of the latest SPEAR data. Through a shift in interpretative practice reminiscent of that involved in the neutral current discovery, he found that it was possible to exhibit evidence for the existence of naked charm.[31] Francois Pierre, a visitor to SLAC from Saclay, France independently reached the same conclusion.[32] Examining the masses of $K\pi$ and $K\pi\pi\pi$ combinations, Goldhaber and Pierre found the narrow peaks in event rates at around 1.87 GeV shown in Figure 9.7.[33] These peaks were identified as being due to the production of a long-lived neutral meson, D^0, decaying to the observed combinations of kaons and pions. In reporting the existence of the D^0, the SLAC-LBL group commented: 'The narrow width of this state . . . . and the fact that the decays we observe involve kaons form a pattern of observation that would be expected for a state possessing the proposed new quantum number charm'.[34]

The paper reporting the existence of the D^0 was submitted for publication in June 1976. It was quickly followed by a second paper reporting the observation of narrow charged state D^+ at a similar mass but decaying to $K\pi\pi$.[35] The D^0 and D^+ mesons had just the properties expected of naked-charm states and their discovery put the nature of the new particles effectively beyond doubt.[36] By the time of the Tbilisi HEP Conference in July 1976 there was a firm consensus in favour of the existence of charm.[37] The critical phase of the November Revolution was over.

One final theoretical development deserves mention here. Even before the SPEAR paper on the discovery of the D meson was submitted for publication, three Harvard theorists had analysed its reported properties in detail in order to argue that it was indeed a manifestation of GIM charm.[38] The three theorists were Sheldon

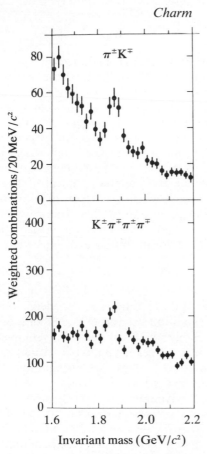

Figure 9.7. D⁰ peaks in the SPEAR data.

Glashow, Alvaro De Rújula and Howard Georgi, and they argued that the properties of the D were just those they had predicted in a paper published in 1975.[39] Their 1975 paper, though, had a much wider significance than simply the prediction of the properties of naked charm. In it they extended the use of perturbative QCD beyond heavy quark systems (hidden and naked charm) to the analysis of light quark systems and the spectroscopy of the old hadrons. In so doing, they set the scene for a rejuvenation of the existing constituent quark model tradition. I will discuss the development of the new, improved CQM in Chapter 11, but it is proper to mention it here, since the new quark spectroscopy was itself an important contribution to the November Revolution.

269

9.5 A New World

The charm explanation of the new particles constituted a nexus of all the new gauge-theory traditions of the 1970s. Through the spectacles of gauge theory (particularly, the charmonium model) the new particles were seen to be the simplest of hadronic composites. Experiment and theory were mutually supporting – theory asked questions which experiment could answer, whilst in turn experimental data provided (apparently) solid ground from which the theory could be extended. This marriage of theory and experiment, consummated by the discovery of the Ds and the establishment of charm, transformed the face of HEP in ways which I will indicate below and elaborate upon in the following chapters. The transformation necessarily involved the simultaneous reorganisation of theoretical and experimental perspectives but, for the purposes of exposition, let us consider theory first.

Theory

I have discussed the evolution of the traditions of HEP theory which entered into the November Revolution in terms of the elaboration of two analogies: a composite analogy which saw hadrons as composite systems and which drew upon resources from atomic, nuclear, and solid-state physics; and a theoretical analogy which sought to model gauge theories of the weak and strong interactions upon the development of quantum electrodynamics. The most direct consequence of the November Revolution was the transformation of the status of the former analogy: hadrons were no longer said to be 'like' composite systems – from 1976 they just *were* composites of quarks. The establishment of charm, and especially the success of the charmonium model, made quarks real. The existence and properties of the hidden-charm states were the exemplary demonstration of this fact. The force of the demonstration derived from the structure of the charmonium model: in the model, the charmed quarks had just the same status as the electrons and positrons in the routine analysis of the positronium atom, and physicists were completely accustomed to seeing electrons and positrons as real. Of course, this demonstration applied, strictly speaking, only to charmed quarks, but in the contemporary theoretical context charmed quarks were simply a new flavour of the old quarks, and the demonstration of reality spilled over to the latter as well.[40]

As far as gauge theory was concerned, matters were not so clear cut. Two gauge theories of quark interactions were at stake – electroweak theory and QCD – and the implications of charm differed for each. For electroweak theory the implication of charm was simple: the Weinberg–Salam plus GIM model fitted all the contemporary data. After the discovery of the weak neutral-current charmed hadrons had been the piece of the jigsaw which was conspicuously missing. Now it had been found. Weinberg–Salam plus GIM became the standard model, upon which all subsequent experimental strategies were, in one way or another, predicated.[41] For QCD the situation was more complicated. Since the confinement problem was unsolved, perturbative QCD entered into the charmonium model, and its extension to quark spectroscopy in general, in a pragmatic *ad hoc* fashion. The proof of the pudding was in the eating, and the value of QCD in this respect was not clearly established in 1976. Thus, while electroweak theories had become central to theoretical and experimental practice in weak-interaction physics by 1976, the same could not be said of QCD and strong-interaction physics. Charm was the lever that turned the world, but by itself it was insufficient to establish fully the gauge-theory element of the new physics.

Experiment

Experimental perspectives were restructured rather drastically during the November Revolution. First the drama of the *R*-crisis and its resolution thrust e^+e^- annihilation to the forefront of physics interest. Until the beginning of the *R*-crisis, e^+e^- colliders had produced few surprises for particle physicists and little data to stimulate the major traditions. From the 11 November 1974 to the end of the decade their data were perceived to be uniquely informative.[42] Other developments were equally straightforward. Whilst e^+e^- machines had the advantage in producing a very clear signal of the existence of massive long-lived particles, other machines had the advantage in energy, intensity and variety of quantum numbers which could be produced. Programmes of experiments designed to produce and investigate the properties of charmed particles got under way in 1975 at all of the major centres of HEP experiment. Experiments using photon, neutrino and hadron beams were quickly planned, and continued throughout the decade.[43] Finally, the establishment of charm, via its relationship to the standard electroweak model, also confirmed neutrino experiments in their new-found high status consequent upon the 1970s studies of scaling and neutral currents.

Thus three general categories of experiment moved to the centre of attention during the period from 1974 to 1976: e^+e^- annihilation; neutrino scattering; and general investigations of heavy particle production using lepton or hadron beams. The experimental tradition which was least affected by this transformation was that which was most productive: high-energy hadron scattering at PSS. Certainly charm searches did get under way at these laboratories, but these in no sense provided for full utilisation of the available facilities. In hadronic scattering the old-physics traditions of experiment persisted, and as long as these experimental traditions persisted, so too did the old theoretical traditions which fed upon them. In the event, though, the old-physics traditions of experiment soon went into decline. The November Revolution was a showcase for QCD and encouraged further elaboration of the theory in its application to scaling and hard scattering as well as to hadron spectroscopy. The former application served in turn to reinforce existing experimental programmes of hard-scattering experiments at PSS. By the late 1970s these had grown and diversified to such an extent that the experimental HEP community could fully occupy itself with new-physics concerns. The penetration of experimental practice by QCD constituted the last act in the establishment of the new physics, and is the subject of Chapter 11.

Charm and Social Unity

In closing this chapter, one point remains to be emphasised. It was no coincidence that the charm explanation of the new particles triumphed and that charm's protagonists succeeded in contributing to so many traditions of HEP theory and experiment. The charm explanation sustained a multifaceted symbiosis of practice, wherein gauge theorists, hadron spectroscopists and experimenters in a whole range of traditions could draw upon one another's research for justification and subject matter for future practice. And within this symbiosis the charm explanation was itself sustained. By 1976, charm was central to such a range of practice that arguments could always be marshalled in favour of seeing misfits between prediction and data as important results rather than as serious problems. That, for example, the hidden-charm η_c and η_c' particles first appeared at the wrong masses and then disappeared caused few doubts, if any, over the validity of charm. Such discrepant phenomena were simply seen to call for further theoretical and experimental work, which would itself command the attention of a wide audience. Alternative models of the new particles did not inspire such dynamic acquiescence. For reasons I have analysed elsewhere, these models neither unified theoretical traditions nor penetrated experimental practice, and their empirical misfits were seen as straightforward counter-arguments by the HEP community at large.[44]

The triumph of charm, then, and the eclipse of its rivals, should not be seen in terms of static comparisons between predictions and data. At no point, before 1976 or after, was charm proved to be right or its rivals wrong. The key to charm's success lay in the social and conceptual unification of HEP practice which was achieved during the November Revolution. And, as we shall see in the following chapters, progressive social unification in the common context of gauge theory was intrinsic to the establishment of the new physics in its entirety.

NOTES AND REFERENCES

1   For more detailed analysis and documentation of the developments discussed here, see Pickering (1978, 1981c). Popular accounts have been given by Drell (1975), Glashow (1975) and Schwitters (1977). For technical reviews from a variety of perspectives, see Goldhaber (1977), Feldman and Perl (1977), Appelquist, Barnett and Lane (1978), Sucher (1978) and Goldhaber and Wiss (1980).
2   Ellis, interview.
3   For the CEA data, see Strauch (1974). For a contemporary theoretical response, see Bjorken (1974).
4   Richter (1974, 41, Fig. 4).
5   Ellis (1974, 20, 30).
6   Aubert *et al.* (1974).
7   Augustin *et al.* (1974). Alerted to the discovery, the experimenters at ADONE were quick to confirm the observation of the J-psi in e^+e^- annihilation: Bacci *et al.* (1974). (This involved operating ADONE at just above its maximum design energy of 3.1 GeV; the psi-prime (see below) remained inaccesible at ADONE). The J-psi was also sighted at DORIS in its first physics runs: Braunschweig *et al.* (1974).
8   For a detailed and documented account of the events leading up to the Brookhaven experiment and of the experiment itself, see Ting's Nobel lecture, Ting (1977). See also Ting (1976).
9   The simultaneous announcement was not a coincidence. It was precipitated by mutual revelations during Ting's visit to SLAC for a meeting of the SLAC Program Advisory Committee on 11 November 1974 (see Ting 1976, 127, Goldhaber 1976, 140 and Richter 1976a, 147). It should be noted that there was a feeling in certain quarters of the HEP community that perhaps the Brookhaven and SLAC discoveries were not truly independent. Certainly the MIT group had evidence in hand for the existence of the J-psi significantly earlier than the SPEAR group. The suspicion that the latter group had prior information on the findings of the former is hinted at in letters to *Science* by Ting and Deutsch (1975) of MIT and equally obliquely rejected by SLAC's Director, Panofsky (1975). It seems impossible to document the truth or falsity of such suspicions.
10 Schwitters (1975, 12, Fig. 8).
11 The discovery of the psi can be seen as an instance of the 'tuning' of experimental techniques (see Chapter 1) to credible phenomena. According to Gerson Goldhaber (1976) – the leader of the Lawrence Berkeley Laboratory group at SPEAR – the process which culminated in the discovery began in early 1974, with an anomalously high cross-sec-

tion measurement at 3.2 GeV (30% above the measured cross-sections at neighbouring energies). This did not seem terribly important, but by mid-October 1974 further examination of the data had revealed another anomaly at 3.1 GeV. Eight experimental runs had been made at 3.1 GeV: 6 of them had revealed normal cross-sections, while 2 revealed much higher cross-sections (by factors of 3 and 5). At this point, 'One obvious and simple cause which we considered was that something had gone wrong with our detector' (Goldhaber 1976, 135). However, when the data from the anomalous runs were scrutinised more closely, an excess of kaon production was observed. 'This was the last straw: an increase in the cross section and an increase in strange particle production – just what the charm advocates had been talking about' (see Section 4 below for the connection between charm and strange particles). Although it later appeared that the increase in kaon production was 'part misidentification and part a statistical fluctuation' (Goldhaber 1976, 136) it was sufficient to encourage Goldhaber to suggest more measurements in the region of 3.1 GeV. During the summer of 1974 SPEAR had been shut down for three months for conversion to operation in the 5 to 9 GeV energy range (SPEAR II). With some difficulty, arrangements were made to lower the energy again, and on 9 November 1974 more data began to be taken around 3.1 GeV. Monitoring the beam energy very precisely, the experimenters obtained curves like that of Figure 9.3. These they regarded as manifestations of a genuine phenomenon, and accordingly put aside their worries about detector performance. They had not proved that the detector was working perfectly; they assumed this because it produced credible evidence.

12   Abrams *et al.* (1974). No evidence for the psi-prime was observed in Ting's experiment.

13   Schwitters (1975, 12, Fig. 8).

14   Early evidence in favour of the hadronic nature of the J-psi came from the observation at Fermilab that it was photoproduced at typical hadronic rates: Knapp *et al.* (1975).

15   Gaillard, Lee and Rosner (1975). See also Glashow (1974).

16   Gaillard, Lee and Rosner (1975, 300). This comment was part of a post-psi addendum to the published version of their paper.

17   Appelquist and Politzer (1975).

18   For early articulations of the model at Harvard, see De Rújula and Glashow (1975) and Appelquist, De Rújula, Politzer and Glashow (1975). Many groups developed the model along similar lines. Amongst them, the Cornell group was most influential: Eichten, Gottfried, Kinoshita, Kogut, Lane, and Yan (1975). During 1975, Appelquist and Politzer (1975) was cited 108 times in the HEP literature. De Rújula and Glashow (1975) received 92 citations, and Gaillard, Lee and Rosner (1975) 105 citations. I thank Edward Nadel for supplying these figures and those quoted in note 42 below; a detailed co-citation analysis of the charm literature is currently in progress.

19   Deutsch (1951).

20   Annihilation to a single gluon was impossible because the gluons were supposed to be coloured while the hidden-charm hadrons were colour-

less. Whether two or three gluons were involved depended on the quantum numbers of the hidden-charm state in question.

21 Appelquist, De Rújula, Politzer and Glashow (1975, 365, Fig. 1). The labelling of the ψ, ψ', χ, η_c and η_c' states has been added for convenience of exposition.

22 This is in contrast to the Coulomb force which diminished as the inverse square of the distance separating a pair of charged particles. The choice of a constant interquark force was supported by theoretical prejudice. In 1969 Nambu proposed that hadrons have a one-dimensional string-like structure, the quarks being situated at the ends of the strings. In developing this idea, Nambu had drawn upon his expertise in solid-state physics, wherein magnetic flux was thought of as trapped in superconductors in string-like vertices. This analogy was extended in a gauge theory context by H.B.Nielsen and P.Oleson at Copenhagen and others. It was also shown that the dual resonance model (an extension of the S-matrix programme) could be reformulated as a field theory of strings. In all of these theories, the energy of the strings was proportional to their length, and a constant interquark force immediately followed. (For a review of this work see Rebbi 1974). The existence of traditions of work on string theory leant credibility to Wilson's lattice-expansion approach to QCD, which itself gave rise to string-like structures of gauge fields linking quarks (see Chapter 7, note 50).

23 The labels E_1 and M_1 refer to different types of electromagnetic transitions, 'electric-dipole' and 'magnetic-dipole' respectively. Note that to get to the single predicted state with quantum numbers 1^{+-} required emission of two photons. Such a transition was not expected to be experimentally observable, and I will ignore it here.

24 See for example the Plenary Report of DORIS experimenter B.Wiik (1976) to the Tbilisi Conference in July 1976. The Tbilisi Conference was the next in the Rochester series after London. Wiik's review contains full references to the experimental literature.

25 This point is best made comparatively (see Pickering 1978, 1981c for more details). In the earliest days of the new particles two theoretical models were considered most promising, charm and 'colour'. The colour model asserted that, unlike conventional hadrons, the psis were coloured particles: they were excitations of the colour quantum number, rather than manifestations of a new quark flavour. One prediction of the colour model was that charged partners of the psi should exist (coloured versions of the ρ) while in the charmonium model only electrically neutral psis were possible. Experimentally the charged psis, $\psi^{\pm}$, were expected to be detectable via the single pion decay of the psi-prime, $\psi' \to \pi^{\mp} \psi^{\pm}$. A 1975 analysis of the DORIS data led to the conclusion that if such decays took place they constituted less than 5% of all ψ' decays (Wiik 1975, 76). Thereafter searches for charged psis were discontinued. This is to be contrasted with the determined and repeated programmes of experiment which led up to the observation of the χ and η_c states. The first such state was reported from DORIS in July 1975 and by August 1975 three had been reported. This was sufficient to encourage Bjorken (1975, 991) to describe charm and the charmonium model as 'standards

of reference which an experimentalist will naturally use to interpret his data'.

The search for the D mesons at SPEAR (see below) contrasted even more markedly with that for charged psis, in that successive failures to find evidence for the existence of Ds were taken as spurs to further detailed investigations and not as evidence that charm was wrong. It is impossible to say what would have happened if experimenters had pursued the predictions of colour with the same tenacity they devoted to charm. (For a review of the technical literature on colour, see Greenberg and Nelson 1977).

26　The experiment which established the non-existence of the η_c and η_c' at their hitherto accepted masses was the so-called 'Crystal Ball' experiment at SPEAR. The Crystal Ball was a massive array of sodium iodide photon detectors. Its construction was proposed in 1974 but only approved in 1975. At that time Elliott Bloom, the spokesman for the Crystal Ball collaboration, 'pointed out how useful the Ball would be in looking for the sharp γ [photon] lines just predicted by the charmonium model' (Bloom: private communication to the author). His arguments were accepted, and four years (and $4 million) later the first results were announced at several HEP conferences (see, for example, Kiesling 1979). These results confirmed the existence of the 3 previously reported χ states, but no sign of the 2 η_cs was to be found. As it happened this did not displease HEP theorists. A series of sophisticated charmonium analyses had made it increasingly apparent that it was difficult to reconcile the model to the reported properties of the η_c and η_c' (see Shifman et al. 1978 and Novikov et al. 1978). Their disappearance therefore came as 'some relief to afflicted theorists' (Davier 1979, 200). The theorists did, however, still want the η_cs to exist somewhere, preferably closer to the predicted masses than the original candidates, and the Crystal Ball experimenters took it upon themselves to find them:

> The disappearance of the X (2820) and chi (3455) is certainly welcome by the proponents of the simpler charmonium model, given the well-known difficulties of the model ... The burden of finding the theoretically necessary pseudoscalar partners of the J-psi and psi-prime is still on the experimenters. If the model predictions of the radiative decays of J-psi and psi-prime into η_c and η_c' are roughly correct *and* the level splittings between J-psi-η_c and psi-prime-η_c' are no smaller than ~ 30 MeV ... there is a fair chance that these states will be found in the Crystal Ball experiment (Kiesling 1979, 302).

By the summer of 1979 there were signs of a new candidate emerging for the η_c from the Crystal Ball data, at a mass of 2980 MeV – 160 MeV heavier than the previous candidate, at roughly the mass expected from the charmonium model (Kiesling 1979, 302).

27　This remained the case even when the strong interaction was suppressed by a super-Zweig rule.

28　As usual neutrino experiments were seen as an especially clean source of information on weak interactions, and several suggestive reports from Brookhaven, CERN and Fermilab were published during 1975 and 1976:

Benvenuti *et al.* (1975 a, b, c, d), Cazzoli *et al.* (1975), Deden *et al.* (1975b), Barish *et al.* (1976), Blietschau *et al.* (1976), Burhop *et al.* (1976) and von Krogh *et al.* (1976). All of these reports could be seen as evidence in favour of the existence of charm, but none were entirely convincing. The reports could be divided into two classes. In one class were a handful of events recorded in bubble chambers (or photographic emulsions: Burhop *et al.* 1976). Here events could be reconstructed from the observed tracks and ascribed to charmed hadrons with the expected masses and weak decay paths. The first member of this class, Cazzoli *et al.* (1975) came from a Brookhaven neutrino exposure. Although each event was suggestive, their combined statistical significance was slight.

Events in the second class were more plentiful but their interpretation in terms of charm was indirect. The earliest members of this class were the 'dimuon' events observed in the HPWF electronic neutrino experiment at Fermilab (Benvenuti *et al.* 1975). These were events in which two muons (of opposite charge) were produced: one muon could be routinely ascribed to conversion of an incoming neutrino in a charged-current interaction, the other to the weak decay of a charmed hadron. Two such events were reported to the London Conference in 1974 (Rubbia 1974) and 73 had been collected by November 1975 (when the HPWF apparatus was dismantled for improvements). Unfortunately, it was impossible to calculate the masses of the decaying hadrons from the HPWF data, and neither was it possible to ascribe the dimuons definitively to charm. For a popular account of the dimuon discovery, see Cline, Mann and Rubbia (1976).

The greatest difficulty which physicists experienced in ascribing the early neutrino data to charm arose from the conflict with the non-appearance of naked-charm in e^+e^- experiments. This conflict was resolved in mid-1976 (see below) and neutrino experiments then became seen as important sources of genuine information on charm. A similar comment applies to photoproduction experiments. The earliest member of this category was a Fermilab observation of a candidate charmed antibaryon, Knapp *et al.* (1976). For a review of the many charm searches (most unsuccessful) in all types of experiment up to February 1977, see Particle Data Group (1977).

29 Boyarski *et al.* (1975).
30 Quoted in *New Scientist* (1976a, 440). This article is a report on an HEP meeting held at the Rutherford Laboratory in February 1976 at which the unsuccessful searches mentioned above were announced.
31 The key signal for the production of charm in e^+e^- annihilation was the expected rise in the proportion of kaons produced. Early analysis of the SPEAR data had failed to show any such rise. After his meeting with Glashow in Madison, Goldhaber proceeded as follows (interview):

> I dropped what I was doing and decided I'm going to look for charmed particles or find out why they're not there ... I really worked on it hard and three days later I found charmed mesons. The data is all there but you have to know how to look. The main breakthrough for me was that I decided to be less timid. People had taken the assumption: well, we don't know if something is a K so

we'll better be safe and only call something a K if you're absolutely sure – and by being absolutely sure one sort of killed the statistics. I decided to push that limit a bit further and accept events where they were likely to be Ks but where there wasn't absolute proof that they're Ks (from time-of-flight identification). I decided to push that a little bit further and this was the thing that led me to find the charmed mesons.

With the evidence in hand, Goldhaber telephoned Glashow who announced the discovery to the world through an article in *The New York Times*. Details of the procedure eventually used to discriminate between pions and kaons in the SPEAR data were given in the more routine announcement by Goldhaber, Pierre *et al.* (1976). Here it is clear that the ultimate justification for the discrimination criteria adopted was the production of credible phenomena: the existence of charmed mesons. This episode is a further example of the stabilisation of experimental procedures against theoretical conceptions of the phenomena at issue. (In this context, is also interesting to note the role played by kaon misidentification in the discovery of the J-psi at SPEAR: see note 11 above).

32 See Goldhaber and Pierre (1976).

33 Goldhaber, Pierre *et al.* (1976, 256, Figs 1 (e) and (f)).

34 Goldhaber, Pierre *et al.* (1976, 258–9).

35 Peruzzi *et al.* (1976).

36 Further support for charm came from the observation that the D mesons appeared to be produced in e^+e^- annihilation in association with heavier particles, D*, at a mass of around 2010 MeV: $e^+e^- \rightarrow \gamma \rightarrow D\bar{D}^*$ (or $\bar{D}D^*$). It was subsequently observed that the D* particles decayed to Ds by pion emisson: $D^{*+} \rightarrow D^0\pi^+$ (Feldman *et al.* 1977) and that the spins of the D and D* were most probably 0 and 1 respectively (Nguyen *et al.* 1977). This pattern was expected in the charm model where the D* was seen as a spin 1 $c\bar{q}$ excited state of the spinless D (see De Rújula, Georgi and Glashow 1976).

Once the existence and naked-charm interpretation of the D and D* had been established, an explanation for the complicated bump structure in *R* between energics of 4 and 5 GeV (Figure 9.4) became obvious. The bumps were due to production of psi-like states above the charm threshold: hidden-charm states which were sufficiently massive to be able to decay into a pair of naked-charm mesons: $\psi'' \rightarrow D\bar{D}$ etc. Such decays did not involve the annihilation of the $c\bar{c}$ constituent quarks; thus they were not suppressed by the Zweig rule; and thus the corresponding peaks in *R* were relatively broad, as observed.

37 See, for example, the Plenary Reports given at Tbilisi by Wiik (1976) and De Rújula (1976).

38 De Rújula, Georgi and Glashow (1976).

39 De Rújula, Georgi and Glashow (1975).

40 The transformation of the status of quarks, and the importance of the new particles in establishing their reality, is clearly evident in technical and popular reviews produced by particle physicists from 1976 onwards: see, for example, Drell (1977, 1978), Ne'eman (1978) and Weisskopf

(1978). A neat illustration of the key role played by the new particles during the November Revolution is given by the frequent references to their significance as the 'hydrogen atom' of strong-interaction physics. In effect, this reference was a *double entendre*: on the one hand, by explicit construction, the charmonium model *was* the hydrogen (or positronium) atom transplanted to strong-interaction physics; on the other hand, there was the hope, at least, that the properties of the new particles would prove amenable to accurate perturbative calculations in QCD, and would hence vindicate QCD in the same way as the properties of the hydrogen atom had vindicated QED. For an example of this line of reasoning, see Eichten *et al.* (1975, 371). Five years later, one of the authors of this paper readdressed himself to the question 'Are They the Hydrogen Atoms of Strong Interaction Physics?' and concluded 'We have discovered not just one, but two hydrogen atoms of hadronic spectroscopy!' (Gottfried 1981, 148). The two 'hydrogen atoms' were the hidden-charm states and the 'upsilons' (see Chapter 10).

41 For quantitative analyses of the early impact of the new particles on the weak-interaction literature, see Sullivan, Koester, White and Kern (1980) and Koester, Sullivan and White (1982).

42 From their analysis of the HEP literature, Irvine and Martin (1983a, 22 3) conclude that between 1973 and 1976, 8 experimental publications were cited more than 100 times within a year. Of these no less than 5 originated at SPEAR. The reports in question were: the discovery of the ψ (316 citations during 1975); the discovery of the ψ' (227 citations during 1975); the discovery of the charmed mesons D and D* (see note 36 above); and the discovery of the heavy lepton τ, discussed in the following chapter. Of the three other highly cited experimental papers, two reported the discovery of the J at Brookhaven and the confirmation of the J-ψ at ADONE. The remaining paper reported that the proton–proton total cross-section rose with energy throughout the energy range opened up by the CERN ISR; this was a result of major importance for workers in the Regge tradition.

43 For the earliest photoproduction experiments see notes 14 and 28 above; for neutrino experiments, see note 28. The earliest relevant hadron-beam experiment was of course the discovery experiment at Brookhaven. Quantitative data on the growth of these experimental programmes is given in Chapter 12.

44 See the discussion of the colour model in Pickering (1981c).

10

THE STANDARD MODEL OF
ELECTROWEAK INTERACTIONS

This chapter reviews the development of electroweak physics in the latter half of the 1970s. Section 1 chronicles the discovery of new quarks and leptons. These required the addition of new families to the ontology of the Weinberg–Salam model, but could be accommodated without change to its basic structure. (Theoretical investigations of the new quark systems also drew upon QCD and are included in Section 1 for convenience.) Section 2 summarises developments in the main-line of empirical electroweak research – neutrino experiment – and reports on novel investigations of the weak neutral-current interactions of electrons. The upshot of all of this work was that by 1979 consensus had been established on the validity of the standard, Weinberg–Salam, electroweak model (with an appropriately expanded set of fundamental particles). The emergence of this consensus was, however, not entirely smooth. Data on both neutrino scattering and neutral-current electron interactions at times threatened the electroweak element of Bjorken's 'new orthodoxy'. Section 2 pays particular attention to the recalcitrant data, and Section 3 discusses how they were brought to heel.

10.1 More New Quarks and Leptons
In 1964 the motivation for the invention of charm had been to ensure a symmetry between quarks and leptons, the four quarks (u, d, s, c) mirroring the four leptons (e, v_e, μ, v_μ). With the establishment of charm during the November Revolution it appeared for a while that such a symmetry was indeed manifest in nature, but this happy situation did not obtain for long. Soon a new lepton was discovered, and then another quark. Here I will outline the circumstances surrounding these discoveries, and their impact upon HEP theory.

The Mini-R-Crisis and the Tau Lepton
The strong form of the R-crisis – the apparent linear increase of R with s – was banished by the discovery of the new particles, never to return. But a weaker form of the crisis was to remain for some time. At around 5 GeV – above the spikes and bumps ascribable to the production of the psi and its relatives – R was seen to level out as

expected. Unfortunately it levelled out at the wrong value: around 5, as against the value of $3\frac{1}{3}$ expected for 4 flavours and 3 colours of quarks (see Figure 9.4). In QCD R was supposed to reach its asymptotic value from above, but the gap was too large for gauge theorists' comfort and showed no sign of decreasing with energy. This discrepancy was the origin of the 'mini-R-crisis', manifested in a hesitancy to accept the charm interpretation of the new particles, and the construction of alternative heavy quark models.[1] These models differed from charm in the number of quarks and the relations between them, thus making possible different predictions for R. An alternative explanation of the excessive magnitude of R was also available, however, and it was the latter which won the day.

The alternative was that at least part of the excess R was due not to yet more quarks, but to the production of new, heavy, leptons. In 1974 the existence of only four leptons was known: the two charged leptons e^- and μ^- and their neutral, massless counterparts v_e and v_μ. But there was no theoretical reason whatsoever for believing these to be the only leptons to exist, and experimenters felt free to entertain the possibility of further leptons. To explain the previous non-observation of such particles the only requirement was that they be heavy, relative to the electron and muon. If one assumed that charged heavy leptons had properties analogous to those of the electron and muon, then they would be pair-produced in e^+e^- annihilation, just like electrons and muons themselves.

In 1974 the lower limit on the mass of new charged heavy leptons was 1 GeV, set by experiments at the ADONE e^+e^- collider. SPEAR opened up a new energy range, and Martin Perl, the leader of one of the experimental groups there, set out to conduct a heavy lepton search.[2] Perl had already looked for evidence of heavy leptons in the data from SLAC, and now he approached the SPEAR data from the same perspective.[3] It was clear that if heavy leptons with masses greater than 1 GeV were produced at SPEAR, they would decay before they reached the detectors, and therefore their existence would have to be inferred from the decay products. The characteristic signal for which Perl looked was the presence of 'μe events'. These were events in which only two particles were detected, an electron and a muon of opposite charge. If a heavy lepton, 'τ' (tau), was being pair-produced, such events would arise from a sequence like

$$e^+e^- \rightarrow \tau^+\tau^-$$
$$\quad\quad\quad \longrightarrow \mu^- v_\tau \bar{v}_\mu$$
$$\quad\quad\quad \longrightarrow e^+ \bar{v}_\tau v_e,$$

the neutrinos escaping undetected.

By late 1974 Perl had found around 20 μe events. He recalls that he experienced some difficulty in convincing the rest of the collaboration at SPEAR of the significance of this, but in 1975 a paper was published.[4] The difficulty which Perl experienced in convincing first his colleagues and later the HEP community in general derived from possible alternative sources of μe events. In particular, there was active speculation that the events were due solely to charm, being manifestations of the pair-production of D mesons. A possible process was

$$e^+e^- \rightarrow D^+D^-$$
$$\quad\quad\quad\quad \rightarrow \mu^- \bar{v}_\mu + \text{hadrons}$$
$$\quad\quad\quad\quad \rightarrow e^+ v_e + \text{hadrons},$$

the hadrons (as well as the neutrinos) escaping undetected in one way or another. Being 'close to the data', Perl felt that the heavy lepton interpretation was 'a natural', while the D-meson interpretation was 'very contrived'.[5] But to those further away, especially to the voluble gauge theorists then looking anxiously to SPEAR for evidence of naked charm, matters did not appear to be so clear cut.

Perl's group reported more data on their discovery in 1976 but, as DESY physicist Bjorn Wiik reported at the Tbilisi Conference, several groups had looked for evidence of the tau at DORIS without success.[6] This was the low point for the heavy lepton, and fortunately for Perl's morale things soon began to improve. In 1976, with better muon detection, a large number of muon hadron events were discovered at SPEAR which were readily interpreted in terms of a heavy lepton. In 1977 the first positive results on μe and μ-hadron events were reported from DORIS, and in 1978 e-hadron events were reported first from DORIS and then from SPEAR.[7] All of these experiments were consistent with the production of a heavy lepton having a mass of around 1780 MeV. Thus, by 1978, three years after Perl had first announced his discovery, the tau was generally acknowledged to exist.

The establishment of the tau laid finally to rest the remnants of the *R*-crisis. Pair-production of taus contributed 1 to the measured value of *R*, leaving roughly 4 to be accounted for by production of quarks. Four flavours and three colours could account for an asymptotic value of $3\frac{1}{3}$ – which was anyway supposed to be approached from above – and theorists were no longer disposed to debate the discrepancy. In retrospect, the tau came to be seen as a complicating factor which, by making its appearance in just the same energy range as the D mesons, had hindered their identification.[8] However, in solving one problem the tau created another: the newly achieved symmetry between the four quarks and four leptons was once more broken. The tau and its associated neutrino, ν_τ, brought the total number of leptons to six (the existence of the tau-neutrino was not experimentally established, but was generally taken for granted in default of counterarguments). The existence of two new flavours of quark was thus desirable, and experiment was not long in providing one, at least.

Upsilons and the b-Quark

If two flavours of quark had gone hitherto undetected it was reasonable to suppose that this was because the quarks themselves, and hence the hadrons containing them, were very heavy. The first place one might think of looking for these flavours was in e^+e^- annihilation, where charm had first manifested itself so clearly. And, quite independently of any promptings from the discovery of the tau, the experimenters at SPEAR and DORIS proceeded to do this. They found no more narrow peaks in the energy range accessible to them, and neither did they find any upward shifts in R. It was thus evident that production of new flavours, if they existed, must take place at energies higher than those available at SPEAR and DORIS. Such energies were available at proton accelerators. The J-psi had, one recalls, been discovered in e^+e^- production at the Brookhaven AGS, as well as in e^+e^- annihilation at SPEAR. The proton beam at Brookhaven had a maximum energy of 31 GeV, and there was no hope of finding extremely massive new flavours there, but at Fermilab 500-GeV beams were available – ample for the production of new heavy states.

The group which set out to exploit the opportunity offered by Fermilab was a Columbia University–State University of New York at Stony Brook–Fermilab collaboration, led by Leon Lederman of Columbia. We have already encountered Lederman in Section 5.5 in connection with his 1968 experiment detecting $\mu^+\mu^-$ pair-production at the Brookhaven AGS. In that experiment, the cross-section for muon production was observed to fall-off smoothly with the mass of the pair, except for a broad bump or shoulder between 3 and 5 GeV. The smooth fall-off could be explained in terms of partons (the Drell–Yan model) but the bump remained enigmatic. Further investigation of the bump through observation of lepton-pairs was the initial aim of Lederman's Fermilab experiment, but this objective had to be reconsidered following the discovery of J-psi. Ting's 1974 Brookhaven experiment was very similar to Lederman's 1968 experiment (except that the former detected e^+e^- pairs and the latter $\mu^+\mu^-$ pairs) and the obvious inference from Ting's observation of the J was that the bump in the 1968 data was nothing more than J seen

283

through the distorting lens of (relatively) crude apparatus. And, as Lederman later wrote:

> Now that the mystery of the shoulder had been solved, we decided to use the new Fermilab accelerator to look for resonances in the unexplored mass range above 5 GeV. In 1975 and 1976 we observed hundreds of events in three lepton-pair runs. . . . This time we could monitor the distorting effects of our apparatus by examining how it altered the J-psi resonance, which we could not have done in 1968. We also had years of experience with muon pairs and of progress in detector developments that we could put to good use.[9]

The results of a preliminary analysis of the Fermilab data were submitted for publication in early 1976.[10] Twenty-seven e^+e^- pairs had been observed with effective masses between 5.5 and 10.0 GeV, of which twelve clustered together between 5.8 and 6.2 GeV. Following Ting's precedent, the experimenters tentatively identified the cluster as evidence for the existence of a new long-lived particle with a mass around 6 GeV, and suggested that it should be named 'upsilon' (Υ). In 1977 changes were made to the apparatus and more data were taken. Now the clustering around 6 GeV disappeared, but a new peak appeared in the $\mu^+\mu^-$ cross-section at around 9.5 GeV. The experimenters again ascribed the peak to production of a long-lived, heavy particle, and again suggested that it be called upsilon. An account of the discovery was submitted for publication on 1 July 1977; Figure 10.1 reproduces the data on which the claim was based.[11]

The 9.5 GeV upsilon report was based on observation of many more lepton pairs than its ill-fated 6 GeV predecessor (9000 as against 27 pairs) and was widely regarded as conclusive. The discovery paper made no attempt to give a theoretical interpretation of the upsilon, but the HEP community showed little hesitation over what such an interpretation should be. As the headline of an American Institute of Physics News Release of Friday, 5 August 1977 put it: 'Particle Discovery Believed to Indicate Existence of Fifth Quark'. The Release cited Lederman as explaining that 'as a result of particle discoveries during the past few years and in context with the present state of quark theory, there is a widespread consensus in the scientific community that associates with the discovery the existence of a new quark. Such was the general reaction of excited physicists, he [Lederman] added, when the results were presented at the annual meeting of the European Physical Society in Budapest on July 8'.

In short, the discovery of the upsilon provoked neither the intensity nor the diversity of theoretical speculation which had followed the discovery of the J-psi less than three years earlier. The saga of the psis

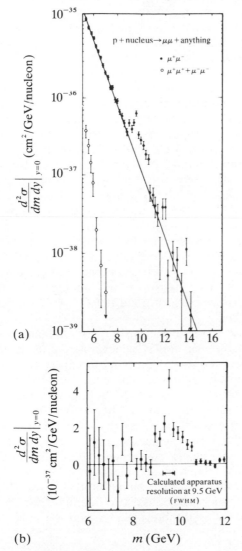

Figure 10.1. Cross-sections ($d^2\sigma/dmdy$) for production of muon pairs as a function of their effective mass (m): (a) full cross-sections for production of opposite-sign dimuons (solid points) and like-sign dimuons (open-points); the solid line fits the smooth exponential fall-off of the data except in the region around 10 GeV, where the upsilon peak stands out. (b) Cross-section differences between the exponential fall-off and the measured data, showing the upsilon peak more clearly.

285

had accustomed the HEP community to the discovery of new families of massive hadrons. From the traumas of the November Revolution a set of techniques for handling the psis had emerged, and these were ready to be transplanted *en masse* to the explanation and analysis of the properties of the upsilon. The upsilon was assumed to be a $q\bar{q}$ composite of a new heavy quark carrying a new flavour. This was a direct extension of the charmonium model image of the hidden-charm psis, and theorists immediately began making detailed calculations of the upsilon's properties on this basis.[12] One conclusion which followed from the charmonium-style analysis was that there should be not one upsilon but several (analogous to the psi and psi-prime). Support for this view was shortly forthcoming from Lederman's experiment at Fermilab.[13] Having taken more data to improve the precision of their measurements, and using computer simulations of their apparatus to compensate for distortions, Lederman's group inferred that at least three distinct particles, each less than 100 MeV wide, were being produced in the upsilon region. The upsilon itself, analogous to the J-psi, was determined to have a mass of 9.4 GeV, and two further narrow states, analogues of the psi-prime, were found to have masses around 10.0 (Υ') and 10.4 GeV (Υ'').

The Fermilab data also served to clarify the quantum numbers of the new quarks. As we have seen, in 1977 evidence for the tau lepton was mounting, and this predisposed theorists to expect two new species of quark. The most popular quantum number assignment of the expected quarks were known as t and b (for 'top' and 'bottom', or 'truth' and 'beauty') and the question most debated after the upsilon discovery was whether it was a $t\bar{t}$ or $b\bar{b}$ state. t and b quarks were assigned different electric charges, $+\frac{2}{3}$ and $-\frac{1}{3}$ respectively, leading to different predictions for the electromagnetic decay rates of the respective composites. The relative magnitudes of the muon yields at the Υ, Υ' and Υ'' peaks favoured the assumption that it was the b rather than the t quark which had been discovered: hidden-t states were therefore presumably even more massive than the hidden-b upsilons.

Upsilons and e^+e^- Annihilation

Lederman's group at Fermilab had claimed the prize of discovering the upsilons, but the November Revolution had taught physicists that e^+e^- colliders rather than proton synchrotrons were the ideal tools for investigating charmonium-type states. Since they decayed to lepton-pairs the upsilons were assumed, like the psis, to have the quantum numbers of the photon; they should therefore be directly produced in e^+e^- annihilation, free from the hadronic background typical of PS experiments. In 1977 the world's highest energy e^+e^- colliders were SPEAR and DORIS, neither of which was designed to operate at sufficient energies to produce upsilons. The SPEAR authorities decided that it was impractical to boost their machine to reach the upsilon region; but at DESY work got under way to operate DORIS in a new configuration above its design energy,[14] and in 1978 sightings of the Υ and Υ' were reported.[15] The DORIS measurements gave precise values for the masses of the Υ and Υ' and confirmed that their lifetimes favoured a hidden-b interpretation. Furthermore, although only two upsilon states were seen at DORIS, the mass values obtained, taken in conjunction with the Fermilab data, reinforced the conclusion of Lederman's group that there were indeed three upsilon states, a conclusion which had previously been debatable. It appeared that the Υ'' should have a mass of around 10.38 GeV, unfortunately beyond DORIS's reach.

In November 1978 the new e^+e^- collider PETRA came into operation alongside DORIS at DESY. PETRA was designed to attain CM energies of up to 40 GeV—well above the expected mass of the Υ''. But the Υ'' remained invisible in Hamburg. The operating characteristics of e^+e^- colliders are such that their experimental efficiency falls off rapidly below their maximum energy, and the Υ'' mass was well below PETRA's optimum range. Thus while the Υ'' was too massive to be produced at DORIS it was too light to be seen at PETRA.

DESY's misfortune was good news for Cornell, where an intermediate energy e^+e^- collider came into operation in late 1979 and was immediately able to establish a monopoly of experimental upsilon physics. The Cornell Electron Storage Rings (CESR – pronounced Caesar) were designed to span the energy range from 4 GeV to 16 GeV, and the upsilons fell nicely within this. Authorisation for the construction of CESR was given in late 1977, shortly after the upsilon discovery at Fermilab. Experiments aimed at detailed investigations of the upsilons at CESR entered the planning stage even before construction of the machine had begun,[16] and once the machine was in operation it was exclusively devoted to upsilon physics.

The experimental programme at CESR mirrored the earlier investigations of charm at SPEAR and DORIS. Within the first few weeks of running, peaks corresponding to the Υ, Υ' and Υ'' were identified in the electron–positron annihilation cross-section. In March 1980 a fourth peak, Υ''', was observed at a mass of around 10.55 GeV. This peak was broader than the lower mass upsilons, and was ascribed to a $b\bar{b}$ state sufficiently heavy to decay into naked-b mesons (B). Subsequent experiments at CESR explored the decay modes of the four upsilons in increasingly fine detail, and established that the weak decays of the B mesons had the characteristics expected in the standard model.[17] In experiment as well as theory, the story of the upsilons and the b quark replayed that of the psis and charm.

The t Quark

The discovery of the upsilons and their interpretation as manifestations of the b quark left the HEP community with a lop-sided ontology. There were six leptons (including the hypothetical tau-neutrino) but only five quarks. What was needed was the sixth, t, quark, and the machine which was expected to provide it was PETRA. From 1979 onwards, the PETRA experimenters searched for the new quark over the accessible energy range, looking both for narrow hidden-t spikes and for the rise in R which would signal the production of naked-t states, but to no avail.[18] In 1980 the new e^+e^- collider, PEP, came into operation at SLAC covering a similar energy range of energies to PETRA, and again no sign of the t quark was seen.

The failure to find the t quark was not regarded in HEP circles as a major setback. No theoretical predictions were frustrated in the t quark's non-appearance, since its mass and the masses of hidden- and naked-t hadrons were unconstrained by theory. It was therefore possible to believe that the t quark existed, but required more energy than that available at PETRA and PEP for its production. This was plausible but disappointing. The conceptual machinery already developed for the analysis of t-quark systems remained on the theoretical shelf,[19] where it seemed likely to remain pending the construction of the next generation of accelerators and colliders. In the interim, most theorists decided to award themselves the t quark by default, and the standard ontology of HEP in the late 1970s was 6 quarks and 6 leptons: e, v_e, μ, v_μ, τ, v_τ and u, d, s, c, b, t.[20]

The New Standard Model

In comparison with the impact of the psis, the discovery of the tau and the upsilons generated little excitement within HEP. The November Revolution had seen a decisive shift in communal practice from the old to the new physics, and the effect of the later discoveries was simply to reinforce that shift. In the post-revolutionary context new quarks and leptons could be accommodated as a matter of routine. The discovery of the tau and the upsilon did of course mark out distinctive new fields of experimental practice – primarily at electron–positron colliders but also, for the upsilons, at high energy PSs (Fermilab and, later, the CERN SPS) and at the CERN ISR.[21] But data from these new fields of experiment served only to reinforce the patterns of theorising established in the November Revolution rather than to redirect theorists' energies.

As far as the strong interactions were concerned, QCD was flavour-blind and new leptons or quark-flavours were, at most, peripheral to the development of the theory. Experiments on upsilon spectroscopy supported extensions of the QCD-inspired charmonium model, but the main lines of QCD development lay elsewhere (see Chapter 11). In respect of the electroweak interactions, an ontology of six quarks and six leptons was easily assimilable to the SU(2) × U(1) group structure of the standard Weinberg–Salam model; all that was required was the inclusion of two new families of fundamental particles (v_τ, τ^-) and (t, b) beside the existing families of Table 8.1. Thus the new quarks and leptons simply went with the flow that charm had established, and their impact was correspondingly less. Harvard theorist Sidney Coleman summed this up when he wrote in 1979, 'Strictly speaking, the current theory is a generalisation of the 1971 theory [Weinberg–Salam plus GIM]; it contains additional entities, generalisations of charmed particles. Such an extension is a minimal modification of the theory, and I do not think anyone would be particularly disturbed if new experiments required further extensions of the same sort.'[22]

Nevertheless, several novel lines of theorising did spring up, grounded in the expanded ontology of the new standard model, of which I will discuss the proposal with the most immediate pheno- menological relevance.[23] The discovery of the b quark and presump- tion of the t brought to particle physicists' attention a 1973 publication from two Japanese theorists, M.Kobayashi and K.Mas- kawa.[24] These authors addressed themselves to the phenomenon of CP-violation in the weak interactions (see Section 3.3) and argued that it could be most easily accommodated in electroweak theory by extending the set of quark flavours from four to six. Kobayashi and Maskawa pointed out that in a 6-quark model there was an extra free parameter which, if suitably chosen, would correspond to the observed extremely weak CP-violating effects. In 1973, this explana- tion was just one of many but, when the upsilons were discovered, it began to receive a great deal of attention. Steven Weinberg, for instance, reviewing the theoretical status of the weak interactions at the 1978 Tokyo Conference, pointed to the utility of the Kobayashi– Maskawa idea.[25] CP-violation assumed major significance in the late 1970s in the context of cosmological speculations based upon grand unified theories. These are discussed in Chapter 13, and we will encounter the 6-quark idea again then. Meanwhile, to sum up the developments reviewed in the present section, the ontological developments in the post-charm era were these:

(1) The existing set of four leptons – e, v_e, μ, v_μ – was augmented to six by the discovery of the heavy lepton, tau, and the presumption of an associated tau-neutrino.

(2) The existing set of four quarks – u, d, s, c – was augmented to six by the discovery of the b quark and the presumption of the t quark.

(3) The resulting 6-lepton, 6-quark ontology blended easily with the developing gauge-theoretical orthodoxy: the existence of six quark flavours and six leptons was immaterial to QCD, and created no problems for the Weinberg–Salam electroweak gauge theory.

10.2 The Slaying of Mutants

For the reasons outlined in Chapter 9, by the mid 1970s neutrino-beam experiments had assumed major significance in the experimental programmes conducted at high energy PSS. Data from these experiments were held to be of immediate relevance both for electroweak theory and for QCD. I will discuss their bearing on the former here, and postpone discussion of the latter to the following chapter.

Electroweak models made detailed predictions concerning the neutral- and charged-current interactions of neutrinos, and experimenters set out to see how these fared empirically. Of greatest interest, following the establishment of charm, was the question of the validity of the Weinberg–Salam model. The Gargamelle experiment at CERN and the first-generation electronic experiments at Fermilab had shown that the neutral-current event rate was roughly as predicted, but much remained to be done before the model could be said to be confirmed. For example, the model predicted that parity should be violated to a specified extent in neutral-current interactions, and the relatively crude discovery experiments were insensitive to parity violation. Succeeding generations of neutrino experiments aimed to remedy this; larger and more sophisticated detectors were built in order to measure parity-violating neutral-current effects as accurately as possible. The second and third generation experiments also yielded new and more precise data on charged-current neutrino interactions which again could be compared with the predictions of the Weinberg–Salam model. And while neutrino experiments dominated electroweak physics, they were not alone in probing the validity of the model. Several ingenious groups of experimenters set out to look for parity-violating neutral-current effects in *electron–*

hadron interactions. Their findings initially threatened the foundations of the new orthodoxy, and are discussed in some detail below. However, by 1979 consensus reigned within HEP on the validity of the Weinberg–Salam model, and in that year Weinberg, Salam and Glashow shared the Nobel Prize for Physics for their seminal theoretical work on electroweak unification.

Writing in commemoration of the Nobel award, Sidney Coleman summarised the development of electroweak physics through the 1970s thus:

> In 1973, experiments at CERN and Fermilab detected neutral current events . . . . of a form and magnitude consistent with the theory. The next 5 years were a confusing period of exhilaration and disappointment, alarms and excursions. Experiment confirmed the theory; experiment denied the theory. Enormous theoretical effort was devoted to producing grotesque mutant versions of the theory consistent with the new experimental results; the new experiments were shown to be in error; the mutants were slain. In the last few years though, the experimental situation seems to have stabilised in agreement with the original 1971 version of the theory. The Weinberg–Salam model is now the standard theory of the weak interactions.[26]

There would be little point in discussing each post-charm experiment which bore upon the standard model. Instead, I will first review the overall experimental situation as it appeared in the tranquility of 1979, and then discuss the more important of Coleman's mutants which were slain along the way.

An authoritative review of the experimental data on neutral currents was given by CERN neutrino experimenter F.Dydak at the European Physical Society Conference held at CERN from 27 June to 4 July 1979. By this time a standard format had been developed for comparison of data with unified models. In general model-independent terms, neutral-current effects in neutrino-electron, neutrino-hadron, and electron-hadron scattering could be represented in terms of ten phenomenological parameters, whilst in the standard model there was only a single free parameter: $\sin^2\theta_w$, where θ_w was known as the electroweak mixing angle or, eponymously, the Weinberg angle. Dydak presented the experimental values, extracted mainly from neutrino experiments at CERN and Fermilab, of the eight phenomenological parameters which had then been measured, and showed that within the quoted errors each of these parameters agreed with the predictions of the standard model if $\sin^2\theta_w = 0.23$ – no mean achievement. One further number could also be extracted from the data, a parameter conventionally denoted as 'ρ'. In the Weinberg–Salam model ρ was predicted to be unity, but this depended upon the use of a minimal choice of Higgs particles in the spontaneous symmetry-breaking mechanism. More complicated choices were possible and implied different values of ρ. The best fit to the world's data, according to Dydak, was $\rho = 1.00 \pm 0.03$, so even the Higgs sector of the standard model appeared to stand up. Dydak concluded his review of the contemporary experimental situation as follows:

(i) All (but two) more recent experimental results are consistent with one another.

(ii) The Weinberg–Salam model gives a good description of all (but two) experimental results, with only one free parameter: $\sin^2 \theta_w = 0.230 \pm 0.015$.

(iii) Eight out of the ten phenomenological parameters governing ve, eq and vq scattering are determined from data. All of them are consistent with the predictions of the Weinberg–Salam model with $\sin^2 \theta_w = 0.23$.

(iv) The experimental precision has reached such a level that theoretical uncertainties in the data analysis can no longer be ignored.[27]

Thus, not only was the unified gauge theory of the electroweak interactions a renormalisable field theory like QED; in 1979 it appeared that when combined with the Weinberg–Salam choice of gauge group and symmetry breaking, gauge theory could also emulate QED in another important respect: its predictions always came out right.

But what of the two experiments parenthetically noted by Dydak as being in disagreement with the standard model? This brings us to mutants and their slaying. According to Coleman, 'this happened at least three times: with high-y anomalies, with trimuons, and with atomic parity violation'.[28] Of the three, the rise and fall of the high-y anomaly and trimuons were rather straightforward and will be dealt with briefly. The atomic parity violation experiments are those referred to by Dydak as still being in disagreement with the standard model in 1979. The fluctuation in the perceived significance of these experiments is more interesting, and will be dealt with at greater length.

Chicago Mutants Slain in Geneva

First the high-y anomaly and trimuons. These phenomena were born just outside Chicago and died near Geneva. Their existence was first reported in charged-current neutrino scattering by the Harvard–Pennsylvania–Wisconsin–Fermilab (HPWF) collaboration in their pioneering neutrino experiment at Fermilab. HPWF were one of the first groups of experimenters to explore the new energy range of neutrino physics opened up by the 500 GeV Fermilab accelerator, and in 1974 were the first to confirm the Gargamelle neutral-current discovery. In 1974 they also reported the existence of the high-y anomaly, and further details followed in two 1976 publications.[29] The anomaly itself was an excess of events relative to standard model expectations in the high-y region of charged-current antineutrino scattering (y is a kinematic variable: $y = E_v - E_\mu / E_v$, where E_v and E_μ represent the energies of the incoming neutrino and outgoing muon respectively). The details of the anomaly are not important here; all that need be said is that its explanation appeared to require the existence of new types of quark not included in the standard four-quark (or six-quark) picture. New quarks, and new particles in general, being the order of the day during this period, a considerable amount of theoretical work was expended upon the explanation of the anomaly.[30] But this mutant theorising was brought to a more-or-less abrupt halt in mid-1977 as data began to emerge from the CERN–Dortmund–Heidelberg–Saclay (CDHS) neutrino experiment at the new CERN accelerator, the 400 GeV SPS. The first publication from the CDHS group, dated July 1977, was entitled 'Is there a high-y anomaly in antineutrino physics?' – a question answered in the negative in the paper's abstract, which read:

> We have analysed data taken in the CERN narrow-band neutrino and antineutrino beams with regard to the "high-y anomaly" observed by preceding experiments at FNAL [Fermilab]. At neutrino energies between 30 GeV and 200 GeV, the $\bar{v}$ and v charged-current cross-section ratios and muon-inelasticity distributions disagree with earlier results. In particular there is no evidence for energy dependent effects in the antineutrino data which constitute an important aspect of the alleged anomaly.[31]

The technical superiority of the second-generation CDHS experiment over the pioneering HPWF experiment was commonly agreed within the HEP community, and although it was possible to argue, for instance, that CDHS had not yet fully de-bugged their procedures, the above paper dealt a body-blow to the high-y anomaly from which it never recovered. It was not sighted again.

Similar comments apply to trimuons. The trimuon phenomena were first reported by the HPWF team in March 1977.[32] Using a new detector HPWF had observed six events in which an incoming neutrino led to the production of three muons in the final state. They argued that these trimuon events could not be understood within the standard picture and that, in this case, they were better explained by the existence of new heavy leptons (*not* Perl's tau). HPWF also identified a significant fraction of the dimuon events which they had previously observed with this source.[33] The statistical significance of six trimuon events was not great, but HPWF argued that conventional processes were inadequate to explain them. Again theorists took up the challenge, fitting the data in terms of cascades of new heavy leptons.

The CDHS collaboration once more dampened theorists' ardour. Their first publication on the subject – dated 5 August 1977 – noted that they had observed two trimuon events in neutrino interactions and one in an antineutrino exposure, but they declined either to unequivocally associate them with any new phenomena, or to comment on the relation of their data with that of HPWF (and of a Caltech experiment at Fermilab which had independently found two trimuon events)[34] saying 'the number of events are small, and the beams different, so that we feel that the differences may not be significant'.[35] The overall effect of the CDHS publication was to damn the anomalous trimuons with faint praise. No more anomalous events were subsequently reported, and the stream of unconventional phenomenology dried up from lack of empirical support. Like the high-y anomaly, the trimuon events had no more than a temporary impact on the ontology and group structure of the standard model. The two mutants from Fermilab were slain at CERN.

Mutants From Washington and Oxford . . .
We now come to 'atomic parity violation', the third major anomaly which challenged the standard model, and the only one which, *pace* Coleman, could be argued to be still alive in 1979 (if one cared to make this argument, which few particle physicists did). The rationale behind the experiments at issue went as follows. In the Weinberg–Salam model, neutral-current effects were mediated by the Z^0 IVB. This coupled to neutrinos, giving rise to neutral-current neutrino–hadron scattering. It also coupled to electrons, and was therefore expected to give rise to neutral-current effects in the interactions of *electrons* with hadrons. In this case, though, the effects were expected to be very small in comparison with the dominant, electromagnetic, interactions of electrons, and the task of detecting them was expected to be correspondingly difficult. The way to surmount this hurdle, experimenters reasoned, was to exploit an important difference between the weak and electromagnetic interactions: the latter conserved parity while the former, at least in the Weinberg–Salam model, did not. Thus if a parity-violating effect could be observed in electron–hadron interactions it could be uniquely ascribed to the

weak interaction, and measurement of its magnitude would serve as a test of the Weinberg–Salam model.

The detection of the parity-violating electron–hadron interactions expected in the Weinberg–Salam model required measurements of a precision far in excess of typical HEP experiments, and few particle physicists took the possibility seriously (for the exception which proved the rule, see below). Atomic physicists, in contrast, routinely performed extremely precise measurements, and they were the first to take up the challenge of testing the Weinberg–Salam model's predictions concerning electron neutral-current interactions.

In the post-revolutionary era, the first two atomic-physics experiments to get under way were performed by groups at the Universities of Oxford and Washington (Seattle). Both groups employed a similar procedure, shining a beam of polarised laser light through bismuth vapour maintained at high temperature (1500°C) and aiming to measure any rotation in the plane of polarisation of the transmitted rays. The point of the exercise was that if the only interactions between the atomic electrons and nuclei of bismuth were parity-conserving then there would exist no preferred direction in space, and hence no rotation of the plane of polarisation. If, on the other hand, there were some parity-violating component in the electron–nucleon interaction then a finite rotation of the plane of polarisation should be observable. The magnitude of this rotation would provide a measure of the magnitude of parity-violation by the weak neutral current, and this could be compared with that predicted by the Weinberg–Salam model.[36] The expected rotation in the Washington and Oxford experiments was very small, around 1.5×10^{-7} radians, but the experimenters felt confident that they could detect this.[37]

The first results from the Washington and Oxford experiments appeared in a joint publication in December 1976.[38] Commenting that 'Although our experiments are continuing we feel that there is sufficient interest to justify an interim report', the experimenters reported the following values for the parameter determining the extent of parity violation: $-8 \pm 3 \times 10^{-8}$ (Washington) and $+10 \pm 8 \times 10^{-8}$ (Oxford). These values were to be compared with the considerably larger prediction of the standard model, which gave a value of roughly -3×10^{-7}. As the publication stated: 'we conclude from the two experiments that the optical rotation in bismuth, if it exists is smaller than the values . . . . predicted by the Weinberg–Salam model plus the atomic central field approximation'. The *caveat* to this conclusion was important. Bismuth had been chosen for the experiment because relatively large effects were expected for heavy atoms, but when the effect failed to materialise a drawback of the choice became apparent. To go from calculations of the primitive neutral-current interaction of electrons with nucleons to predictions of optical rotation in a real atomic system it was necessary to know the electron wave-functions, and in a multi-electron atom like bismuth these could only be calculated approximately. Furthermore, these were novel experiments and it was hard to say in advance how adequate such approximations would be for the desired purpose. Thus in interpreting their results as a contradiction of the Weinberg–Salam model the experimenters were going out on a limb of *atomic* theory. Against this they noted that four independent theoretical calculations of the electron wave functions had been made, and that the results of these calculations agreed with one another to within twenty-five per cent. This degree of agreement the experimenters found 'very encouraging' although they conceded that 'Lack of experience of this type of calculation means that more theoretical work is required before we can say whether or not the neglected many-body effects in the atomic calculation would make R [the parity-violating parameter] consistent with the present experimental limits'.

The Washington–Oxford publication spread confusion in the relevant circles of HEP. High-energy physicists were not expert in the kind of atomic physics calculation involved in the interpretation of the experiment and, faced with the assertions of the atomic experts, were simply left with the options of taking them or leaving them. In November 1976 the HPWF collaboration at Fermilab had reported the first experimental confirmation of the predicted parity-violation in neutral-current neutrino scattering.[39] And thus, if one accepted the Washington–Oxford result, the obvious conclusion was that neutral-current effects violated parity conservation in neutrino interactions and conserved parity in electron interactions. This conclusion would rule out the Weinberg–Salam model but, as HEP theorist Frank Close put it:

> Whether such a possibility could be incorporated into the unification ideas is not clear. It also isn't clear, yet, if we have to worry. However, the clear blue sky of summer now has a cloud in it. We wait to see if it heralds a storm.[40]

In December 1976 Patrick Sandars, the leader of the Oxford group, was quoted as believing that 'six months . . . should be sufficient time to bring the situation to clear confrontation or resolution'.[41] On 31 March 1977 he wrote that the Oxford and Washington experiments

> together provide convincing evidence that there are no rotations of the magnitude required by the 1967 Weinberg–Salam theory combined with the atomic calculation. The limit obtained by

combining the results of the two experiments is at least a factor three below the prediction. Unfortunately one cannot at present rule out the possibility that the source of the disagreement lies in the atomic calculation which is used to relate the Weinberg–Salam theory to the predicted angle. Such a large discrepancy seems unlikely, but strenuous efforts to improve the calculation are under way.

If the atomic theory survives intact, or if improved experiments still see nothing, the implications for our current ideas on the weak and electromagnetic interactions will be very considerable.[42]

By this time it had become clear that although a null-result in the atomic physics experiments would contradict the Weinberg–Salam model it would not necessarily overthrow the unified gauge-theory approach, and Steven Weinberg was quoted as saying that he had 'an open mind on the atomic physics calculations' and was willing to embrace 'an attractive class of theories which are not radical departures from the original model'.[43] Such non-radical models typically used additional IVBs or neutral heavy leptons to make the required difference between neutral-current effects in neutrino scattering and atomic physics experiments.[44]

In September 1977 the Washington and Oxford groups published their latest findings in *Physical Review Letters*.[45] They had continued to find no optical rotation, and described two 'hybrid' unified electroweak models, which used neutral heavy leptons to accommodate the divergence with the findings of high-energy neutrino scattering. Sandars gave a status report on the atomic physics experiments to the Symposium on Lepton and Photon Interactions at High Energies, held in Hamburg, 25–31 August 1977, and the general response to this report was summarised by David Miller (a neutrino experimenter from University College London) as follows:

S. Weinberg and others discussed the meaning of these results. It seems that $su(2) \times u(1)$ is to the weak interaction what the naive quark–parton model has been to QCD; a first approximation which has fitted a surprisingly large amount of data. Now it will be necessary to enlarge the model to accommodate the new quarks and leptons [the b quark and the tau], the absence of atomic neutral currents and perhaps also whatever it is that is causing trimuon events.[46]

In fact, as discussed earlier, the b quark and the tau heavy lepton required no modification of the $su(2) \times u(1)$ group structure. The atomic physics results, however, did require such a modification, and it is clear that in late 1977 particle physicists were quite prepared to accept this. The storm which Frank Close had glimpsed had materialised and was threatening to wash away the basic Weinberg–Salam model, although not the gauge-theory enterprise itself. It is interesting to note that at this time theorists were seeking to explain in a single new model not only the null-result of the atomic physics experiment, but also the HPWF trimuons and, until the CDHS data, the high-y anomaly. These were Coleman's three mutants and at one time, around mid-1977, they promised to provide the basis for a new fundamental group structure for the electroweak interactions. But this new structure was not to be. In the event the standard model emerged unscathed. We have seen the fate of the high-y anomaly and trimuons, and now we can look at the more complex demise of the atomic physics anomaly.

... Slain at Stanford

There were two ways of looking for neutral-current effects in the weak interactions of electrons. One way was the atomic-physicist's bench-top approach. The other was the particle physicist's way, using high-energy electron beams and all of the paraphernalia of HEP experiment. The idea here was to measure the scattering of polarised electron beams on nucleon targets and to compare cross-sections when the electron-spins were aligned parallel and antiparallel to their direction of motion. Any dependence of cross-sections upon the electron spin orientation would point immediately to a parity-violating neutral-current effect, and hence such measurements would serve to test the Weinberg–Salam model.

The only problem with the HEP approach to the investigation of the neutral-current interaction of electrons was that it appeared to be technically impossible. The expected spin-dependence of electron-nucleon scattering cross-sections was much smaller than the experimental uncertainty of a typical HEP measurement. But one physicist refused to subscribe to such defeatist arguments, Charles Prescott of SLAC. As early as the 1960s, Prescott had begun exploring the possibility of detecting parity-violating interactions in polarisation measurements, and in the 1970s he was at the heart of developments which culminated in the SLAC experiment code-named E122. Performed by a collaboration of physicists from SLAC and Yale in early 1978, experiment E122 contributed more than any other to the establishment of the Weinberg–Salam model as the standard model of electroweak interactions. To search for parity violation in electron scattering at the level of accuracy required to test the Weinberg–Salam model, the SLAC–Yale collaboration assembled a unique high-intensity source of polarised electrons and unprecedentedly precise techniques of beam monitoring and particle detection.[47] Their experiment took data from January to May 1978. The first results were announced in a SLAC seminar on 12 June 1978 and were published one month later.[48] The experimenters had found what they were looking for: as expected in the Weinberg–Salam model, the scattering cross-sections for electrons polarised parallel and antiparallel to the beam direction differed by around 1 part in 10^4 (measured to a precision of roughly twenty per cent).

Data from SLAC were something particle physicists were used to dealing with, unlike bench-top atomic-physics experiments, and at once the Washington–Oxford results began to recede in perceived significance. Theorists manifested a new disinclination to tinker further with the Weinberg–Salam model, a position which was reinforced when the Washington and Oxford groups found themselves contradicted in their own backyard. In mid-1978 a group of Soviet atomic physicists from Novosibirsk entered the fray, with their first measurements on bismuth at exactly the same wavelength as the Oxford group, measurements which *agreed* with the prediction of the standard model.[49] The details of the Soviet experiment were not known to Western physicists, making a considered evaluation of its result problematic. But the discrepancy amongst the reported measurements on bismuth was enough to enable high-energy physicists to argue that, in one way or another, the atomic physics experiments were very difficult and should not, for the time being, be taken too seriously.

The extent to which the balance had shifted away from the Washington–Oxford results was evident in Dydak's 1979 neutral-current review. His opening remarks on weak electron scattering were:

> Very recently, substantial progress has been made in the understanding of eq scattering via weak neutral current. This progress stems mainly from new results on parity-violating effects in the inelastic scattering of polarized electrons at SLAC, but also from new results on parity-violating effects in optical transitions of heavy atoms.[50]

Of the atomic physics experiments in particular, Dydak commented:

> The experimental situation on parity violations in atoms is not satisfactory. The experiments carried out at Seattle and Oxford . . . showed essentially null results. A Novosibirsk team, however, reported evidence for a non-zero result in Bi, being consistent with the standard model prediction. Barkov reported to this conference a new result of this team. The ratio of the

experimental to the theoretical rotation (standard model with $\sin^2\theta_w = 0.23$) is 1.07 ± 0.14, which is clearly incompatible with parity conservation. Note that the Novosibirsk experiment is carried out on the same optical transition as the Oxford experiment, with conflicting results.

Recently, another experiment carried out at Berkeley also reports parity violation observed in atomic thallium, although the effect has only a 2σ [2 standard deviation] significance.[51]

Dydak's goal in this part of his talk was to determine the best values of the phenomenological parameters describing neutral-current electron scattering, for comparison with the standard model predictions. On this, he commented:

It is difficult to choose between the conflicting experimental results in order to determine the eq coupling constants. Tentatively, we go along with the positive results from the Novosibirsk and Berkeley groups and hope that the future development will justify this step (it cannot be justified at present, on clear-cut experimental grounds).[52]

Having decided not to take into account the Washington–Oxford results, Dydak concluded from the Novosibirsk–Berkeley data that parity violation in atomic physics was as predicted in the standard model.

10.3 The Standard Model Established:
Unification, Social and Conceptual

By the summer of 1979 all of the anomalous storm clouds which threatened the standard electroweak model had been dispelled to the satisfaction of the HEP community at large (if not, perhaps, to the authors of the anomalous data). The mutants had been slain. Weinberg, Salam and Glashow shared the 1979 Nobel Prize for their part in an achievement which marked, in the words of Sussex HEP theorist Norman Dombey, 'the end of the second phase of post-war theoretical physics'.[53] The first phase had been the development of QED as a renormalisable and phenomenologically accurate field theory of the electromagnetic interactions. The second phase had been the extension of field theory to the weak interactions, and the unification of the weak and electromagnetic forces in an electroweak gauge theory which was again both renormalisable and phenomenologically accurate. One theoretical element of the new physics (the other being QCD) – one element of Bjorken's 'new orthodoxy' – had been established. As we shall see in Chapter 12, this had far-reaching consequences for the subsequent development of HEP: several major investment decisions of the late 1970s were structured around the world-view provided by electroweak theory. It is appropriate, therefore, to close this chapter with some discussion of how consensus on the standard model was achieved.

In retrospect, it is easy to gloss the triumph of the standard model in the idiom of the 'scientist's account': the Weinberg–Salam model, with an appropriate complement of quarks and leptons, made predictions which were verified by the facts.[54] But missing from this gloss, as usual, is the element of choice. In assenting to the validity of the standard model, particle physicists chose to accept certain experimental reports and to reject others. This element of choice was most conspicuous in the communal change of heart over the Washington–Oxford atomic-physics experiments and I will focus on that episode here. We saw in the preceding section that in 1977 many physicists were prepared to accept the null-results of the Washington and Oxford experiments and to construct new electroweak models to explain them. We also saw that by 1979 attitudes had hardened. In the wake of experiment E122, the Washington–Oxford results had come to be regarded as unreliable. In analysing this sequence, it is important to recognise that between 1977 and 1979 there had been no *intrinsic* change in the status of the Washington–Oxford experiments. No data were withdrawn, and no fatal flaws in the experimental practice of either group had been proposed.[55] What had changed was the *context* within which the data were assessed. Crucial to this change of context were the results of experiment E122 at SLAC. In its own way E122 was just as innovatory as the Washington–Oxford experiments and its findings were, in principle, just as open to challenge.[56] But particle physicists *chose* to accept the results of the SLAC experiment, *chose* to interpret them in terms of the standard model (rather than some alternative which might reconcile them with the atomic-physics results) and therefore *chose* to regard the Washington–Oxford experiments as somehow defective in performance or interpretation.[57]

Thus the development of electroweak physics in the late 1970s diverged markedly from the adverserial image of theory and experiment implicit in the 'scientist's account'. Experimental facts did not exert a decisive and implacable influence on theory. The relation was two way, with theory acting, in certain circumstances, as the yardstick for the acceptability of experimental data. And, as I have argued before, this two-way relationship can be understood in terms of the dynamics of research practice. The standard electroweak model unified not only the weak and electromagnetic interactions; it served also to unify research practice within otherwise diverse traditions of HEP theory and experiment. The standard model was the

301

medium of a social and conceptual symbiosis wherein gauge theorists and experimenters working on neutrino scattering, electron scattering and electron–positron annihilation could draw upon the products of one another's labour for justification and subject matter. Simultaneously, the standard electroweak world-view was the product of that symbiosis, in the sense that resources from all of the traditions it interlinked could be mobilised in arguments for its defence. Worldview and traditions of practice stood or fell together. Matched against the mighty traditions of HEP, the handful of atomic physicists at Washington and Oxford stood little chance. Their results, which potentially represented a problem for electroweak gauge theory, were returned to them by the HEP community as a problem for atomic physics. And, of course, since atomic physicists around the world could not agree amongst themselves, there was little hope that they could mobilise sufficient resources to send the problem back whence it had originated – to particle physics. The standard model, a central pillar of the new orthodoxy, was established through a process that was *at once* conceptual and social: the two cannot be separated.

NOTES AND REFERENCES

1   Numerous private communications, especially from SLAC.
2   For a fully-documented semi-popular account of the developments discussed here, see Perl (1978). For an extended popular account see Perl and Kirk (1978). For technical details, see Feldman and Perl (1977) and Perl (1980).
3   See Perl (1978).
4   Perl *et al.* (1975). The policy at SLAC was only to publish results if the experimental group were unanimous upon their validity.
5   Perl, private communication.
6   Perl (1976), Wiik (1976).
7   See Perl (1978) for access to the literature.
8   Private communications from SLAC.
9   Lederman (1978, 64). This article provides a popular account of the Fermilab discovery.
10 Hom *et al.* (1976).
11 Herb *et al.* (1977, 254, Fig. 3). Support for this report was shortly forthcoming from another Fermilab experiment performed by a Northeastern University – University of Washington – Tufts University collaboration: Garelick *et al.* (1978).
12 For a review of the redeployment of charmonium technology in the analysis of the upsilons, see Jackson, Quigg and Rosner (1979). As one would expect, many of the theorists who made such analyses of the upsilons were those who had contributed to the development of the charmonium model. Rosner fell in this category, as did the Cornell group (see Chapter 9, note 18, Eichten and Gottfried 1977, and Eichten, Gottfried, Kinoshita, Lane and Yan 1980). Having applied a similar

model to the psis and upsilons, several theorists went one step further and developed a general 'quarkonium' analysis, applicable to any hidden-flavour system of heavy quarks: see Quigg and Rosner (1979), Krammer and Krasemann (1979) and for a popular account, Bloom and Feldman (1982). The principal referent of the quarkonium studies were the hoped-for manifestations of the t quark (below). Note that the theorists most active in extending the charmonium model were all closely involved with experimental groups working on upsilon physics. Quigg, for example, was the leader of the theory group at Fermilab when the upsilon was discovered. Krammer and Krasemann were members of the DESY theory group (see the discussion of experiments at DORIS and PETRA below) and Eichten *et al.* were based at Cornell University (see the discussion of the CESR experimental programme below).

13 Innes *et al.* (1977), Ueno *et al.* (1979).

14 In its original configuration DORIS was a two-ring machine in which electrons and positrons circulated separately. In its new configuration only a single ring was used, filled with counter-rotating e^+ and e^- beams. More technical modifications were also made and additional power supplies, intended for the new PETRA e^+e^- machine, were borrowed: see *CERN Courier* (1978).

15 Berger *et al.* (1978), Darden *et al.* (1978a,b), and Bienlein *et al.* (1978).

16 See *CERN Courier* (1977b).

17 For an extensive account of the experimental programme at CESR, see Franzini and Lee-Franzini (1982).

18 See Duinker (1982).

19 See the discussion of quarkonium studies in note 12 above.

20 Alternatives to the six-quark scheme were proposed in the literature (for a review, see Harari 1978) but none of these had any special empirical advantage. I will not discuss them further.

21 The first sightings of the upsilons at the ISR were reported by three groups at the Tokyo HEP Conference in August 1978: Camilleri (1979), Mannelli (1979) and Newman (1979).

22 Coleman (1979, note 21).

23 Another line which should be noted concerned the so-called 'mass hierarchy' problem. This dated back to the discovery of the muon. The muon appeared to be identical to the electron in all respects except for its mass. Theorists wondered why this should be, but found no convincing solution. The discovery of more quarks and leptons exacerbated the problem. Within the standard model, the lepton families (v_μ, μ^-) and (v_τ, τ^-) appeared as exact replicas of the (v_e, e^-) family, except that the charged leptons were of progressively higher mass. Similarly the quark families (c, s) and (t, b) were carbon copies at higher mass of the (u, d) family. Many conjectures were proposed to explain this mass hierarchy of otherwise identical families, but none of them received significant support from experiment. One popular conjecture was that quarks and leptons were not truly fundamental entities but were themselves composite. For a technical review of work on this recycling of the atomist analogy, see Kalman (1981); for a popular account, see Harari (1983).

24 Kobayashi and Maskawa (1973).
25 Weinberg (1979).
26 Coleman (1979, 1292).
27 Dydak (1979, 47). For later detailed reviews of the experimental status of the Weinberg–Salam model, see Kim, Langacker, Levine and Williams (1981), Hung and Sakurai (1981) and Myatt (1982).
28 Coleman (1979, note 20).
29 B. Aubert et al. (1974c), Benvenuti et al. (1976a,b).
30 Although they were subsequently dismissed as illusory, the high-y anomaly and trimuons both had considerable short-term impact upon HEP practice. At least four experimental papers from Fermilab on these phenomena achieved more than 50 citations in the HEP literature within a single year (for comparison, only 57 experimental reports in all achieved this citation rate between 1969 and 1978): see Martin and Irvine (1983, 37–8).
31 Holder et al. (1977a, 433).
32 Benvenuti et al. (1977a,b).
33 For the dimuon events and their relevance to charm see Chapter 9, note 28.
34 Barish et al. (1977).
35 Holder et al (1977b). The apparent discrepancy between the CERN and Fermilab trimuon results led to a series of 'beam dump' neutrino experiments at both laboratories. For early descriptions of these experiments, discussions of the theoretical implications of their findings, and references to the original literature, see Miller (1978a,b) and Morgan (1978).
36 For technical details, see Brodsky and Karl (1976).
37 Sandars (1977) provides a good popular account of the Oxford and Washington experiments.
38 Baird et al. (1976).
39 Benvenuti et al. (1976c).
40 Close (1976, 506).
41 New Scientist (1976b).
42 Sandars (1977, 766).
43 Walgate (1977).
44 Amongst the authors of such models was Salam himself: see Pati, Rajpoot and Salam (1977).
45 Lewis et al. (1977), Baird et al. (1977).
46 Miller (1977, 288).
47 A discussion of the history of experiment E122, and of Prescott's contribution to it, will serve to illustrate the structuring of experimental practice by expertise and context (the following account is based upon interviews with Prescott and R.E.Taylor; quotations are from the former).
  Prescott trained in experimental HEP at Caltech, where he obtained a PhD in 1966 and where he remained as a post-doctoral researcher until 1970. His early work was on low-energy photoproduction experiments using the small Caltech electron synchrotron. One topic with which he was involved was the measurement of proton polarisation in such

experiments, and it occurred to him that polarisation measurements could provide a route to the detection of parity-violating effects. He recalls that he became 'obsessed' with looking for such effects in electron–proton scattering. While still at Caltech he discussed plans to do so at SLAC with Richard Taylor, the leader of one of the SLAC experimental groups. Taylor was not enthusiastic: Prescott's proposed experiment appeared to be difficult and time consuming, and the SLAC deep-inelastic scattering programme was then in full swing.

In 1970 Prescott left Caltech to spend a year at the University of California at Santa Cruz before joining Taylor's group at SLAC. In 1971 he attended a talk given by Yale experimenter Vernon Hughes. Hughes described a source of polarised electrons which had been constructed at his home university. Hughes and his colleagues proposed to bring the source to SLAC where the polarised electrons would be accelerated in the 22 GeV accelerator and directed upon a polarised hydrogen target. Cross-section measurements of electron–proton scattering for different spin orientations would then provide information on the spin-dependence of the deep-inelastic structure functions (Baum *et al.* 1971). At Hughes' talk, Prescott realised that 'that was the way to do it': measurements of polarisation-dependent cross-section *differences* using a polarised electron beam deriving from the Yale source, incident upon an unpolarised target, would indicate a parity-violating effect.

Prescott's realisation culminated in 1972 in a formal proposal from a SLAC-Yale collaboration to do such an experiment at SLAC (Atwood *et al.* 1972). It is important to stress that at this time the experimenters were not thinking in terms of the parity-violating neutral-current effects expected in unified electroweak theories. These theories had had little overall impact upon the experimental community by 1972 and, in any event, it was later calculated that it would have taken around twenty-seven years to gather sufficient data with the Yale source to measure the very small effects predicted by unified models. Instead, the experimenters took the pragmatic view that parity-conservation in purely electromagnetic electron–proton interactions had never been closely investigated at SLAC energies and that here was a chance to do so. Inasmuch as they sought theoretical justification they argued by analogy from the weak to the electromagnetic interaction: the two interactions were similar, the weak interactions failed to conserve parity, so perhaps the electromagnetic did likewise (Atwood *et al.* 1972, 5–6).

The SLAC-Yale proposal, code-named E95, was approved by the SLAC Program Advisory Committee and the experiment went ahead. However, in 1973 and 1974 data on neutral currents in neutrino scattering began to emerge from CERN and Fermilab, and the experimenters realised the potential significance of their work within the context of electroweak theory. The problem they then encountered was one of beam intensity. E95 was designed to measure asymmetries as small as 1 part in 10^4, while any meaningful test of the Weinberg–Salam model required an accuracy of 1 part in 10^5 (several theoretical calculations of the expected effects were made from 1972 onwards: for one such calculation and references to earlier work, see Cahn and Gilman 1978).

To improve their accuracy by a factor of ten the experimenters needed to measure one hundred times more events. To run the experiment with the existing beam for 100 times as long as planned was impossible, while it appeared that to improve the intensity of the Yale source by more than a factor of three was out of the question.

Here microsocial factors intervened. Prescott set out to look for a new and more powerful source of polarised electrons, and he was fortunate to find appropriate expertise resident at SLAC – in the shape of his colleagues Edward Garwin and Charles Sinclair. Garwin was a solid-state physicist, familiar with progress made elsewhere in the early 1970s on the production of polarised electrons from semiconductors (using surface-treated gallium arsenide, GaAs). The feasibility of producing polarised electrons in this way was reported in 1974 by a group of physicists at Zurich with whom Garwin was in close touch (see Garwin, Pierce and Siegmann 1974). Polarised electrons were emitted from GaAs exposed to polarised laser light, and Sinclair was a laser expert. It was Sinclair who eventually built the laser for the improved electron source, and who contributed many optical techniques to E122 (including the use of 'Pockel's cells' to reverse the polarisation of the light, and hence that of the electron beam itself).

By the end of 1974 Prescott, Garwin and Sinclair believed that they could develop a GaAs source which would provide sufficient intensity of polarised electrons to make a test of unified electroweak models feasible, and a formal proposal from the SLAC-Yale collaboration went before the SLAC Program Advisory Committee in May 1975 (Baum *et al.* 1975). The proposal was approved, and Prescott, Garwin and Sinclair then spent three years in developing an operational GaAs source for the new experiment, E122. At the same time, a great deal of attention was given to more conventional aspects of the experiment, such as beam monitoring and detection equipment, in order to reach the required precision. During this time, too, experiment E95 with the original Yale source was in progress (Alguard *et al.* 1976, 1978). E95 failed to detect any parity-violating effects, but as Prescott later put it: 'it was the techniques developed in the Yale experiment [E95] which made it [E122] work'. Thus, when measurements at E122 got under way, the SLAC-Yale group had at their disposal unprecedented technical resources and unprecedented expertise in their use.

48 Prescott *et al.* (1978, 1979).
49 Barkov and Zolotorev (1978, 1979).
50 Dydak (1979, 32).
51 Dydak (1979, 35). For the Berkeley experiment, see Conti *et al.* (1979) and Bucksbaum *et al.* (1981).
52 Dydak (1979, 35).
53 Dombey (1979, 131).
54 For typical 'scientist's accounts' see Coleman (1979) and Dombey (1979).
55 In 1981 the Washington group recanted. They published measurements which 'reveal a well-resolved optical rotation that agrees in sign and approximate magnitude with recent calculations of the effect in bismuth

based upon the Weinberg–Salam theory' (Hollister *et al.* 1981). It is interesting to note that this new-found agreement did not come about through improvements in atomic theory (the element of the original Washington–Oxford reports which had been held to be most suspect). More sophisticated atomic calculations performed between 1979 and 1981 did suffice to reduce the expected value of R, the parity-violating parameter, by a factor of around three. But that was still insufficient to bring the 1977 Washington–Oxford results into line with the Weinberg–Salam model. Instead, the 1981 Washington recantation was based upon new data taken with new apparatus, and it remained the case that no defect of interpretation or performance of their 1977 experiment had been identified. Significantly, though, the Washington group sought to legitimate their new positive findings by casting doubt upon the earlier reports from atomic-physics experiments, including their own. In the introductory remarks of their paper (1981, 643) they stated:

> Since the time of our earliest measurements upon this bismuth line, which were not mutually consistent, we have added a new laser, improving the optics, and included far more extensive systematic checks . . .

> Although a PNC [parity-nonconserving] neutral-current interaction between electrons and nucleons in agreement with the Weinberg–Salam theory has been observed in high-energy electron scattering, the situation in atoms is unclear. Our experiment and the bismuth optical-rotation experiments by three other groups [Oxford, and two Russian groups] have yielded results with significant mutual discrepancies far larger than quoted errors. The overall evidence possibly favours some PNC effect on bismuth, and similar evidence about thallium comes from measurements at Berkeley of circular dichroism in thallium vapour.

For a review of all of the atomic-physics neutral-current experiments performed up to 1980, see Commins and Bucksbaum (1980).

56 The status of the findings of experiment E122 could be regarded as ambivalent for an unusual reason. In discussing E122 with SLAC experimenters who had *not* worked on it, I frequently encountered comments like 'no-one really knows what went on in that experiment'. I should stress that the intention of such comments was not to impugn the integrity of the E122 experimenters; it was rather to recognise that the performance of E122 (and E95) had involved the development of specialised techniques and procedures foreign to the rest of experimental HEP. No-one outside the SLAC-Yale collaboration had the required expertise for informed comment or criticism of the performance of E122: if the SLAC-Yale experimenters had made a 'mistake', only they could find it. It is also relevant to note that it was very difficult to attempt to check E122's findings outside SLAC. Elsewhere, high-energy charged leptons (electrons and muons) were only available as secondary beams at PSS which were not as intense nor subject to the same precise control as the SLAC primary electron beam. In 1983, a high-energy muon experiment at CERN did report similar findings to E122, but within a large margin of error: Argento *et al.* (1983) reported a measurement

giving $\sin^2\theta_w = 0.19 \pm 0.11$, consistent with the accepted value of 0.23, but by no means as decisively so as the SLAC result. Although highly regarded within the HEP community, the findings of E122 hardly conformed to the philosopher's paradigm of intersubjectively observable empirical fact.

57 The general transformation in the significance ascribed to the Washington–Oxford experiments can be exemplified in the work of Abdus Salam. In late 1977, the atomic-physics experiments held the stage alone – E122 had yet to get under way – and Salam was co-author of a paper in which the non-observation of parity violation at Washington and Oxford was used to motivate the development of an unconventional electroweak model (Pati, Rajpoot and Salam 1977). Two years later, in his concluding address to the European Physical Society Conference, Salam had this to say: 'After the beautiful presentations of Dydak and of Prescott [on the SLAC experiment], there is little that I can add about the agreement of the SU(2) × U(1) theory . . . with all the currently measured weak and electromagnetic phenomena below 100 GeV or so'. He qualified this statement in a footnote which remarked that the Novosibirsk atomic physics results depended upon 'the atomic theory of Khriplovich et al. for the complicated bismuth atom. Since the Oxford group contest (among other things) this atomic theory . . . the issue of atomic parity violation is a problem for atomic physicists, rather than a problem for particle physics' (Salam 1979, 855 and note 2). Thus the Washington–Oxford experiments were shuffled from prominence to the sidelines of HEP and beyond – to atomic physics, in fact. As Salam put it in October 1979, the only conflict with the standard model came from 'one unfortunate set of experiments' at Washington and Oxford, adding in a humorous aside, 'a few years ago one should have said an unfortunate theory' (A. Salam, speaking at the Symposium in Honour of Professor Nicholas Kemmer, Edinburgh, 1 October 1979, unpublished).

11

QCD IN PRACTICE

Although it is still early days, great optimism abounds that we at last
have a theory of strong interactions as well as weak-electromagnetic. All of these ideas have grown out of the realization that
quarks underwrite the phenomena of the universe, and today
practically all high-energy physicists accept this (indeed, very
complicated experiments are designed, and their results analysed, with the quark–parton model being used as an axiom).
FRANK CLOSE, 1978[1]

Spin one-half quarks with flavor and color, and their associated
flavorless gluons with color, are the fundamental objects in the
description of hadrons. The local color-SU(3) invariant field
theory of quarks and gluons, QCD, is a good candidate for the
description of hadrons. The replacement of naive guesses for
quark dynamics by calculations based on QCD seems likely to
transform the quark model, already successful in a large range of
areas in elementary particle physics, into a theory which can
unify the many disparate applications of the quark model,
foremost of which are the applications to hadron spectroscopy
via the constituent quark model and applications to high energy
inclusive collisions via the quark–parton model.
O.W.GREENBERG, 1978[2]

It is now fashionable to attach to everything the label 'predicted by
QCD'. In fact, despite favourable auguries, the confinement
problem has not yet been solved and there are *no* rigorous results
from QCD for hadron spectroscopy. ANTHONY HEY, 1979[3]

The use of Quantum Chromodynamics (QCD) in treating the
hadronic world has become an overwhelming trend in particle
physics . . . Perhaps it is for the first time in the history of physics
that a theory which is neither precisely defined nor proved to
have the right to exist as a consistent theory has become so
popular. DOKSHITZER, DYAKONOV AND TROYAN, 1979[4]

309

11.1 Phenomenological QCD – An Overview

We now come to the second great theoretical pillar of the new physics: quantum chromodynamics, the gauge theory of the strong interaction. In the period following the November Revolution, QCD rapidly came to dominate theoretical and experimental perspectives on strong-interaction physics. The present chapter reviews how this dominance was achieved. The story is a complicated one and, for simplicity of exposition, documentation of the extent of QCD dominance will be postponed to Chapter 12.

In the latter half of the 1970s, theorists elaborated QCD into three distinct traditions; one purely theoretical and two phenomenological. The theoretical tradition centred upon various nonperturbative approaches to the confinement problem. This work was not perceived as having phenomenological implications during the 1970s (neither was an agreed proof of confinement forthcoming) and I will not discuss it further here.[5] I will concentrate instead upon the two QCD-based phenomenological traditions. These related to hadron spectroscopy and hard-scattering phenomena, and were, respectively, extensions of the charmonium model and the parton model. Theorists in both traditions drew upon perturbative QCD calculations, analysing existing phenomena and predicting the existence of new ones. The development of these traditions thus served to set up a cross-cutting network of symbioses between theory and experiment. Within all three major areas of HEP experiment – electron–positron, lepton–hadron and hadron–hadron physics – traditions existed devoted to the exploration of spectroscopic and hard-scattering phenomena. Data from these traditions were seen as both justification and subject matter for phenomenological QCD analyses. In turn, these analyses generated a context for further growth of the relevant experimental programmes. And the upshot of this symbiotic spiral was that, by the end of the 1970s, experimental practice in HEP was almost entirely centred upon phenomena of interest to QCD theorists.[6] Data on phenomena like hadronic soft-scattering, for which QCD theorists could not supply a ready-made analysis, were no longer generated in significant quantities. Proponents of alternative theories of the strong interactions – principally Regge theorists – were starved of relevant data: the world described by their analyses had become invisible in experimental practice. HEP experimenters had come to inhabit the phenomenal world of QCD and, in effect, obliged recalcitrant theorists to do likewise.

In the following sections we can examine various aspects of the QCD takeover of strong-interaction physics, focusing upon hadron spectroscopy and hard-scattering processes in turn. First, though, one general point can usefully be made. In discussing the establishment of electrowcak gauge theory I went to some lengths to expose the inadequacy of the 'scientist's account' of events. In the present chapter, a similar effort would be superfluous. Although enthusiastic theorists did, from time to time, advertise future data as decisive tests of QCD, when the data arrived matters were seen to be more complex. Never, during the 1970s, was QCD alone brought to bear upon existing data.[7] Thus the story told here does not even begin to resemble the adversarial confrontation of theory with experimental facts which typifies the 'scientist's account'. Instead we will be dealing with what can be best described as the construction of the QCD *world-view*. Within this world-view, a particular set of natural phenomena were seen to be especially meaningful and informative, and these were phenomena conceptualised by the conjunction of perturbative QCD calculations and a variety of models and assumptions (different models and assumptions being appropriate to different phenomena within the set).

This QCD-based world-view was sustained by the cross-cutting symbioses of theoretical and experimental practice discussed above, and it was appropriately resilient. In a given phenomenal context, mismatches could, and did, arise between data and the predictions of particular QCD-based models. But the overall context, constituted within the world-view, was such that these mismatches could be represented more readily as subject matter for further research, than as counterarguments to QCD itself. I mention this now because the exposition which follows is linear, examining a sequence of applications of QCD one by one. Taken individually, each application had only limited persuasive power. To appreciate the grip of the QCD-based world view upon HEP practice in the late 1970s, it is important to bear in mind that the episodes discussed below occurred in parallel, generating mutually reinforcing contexts. Once the QCD bandwagon had begun to roll it is hard to imagine what could have stopped it.

11.2 QCD and Spectroscopy
In the latter half of the 1970s QCD-based models were applied to all areas of hadron spectroscopy. These models were variants on the charmonium theme, suitably tailored for particular contexts. Quarks were assumed to move under the influence of long-range confining forces (described by models which could not be derived from QCD) while their short-range interactions were taken to be those given by single gluon exchange (the first-order perturbative approximation to

311

QCD). Three areas of spectroscopic physics can be distinguished in the period under consideration – the study of the new particles, of 'baryonium' and of conventional hadrons – and I will discuss these in turn.

The New Particles

The first spectroscopic application of QCD was to the analysis of data on the new particles, the J-psi and its relatives, and little needs to be added to the discussion of Chapter 9 to cover subsequent developments. We saw in Chapter 9 that, during the November Revolution, data on the new particles assumed enormous significance within HEP. The established theoretical resource was the charmonium model which drew directly upon the asymptotic freedom property of QCD (in conjunction with a confining potential which did not follow directly from QCD). The charmonium model was at the heart of phenomenological work on both hidden- and naked-charm states, as it was of the experimental programmes which sought to investigate their properties. In the post-revolutionary era, investigation of the properties of the psis and their relatives remained the primary focus of research at the low-energy e^+e^- colliders, SPEAR and DORIS, and charmonium models remained the primary interpretative resource. With the discovery of the upsilons in 1977 yet another field for charmonium-style phenomenology opened up, fed, from 1979 onwards, by data from the CESR e^+e^- collider operating as an upsilon factory. Experiments with lepton and hadron beams at the high-energy PSS also yielded a growing supply of data on c- and b-quark composites – more grist to the QCD phenomenologist's mill. As we saw in Section 9.4, mismatches did develop from time to time between theorists' expectations and experimenters' findings – the disappearing η_c and η_c' being the most conspicuous case in point. But in an overall context of thriving traditions of new-physics theory and experiment, these mismatches were seen as interesting results, not serious problems; as justifications for further research and not for downing tools (experimental or theoretical).

Baryonium

Baryonium was the generic name given in the mid-1970s to hypothetical hadrons containing two quarks and two antiquarks ($qq\bar{q}\bar{q}$). Such entities occupied an almost paradoxical place within hadron spectroscopy. The major theoretical advance of the late 1950s and early 1960s was the grouping of the known hadrons into the Eightfold Way families: singlets, octets and decuplets. In 1964, the constituent quark model took off from this classification, explaining the observed multiplet structure in terms of the quark–antiquark ($q\bar{q}$)

composition of mesons and the three-quark (qqq) composition of hadrons. Only singlets, octets and decuplets of hadrons could be formed from these combinations, and it was held to be a success of the CQM that throughout the 1960s and early 1970s all of the observed hadrons could be fitted into such families. Experiment did not significantly challenge the CQM assumption that only qq̄ and qqq combinations existed in nature.[8] In 1965, Nambu proposed the idea of colour to explain why this was so. If quarks came in three colours then, Nambu argued, they would tend to bind together in colourless combinations. And the simplest colourless combinations which could be formed from tri-coloured quarks were just qq̄ and qqq: the standard CQM states. But here the element of incipient paradox arose. qq̄ and qqq were by no means the *only* colourless multiquark states which could be constructed. Colourless hadrons could also be put together from an infinite variety of more complex combinations: qqq̄q̄, qqqq̄q̄, and so on. Nambu's argument, if taken seriously, implied that such 'exotic' multiquark hadrons should also exist, particles lying outside the Eightfold Way classification. Thus having tried to explain the success of the CQM, Nambu arrived at a point where evidence which had originally counted in the model's favour – the absence of exotic hadrons – stood itself in need of explanation.[9]

During the 1960s and early 1970s Nambu's work on colour was outside the main line of CQM development. But it assumed central importance with the advent of QCD, the field theory of colour. Nambu's sketchy argument that the observed hadrons should be colourless became a key tenet of QCD theorists. In default of a satisfactory theoretical analysis of quark binding, QCD theorists defended the doctrine of confinement: for some unstated and non-perturbative reason, the properties of QCD were such that only colourless combinations of quarks could exist as physical particles. In QCD, as in the more primitive CQM formulations, there appeared to be no particular reason why the only relevant colourless combinations should be qq̄ and qqq. QCD theorists therefore expected the existence of exotic multiquark hadrons, and the properties of such particles were worked out in some detail in several QCD-inspired composite models.

The rise of theoretical interest in exotic hadrons generated a context in which novel experimental traditions could flourish. In the latter half of the 1970s these traditions focused upon the search for qqq̄q̄ baryonium states, which promised an especially clear experimental signal. Since baryonium decays involving qq̄ annihilation were expected to be subject to the usual Zweig rule suppression, these

313

particles were predicted to be narrow and to decay predominantly to baryon–antibaryon pairs (hence the name baryonium) as shown in Figure 11.1. This decay mode was quite untypical of conventional hadrons, and offered an easy way to distinguish between baryonium states, if they existed, and more mundane entities.[10]

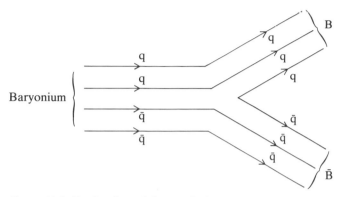

Figure 11.1. Zweig-allowed decay of a baryonium state into a baryon–antibaryon (BB̄) pair.

By around 1976 experimental baryonium physics had become an established research programme at the major HEP laboratories, but findings within this programme were indecisive. Despite, or perhaps because of, the enthusiasm with which HEP experimenters embarked upon the hunt, the empirical status of baryonium remained unclear. Many early reported sightings went unconfirmed in the next generation of experiments, in a pattern which was repeated up to the end of the decade.[11] At the 1979 CERN conference, for example, Heidelberg experimenter B.Povh summarised the situation thus:

> When accepting this invitation to talk about the present experimental status of the search for baryonium I was quite sure that I would be able to give a definite answer as to the existence or nonexistence of narrow baryonium states. But some of the hurried experiments trying to confirm their existence and possibly to find new ones failed to show any indication of the baryonium. The second generation experiments just being started will do, I hope, a careful job in scanning the NN̄ system for narrow states. But at present the situation in baryonium physics is still as confusing as it used to be.[12]

In the late 1970s whether baryonium states actually existed was open to doubt. But what was not open to doubt was the existence of an experimental tradition devoted to baryonium physics and structured around expectations deriving from QCD. Irrespective of the experimenters' long-term success or failure in establishing the existence of baryonium, QCD was central to practice in this tradition. Baryonium experiments were unequivocally part of the new physics, even though their relation to QCD-inspired models could not be expressed within the conventional language of confirmation or falsification.[13]

The Constituent Quark Model Reborn

In the latter half of the 1970s the principal points of contact between HEP experiment and QCD-inspired spectroscopic models were baryonium physics and the study of heavy-quark systems, but it would be inappropriate to close this section without some discussion of the relation between QCD and the spectroscopy of light-quark systems: the conventional hadrons, built from mundane u, d and s quarks in mundane $q\bar{q}$ and qqq combinations. The first point to note is that by the mid-1970s experimental research on conventional resonances was in decline.[14] Experimenters had pursued such research in increasingly fine detail for more than fifteen years and, as discussed below, there remained an active CQM phenomenological tradition devoted to the analysis of the data which had been amassed. But the shift in theoretical context which took place during the November Revolution left the CQM on the sidelines, and many resonance experimenters sought to redeploy their skills in areas of greater perceived significance (the study of charmed particles, for example, or baryonium). This withdrawal of experimental support, coupled with the overall post-revolutionary decline of interest in resonance physics, left CQM modellers in a difficult position. There was no immediate shortage of subject matter for their research – the problems of conventional resonance physics had by no means all been solved – but the audience for such research was fast disappearing (amongst theorists as well as experimenters). One possible response to this would have been for CQM modellers to abandon their trade, seeking as best they could to find a niche in the new physics.[15] But a less drastic option was also open: to incorporate QCD with the CQM, and hence to align resonance phenomenology with the new physics.

The decisive step towards realignment of the CQM from the old to the new physics came from gauge theorists. In 1975, at the height of the November Revolution, theorists at Harvard and MIT extended the basic axioms of the charmonium model to the analysis of the conventional hadrons.[16] They began by assuming that for light-quark as well as heavy-quark systems the long-range confining force was independent of quark flavours and spins. This assumption immediately implied the exact su(6) hadronic symmetry which was the conventional starting point for CQM analyses. However, the Harvard and MIT theorists then departed from the standard CQM by assuming that the origin of su(3) symmetry-breaking lay solely in the quark masses (the s quark being heavier than the u or d quarks). This was in line with the 'flavour-blindness' of QCD. When this assumption was conjoined with the further assumption that all short-range interquark forces followed from single gluon exchange, a relatively well defined pattern of su(6) symmetry-breaking was predicted. The pattern incorporated several mass-formulae already established in the CQM and offered more besides. The QCD-inspired model explained, for example, the magnitudes and signs of the Δ–N, $\rho - \pi$ and Σ–Λ mass-differences, quantities which could only be reproduced in conventional CQM analyses by the adjustment of free parameters.

In the sense that the new spectroscopic models offered a calculation of previously arbitrary parameters, they could be seen as an advance over earlier formulations of the CQM. Certainly in 1975 the Harvard group saw things that way: 'We can see a time coming', they wrote, 'when the subject of hadron spectroscopy as it is now known will be generally recognized to be interesting, but no longer truly fundamental. Hadron masses, widths and cross sections may soon be "understood" if not precisely calculable'.[17] Given the tide then running in favour of gauge theory and the new physics, many particle physicists were disposed to agree. But work remained to be done. The sophistication of the early QCD-inspired spectroscopic models was on a par with that of CQM analyses a decade earlier. The route to improvement was plain: to rerun the historical development of the CQM, beginning this time with the assumptions of perturbative QCD. In the mid-1970s gauge theorists had neither the expertise nor the inclination to carry this programme through. Established proponents of the CQM, in contrast, had both. First into the water were quark modellers Nathan Isgur and Gabriel Karl, working at the Canadian Universities of Toronto and Guelph respectively.[18] Beginning in 1977, they published a long series of papers on the QCD-based CQM analysis of hadron spectroscopy.[19] Other quark modellers followed suit.[20] The upshot of this outburst of CQM activity was that a number of features of the hadron spectrum which had been previously described by arbitrary parameters (or not described at all) were now described by symmetry-breaking parameters of the form suggested by perturbative QCD. If one accepted the basic assumptions of the QCD approach, then what had once been arbitrary facts could now be understood as manifestations of an underlying theoretical structure.

316

Quantum Chromodynamics in Practice

The rejigged CQM thus constituted a new approach to conventional hadron spectroscopy. It served, moreover, to link work in this area with the wider context of the new physics. Once they had espoused the tenets of perturbative QCD, quark modellers could cite the programmes of new-physics theory and experiment in justification of their own work (and vice versa). And, for these reasons, the advent of QCD was regarded by quark modellers in the late 1970s as an unequivocal advance. Greenberg, for example, one of the pioneers of the CQM, spoke for most quark modellers when he described the work of Isgur and Karl as 'a striking success of QCD applied to the constituent quark model', and 'important progress towards a fundamental theory of baryon spectroscopy'.[21]

Thus, as the 1970s drew to a close, the CQM moved from the old physics to the new, to the mutual satisfaction of quark modellers and new physicists in general. Ironically, the section of the HEP community least impressed by the rejuvenation of the CQM were the experimenters. They had moved on from conventional resonance physics and showed little inclination to return, despite the newly fashioned relationship between the area and QCD. Contemporary theoretical work in the QCD-inspired CQM resembled that which immediately preceded the advent of QCD, in that it concerned fine details of the hadron spectrum rather than striking and significant new phenomena. In general, experimenters subscribed to the gauge theorists' sentiment expressed above – hadron spectroscopy was 'interesting, but no longer truly fundamental' – and their attention remained fixed elsewhere.[22]

11.3   QCD and Hard Scattering

The spectroscopic developments just reviewed were important in the post-revolutionary career of QCD, but in themselves they were insufficient to guarantee its establishment. They covered only a limited domain of experimental practice, and left the rest untouched. The reason for this was as follows. The standard experimental technique for resonance spectroscopy was the bombardment of fixed targets by PS hadron beams. Typical resonances decayed into *a few* lower-mass (quasi)stable hadrons; these were detected, and measurements upon them yielded, through relativistic kinematics, the properties of the resonances themselves. Now, the typical yield of detected particles (generally pions) per collision in HEP experiments was found to increase with the energy of the beam – to such an extent that at the high-energy machines of the 1970s (the Fermilab PS and the CERN SPS and ISR) the decay products of low-mass resonances were swamped by an enormous background of unrelated particles.

317

The high-energy machines were therefore not, in general, the ideal tools for hadron spectroscopy; for the spectroscopy of the conventional hadrons and baryonium, the old low-energy machines – such as the CERN PS and Brookhaven AGS – were not only adequate, they were technically to be preferred.

Certainly Fermilab, the SPS and the ISR were the only machines with sufficient energy to investigate the production and spectroscopy of charm and b-quark systems in purely hadronic processes but, even so, their unique possibilities for the experimental investigation of multiparticle production at the highest energies were left largely untouched by the spectroscopic development of QCD. Similar comments apply to other areas of experiment. In comparison with hadronic experiments, fixed-target experiments with lepton beams (electrons at SLAC, muons and neutrinos at the PSS) were highly inefficient sources of spectroscopic data; from the point of view of the strong interactions, their principal interest lay in the investigation of hard-scattering scaling phenomena, again manifested in multiparticle production rather than resonance processes. Of course, the electron–positron machines were in many respects the ideal tools for the examination of charm and b-quark systems, but here, too, QCD-spectroscopy had little to say about many possible experiments. This lack was strongly felt when PETRA failed to discover the t quark. The energy range accessible at PETRA lay above the mass of the upsilons and, in this 'desert', spectroscopy offered no guide to the search for interesting phenomena.

QCD-spectroscopy thus engaged only peripherally with experimental programmes devoted to high-energy hadron–hadron, lepton–hadron and lepton–lepton (e^+e^-) physics. However, in parallel with spectroscopic developments, there grew up new traditions of QCD-based hard-scattering phenomenology. These took over from the parton-model traditions discussed in Chapter 5, and became central to the high-energy experimental programmes we first met there. QCD theorists drew support from those programmes, and at the same time their work served to constitute the subject matter for succeeding generations of experiment. Through the 1970s the traditions of hard-scattering experiments grew to dominate high-energy research at the major HEP laboratories, completing the gauge-theory take-over of experimental practice and setting the seal on the establishment of the new physics.

In what follows, I will discuss the rise of hard-scattering experiment under three headings. The first is lepton–hadron scattering. This was the paradigmatic reference point for QCD. QCD predicted deviations from the exact scaling of the parton model, and observation of such deviations was expected conclusively to demonstrate the validity of QCD. We shall see that, in the event, developments were not so simple. Under the second heading I will discuss electron–positron annihilation. Here the gluons of QCD were used to fill the desert above the upsilons, giving meaning and significance to research at PETRA and PEP. Finally I will discuss the evolution of hard-scattering traditions of hadronic experiments at CERN and Fermilab. These traditions had the least clear-cut relation to QCD, but many experimenters (and theorists) worked to progressively reduce the ambiguity.

The emphasis in what follows is thus on the interrelation of experiment and theory, but one purely theoretical development should be mentioned in advance. The takeover of the parton model by QCD theorists, and the consequent permeation of traditions of hard-scattering experiment by QCD considerations, took place in two distinct stages. The first phase began in 1973 with the discovery of the asymptotic freedom of massless gauge theory. Using the techniques of the operator product expansion and the renormalisation group, theorists argued that QCD could underwrite the parton-model analysis of the classic hard-scattering processes: deep-inelastic lepton scattering and electron–positron annihilation. For the former, the parton model explained existing observations of scaling and, for the latter, the model predicted the constancy of R. By virtue of its asymptotic freedom, theorists argued that QCD could reproduce these predictions over a limited energy range, but that small corrections to the parton-model results should be visible in measurements over larger ranges. As mentioned in Section 10.1, the QCD prediction for e^+e^- annihilation was that R should asymptotically decrease to a constant value at high energies; more importantly, calculable scaling violations were expected to become visible in high-energy deep-inelastic data.

It is important to emphasise that at this stage (1973 onwards) QCD only underwrote a limited subset of parton-model predictions. For technical reasons, the operator product expansion/renormalisation group approach to QCD was not held to be valid for the analysis of the other processes to which the parton model had been applied: lepton-pair production and hard-scattering in hadron–hadron collisions. Neither was this approach held to apply to the analysis of the details of hadronic final states in deep-inelastic scattering or electron–positron annihilation.[23] Thus the advent of QCD served to divide the phenomena covered by the parton model into two classes: those which were amenable to QCD analysis using existing theoretical techniques, and those which were not. As one might expect, theorists initially sought to refine their analysis of the former class of processes.

The latter class remained outside the ambit of QCD, and the experimental traditions devoted to their investigation continued to evolve within the conceptual framework supplied by the parton model. Progressively, though, increasing numbers of theorists began to search for new techniques which would facilitate the extension of perturbative QCD to cover the entire range of parton-model applications. This search culminated in late 1977 and early 1978 in an outburst of theoretical activity centring upon a new calculational scheme for QCD. Within the new approach, QCD was indeed seen to be applicable to all of the processes routinely described by the parton model, and thus 1978 saw the second and final phase of the QCD-takeover of hard-scattering physics.

The new calculational scheme was the 'intuitive perturbation theory' (IPT) approach.[24] Its departure from the original 'formal' approach to QCD is most easily characterised in terms of how theorists coped with gluons. Gluons having specified interactions with one another and with quarks were the distinguishing feature of QCD. In the formal approach to QCD, gluons appeared in simple perturbative diagrams containing a few closed loops; the renormalisation group β-function was evaluated from these; and predictions of physical quantities followed from formal field-theory manipulations. In the IPT approach, theorists computed infinite (but restricted) sets of diagrams in which quarks emitted arbitrary numbers of gluons. In making these computations, theorists leaned heavily upon the similarity between QCD and QED. A great deal of sophisticated theoretical work had been done in the 1960s and 1970s on the evaluation of QED diagrams containing arbitrary numbers of photons, and the development of the IPT approach to QCD was one of analogical importation of established techniques and results from QED (taking account, of course, of the differences in group structure of the two theories).[25]

Once the trick of taking over resources from QED had been spotted many theorists contributed to the development of the IPT scheme for QCD,[26] and by mid-1978 the following results were generally accepted. The gluons appearing in perturbation theory could be divided into three classes: 'soft', 'collinear' and 'hard'. Soft gluons were those emitted with low energy and momentum; collinear gluons were of high momentum, travelling parallel to the emitting quark; and hard gluons were those of high transverse momentum relative to the emitting quark. In perturbation theory, the effects of soft and collinear gluons could be taken into account by redefinition of the structure functions used in parton-model analyses. The redefined structure functions incorporated the scaling deviations already predicted by the formal approach to QCD, and thus all of the existing formal predictions could be rederived in the IPT approach. More importantly, in the latter scheme the redefined structure functions were seen to be *universal*: they could be used in all of the standard applications of the parton model, and not just in those to which formal considerations applied. Thus the development of the IPT analysis completed the QCD takeover of the parton model: all of the basic parton results could be rederived up to small and calculable predicted deviations from exact scaling behaviour.

Beyond this extension of theoretical range, the IPT approach to QCD brought with it one further bonus. In perturbation theory, the third class of gluons – hard gluons – could *not* be accommodated by redefinition of structure functions. Instead, emission of hard gluons was expected to lead to novel physical phenomena, quite distinct from those described in the parton model. Such phenomena could be calculated from simple first-order QCD diagrams, and the predictions which followed were welcomed by HEP experimenters. Hard-gluon phenomena constituted a new and significant focus for empirical research and, as we shall see, experimenters structured their practice accordingly.

Lepton–Hadron Scattering and Scaling Violation

QCD was born in 1973, when massless gauge theory was shown to be asymptotically free. Asymptotic freedom implied that, over a limited energy range, scaling in lepton–hadron scattering should appear to be exact, as reported from SLAC and as described by the quark–parton model. Over large energy ranges, calculable violations of scaling were predicted, and this prediction was regarded as the potential acid-test of the validity of QCD. The expected scaling violations took the form of corrections to the quark–parton model results varying as the logarithm of q^2/Λ^2, where q^2 was the square of the momentum transfer between the lepton and hadron systems in lepton–hadron scattering, and Λ^2 was a scale parameter. Λ was unspecified by QCD, but the early observations of scaling implied that it must be of the order of 1 GeV. This in turn implied that, since a logarithm is a slowly varying function of its argument, experiments over a large range of q^2 would be necessary to distinguish between the predictions of the parton model and QCD.

Experiment. As we have seen, the parton model was already a standard resource in the planning of lepton–hadron scattering experiments by 1973, and with the advent of QCD the experimenters acquired a new goal: measurements of sufficient precision to establish the nature of *departures* from the parton-model predictions. In 1975

321

data on deep-inelastic electron scattering were forthcoming from SLAC which appeared to show that scaling was not exact. In the same year a higher energy experiment using a secondary muon beam at Fermilab showed a similar pattern of variation.[27] SLAC, however, lacked the range of q^2, and Fermilab lacked the precision required to determine the precise form of the scaling violations. More experiment was clearly in order, and since lepton–hadron scattering was now a well established tradition – particularly neutrino–hadron scattering which was also seen as the key to weak-interaction physics – it was shortly forthcoming. Rather than attempt to discuss individual experiments, I want now to outline the situation with respect to scaling violations as it appeared at the Tokyo HEP Conference in August 1978 (the nineteenth in the Rochester series). The pattern which had by then emerged persisted to the end of the decade and beyond.

The status of QCD in its many putative applications to hard-scattering phenomena was reviewed at Tokyo in a plenary session by theorist R.D.Field of Caltech.[28] Field first compared the predictions of QCD with the existing electron–proton and muon–proton scattering data from SLAC and Fermilab. QCD gave expressions for the asymptotic form of the structure functions in terms of the single, *a priori* unknown, scale parameter Λ. Field displayed detailed QCD fits of the data, from which the Λ-parameter was determined to be in the vicinity of 0.5 GeV. These fits are reproduced here as Figure 11.2.[29] Scaling violations are manifest there as a dependence of the structure function vW_2 upon q^2 at fixed values of x ($= 1/\omega$): if scaling were exact, there would be no such dependence. As can be seen from the solid curve drawn through the data, the observed violation of scaling could be easily accommodated by a suitable choice of Λ. Field commented, 'the data show a decrease at large x and an increase at small x as Q^2 increases in precisely the manner expected from QCD'.[30]

Field then went on to discuss recent measurements of scaling violations in neutrino and antineutrino scattering made by the BEBC (Big European Bubble Chamber) collaboration working at the CERN SPS. Here again the structure functions were found to be compatible with QCD for a value of Λ in the vicinity of 0.5 GeV. Furthermore, the CERN data were sufficiently precise to allow meaningful calculations of the 'moments' of the structure functions. The moments were variously weighted integrals of the structure functions, and their ratios could be predicted from QCD with no free parameters whatsoever.[31] At Tokyo Field displayed a plot of the CERN moment ratios – reproduced here as Figure 11.3 – which showed a remarkable degree of agreement between theory and experiment.[32] The slopes of

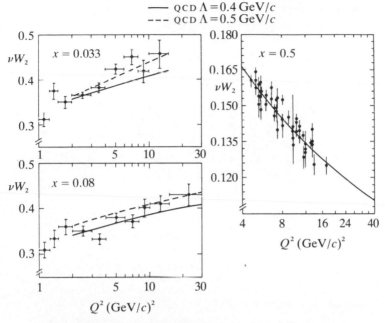

— QCD Λ = 0.4 GeV/c
--- QCD Λ = 0.5 GeV/c

Figure 11.2. Scale breaking in inelastic electron and muon scattering: νW_2 as a function of momentum transfer (Q^2) at fixed values of x. The solid and broken curves represent the QCD predictions for different values of Λ.

the straight lines fitted to the data in Fig. 11.3 were given in QCD by the anomalous dimensions of the relevant operators (Section 7.1) and, as Field noted:

> The anomalous dimensions . . . predicted by QCD are in agreement with the data at about the 10% level. This is one of the most impressive tests of QCD to date. Remember the naive parton model would predict that all the data along the lines in Figure [11.3] should lie at one point (*i.e.* no Q^2 dependence).[33]

Theory. On the face of it, the one-parameter and zero-parameter QCD fits to a whole range of data on electron, muon and neutrino scattering were impressive and convincing. But they were not hailed as a demonstration of the validity of QCD. As theorist Frank Close reported on the Tokyo Conference: 'Although the emerging data in lepton and hadron interactions are all consistent with QCD, no conclusive tests that would confirm or deny the theory are yet available'. The reason for this subdued response was that 'theorists are still uncovering some of the rich subtleties in the theory and [at

323

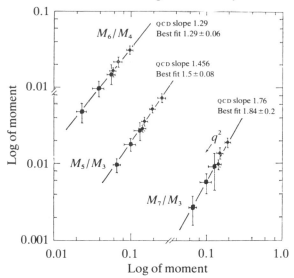

Figure 11.3. Moment ratios in neutrino scattering. The solid lines are the QCD predictions.

Tokyo] there was much discussion as to the understanding we have of the theory and what we must do to really test it in a severe way'.[34]

While HEP experimenters had been busy collecting data which supported QCD, theorists had been engaged in cutting the ground from beneath their own feet. To understand this one has to recognise that the early predictions of scaling violations were truly asymptotic: they were expected to apply only at infinitely high energy and momentum transfer. The question of pressing interest in phenomenological application was how close were currently accessible energy regimes to effective asymptopia. To answer this it was necessary to compute the expected scaling violations beyond the leading-order approximation which dominated only asymptotically. Calculations, and estimations where calculations were impossible, were under way in 1978 – a few had been completed – and revealed that, if the Λ-parameter were indeed around 0.5 GeV, higher-order effects were *not* negligible. These higher-order effects should properly be included in fits to data and, as a result, the existing data were possibly inadequate to provide a good test of QCD.

Perhaps the most accurate statement of the theoretical position in 1978 is that it was confused. Calculation of higher-order corrections to scaling violations were highly complex, and experts disagreed amongst themselves. The results of a careful computation by a collaboration of Fermilab theorists were reported by one of them, T.Muta, at Tokyo. The Fermilab group had endeavoured to compare their results with those of other authors (a process complicated by the variety of different computational procedures employed) and Muta had this to say:

> By carefully taking into account the difference of schemes we find that our result is consistent with that of AEM [Altarelli, Ellis and Martinelli], but disagrees in a minor term with DGP [De Rújula, Georgi and Politzer] . . . After taking into account the difference of the ways to handle mass singularities we find that our result agrees with Kingsley and AEM, but disagrees with Ahmed–Ross, Witten, Calvo and Hinchcliffe–Llewellyn-Smith.[35]

During 1978 and 1979 these differences (which stemmed from the complexity of calculation rather than doctrinal schisms) and others were more-or-less sorted out, and the situation which emerged was broadly as follows. The QCD predictions continued to fit the data, including new measurements from CERN. Various counter-arguments could however be proposed. Many terms were agreed to contribute non-asymptotically to the QCD predictions of scaling violation, and not all of these had been computed. Some of the omitted terms could in principle be computed perturbatively, given sufficient theoretical effort. Others arose from confinement effects. These were intrinsically nonperturbative, and could only be estimated in models. Thus it was possible to maintain that the agreement between existing QCD calculations and the scaling-violation data was accidental: that the results of a complete calculation, including all of the perturbative and nonperturbative terms agreed to be relevant at contemporary energies, might be quite different from present approximations. Conversely, given the complexity of the situation, it was conceivable that any renormalisable theory would lead to similar predictions in the domain of observations; and so on. The arguments for and against QCD from the data became highly technical – certainly not as clear-cut as QCD enthusiasts had once hoped – and I will not go into them further here.[36]

Thus, even as HEP entered the 1980s, particle physicists were unable to convince themselves that the scaling violations seen in deep-inelastic scattering demonstrated the validity of QCD. But this was irrelevant to the continuing centrality of QCD to this area of practice. Theorists made sophisticated perturbative calculations of non-asymptotic contributions to scaling violation; more precise determination of the scale-violating parameters lent point to second and third generation experiments – particularly with the high quality neutrino and muon beams which became available at the CERN SPS; and phenomenologists maintained a satisfying symbiosis with both groups. The validity of QCD may not have been proved according to the canons of deductive logic, but QCD was none the less solidly entrenched as the central resource in this branch of physics.[37]

Electron–Positron Annihilation and Jets

In late 1978 the new electron-positron collider, PETRA, came into operation at DESY. In 1980, experiment began at a similar machine, PEP, at SLAC. PETRA and PEP were both high-energy facilities, designed to attain centre-of-mass energies in the vicinity of 40 GeV. They opened up a large range of energies above SPEAR and DORIS, and experimenters were keen to exploit this. The question was: what to look for? An obvious target was the t-quark, but, as noted above, this declined to manifest itself.[38] Failing that, QCD theorists urged the search for 'three-jet' events, and the experimenters gratefully responded. I will first explain the concept of a jet, and then turn to the three-jet events themselves.

We saw in Section 5.5 that one of the earliest applications of the parton model was to electron–positron annihilation. Here the virtual photon formed in the annihilation process was supposed to rematerialise as a quark–antiquark pair. The quark and antiquark were then assumed to somehow 'dress' themselves to form normal hadrons. This line of reasoning led to predictions for the total hadronic cross-section in e^+e^- annihilation (and hence to the R-crisis and its resolution, as discussed in Chapter 9). But not all parton-modellers were content with predicting total cross-sections. In 1969 and 1970 several attempts were made to predict details of the configuration of hadrons produced in e^+e^- annihilation.[39] The common feature of these attempts was that quarks were expected to dress up as 'jets' of hadrons: sprays of particles all travelling in the general direction of the parent quark (or antiquark) and having only a small component of momentum transverse to that direction.[40] Each quark would convert to a jet, and the hadrons resulting from e^+e^- annihilation were therefore expected to show a two-jet structure. This is schematically represented in Figure 11.4.

Much of the early development of the parton model was carried out by members of the SLAC theory group, and in 1975 the SPEAR experimenters repaid them with the observation of two-jet events. Gail Hanson found from an analysis of high-energy SPEAR data that hadrons tended to emerge from the e^+e^- interaction region in two back-to-back jets.[41] Furthermore, the cross-section for jet production

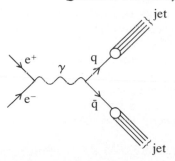

Figure 11.4. Parton-model diagram for two-jet
production in electron–positron annihilation.

was found to have an interesting dependence upon the angle between
the jet and beam axis. This dependence was that given by QED for the
production of a pair of spin-one-half particles – such as quarks – despite the fact that the majority of the detected particles were spin-zero
pions. Models which explained such behaviour without resort to the
quark–parton picture were conceivable but, at the height of the
November Revolution, they attracted few adherents.[42]

Early discussions of jets, then, were couched in terms of the
quark–parton model. But jet physics, like the study of scaling, was
transformed by the advent of QCD. In 1976, three CERN theorists,
John Ellis, Mary Gaillard and Graham Ross, proposed that three-jet
as well as two-jet events should be visible in e^+e^- annihilation.[43] They
argued that in QCD three-jet events would arise when one of the
primary quarks emitted a 'hard' gluon – a gluon carrying a large
momentum transverse to that of the emitting quark. The gluon would
then convert into a third jet of hadrons, as indicated in Figure 11.5.[44]

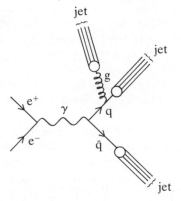

Figure 11.5. Three-jet production in perturbative QCD.

The rate of emission of hard gluons was determined by α_s, the effective QCD quark–gluon coupling constant. This was considerably less than unity in the high-energy regime where perturbative consideration were applicable, and thus three-jet events were expected to occur less frequently than two-jet events. Furthermore, although e^+e^- annihilation was expected to show signs of three-jet formation in the upsilon region, clear evidence for three-jet events was only expected at the higher energies available at PETRA (and, later, PEP).[45]

The PETRA experimenters took careful note of these predictions and, by the time experiment got under way at the new machine, computer programmes for the jet-analysis of data were already well developed.[46] This observation bears emphasis, since it illustrates very clearly the centrality to practice in e^+e^- experiment achieved by QCD in the late 1970s. When PETRA became operational in late 1978, the overall pattern of experiment at electron–positron machines had become more-or-less standardised. Modelling their practice on the pioneering Mark I detector at SPEAR, experimenters aimed to surround the electron–positron interaction region as completely as possible with counters and thus, ideally, to detect every particle which emerged.[47] The subtlety of the experimenter's art lay, therefore, as much in data analysis as in data collection. Now, although it is easy to describe a two- or three-jet event, identification of such phenomena from raw data was another matter. Particles had to be identified, their momenta measured, possible jet axes adjusted, momenta along and transverse to these axes computed, and so on. This process was relatively simple for two-jet events where the experimenters could assume the existence of a single jet axis (the two jets emerging back-to-back) but was extremely complex for three-jet events in which there were three independent and *a priori* unknown axes. Given the expenditure of analytical effort involved, it seems highly unlikely that three-jet events would ever have become a topic for experimental investigation in the absence of prompting from QCD enthusiasts. Certainly they were not the kind of thing one would stumble upon from a pragmatic sorting of data: as PETRA experimenter Bjorn Wiik put it, 'one would never see gluon jets without deliberate treatment'.[48]

Prompted, then, by QCD theorists, the PETRA experimenters searched for three-jet events, and the first relevant data were released at HEP conferences in mid-1979.[49] Four early candidate three-jet events are reproduced in Figure 11.6.[50] The solid lines shown there represent the trajectories of charged particles leaving the e^+e^- interaction region. The length of each line is proportional to the measured energy of the particle in question. A tendency of the particles to group into three sprays is fairly apparent for each event, and the dotted lines correspond to the jet axes as computed by the experimenters.

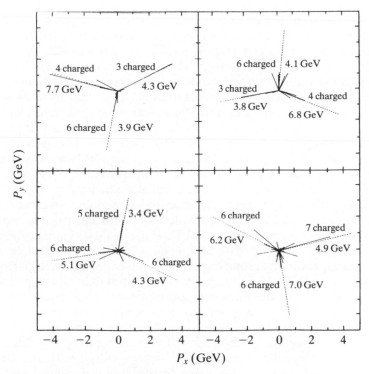

Figure 11.6. Three-jet events from PETRA. P_x and P_y are particle momenta relative to a conventional set of coordinates.

The PETRA observations were encouraging to theorists and experimenters alike: to theorists they represented support for the underlying QCD picture of hard-gluon emission; to experimenters they represented a manifestation of a novel phenomenon to be investigated further. But they were not regarded as conclusive proof of the validity of QCD within the HEP community (the popular press saw matters differently).[51] Many obstacles remained before any such claim could be made. Most obviously, it was impossible to distinguish on an event-by-event basis between genuine three-jet events and statistical fluctuations in hadron directions and energies: even if hadrons were actually emerging at random from the interaction region, occasionally they would clump together in three jets.

To counter such arguments, the PETRA experimenters performed sophisticated analyses of large samples of events, including complex 'Monte Carlo' simulations of their apparatus. This work showed that the rate and overall characteristics of the three-jet events were consistent with expectations deriving from QCD – but only at the expense of introducing extraneous theoretical considerations. For example, while QCD offered predictions concerning hard-gluon emission, it offered no guide as to how the gluon or quarks would dress themselves as hadron jets. The dressing mechanism was clearly related to the confinement problem, and no solution to this in QCD was available. The experimenters filled this gap either by looking at the broad features of their data which were expected to be insensitive to the dressing mechanism[52] or by describing the mechanism in terms of 'fragmentation functions'. These functions were analogues of the structure functions used in the analysis of deep-inelastic scattering, and were used to parametrise the non-perturbative dressing of a gluon (or quark) of a given momentum into a hadron jet of specified characteristics. Fragmentation functions were either pragmatically constructed to fit the overall characteristics of the data or grounded in some simple confinement model. Their use facilitated more detailed analyses of the e^+e^- data and made it possible, for example, to extract quantitative estimates of α_s, the QCD coupling constant, from three-jet event rates. But such estimates proved to be sensitive to the chosen form of the fragmentation functions. They could not be regarded as direct measurements in themselves, and nor could the observations be seen as unproblematic confirmations of QCD.[53]

Thus, in the late 1970s the situation in high-energy e^+e^- experiment resembled that in deep-inelastic lepton–hadron scattering. QCD alone could not be used to make detailed and well defined predictions. But taken in conjunction with various models and assumptions, perturbative QCD could be used to motivate a world-view – this time centring on three-jet phenomena. Experimenters had the tools to explore these phenomena, and the data they generated fuelled further phenomenological work. QCD was central to theoretical and experimental practice in high-energy e^+e^- physics, as elsewhere.

Hard Scattering in Hadron–Hadron Interactions

In Section 5.5 we discussed two classes of phenomena in hadron–hadron physics which were analysed in terms of the parton model: lepton-pair production and high transverse-momentum hadron production. In this section we can review how knowledge of these phenomena developed during the 1970s.

Lepton-Pair Production. In the 1960s, theoretical interest in inclusive lepton-pair production – the process $pp \rightarrow \mu^+ \mu^- X$ – centred on the possible detection of weak interaction IVBs. As such, the experiments which were performed were failures. However, the arrival of the quark–parton model in the late 1960s constituted a new context in which experimental research could flourish. Early formulations of the parton model could describe the overall features of the existing data, and inclusive lepton-pair production became known as the Drell–Yan process after the inventors of the most influential parton-model analysis. Investigation of hadronic lepton-pair production received a further boost during the November Revolution, with the discovery of the J-psi at Brookhaven. Likewise, the discovery of the upsilon in a lepton-pair experiment at Fermilab in 1977 did little to dampen experimenters' enthusiasm for this process. The narrow psi and upsilon peaks in the lepton-pair cross-sections were, of course, subject to a different theoretical analysis from the Drell–Yan continuum. But data on peaks and continuum typically derived from the same experiments, and the two theoretical contexts fostered the growth of a unitary tradition of experiment. Lepton-pair experiments flourished accordingly during the 1970s, principally at the high-energy machines: the CERN ISR and the Fermilab and Serpukhov PSs.[54] Initially the parton model provided the interpretative context for studies of the Drell–Yan continuum, but in 1978 the new IPT approach to QCD served both to reinforce that context and to extend it significantly. According to QCD the basic diagram contributing to lepton-pair production was just the standard parton-model diagram, Figure 11.7(a). However, QCD predicted the existence of corrections to this diagram, coming from hard-gluon emission. The first-order corrections (suppressed by a factor of α_s) took the form shown in Figure 11.7(b).

Hard gluons, by definition, were those emitted at large transverse-momentum to the parent quarks. The diagrams of Figure 11.7(b) therefore translated into the prediction that lepton-pairs should be produced at high transverse-momentum (p_T) relative to the beam-target axis. By mid-1978 this prediction had become a central focus of HEP interest as a new test of QCD, and appeared to be well satisfied by the available data. At the Tokyo conference, Wisconsin theorist F.Halzen reported that a qualitative fit to the lepton-pair data could be obtained from QCD which was a 'definite success', and that the agreement could be made quantitative by additional assumptions.[55] As in the case of scaling violations, however, elements extraneous to QCD intruded upon the comparison with data. Even the qualitative description of lepton-pair production entailed assumptions about unmeasured gluon structure functions, and the quantitative fit

331

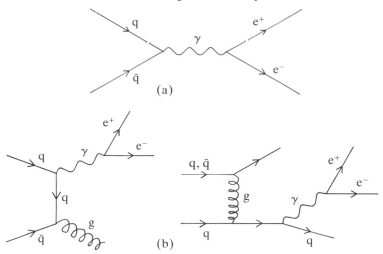

Figure 11.7. Lepton-pair production in hadronic collisions (only the interacting quarks are shown): (a) lowest order QCD diagram; (b) first-order perturbative corrections from hard-gluon exchange.

depended upon specification of the p_T distribution of quarks within hadrons. This distribution was related, via the Uncertainty Principle, to the confinement of quarks within a region of finite size and, as usual, QCD theorists had nothing to say on such topics. After Halzen's talk the situation became even more complicated with the accumulation of more data, and the realisation that, as in the analysis of scaling, many perturbative and nonperturbative effects had been unjustifiably neglected in existing analyses. This new test of QCD, like the classic one of scaling violation, proved to be not a very good test at all: the observation of high-p_T lepton pairs could not be said to confirm QCD. Nevertheless, the IPT approach did constitute the observation of such pairs as a significant phenomenon for experimental investigation, and thus served to reinforce and structure the subsequent development of work in this tradition.[56]

Hadronic Hard-Scattering. We now come to the last of the parton-model-related experimental traditions: the study of large-p_T hadron (LPTH) production in hadronic interactions. The LPTH tradition was born in 1972, with three reports from the CERN ISR that an excess had been observed of large-p_T hadrons relative to established soft-scattering expectations. As we saw in Section 5.5, the excess was immediately ascribed by parton modellers to the hard-scattering of hadronic constituents, and the parton model provided the context for subsequent generations of experiment. During the 1970s developments in this area (theoretical and experimental) were complex and often confusing. I will only sketch out the main lines.

From a theoretical perspective, LPTH phenomena were the least conducive to parton-model analysis. In LPTH, unlike the other fields of hard-scattering experiment, there was no photon (or IVB) intermediary of known properties. LPTH therefore involved strong-interaction dynamics at an immediate level: if partons behaved as quasi-free hadronic constituents, as assumed in other applications of the model, there would be no LPTH at all. Theorists faced up to this by attempting to guess the basic form of hard-scattering parton–parton interactions. One of the most influential of the early guesses came from SLAC theorists Blankenbecler, Brodsky and Gunion, and was embodied in their Constituent Interchange Model (CIM).[57] The central idea of the CIM was that mesons as well as quarks were active constituents of hadrons. The basic LPTH process was assumed to be quark exchange in quark–meson scattering: $\pi^+ + d \rightarrow \pi^0 + u$, for example. Since mesons were routinely supposed to be $q\bar{q}$ composites, the CIM approach to LPTH was rather contrived, but it had the great merit that it worked. In particular, the initial ISR measurements had shown that inclusive cross-sections for processes like $pp \rightarrow \pi^0 X$ had a $1/p_T^8$ dependence, and this behaviour followed immediately from the CIM by dimensional analysis.

In the wake of the ISR discoveries, and in a context defined by elaborations of the CIM and other parton models, experimenters at the ISR and Fermilab set out to explore LPTH further. One important aim of this work was to elucidate the structure of LPTH final states: if one particle emerged with high p_T from the interaction region, what about the others? Did they cluster around the high-p_T particle; did another high-p_T particle emerge in the opposite direction as well; or was momentum balanced in a more subtle fashion?[58] Progress here was crucially dependent upon developments in detector technology (see Chapter 12) and was relatively slow. However, by the mid-1970s the generally accepted picture was that large-p_T hadrons were typically members of jets, similar to those identified in e^+e^- annihilation. This picture received a major boost in 1977 with the first reports from 'calorimeter' experiments at Fermilab.[59] Earlier experiments had triggered their detectors on single high-p_T particles, but these experiments used a jet trigger: as many particles as possible were caught in a large 'calorimeter' detector, and the electronics were arranged to count events in which jets of particles carried large p_T in aggregate.[60] One of the major findings of the Fermilab calorimeter groups was that jet events occurred around 100 times more frequently than production of a single particle carrying the same total p_T. This was both encouraging to experimenters, who liked to study as large cross-sections as possible, and confirmed what was already suspected: that jet phenomena were the basis of LPTH, and that only rarely was the jet dominated by a single high-p_T particle.[61]

By 1978, then, LPTH production had come to be seen as another instance of the jet phenomena already identified in e^+e^- physics. At the same time, theorists were using the IPT approach to extend QCD to cover LPTH, and the jet picture of the latter was highly congenial to this exercise. The overall characteristics of LPTH were held to support the extended QCD analysis; the latter in turn provided a context for further LPTH studies. One clear example of this structuring role of QCD concerned the LPTH single-particle inclusive data. In QCD the primitive hard-scattering interaction was the exchange of a hard gluon. This was predicted to lead to a $1/p_T^4$ dependence of inclusive cross-sections. Unfortunately, as noted above, the early data showed a $1/p_T^8$ dependence instead. This was originally held to rule out gluon exchange as the basic mechanism of LPTH, and to favour parton models of the CIM type. But such simple arguments were no longer in vogue in the late 1970s. QCD had already come to dominate many experimental traditions, and theorists now sought to reconcile its predictions with the recalcitrant single-particle data. Extensive phenomenological analyses were made, employing quark–gluon and gluon–gluon interactions as the basic source of LPTH, augmented by the full apparatus of structure functions, fragmentation functions and scale-violating terms (perturbative and nonperturbative). Thus armed, theorists were able to defeat the $1/p_T^8$ fall-off. It was shown that for modest p_T (i.e. for p_T values at which measurements had been made) QCD could be reconciled to $1/p_T^8$ behaviour. At higher p_T, though, a $1/p_T^4$ dependence was still expected to manifest itself, and this encouraged the experimenters to devote the required effort to measuring the relevant extremely small cross-sections. In the late 1970s, new results from the ISR showed that indeed the fall-off of single-particle cross-sections did become less steep as higher and higher p_T ranges were explored. At the highest p_T, the data suggested that a $1/p_T^4$ behaviour might be becoming apparent, and this was regarded as a new justification for the QCD picture.[62]

The determined pursuit of single particles to extreme p_T values was a simple instance of the permeation of QCD considerations into experimental LPTH physics in the late 1970s. Many such instances could be given, especially in respect of jets, but discussion of these would require extensive technical background. Here, instead, is a quotation from Pierre Darriulat, a leading ISR experimenter, taken from his 1980 review of LPTH research:

[T]he difficult LPTH experiments, with complicated multibody final states, require guidance to suggest relevant measurements and efficient modes of data presentation. Such guidance is presently best provided by the QCD approach, which is the most successful at describing existing data and possesses a strong predictive power. The hope of reaching in the QCD framework a coherent and unified description of strong interaction dynamics is an important stimulus to encourage further experimental efforts.[63]

Thus, by 1980, QCD had become both the reason for, and the guiding light of, LPTH experiment. The world-view within which LPTH experimenters conducted their practice was unequivocally that proferred by QCD theorists. LPTH experiment had taken its place amongst the new-physics hard-scattering traditions. As we shall see in the next chapter, these traditions, together with those devoted to QCD-related hadron spectroscopy, made up an overwhelming proportion of contemporary experimental programmes. Even more strikingly, the future of experimental strong-interaction physics was entirely conceived in terms of the elaboration of the new-physics traditions. The HEP community had succumbed *en masse* to the QCD world view: strong-interaction physics had become synonymous with the study of quark–gluon interactions.

NOTES AND REFERENCES

1   Close (1979, vi).
2   Greenberg (1978, 381).
3   Hey (1979, 523).
4   Dokshitzer, Dyakonov and Troyan (1980, 271).
5   The origins of the nonperturbative attack on QCD were outlined in Chapter 7, note 50. For reviews of some subsequent developments in this area, see Gervais and Neveu (1976), Marciano and Pagels (1978) and Gross, Pisarski and Yaffe (1981). My reason for omitting further discussion of this work is that I am particularly concerned here with the relation between QCD and experimental practice as it evolved during the 1970s. As stated in the text, the non-perturbative tradition had no direct bearing upon this relation: the quark confinement problem remained unsolved and no questions were raised which experimenters could seek to answer. Nonperturbative QCD did, however, have an indirect bearing which should be noted. In parallel with the growth of phenomenological applications of QCD, the development of nonperturbative analyses came to be seen as increasingly fundamental and important. Conversely, nonperturbative analyses of QCD promised to legitimate phenomenological applications of the theory. Thus nonperturbative analyses contributed to a single symbiotic constellation of practice embracing pure

theory, phenomenology and experiment: all three components grew together, feeding off one another.

In the early 1980s, the nonperturbative lattice approach to QCD was brought to the point where meaningful predictions could be extracted, and these are briefly discussed in note 13 below.

6   Strictly speaking, the phenomena in question were those of interest to electroweak gauge theorists as well as QCD theorists. However, the principal experimental tradition central to electroweak theory – lepton–hadron scattering – was also part of the constellation supporting QCD. Experimental lepton–hadron physics was therefore structured around considerations deriving both from electroweak theory and QCD. More generally, it can be noted that electroweak theory was tangled up in the same cross-cutting symbiosis with experiment as QCD: the historical processes promoting the rise of QCD were, then, reinforced by the electroweak-theory tradition (and vice versa). Beyond this, it can be noted that QCD and electroweak theory were synthesised within another tradition which flourished in the late 1970s: grand unified theory (see Chapter 13). So, to be explicit, a full analysis of the triumph of gauge theory should *simultaneously* take account of, at least, three QCD traditions (one nonperturbative, two phenomenological), electroweak theory, grand unified theory and the experimental traditions with which they were variously interwoven. In this sense, the triumph of QCD was both more complicated and more overdetermined than stated below. My aim in the text is to keep the discussion as simple as possible while illustrating the overall general pattern.

7   By 'QCD alone' I refer here to results derived directly from the QCD Lagrangian using agreed field-theory techniques without extraneous assumptions.

8   See the discussion of the hunt for the Z resonances (exotic baryons) given in Chapter 4, note 53.

9   The existence of exotic hadrons could also be motivated from 'duality' arguments. The duality principle asserted that resonance production and Regge-pole exchange provided equivalent descriptions of hadronic scattering. Relations between Regge trajectories could be deduced from this principle (which was espoused in the late 1960s by both bootstrappers and quark modellers) and the existence of baryonium trajectories appeared to be required for a self-consistent description of baryon–antibaryon scattering (Rosner 1968: for a review of the duality approach to baryonium, see Rosner 1974, 267–72. Rosner here refers to baryonium states as 'gallons' – see his note on p. 270 for an explanation of this).

10   For a review of theoretical analyses of baryonium, see Montanet, Rossi and Veneziano (1980, 149–200). This article discusses no less than six different models, all of which led to predictions of baryonium states: the 'QCD inspired dual topological picture'; the 'MIT-bag model'; the 'colour chemistry model'; the 'nuclear physics approach'; the 'topological bootstrap model'; and the 'large N_c approach to baryons in QCD'. I will give a brief outline of the MIT-bag model in order to give some feeling for the details of theoretical work in this area, but first it is interesting to

note that two of the approaches listed above were independent of QCD: the nuclear physics approach and the topological bootstrap. The former was an extension of the routine nuclear physics approach to nucleon–nucleon and nucleon–antinucleon interactions; the latter was the 1970s version of the S-matrix bootstrap (still advocated and elaborated by Geoffrey Chew). The topological bootstrap incorporated duality (see note 9 above) and hence led directly to the prediction of exotic states. It was Chew who coined the name 'baryonium' in 1976 (see Chew and Rosenzweig 1978, 320: I thank B. Nicolescu for pointing this out to me). It should, however, be noted that the bootstrap commanded very little active support in the mid-1970s; it was undoubtedly QCD-inspired theoretical practice which created the context in which experimental baryonium physics flourished. For a review of work on the topological bootstrap, see Chew and Rosenzweig (1978).

The MIT bag model was first proposed by a group of MIT theorists in early 1974 (Chodos, Jaffe, Johnson, Thorn and Weisskopf 1974). In the model, hadrons were represented as bubbles of vapour within a liquid. Quarks were confined within the bubbles (bags), and interactions between quarks were described by single gluon exchange, the first-order perturbative approximation to QCD. The MIT group constructed a simple calculational scheme embodying this image, and in subsequent years the bag model was extended by theorists at MIT and elsewhere to cover all areas of hadron spectroscopy: heavy-quark systems (containing b or c quarks) and light-quark systems (conventional hadrons) as well as to baryonium. For the latter, see Jaffe and Johnson (1977), Jaffe (1977a,b). For technical reviews of bag physics see Hasenfratz and Kuti (1978) and Squires (1979); for a popular review see Johnson (1979).

11 The history of the S(1936) meson, the earliest and most promising baryonium candidate, will serve as an instance of the general pattern. The existence of the S was first reported in 1974 (Carroll *et al.* 1974). In an experiment at the Brookhaven AGS a large, narrow peak was observed in total cross-section measurements on proton–antiproton and antiproton–deuteron interactions (the peak corresponded to a resonance of mass 1932 ± 2 MeV and width 9 ± 4 MeV). The S had just the properties expected for a baryonium state, and new experiments got under way (at the CERN PS, the Brookhaven AGS and elsewhere) to explore these in more detail. By April 1977 three experiments had reported data confirming the 1974 sighting and 'it seemed reasonable to assume . . . that the S(1936) was well established' (Montanet 1979, I-3). But from then on the situation became more complicated. Of the 9 further experiments which reported investigations of the S up to the end of the decade, 5 reported positive findings and 4 reported that they could find no trace of it. Moreover, the positive reports were not entirely consistent with one another: in particular, two experiments reported widths of 23 ± 6 MeV and 80 ± 20 MeV respectively for the S – much larger than those reported from other experiments. Different experiments had employed different apparatus and detectors to study different production and decay channels, and by the late 1970s it had become difficult, to say the least, to argue convincingly from the overall data that

the S did or did not exist. Taking all experiments into account it was possible to maintain that the S gave a consistent explanation of observations (see Montanet 1979, I-5); alternatively it was possible to accept the non-existence of the S and to look for faults in the experiments which had reported its presence (this was the strategy of the experimenters who had looked in most detail for the S but without success: see Tripp 1979). For a documented history of the S(1936) and other candidate baryonium states, see Montanet, Rossi and Veneziano (1980, 201–222). For detailed discussions of experiments in progress in the late 1970s, see Nicolescu, Richard and Vinh Mau (1979) (this volume also includes contributions on contemporary theoretical approaches to baryonium physics).

12 Povh (1969, 604).

13 Another class of hypothetical hadrons deserves to be mentioned: glueballs. These were hadrons containing no quarks at all but only gluons. That glueballs should exist followed from the colour confinement dogma of QCD: gluons were themselves coloured, but colourless combinations of two or more gluons could be constructed and, according to the dogma, should exist as observable particles. The prediction of glueballs was QCD-specific (in the sense that many quark models could be devised, but that only QCD-inspired models included coloured gluons as well) and the expected properties of glueballs could be calculated in most QCD-inspired spectroscopic models, but they attracted little interest relative to baryonium until the end of the 1970s. In the early 1980s glueballs assumed a more prominent position in HEP, for two reasons. First, this was the time when lattice gauge theory began to bear fruit. The idea of quantising QCD on a discrete lattice of space–time points had been originally suggested by Kenneth Wilson (1974) as an approximate way of extracting the spectroscopic predictions of the theory. In practice, lattice calculations proved to require extensive computer facilities and the development of new computing techniques (largely taken over from solid-state physics). This work was brought to the point where phenomenological comparisons between calculations and data could be said to be meaningful only in the late 1970s and early 1980s. At this stage, lattice theorists sought only to reproduce the accepted overall systematics of conventional hadron spectroscopy, and calculations of complex multiquark states were beyond them. But analysis of glueball spectroscopy was possible, and represented an area in which lattice calculations could make distinctive contact with experimental data. Lattice theorists emphasised the significance of glueballs accordingly, thus strengthening the support for contemporary glueball experiment.

The second development bearing on the popularity of glueball physics was the tentative identification in 1980 of an experimental candidate. Glueballs were not expected to have as distinctive production and decay modes as baryonium states, but QCD theorists argued that a 1440 MeV particle recently observed in J-psi decays at SPEAR had been misidentified as conventional meson and was better described as a glueball (Chanowitz 1980). The glueball identification was by no means

338

unambiguous, but theorists outlined programmes of e^+e^- and $p\bar{p}$ experiments which might support their position, delineating more sharply the context for future empirical research.

For technical reviews of developments in lattice gauge theory, see Rebbi (1980) and Creutz (1981); for a popular review, see Rebbi (1983). For corresponding accounts of glueball physics, see Dalitz (1982, 36–37), Donoghue (1982) and Ishikawa (1983).

14 See Chapter 4, notes 54 and 55.

15 This was the response of many Regge phenomenologists to their own increasing marginality during the late 1970s.

16 The work of the Harvard group, De Rújula, Georgi and Glashow (1975) has already been mentioned in Section 9.4. The MIT group, DeGrand, Jaffe, Johnson and Kiskis (1975), were working in the framework of the bag model (see note 10 above).

17 De Rújula, Georgi and Glashow (1975, 160).

18 Professor Gabriel Karl's early involvement with the quark model offers an especially neat illustration of the importance of expertise and microsocial context in the determination of research strategies (I have no information on Professor Isgur's background). Karl was born in Romania and studied chemistry as an undergraduate at the University of Cluj. In 1961 he emigrated to Canada. He enrolled as a graduate student in physical chemistry at the University of Toronto, where he worked on molecular physics. He was awarded a PhD in 1964, and continued to work on molecular physics until 1966, when, in his own words (Karl 1974, 1):

> . . . I received an NRC postdoctoral fellowship and went to the Department of Theoretical Physics in Oxford. I spent three fruitful years working in the research group of Professor R.H. Dalitz. At this time there was a great deal of excitement about the 'quark' model . . . This model is basically a 'molecular' model for the proton and other simple particles. I worked on this model, since I felt that my expertise with molecules would be useful. In 1969 Les Copley, Edward Obryk and myself found the quark model solution of an old experimental puzzle concerning the angular distribution of pions photoproduced from protons. The interpretation which we proposed was accepted quickly and led soon to work in the same direction by many other people.

Having moved into particle physics by exploiting the composite analogy connecting the quark model with his original field of expertise, molecular physics, Karl continued to work on quark physics after he took up a faculty position at the University of Guelph. His collaboration with Nathan Isgur (which built directly upon his earlier work) was a manifestation of this. Much of this collaborative work was done in England, while Isgur was visiting the theory group at Oxford and Karl the theory group at the nearby Rutherford Laboratory. At both institutions the CQM and QCD were topics of active local interest.

Karl's early work on the CQM, referred to in the above quotation, was reported in two publications: Copley, Karl and Obryk (1969a,b). I am grateful to Professor Karl for the provision of biographical materials.

19   Between 1977 and 1979 Isgur and Karl published seven papers on their QCD-based model; the first was Isgur and Karl (1977).

20   For a review of this work, see Hey (1979). Hey's talk was a summary of phenomenological CQM analyses, and it is significant to note that he divided it into two sections: 'Pre-1978 Status' and 'Post-1978 Developments'. The latter section was devoted exclusively to the work of Isgur, Karl and the other authors who had begun to import resources from QCD into the mainline CQM.

21   Greenberg (1978, 348). All of the major post-revolutionary CQM reviews (see Chapter 4, note 34) saw the advent of QCD as an unequivocal advance. But it should be emphasised that this did not imply that conventional hadron spectroscopy was 'solved'. As Dalitz (1982, 45–6) put it: 'We conclude that there are still some serious problems to be settled before our understanding of the spectrum of even the simplest baryonic excitations can be said to be satisfactory, even though there has been a great deal of progress made on the basis of QCD over the last few years'. But, he continued, 'the qualitative [QCD-inspired] fit to the data is over-all remarkably good. The discrepancies existing tend to be concerned with matters of detail; the general picture works surprisingly well. That is the situation today'.

22   There were some signs of a revival in experimental resonance physics in the early 1980s. For example at SLAC there existed a high precision electronic detector, LASS, well suited to resonance measurements using secondary hadron beams derived from the primary SLAC electron beam. A long-range planning document for SLAC issued in 1980 commented upon the possible future use of LASS thus (SLAC 1980a, 85):

> The advent of SLED [a programme to increase the energy of the linear accelerator] has resulted in increased hadron yields in the existing beam to LASS. Although recent discoveries have somewhat overshadowed this former 'bread and butter' area of research, there are indications that the new theories based on QCD or other models used in explaining the new physics will lead to new predictions based on subtler interpretations of ordinary strong interactions, and thus to a renewed effort in this area of physics.

23   In the jargon of HEP, the standard approach to QCD was only valid for processes which were 'light-cone dominated' – processes which could be represented in terms of products of current operators at light-like separations (see the discussion of the light-cone algebra in Section 7.2). It was generally agreed that only two processes fell in this category: inclusive lepton–hadron scattering, $lN \rightarrow lX$, and inclusive electron–positron annihilation, $e^+e^- \rightarrow X$, where X represents all possible combinations of hadrons. If X were in any way restricted (as in the measurement of an exclusive cross-section , $e^+e^- \rightarrow \pi^+\pi^-$, for example) then light-cone dominance ceased to apply even to these two processes.

24   For technical reviews of the IPT scheme, and access to the original literature, see Llewellyn Smith (1978), Dokshitzer, Dyakonov and Troyan (1979), Ellis and Sachrajda (1980), Buras (1980), Reya (1980), Altarelli (1982) and Wilcek (1982).

Quantum Chromodynamics in Practice

25 The analogical aspect of the development of the IPT approach was neatly summed up by some of the physicists involved as follows (De Rújula, Ellis, *et al*. 1982, 612):

INFALLIBLE PROGRAM TO JUSTIFY YOUR SALARY
1. Take a QED calculation (RUSSIAN[28], if possible) at random
2. Change $\alpha \rightarrow \alpha_s(Q^2)$
3. Change QED into QCD in title*
4. Publish
5. Go to 1
*and the names of the authors

Their note 28 was to Gribov and Lipatov (1972), one of the seminal papers on the relevant techniques in QED. Fuller references to the original QED literature are to be found in the reviews cited in note 24 above.

26 Amongst the seminal 1977 contributions on the IPT approach were Politzer (1977a,b) and Sterman and Weinberg (1977), and some comment on the genesis of these papers is appropriate (the following remarks are based upon interviews with Politzer and Weinberg). The routes followed by Politzer and Weinberg to an IPT-style analysis were similar. Politzer had laid the foundations of the formal approach to QCD with his 1973 discovery of asymptotic freedom. Subsequently he concentrated first on exploring the detailed implications of his discovery for deep-inelastic phenomenology (see note 31 below) and for heavy-quark systems (Section 9.4). Throughout this period he was based at Harvard, but in 1976 he spent a month visiting the University of Chicago. There he came into close contact with HEP experimenters who were interested in QCD predictions for hadronic hard-scattering phenomena at high energies (the experimenters had in mind the Doubler/Saver project at nearby Fermilab, which aimed to produce 1 TeV beams for fixed-target experiments and to realise 2 TeV centre-of-mass energy proton–antiproton collisions). Politzer 'felt embarrassed' at being unable to generate such predictions using formal techniques, and set about making direct perturbative QCD calculations of the Drell–Yan process, $pp \rightarrow \mu^+\mu^-X$. This led to the papers cited above and a continuing programme of research on the IPT approach.

One existing result which Politzer built into his analysis was the Kinoshita–Lee–Nauenberg (KLN) theorem. This theorem had been developed in the early 1960s, and concerned technical problems ('infrared divergences') arising from the masslessness of photons in QED (Kinoshita 1962, Lee and Nauenberg 1964). Politzer argued that the KLN theorem could be used to solve analogous problems arising from the massless gluons of QCD. Weinberg made a similar connection between KLN and QCD in 1977 whilst attending a SLAC seminar on jet production in e^+e^- annihilation (for the relation between jets and QCD, see below. Weinberg's contributions to electroweak theory have been discussed already; for an early contribution to the development of QCD, see Weinberg 1973; and for his work on GUTs, see Chapter 13 below). He then telephoned 'the expert', Kinoshita, for advice. Kinoshita, in turn, recommended that he contact George Sterman at the State University of

New York at Stony Brook, who was already working on the extension of the KLN theorem to QCD (Sterman 1976). Thus began the (telephonic) collaboration between Sterman and Weinberg which resulted in the QCD analysis of jet production in e^+e^- annihilation cited above.

27   For the SLAC data, see Taylor (1975); for Fermilab, see Chang *et al.* (1975).

28   Field (1979). It is worth noting that although Field's talk had the very general title of 'Dynamics of High Energy Reactions' it was exclusively devoted to QCD phenomenology.

29   Field (1979, 748, Fig. 6).

30   Field (1979, 748). In a plenary report on electron, muon and photon scattering experiments, CERN experimenter Erwin Gabathuler (1979, 846) commented: 'At present all results from deep-inelastic scattering from H_2 and D_2 [hydrogen and deuterium] are in excellent agreement with QCD'.

31   The n-th moment was defined as $M_n = \int_0^1 dx \, x^n \, F(x)$. The QCD predictions for the moment ratios were first derived by the discoverers of asymptotic freedom in detailed analyses of deep-inelastic scattering: Gross and Wilcek (1973b, 1974), Georgi and Politzer (1974).

32   Field (1979, 750, Fig. 9).

33   Field (1979, 750).

34   Close (1978, 267).

35   Muta (1979, 235).

36   For reviews, see Gaillard (1979) and De Rújula (1979).

37   A further axis along which QCD penetrated lepton-scattering experiment concerned the structure of the hadronic final states. QCD predicted the production of hadronic jets (see the following section) and experimenters set out to look for them. For early reports from Fermilab and CERN on the observation of jets in neutrino experiments, see Van der Velde *et al.* (1979) and Scott (1979). For a review of the experimental situation, see Renton and Williams (1981).

38   Two further targets for high-energy e^+e^- measurements can be mentioned here. First there was the QCD/parton-model prediction that R should tend to a constant. This was found to be the case and constituted evidence against the existence of t-quark states in the accessible energy range. Secondly, electroweak models predicted the existence of small effects due to Z^0 production: $e^+e^- \rightarrow Z^0 \rightarrow$ lepton pairs or hadrons (note that the Z^0 here is a virtual intermediate particle, its mass was predicted to be above the maximum energy of PETRA and PEP). The experimenters set out to search for such effects, and relatively imprecise data favouring the standard model began to be reported from PETRA in the early 1980s. For a review of this work, see Duinker (1982).

39   Drell, Levy and Yan (1969), Cabibbo, Parisi and Testa (1970), Bjorken and Brodsky (1970).

40   The assumption that quarks dressed to form jets of limited transverse momenta (of the order of a few hundred MeV) was a way of reconciling the parton model with the observation that soft hadronic cross-sections decreased rapidly with transverse momentum.

41 Hanson *et al.* (1975). See also Schwitters *et al.* (1975).

42 In his Nobel Lecture, Burton Richter (1977, 1296) remarked that
    I find it quite remarkable that a collection of hadrons, each of which
    has integral spin, should display all of the angular-distribution
    characteristics that are expected for the production of a pair of
    spin-$\frac{1}{2}$ particles. Such behaviour is possible without assuming the
    existence of quarks . . . but any other explanation seems difficult
    and cumbersome. In my view the observations of these jet
    phenomena in e^+e^- annihilation constitute one of the very
    strongest pieces of evidence for believing that there really is a
    substructure to the hadrons.

43 Ellis, Gaillard and Ross (1976).

44 Note that the work of Ellis, Gaillard and Ross pre-dated the IPT
    approach to QCD. Their prediction of three-jet events was grounded
    in a 'common-sense' reading of Figure 11.5. The advent of the IPT
    scheme served to disarm technical objections to the prediction, and
    hence to reinforce the theoretical context supporting three-jet
    searches.

45 The earliest data supporting the existence of three-jet events in e^+e^-
    annihilation were reported at the 1978 Tokyo Conference from DORIS
    experiments in the upsilon energy region: Spitzer and Alexander (1979)
    (for a review of subsequent developments, see Söding 1979). The
    hidden-b upsilons were expected to decay primarily to three gluons
    (analogously to the decay of psis; see the discussion of the charmonium
    model in Chapter 9.4) which would then materialise as jets. However,
    the total amount of energy to be shared amongst all of the final-state
    hadrons was small relative to typical transverse momenta within jets,
    and the DORIS experimenters were unable to positively identify well
    collimated sprays of hadrons in their data. They were, however, able to
    show that the two-jet structure visible at energies away from the upsilon
    peaks was blurred when the beam energies were set to coincide with the
    upsilon masses. They interpreted the blurring as evidence that three jets
    were being formed instead of two, as expected from QCD. A similar
    conclusion was also reached from experiments on the upsilon at CESR:
    see Franzini and Lee-Franzini (1982, 261–4).

46 This section of the account is based upon an interview with PETRA
    experimenter B.Wiik.

47 For a brief description of the SPEAR Mark I detector and its construc-
    tion, see Richter (1977, 253).

48 Wiik, interview.

49 For fully referenced reviews of the early PETRA three-jet data, see
    Duinker and Luckey (1980), Söding and Wolf (1981), and Duinker
    (1982). For a theorist's review of the data, see Ellis (1981).

50 Söding (1979, 277, Fig. 9).

51 The first announcement of three-jet events was made at an HEP
    conference held at Fermilab in August 1979. In popular accounts, this
    announcement was translated into conclusive proof of the existence of
    gluons. Typical of such presentations was the report from Harold
    Jackson in Washington which appeared in the *Guardian*, 30 August

1979, under the headline 'Gluon Holds It All Together'. 'The hunting of the quark progressed a little further yesterday', revealed Jackson,

> with the announcement by a group of physicists meeting in America that they had identified a new particle which binds the centre of atoms. In an unaccustomed fit of whimsy, they have named it gluon, a fairly precise description of the function it serves in holding the universe together.
>
> The discovery is regarded as an important advance in the efforts to understand the internal structure of atoms. Mr. Don [sic] Lederman, Director of the Fermi National Laboratory in Illinois, said that 'we're beginning to understand how it is all put together'. The existence of the gluon was theoretically predicted 10 years ago but has only been established by the combined efforts of 300 physicists at PETRA . . .

This style of reporting jarred nerves within the European HEP community, as the *New Scientist* (1979) revealed in an unsigned news item entitled 'Do Gluons Really Exist?':

> Newspaper stories of the discovery of the gluon – an elusive particle believed by theorists to hold nuclear matter together – bemused many physicists in Europe, not least because the Americans were proclaiming a discovery made on an accelerator in Hamburg. The announcement of this breakthrough in particle physics research came at the Fermi Laboratory in Batavia, Illinois. But what was all the fuss about? . . .
>
> The researchers do not actually 'see' gluons in their apparatus. At the highest energy at which PETRA runs they find that a small fraction of electron–positron collisions produce three sprays, or 'jets', of particles which all lie in the same plane. The physicists call these 'propeller', or 'Mercedes', events after their three-pronged appearance, and believe that the jets originate as a gluon, quark and antiquark that materialise from the electron–positron annihilation.
>
> According to David Saxon from the UK's Rutherford Laboratory, and a member of the team called TASSO – the first group to report the three-jet events – the results revealed in the US do not represent a conclusive proof of the existence of gluons. The evidence is therefore weak, so why have these results been hailed as so remarkable, particularly in the US?
>
> Answer:
>
> The only conclusion seems to be that American particle physicists are trying hard to keep up the momentum for federal funding for their expensive form of research. The battle is already on for the next generation of accelerators, to go to higher energies, so physicists need to prove that future expenditure is well invested.

52   This was the approach advocated by Sterman and Weinberg (see note 26 above) and subsequently by many authors. Its disadvantage was that by focusing on broad features of the data it made detailed measurements of hadronic final-states irrelevant. Experimenters had the capability to make such measurements, and the need for a finer analytical scheme was therefore pressing.

53   A brief description of the two most popular choices of fragmentation function is given in Branson (1982). The standard choice was that due to Caltech theorists, Field and Feynman (1978). This was a phenomenological function found to provide an adequate description of quark jets in many experiments. Extensions of the Feynman–Field approach to gluon jets in e^+e^- annihilation were given by Hoyer *et al.* (1979) and Ali *et al.* (1980). A more theoretical approach to fragmentation functions was the 'Lund model': Andersson, Gustafson and Sjostrand (1980). In line with theoretical prejudice, these authors assumed that quarks and gluons were confined via string-like structures, and that jet formation was due to progressive breakings of these strings. The Feynman–Field and Lund fragmentation functions were different from one another and led, for example, to quite different estimates of α_s when used as input to three-jet analyses. Branson (1982, 30) argues that the variability in α_s extracted using the two models gives 'a good indication of the model dependence of the result', and goes on to note that 'It would of course be useful to try many possible fragmentation models, but the time required to tune each model to fit the data well has so far proved prohibitive'.

54   For example, at the Tbilisi Conference in 1978, new data on lepton-pair production were reported from 8 experiments: 4 at the ISR, 2 at Fermilab and 2 at Serpukhov (Homma *et al.* 1979, 184–203).

55   Halzen (1979, 215–18).

56   For reviews of subsequent developments, see DeCamp (1982) and Kenyon (1982). DeCamp (1982, 330) noted that when the second-order (two gluon) QCD corrections to the parton model were computed they were found to be of around the same magnitude as the first-order. This result served both to render earlier fits obsolete, and to raise the spectre that higher-order, as yet uncalculated, corrections might also be phenomenologically important.

57   The first presentation of the CIM was Blankenbecler, Brodsky and Gunion (1972). For a review of developments of the CIM and other models of LPTH see Sivers, Brodsky and Blankenbecler (1976), Ellis and Stroynowski (1977) and Michael (1979).

58   Investigations of such topics in high-energy soft scattering were routine in the early 1970s. The ISR and Fermilab experimenters transposed the same set of questions to hard scattering.

59   Bromberg *et al.* (1977, 1978), Corcoran *et al.* (1978).

60   The prototypical calorimeter experiments of the 1970s were the HPWF and Caltech-Fermilab neutrino experiments, also conducted at Fermilab. For a general outline of the principles and advantages of calorimeter detectors, see Chapter 12, note 34.

61   Thus the original ISR experiments on single particle production had observed only a vestige of the underlying jet phenomena. The possibility that concentration on single high-p_T particles might distort perceptions of overall event structures was known as the 'trigger-bias' effect (see Jacob and Landshoff 1976).

62   For a review of these developments and references to the theoretical and experimental literature, see Darriulat (1980, 171–2).

63   Darriulat (1980, 194). The continuation of this quotation (p. 195) may

scrve as an indication of the details of LPTH research in the late 1970s:

> Whenever possible experiments testing a specific aspect of the interaction should be preferred over those involving its whole complexity . . . ratios of inclusive cross-sections measured either with different beams or for different types of produced particles, jet fragmentation studies without reference to the mode of formation, measurements of the transverse momentum imbalance between the two jets independently of their mode of formation, etc. A particularly rich source of information, as yet almost unexploited, is the study of final states containing a large transverse momentum direct photon, the analysis of which [according to QCD] should be greatly simplified.

Full details of the significance of such measurements are given in Darriulat's review. Darriulat (202–7) also lists all of the experimental groups working on LPTH during the 1970s (23 different collaborations), the types of experiment each performed, and the experimental publications produced (73 publications in all, up to 1980).

The rapid penetration of QCD into LPTH experimental practice can be followed by comparing Darriulat (1980) with equivalent reviews published in the previous two years: See, for example, Jacob and Landshoff (1978) and Hansen and Hoyer (1979). For a popular account of the emergent QCD picture of LPTH phenomena, see Jacob and Landshoff (1980).

12

GAUGE THEORY AND EXPERIMENT:
1970–90

The new physics consisted of a set of experimental and theoretical traditions organised around a common gauge-theory world view. Chapters 9 to 11 reviewed the joint establishment of the traditions and world view during the 1970s. The present chapter rounds off the discussion by examining the developments in the same period from a somewhat different perspective. Previous chapters sought to analyse *how* and *why* the new physics came to dominate HEP. Answers were sought in terms of the symbiosis of practice. The present chapter aims to emphasise the *extent* of the dominance achieved by the new physics. To this end, the focus will be upon experiment. We will see that during the 1970s the gauge-theory world-view was progressively built into both the present and future of experimental HEP. Experimenters aligned their communal practice more and more closely to new-physics phenomena, creating an empirical context wherein gauge theory could flourish but the theories of the old physics were doomed to wither away. By 1980, HEP experimenters had effectively defined the elementary particle world to be one of quarks and leptons interacting according to the twin gauge theories, electroweak theory and QCD.

Section 1 presents quantitative data on the development of experimental programmes in HEP from 1970 to 1980. In 1970, there existed only one major new-physics tradition, pursued at a single laboratory – deep-inelastic electron scattering at SLAC; by 1980 new-physics traditions dominated experimental practice at HEP laboratories around the world. Section 2 explores the 1970s transition from old- to new-physics experiment more closely. It emphasises that the transition involved much more than a simple redeployment of 'theory-neutral' detectors. Throughout the 1970s, particle detectors were progressively refined to highlight the rare phenomena of the new physics, and to exclude the old. Two examples are presented, which typify how the new physics was made concrete in detector design. Finally, Section 3 discusses particle physicists' visions of the future of HEP. We will see that, by the late 1970s, planning of new major facilities around the world had come to be predicated upon the gauge-theory world-view. The future of HEP was given over to the new physics.

In what follows, I will not give a comprehensive review of experimental developments at all of the HEP laboratories operating in the 1970s. Instead, I will focus upon the major European laboratory, CERN. A discussion of developments there can serve as an indicator of wider ranging trends, for two reasons. The first and most obvious reason is that throughout the 1970s CERN was the largest and most productive centre of HEP research in the world.[1] The second reason is more subtle. Table 12.1 lists the major facilities for HEP experiments of the 1970s.[2] These can be divided into two categories: electron machines (accelerators and colliders) at SLAC, DESY and Cornell; and proton machines, at CERN, Fermilab, Brookhaven and Serpukhov. We have already seen that the electron machines, especially the five e^+e^- colliders, were essentially new-physics machines from the first. From 1974 onwards, heavy-quark spectroscopy and jet phenomena constituted the principal topics for e^+e^- experiment. The historical development of these programmes has already been discussed, and it would serve no purpose to repeat the discussion here.[3] In contrast, the proton machines of Table 12.1 represented the 'hard case' for the new physics.

Table 12.1. Major HEP accelerators and colliders
in the 1970s.

Lab.	Machines
SLAC	SLAC (e^-), SPEAR (e^+e^-), PEP (e^+e^-)
DESY	DORIS (e^+e^-), PETRA (e^+e^-)
Cornell	CESR (e^+e^-)
CERN	PS (p), SPS (p), ISR (pp)
Fermilab	PS (p)
BNL	AGS (p)
Serpukhov	PS (p)

If e^+e^- colliders were the ideal new-physics machines, then PSs were the ideal old-physics facilities. Fixed-target research at PSs was the driving force behind the development of the old physics in the 1960s, and wide-ranging, highly sophisticated old-physics traditions dominated PS research well into the 1970s. Pss could, furthermore, outgun any contemporary collider: they could support a wider range of experiments via the provision of diverse secondary beams, and data collection rates in fixed-target experiments were inherently superior to those at colliders.[4] In terms of manpower employed, and quantity and variety of data produced, PSs continued to dominate experimental HEP throughout the 1970s.[5] Thus a reorientation of practice from the old to the new physics in PS research was crucial to the establishment of the latter, and that is why I will concentrate upon proton machines here.

I choose to focus upon CERN in particular because of the variety of proton machines assembled there. A discussion of the CERN PS programme can serve also to characterise developments at the comparable Brookhaven AGS; similarly, a discussion of the CERN SPS programme can illuminate parallel developments at the high-energy Fermilab PS.[6] An added bonus in concentrating upon CERN is that it was the home of the ISR – the only proton–proton collider operational during the 1970s. Trends at the ISR will illustrate particle physicists thinking on the pp and p̄p colliders planned for the 1980s.[7] Finally, since the enthusiasm for the new physics was, by the late 1970s, uniformly distributed around the world, future planning at CERN will serve as an instance of the overall trend.

12.1 Experimental Trends
In this section I will review the development from 1970 to 1980 of the experimental programmes at the three CERN big machines, the PS, ISR and SPS, respectively.

Experiments at the CERN PS
The CERN PS, a 28 GeV proton synchrotron, was the workhorse of CERN during the 1970s. It sustained an active programme of HEP experiment, and also fed proton beams to the ISR and the SPS. Figure 12.1 illustrates the evolution of the PS experimental programme from 1970 to 1980. Shown here are the number of experiments 'on the floor' – either being assembled, or taking data – year by year at the PS.[8]

The most conspicuous feature of Figure 12.1 is the overall decline in the PS programme in the latter half of the 1970s. The decline was clearly correlated with the PS's new role, from 1976 onwards, of feeding protons to the SPS, but should not be seen as unproblematic. The balance of activity between the PS and SPS reflected the choice of HEP experimenters to explore new-physics phenomena, most of which were inaccessible at PS energies.[9]

In order to examine the development of the PS programme more closely, Figure 12.1 categorises experiments into old- and new-physics traditions. The former category includes conventional resonance physics, the study of high-energy soft-scattering and studies of the

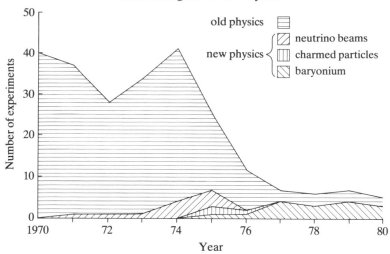

Figure 12.1. Experimental trends at the CERN PS.

weak decays of the ordinary hadrons, and was completely dominant at the PS in 1970. The first trace of the new physics came in 1971 with the start of neutrino experiment using the Gargamelle bubble chamber. In the early 1970s, interest in the Gargamelle experiment converged on investigations of scaling in charged-current events (the parton model) and the search for the neutral current (electroweak theory). Following the 1973 discovery of the neutral current, neutrino physics enjoyed a brief efflorescence at the PS before vanishing in 1976. The cessation of activity was not due to a lack of interest in neutrino physics; in 1976 the large bubble chambers used to detect neutrinos were repositioned in the SPS neutrino beams, where physicists were keen to explore the vistas opened up by higher energies.

Coincidentally with the last year of neutrino experiment, the new physics made two further incursions into the PS programme, in the shape of charm and baryonium experiments. The discovery of the first hidden-charm particle, the J-psi, had been announced in November 1974 and experimenters at CERN (and all over the world) were quick to respond. Three experiments explicitly devised to investigate the properties of charmed particles were mounted at the PS in 1975, and two in 1976. Thereafter charm experiments, like neutrino physics, disappeared from the PS programme, and for the same reason: the higher energies available at the SPS (and ISR) made possible more detailed and extensive investigations than were

conceivable at the PS. Baryonium experiments proved more endur-
ing, and had come to dominate the residual PS programme by the end
of the decade. The qqq̄q̄ baryonium states were expected to have
masses in the same range as conventional resonances, and constituted
the only new-physics phenomenon for which higher-energy machines
had no intrinsic advantage over the PS. Elusive as the baryonium
states were, baryonium physics represented the only PS programme
guaranteed a large audience in the late 1970s.[10]

Experiment at the ISR
In 1971 the Intersecting Storage Rings facility came into operation at
CERN. Fed by beams from the PS, the ISR enabled the study of
proton–proton collisions at centre-of-mass energies up to 62
GeV – the highest energies available anywhere in the world in the
1970s. Figure 12.2 gives a yearly breakdown of the ISR programme.[11]

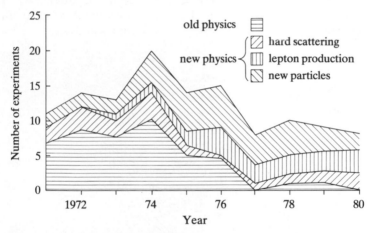

Figure 12.2. Experimental trends at the CERN ISR.

The ISR was conceived in the old-physics era and, as Figure 12.2
shows, old-physics topics (concerning soft hadronic collisions)
dominated the early years of the ISR programme. In the midst of the
November Revolution, however, the new physics moved into the
ascendant, and only a handful of old-physics experiments were
mounted at the ISR between 1977 and 1980.[12] The new-physics
experiments recorded in Figure 12.2 have been crudely sorted into
three categories: the study of purely hadronic hard-scattering
(single-particle inclusive and jet measurements); lepton production
(the detection of single leptons and lepton-pairs); and the study of the

351

new particles (principally psis and upsilons). Experiments on the first two topics were included in the ISR programme from the first: lepton production in hadronic collisions had already been observed and related to the parton model in AGS experiments, and hadronic hard-scattering was discovered in the first round of ISR experiments. The study of new particles grew rapidly after the 1974 discovery of the J-psi, and received a further boost with the 1977 discovery of the upsilon.[13]

The dominance achieved by the new physics in the ISR programme is especially revealing of the overall context of HEP in the late 1970s. As a purely hadronic machine, the ISR was the hardest of hard cases for the new physics. We shall see in the following section that ISR experimenters could only tune their practice to new-physics phenomena at the cost of considerable technical effort and ingenuity. That they invested the required resources amply testifies to the grip established by the gauge-theory world-view in the aftermath of the November Revolution. And, of course, the evolving practice of the ISR experimenters served itself to sustain that world-view.

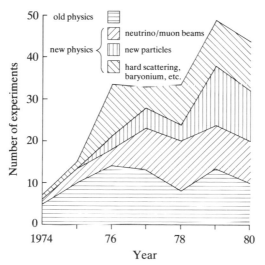

Figure 12.3. Experimental trends at the CERN SPS.

Experiment at the SPS

Figure 12.3 shows a breakdown of experiment at the SPS in the 1970s. The first thing to note about this figure is that it is a composite: the data from 1974 to 1976 do not have the same status as those from 1977 to 1980. Since experiments did not begin at the SPS until late

1976, only the data for the years 1977 to 1980 are directly comparable with those presented for the PS and ISR. However, many experiments were proposed and approved for access to SPS beams before the SPS began running, and the portion of the figure up to 1976 categories such proposals by date of approval. It therefore serves to give some impression of the experiments which physicists wanted to perform at the SPS in those years, and enables us to discuss the development of the SPS programme over a longer period.[14] This development shows similar features to that at the ISR. In the build-up of approved proposals from 1974, the old physics was at first in the ascendant, but by 1976 new-physics proposals outnumbered the old by around three to two. In the first year of operation, the ratio of experiments on the floor remained constant at 3:2 in favour of the new physics, but rose smoothly thereafter to 3:1 in 1980.[15]

As Figure 12.3 indicates, the new-physics programme at the SPS was broadly based and covered all of the phenomena of interest within the gauge-theory framework. The two largest classes of new-physics experiments were those using neutrino beams and investigations of the new (c and b quark) particles, but neither featured to the exclusion of the rest. The wide spread of phenomena covered by the SPS new-physics programme was, of course, correlated with the variety of beams available there. Neutrino beams were used to investigate weak neutral currents, hard-scattering and new particle production;[16] muon beams were principally appropriate to hard-scattering; hadron beams were used in hard-scattering, new particle and baryonium experiments. As far as the new physics was concerned, there was something for everyone. In principle, the SPS was an ideal tool for old-physics experiment – it was originally conceived as such – but in the late 1970s the HEP community chose instead to run it as a new-physics factory.[17] As in the case of the ISR, this entailed the investment of considerable effort and resources in novel detector technology. But again experimenters regarded this as the entrance fee to the new-physics symbiosis, and were happy to pay.[18]

12.2 Theory Incarnate
The old physics was common-sense physics. Experimenters began by mapping out large cross-section processes and then examined them in detail. The new physics was theory-loaded. Experimenters sought to isolate and explore the rare, small cross-section processes of particular significance in the gauge-theory world-view. Some new-physics phenomena were relatively easy to isolate while others were not. A large detector positioned in a lepton beam, for example, poured forth uncontaminated new physics. In hadron–hadron interactions,

though, new-physics phenomena were submerged in a sea of soft-scattering old-physics background, and their detection required considerable effort and ingenuity.

Here again I will focus upon the hard case, and examine two examples of how hadron–hadron experiments were harnessed to the new physics. The first concerns the detection of charmed particles in fixed-target research. This will provide an opportunity to discuss the social dynamics of the transformation from the old to the new physics. We will see how characteristic old-physics resources (of expertise and equipment) were reorganised and redeployed at the forefront of new-physics experiment. The second example concerns detector developments at the ISR collider. These were the developments which identified the ISR as a new-physics machine in the latter half of the 1970s, and which set the trend for collider research in the 1980s.

Charm in Fixed-Target Experiments

During the November Revolution, HEP experimenters all over the world sought to devise ways to investigate the properties of charmed particles using whatever resources they had available. Early work in this area was reviewed in Chapter 9; here I will concentrate upon the programme that emerged at high-energy fixed-target accelerators such as the CERN SPS.

Whilst e^+e^- experiment offered particularly clean data, free of extraneous background, on charmed particles, fixed-target hadron-beam experiment offered much higher data-taking rates. Also, a wide variety of charmed particles were expected to be produced in hadronic experiments, whilst e^+e^- experiments led primarily to data on hidden-charm states having the quantum numbers of the photon (the psis) and on the particular naked-charm states into which they decayed. The attractions of fixed-target charm experiments were obvious; but so were the difficulties involved. The standard way to detect unstable particles was 'bump-hunting' amongst their decay products: parent masses were calculated for combinations of particles emerging from the interaction region, and the clustering of events around a given mass was taken to indicate the existence of an unstable hadron. Unfortunately, in high-energy experiments the number of particles emerging from the interaction region was large – of the order of 10 at SPS energies. Around 3 of these were expected to derive from the decay of a single charmed particle, and the likelihood of picking the appropriate combination of 3 from 10 in a bump-hunting exercise was small. Thus it was clear that while many charmed particles might be formed in fixed-target experiments, straightforward data analysis would be defeated by 'combinatorial background'.[19]

To circumvent this problem, experimenters focused their attention upon naked-charm particles ($c\bar{q}$ or $\bar{c}q$: I will refer to them simply as charmed particles in what follows). The charm quantum number was supposed to be conserved by the strong interactions, so charmed particles could decay only weakly. This implied that they should be relatively long-lived, and theoretical estimates set their lifetimes in the vicinity of 10^{-13} seconds. Within 10^{-13} seconds a relativistic particle could travel around 30 microns (30×10^{-6} metres). And thus, the experimenters reasoned, the observation of tracks of this length would constitute a direct identification of charmed particles, quite free from the problems of combinatorial background. The question which remained was just how to observe such short tracks: no existing detector approached the required spatial resolving power. In the late 1970s at least four different routes to the development of high-resolution charm detectors were pursued, and a discussion of one of them – high-resolution bubble chambers – will serve to illustrate the general strategy.[20]

In 1977, Colin Fisher of the UK Rutherford Laboratory proposed that it might be possible to observe charmed-particle tracks using a bubble chamber operated in an unconventional mode. In conventional bubble chambers, bubbles were allowed to grow so large before being photographed that charmed-particle tracks were unobservable. Fisher argued that, in a suitably designed chamber, the bubbles could be photographed when still sufficiently small that 30-micron tracks would be identifiable. Special optics would also be required in order to record events with high precision. Technical developments went ahead, resulting in the production of LEBC (LExan Bubble Chamber) – a small hydrogen bubble chamber, with a diameter of 20 cm and 4 cm deep.[21]

The first experiment with LEBC was performed in a high-energy pion beam from the CERN SPS in 1979. 48,000 events were recorded in a total of 110,000 pictures. From the 48,000 events 20 events were identified which showed the short tracks expected for charmed particles (as against an expected background of 8 events coming from unusual kaon decays).[22] Unfortunately, it was difficult to argue that the candidate events were definitive observations of charmed particles. The LEBC photographs alone offered no information on the identity and momenta of the decay particles, and it was therefore impossible to compute the mass and charge of the candidate particles: all that was known was that their production and decay had the expected topology – a very short track, decaying to several hadrons.

The obvious next step was to seek more information on the identity and momenta of the decay hadrons. Here the LEBC experimenters executed a neat manoeuvre. In 1974 the construction of a major experimental facility for SPS experiment had been proposed. This was known as the European Hybrid Spectrometer or EHS.[23] The EHS consisted of a bubble chamber, used as the experimental target, followed by an array of highly sophisticated electronic detectors designed to analyse the reaction products which emerged. Conceived in the old-physics era, the original aim of the EHS was to improve upon pioneering Fermilab bubble-chamber measurements of high-energy soft multiparticle interactions.[24] In the late 1970s the EHS was nearing completion, and was in danger of becoming a white elephant – an old-physics facility in a new-physics era.

Enter Fisher and his collaborators. They proposed to use LEBC in conjunction with the partially completed electronic detectors of the EHS. LEBC would record charmed-particle tracks and the EHS electronics would analyse the decay products. The EHS would become a new- rather than an old-physics facility. The proposal was accepted, and the LEBC–EHS experiment ran at CERN in 1980. The combination of detectors proved fruitful and charm events were quickly identified. Figure 12.4(a) reproduces one such event, and Figure 12.4(b) gives a schematic reconstruction.[25] This event was interpreted as the production of a $D^0 \bar{D}^0$ pair of charmed mesons by a 360 GeV pion (the short track entering on the left). Being neutral the D^0s leave no tracks, but decay to charged hadrons at the points marked V_2 and V_4. At V_2 the D^0 decays to $K^- \pi^+ \pi^0 \pi^0$, and at V_4 the $\bar{D}^0$ decays to $K^+ \pi^+ \pi^- \pi^-$. The identities and momenta of the decay particles shown in the figure are those given by the EHS electronic detectors, and imply masses for the D^0 and $\bar{D}^0$ of around 1860 MeV – in agreement with previous measurements at SPEAR.

With the LEBC–EHS experiment, bubble-chamber charm experiments had come of age. Using this high-resolution technique, experimenters could positively identify charmed-particle tracks. Significant technical developments were still in progress,[26] but already experimenters were set for a detailed programme of measurements on charm lifetimes and decay modes, and for searches for hitherto unobserved species of charmed particles. At around the same time other approaches to high-resolution fixed-target experiment were also reaching a viable state. The overall field was regarded as one of great promise for the future, and in 1982 one physicist went so far as to suggest that the 'physics interest for such development is such that . . . the pertinent experimental techniques will determine a large fraction of the research activity with fixed target, very high

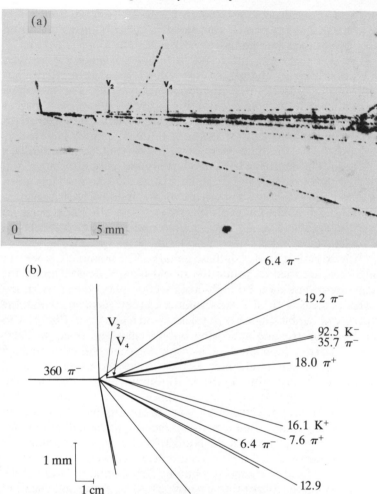

Figure 12.4. (a) Photograph of charmed-particle production in LEBC; (b) schematic reconstruction of the same event (note that the vertical scale has been expanded to clarify the track structure: momenta in GeV/c).

energy proton synchrotrons during the coming decade.'[27] Enough has been said on the technicalities of these developments, and I want now to discuss some overall features of this episode.

The evolution of the high-resolution bubble-chamber approach to charmed-particle detection illustrates well the processes through

357

which the new physics came to permeate high-energy fixed-target experiment. One key point is encapsulated in the following quotation from two of the LEBC team:

> Several years ago the trend in bubble chamber physics was towards bigger and bigger devices intended as general facilities. Most laboratories were abandoning the development of bubble chamber technology. With the exception of a few apparent cranks, bubble chamber physicists were reeducating themselves in counter [electronic] techniques. It was commonly said that, as a useful tool in high energy physics, the bubble chamber was dead. The advent of the small 'disposable' bubble chamber and the use of esoteric high resolution optical techniques has given the bubble chamber a new lease on life. It now [1981] appears to be the best device for the study of short lifetime particles produced with small cross sections and the future is indeed an exciting one.[28]

Little exegesis is needed of these remarks. The bubble chamber had long been regarded as primarily an old-physics device, useful for gathering precise data on high-cross-section phenomena. With the advent of the new physics its usefulness was devalued, and with it the expertise of bubble-chamber physicists – men like Colin Fisher, who had been constructing and using bubble chambers since the early 1960s.[29] The development of high-resolution chambers represented a way in which they could recycle their imperilled expertise into a context at the forefront of HEP. Resources more concrete than expertise were also at stake in this episode. As noted already, the European Hybrid Spectrometer was, in its original form, an extremely sophisticated (and expensive) facility which had no obvious place in the new physics. In the late 1970s, the EHS was in danger of being devalued even before it was completed. However, the substitution of LEBC for the original EHS bubble-chamber target brought the EHS too within the fold of the new physics. An attentive audience and significant topics for the EHS research were thus guaranteed, to the satisfaction of the CERN management, the builders of the EHS and the HEP community in general.

It is straightforward, then, to understand the growth of this branch of fixed-target experimentation. And it is also straightforward to understand its effects upon the phenomenal world of HEP. High-resolution detectors had the prejudices of the new physics built in. They were explicitly designed to measure the very short tracks left by charmed particles, and were useless for the old-physics purposes of their conventional relatives. Their sole aim was to make visible rare new-physics phenomena, and invisible the common phenomena of the old physics. Such techniques were progressively refined throughout HEP in the late 1970s, and their effect was to blot out of existence the phenomenal world of the old physics.[30]

Detector Development at the ISR

The early ISR experimental programme was evenly balanced between the old and new physics. The machine had been conceived in the old-physics era, and such experiments made up the bulk of the programme until 1974; but new-physics-related lepton-production searches were included from the start, and the initial round of ISR experiments saw the discovery of the first LPTH phenomena. As the 1970s drew on, the growth of the new physics around the world provided a context wherein the new-physics component of the ISR programme could prosper, while the old-physics component received little support. I want here to examine the rising dominance of the new physics in respect of the technological developments it entailed. As in the preceding discussion of fixed-target experiments, I will emphasise the extent to which new-physics perspectives were increasingly built into experimental hardware at the ISR.

The differences between old- and new-physics ISR experiments were evident from the first. Figure 12.5 illustrates the detector configurations used in two early ISR experiments.[31] The apparatus of Figure 12.5(a) was used to measure elastic and soft multiparticle scattering. These were classic old-physics topics involving small momentum transfers; the detectors were accordingly placed close to the beam pipes (the ISR protons travel within the pipes in the directions indicated by the arrows, intersecting and interacting in the region between the points marked A1 and A2; the details of the detectors themselves are not important here). The apparatus of Figure 12.5(b) is that used by the CERN–Columbia–Rockefeller collaboration in their discovery of the inclusive production of high-p_T pions. In this view, we are looking along the beam pipes at the interaction region (the oval in the centre of the figure). Note that instead of lying close to the beam pipes, the detectors now stand away from them, along an axis drawn through the interaction region perpendicular to the beam direction. This geometry was chosen precisely to avoid detecting the high flux of softly scattered particles emerging at small angles from the proton–proton collisions; it was a stratagem to cut out as many old-physics events as possible before they even reached the detectors. This stratagem was further reinforced by triggering only on high-momentum particles – the electronics were so contrived that any slow softly scattered particles entering the detector would go unrecorded.

Experiments like that shown in Figure 12.5(b) were the first to observe LPTH phenomena, and, as one ISR experimenter put it:

> The ISR discovery . . . . demonstrated the possibility of exploring the large p_T region with existing machines and immediately triggered a large experimental effort. [However] Instrumentation at the CERN ISR and at the newly operational Fermilab was indeed quite inadequate to tackle LPTH experiments. This lack of experimental preparation [was] a severe handicap and it took several years until adequately instrumented detectors became available.[32]

The problem was that although hard-scattering took place more frequently than had been expected, it still remained a very rare phenomenon (as did lepton production: the following remarks apply to experimentation on all of the new-physics processes). Thus, in order to gather detailed information on processes of interest within a reasonable space of time, experimenters had to construct novel detectors. They reasoned that it was necessary to catch as many of the rare events as possible, which meant surrounding the interaction region with detectors, ensuring that few particles could escape. Figure 12.5(c) shows the apparatus of the CERN–Columbia–Rockefeller collaboration (now joined by a group from Oxford) as it stood from 1976 onwards. This, like Figure 12.5(b), is a view along the beam pipes, and now one sees the interaction region completely surrounded by detectors.

A similar pattern of development was evident throughout the ISR programme. As one further example we can take the sequence illustrated in Figure 12.6. The three configurations depicted are all based upon the use of the Split Field Magnet (SFM) – a major multi-purpose facility, designed to measure particle momenta via observation of their deflections in a magnetic field. Detector configuration (a) was used in the early 1970s by a CERN–Hamburg–Orsay–Vienna collaboration. The figure shows their apparatus as seen from above, looking down on the interaction region. This experiment was designed to study an old-physics topic, resonance production, and the detector arrays were accordingly aligned with the beam pipes. Configuration (b) was used from 1974 onwards by a Liverpool–MIT–Orsay–Scandinavian collaboration to investigate LPTH. Here, as in Figure 12.5(b), the detectors have been positioned at 90° to the interaction region away from the beam pipes. Finally, configuration (c) shows the upgraded SFM facility of the late 1970s. In this arrangement, the interaction region was surrounded on all sides with a rather impressive profusion of detectors, intended to observe as many charmed particle decays as possible.

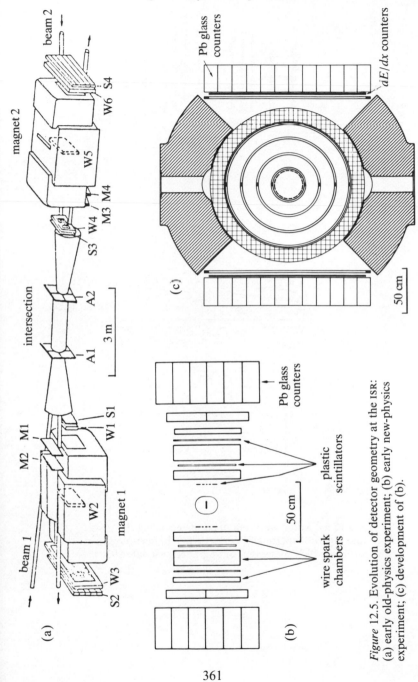

Figure 12.5. Evolution of detector geometry at the ISR: (a) early old-physics experiment; (b) early new-physics experiment; (c) development of (b).

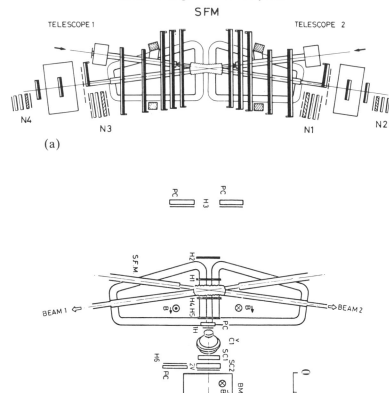

Figure 12.6. Evolution of ISR experiments using the SFM:
(a) old-physics experiment, early 1970s; (b) new-physics
experiment, 1974 onwards; (c) upgraded SFM, late 1970s.

(c)

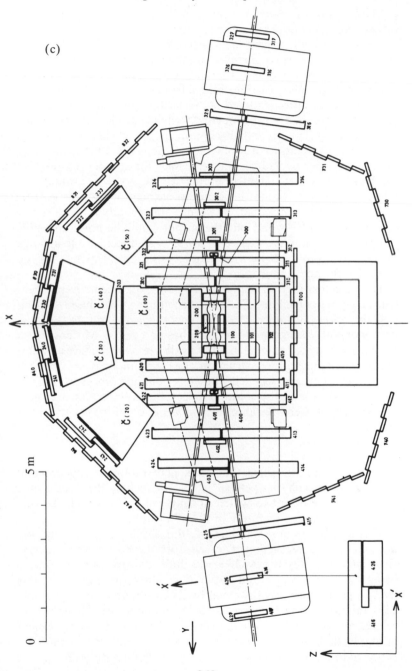

These changes in detector configuration illustrate the penetration of new-physics considerations into experimental practice at the ISR. They do not fully represent the depth of that penetration. The use of complex detector arrays surrounding the interaction region brought with it new problems for ISR experimenters. No longer could they rely upon simple geometry to protect them from the tedious but dominant background of softly scattered hadrons. Instead sophisticated electronic triggers had to be devised, which would record events containing, say, leptons or high-p_T hadrons, but which would consign soft-scattering events to the oblivion they deserved. Furthermore, by this stage triggering had become necessary for another reason. The total proton–proton interaction rate at the ISR was around one million events per second. Large multi-element detectors and their associated electronics could, however, only handle rates of a few events per second. Beyond that, the output signals became impossible to process. Thus, to use these detectors it was imperative to arrange the electronics to reject the vast majority of events, leaving only the rarest for further examination.[33]

The detectors used in the late-1970s ISR experiments were, then, true new-physics facilities. Their construction was optimised to bring out the rare phenomena of the new physics, and to kill the preponderant background of old-physics events. Moreover, so specialised had they become that their sole use lay in the investigation of rare phenomena. By construction, they were impotent to cope with the high event-rates associated with more common processes. The ISR detectors, like the high-resolution detectors developed for fixed-target experiment, were the laboratory incarnation of the gauge-theory world-view.[34]

12.3 Visions

So far we have examined the 1970s penetration of experimental practice by new-physics perspectives. To conclude the discussion I will review the extent to which, during the same period, the HEP community invested its future in the new physics.

Throughout its post-war history, the development of HEP was marked by the construction of accelerators of ever-increasing energy. Even while the latest big machine was under construction, plans were being laid for a bigger and better one. Until the mid-1970s, this leap-frogging cycle was sustained by its own momentum. The only constraints upon machine design were technological and financial; no reference was made to particular needs of the HEP community, except inasmuch that PSS provided the staple diet of the old physics and were expected to continue to do so. As Wolfgang Panofsky, the Director of

Gauge Theory and Experiment

SLAC, wrote in 1974: 'I believe that most elementary particle physicists would agree that the entire field of particle physics has been paced by the opportunities offered by accelerator technologies rather than by the demands of high energy physicists'. However, with the advent of the new physics, perspectives were changing. Panofsky continued:

> Today . . . . I can not be quite so sure whether that is still the case . . . possibly for the next generation of accelerators or storage rings [colliders] history may not repeat: it may be that the open problems of elementary particle physics rather than new technology will determine events. I can say that particle physics today has rarely been in a more 'expectant' condition. Recent results [on LPTH production, neutrino and e^+e^- physics] all cry out for exploitation at high energies and luminosities [interaction rates]. So there is great urgency about the further steps that might be taken in the evolution of accelerators and storage rings.[35]

Panofsky's words were quickly followed by the November Revolution, and many particle physicists came to echo his sentiments. No longer was it sufficient to construct one big PS after another. Panofsky's own laboratory had been in the vanguard of the new physics, and had shown the rival attractions of electron accelerators and electron–positron colliders. Likewise, the ISR programme had demonstrated the potential of proton–proton colliders. Physicists now had a whole range of possible big machines before them. And to determine which to build, and what its specifications should be, they turned to gauge theory for inspiration. QCD was not much use in this respect. At higher energies, QCD theorists predicted more of the same: scaling deviations should become more pronounced; jet phenomena should become more important; new heavy quark systems, if they existed, would look even more like transplanted hydrogen atoms. No new targets were evident for experiment at new machines. But electroweak theory offered a particularly seductive and well-defined target on which to predicate investment decisions: the detection of the intermediate vector bosons (IVBs): W^+, W^- and Z^0. As SLAC theorist James Bjorken put it in a 1977 talk entitled 'Future Accelerators: Physics Issues', 'I find it remarkable (but true) that *nondiscovery* of W and Z . . . would be a more revolutionary development than discovery [of a W and Z] with all the expected properties'.[36] The existence of the IVBs was crucial to unified electroweak theory, and their masses were predicted to lie in the 50 to 100 GeV range.[37] This was beyond the reach of contemporary accelerators, but well within the bounds of technological feasibility.

365

Around the world, IVBS suffused particle physicists' visions of the future. All of the new machines under consideration in the late 1970s were conceived around the requirements of IVB physics. Here we can take developments at CERN as an example of a world-wide phenomenon. At CERN, as discussed below, long-term planning converged upon a giant electron–positron collider. In the short-term, attention focused upon the possibility of running the SPS as a high-energy proton–antiproton collider. A similar plan was hatched at Fermilab, and in what follows I will discuss developments both sides of the Atlantic.

Proton–Antiproton Colliders

In principle, the proton–antiproton (pp̄) collider projects were very simple. Antiprotons had the opposite electric charge to protons. Physicists therefore expected that counter-rotating proton and antiproton beams could be accelerated within high energy PSS, just as counter-rotating beams of electrons and positrons were accelerated in e^+e^- colliders. Where the proton and antiproton beams intersected, ample centre-of-mass energy would be available for IVB production. Thus it appeared possible to use the existing high-energy PSS at CERN and Fermilab to do experimental IVB physics.

In practice, matters were not so clear cut. Aside from the technical difficulties of adapting PSS to a purpose for which they had not been designed, there was the problem of the antiprotons: where were they to be obtained? A solution to this problem was forcefully championed by Italian experimenter Carlo Rubbia (of Harvard and CERN, and a principal of the HPWF experiment at Fermilab). In 1976, together with P. McIntyre and D. Cline, he argued as follows.[38] Antiprotons can be generated (alongside all sorts of different hadrons) by directing a PS beam onto a metal target. These antiprotons are not directly suitable for the desired purpose. There are too few of them to constitute a beam of worthwhile intensity. And they are 'hot' – spread over a large range of energies and directions – while accelerators require 'cool' bunches of particles travelling together within a narrow band of momenta. However, techniques of beam cooling had been under investigation since the late 1960s,[39] and Rubbia and his collaborators argued that these could be developed to cool the hot antiprotons emerging from an experimental target. Successive batches of cooled antiprotons could then be stacked together in a beam sufficiently intense to give a detectable IVB production rate in pp̄ collider experiments.[40]

This proposal promised to bring IVB physics within reach, without requiring the construction of a major new accelerator. Only modifications of existing facilities were required. Thus, on the typical scale of HEP expenditure, it appeared that p̄p colliders would make IVB physics possible 'fast and cheaply', in the words of CERN Director-General Leon Van Hove.[41] The laboratory managements of CERN and Fermilab were appropriately enthusiastic. Both laboratories set about the required technical developments, on which the first Director of Fermilab commented:

> The parallel efforts of CERN and Fermilab to establish proton–antiproton collisions will undoubtedly be seen as another race, and with some justice. One prize may well be the first observation of the Weakon [i.e. $W^\pm$, Z^0].[42]

Thus in the latter half of the 1970s, the new physics was made flesh in the construction of p̄p collider facilities at CERN and Fermilab.[43] Money and effort were devoted to the conversion of the big accelerators to IVB production (to the detriment of fixed-target non-IVB physics at the same machines). And, to step briefly outside our chosen decade, CERN won the race. In 1981 the first 540 GeV p̄p collisions were observed at the SPS. In 1983 reports from CERN of candidate IVB events hit the headlines of the popular press.[44] Only a handful of such events had been found but, in the context of the time, few particle physicists were inclined to dispute that the Ws and Zs of electroweak theory had been discovered. A considerable achievement for European physics, and a disappointment for the Americans, whose own collider remained in the future.[45]

LEP

At CERN the conversion of the SPS to collider physics was fast and cheap, but the SPS collider was not seen as sufficient to support a thriving HEP programme through the 1980s and beyond. The collider was a piece of technological opportunism, designed to achieve the first sightings of the IVBs and to map out their gross features, but with inadequate beam intensities for detailed investigations. Furthermore, the collider was designed to have only two interaction regions, each supporting one major experiment. Whilst these were envisaged to be big experiments, it was clearly impossible for the entire European experimental HEP community to involve itself in them.[46] Something more was needed, and that proved to be LEP, the Large Electron–Positron collider.

From its inception, CERN had been a proton laboratory. The PS had been joined by the ISR and then the SPS. In the early 1970s, it was anticipated that CERN would remain a proton laboratory for the forseeable future. In 1973 a design study got under way for the LSR, the Large Storage Rings. Predictably, the LSR was a super-ISR, designed to store 400 GeV beams from the SPS and to collide them together at 800 GeV centre-of-mass energy. The first design report on the LSR was completed in 1975,[47] but already the effects of the November Revolution were making themselves felt throughout the world of HEP. The machines which came into vogue during the November Revolution were, of course, e^+e^- colliders. And, beyond their established virtues as new-physics facilities, it was clear that e^+e^- machines of sufficient energy would, in principle, be ideal for IVB physics. Given sufficient energy, the neutral Z^0 IVBs could be produced singly in e^+e^- annihilation (Figure 12.7) and detected through their decays to hadrons or leptons. Simply by setting the beam energies appropriately, an e^+e^- machine could be operated as a Z^0 factory – just as SPEAR and DESY were hidden-charm factories, and CESR was an upsilon factory. The charged IVBs, W^+ and W^-, could not be produced singly at e^+e^- machines, but processes like $e^+e^- \rightarrow W^+W^-$ offered a clean experimental source, free from the hadronic background inherent in proton collider experiments. The only question was: was a LEP machine technologically and financially feasible? According to Burton Richter, the architect of SPEAR, the answer was: yes. In March 1976, during a visit to CERN, he sketched out the design of a 200 GeV centre-of-mass energy e^+e^- machine.[48]

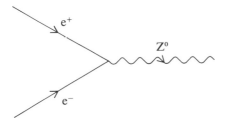

Figure 12.7. Z^0 production in electron–positron annihilation.

With Richter's authority behind it, at CERN the LEP project came rapidly to eclipse the LSR (and other proton facilities then under consideration). CERN was set to become an electron laboratory. In April 1976, a LEP study group was set up at the instigation of John Adams, the Executive Director-General of CERN and eminent machine builder. As chairman of the European Committee for Future Acclerators (ECFA) Guy von Dardel put his weight behind the project. ECFA was an international body, independent of CERN, with members nominated by individual states. Its role in the LEP saga was to provide a vehicle whereby the European HEP community at large could contribute to planning and influence decisions.[49] Several LEP studies and workshops followed, organised by both CERN and ECFA. In May 1977, ECFA recommended that LEP should be considered as 'the prime candidate for a major European project for the 1980's'.[50] In 1979, CERN's policy-making body, the Science Policy Committee, recommended the construction of LEP with the words: 'throughout the history of particle physics it has never before been possible to identify so readily the scientific questions that need to be resolved and also what accelerators need to be built to enable the crucial experiments to be done'.[51] In 1981 the CERN Council, on behalf of the governments of the CERN Member States, concurred: LEP was the machine which would carry European HEP research into the 1990s.[52]

First operation of LEP was envisaged for around 1987. With four to six interaction regions, each feeding a large experimental collaboration, LEP was expected to provide work for a large proportion of the European experimental HEP community.[53] In opting for LEP, European physicists were staking their future on the new physics. Like the e^+e^- machines of the 1970s, LEP was seen as an ideal tool for the investigation of new-physics phenomena (and as effectively useless for the old physics of soft hadronic scattering). Top of the LEP agenda was the search for the IVBs and detailed investigation of their properties. But LEP, like its predecessors, also offered opportunities to pursue to higher energies QCD-related phenomena (hadronic jets, new heavy quark states, and so on). QCD, alongside electroweak theory, permeated the planning for the LEP experimental programme.[54]

The new-physics world-view even entered into engineering schedules for the construction of LEP. The CERN management were keen to see LEP built as soon as possible. But they were acting within tight financial constraints. It was taken for granted that no additional funds would be made available for LEP from the Member States: LEP would have to be financed from within CERN's existing total budget. The construction cost of LEP was estimated at around 1000 million Swiss Francs, and to find this sum the closure of the ISR and cutbacks in the PS and SPS programmes were proposed. To further speed construction schedules, it was also proposed that LEP be built in stages.[55] This was where engineering and financial considerations joined harness with the electroweak IVBs. The first stage of LEP construction, designated Stage $\frac{1}{3}$, was intended to provide beams up to 62 GeV in energy. This was the version of LEP expected to come into service in 1987 as a 'very worth-while Z^0 factory'.[56] At Stage $\frac{1}{3}$, LEP would have insufficient energy to produce W^+W^- pairs, but once the resources of the Z^0 factory had been fully exploited, it was planned to move to Stage 1. Additional power supplies would be used to bring beam energies up to 86 GeV. This would make available centre-of-mass energies at which 'the best present estimates predict W pair production at a good rate'.[57] Finally, once W physics had been fully explored, and technological developments permitting, the Stage 2 transformation of LEP could go ahead. The conventional power supplies of Stage 1 would be replaced with superconducting devices, giving beams of up to 130 GeV in energy. This was expected to open up the energy regime best suited to investigation of the enigmatic Higgs particles, which were part and parcel of the spontaneous symmetry breaking mechanism in electroweak theory.[58]

Thus during the 1970s, the new-physics world-view penetrated every aspect of experimental life at CERN (and at the other major laboratories around the world). The day-to-day practice of HEP experimenters became structured around the investigation of new-physics phenomena, using detectors which embodied the prejudices of gauge theory. The phenomena of the old physics were written out of existence. And in the future, by common consent and international agreement, lay two new-physics machines: the SPS $p\bar{p}$ collider, a fast and cheap route to the IVBs; and LEP, an enormous and expensive e^+e^- collider, a dedicated new-physics facility conceived to be the mainstay of the European experimental programme into the 1990s. In word, deed and financial calculation, HEP experimenters testified that they lived in the world of the new physics.[59]

NOTES AND REFERENCES

1 Relevant quantitative data are given by Martin and Irvine (1983b). They show (p. 23, Table 5) that, between 1969 and 1978, the total number of experimenters working at CERN accounted for 25–32% of the experimental HEP community. Funding for CERN similarly fell in the range 25–32% of world funding for HEP. The next most populous laboratory was Fermilab, housing between 11 and 12% of the experimental HEP community, and absorbing 12–13% of world funding. In the same period, Martin and Irvine find (p. 28, Table 8) that CERN produced 26.5% of the world's experimental HEP publications. In comparison, Fermilab generated 8.5% (13.5% between 1973 and 1978); Brookhaven, 11.5%; SLAC, 15.5%; and DESY, 3%.

2 Most lower-energy facilities were closed down during the decade. Note that Table 12.1 does not include the electron accelerators at DESY or Cornell. In the late 1970s the primary function of both machines became one of feeding the associated e^+e^- colliders, rather than supporting an independent programme of low-energy fixed-target experiment (see Madsen and Standley 1980, 33, 35).

3 Note that colliders provided for small experimental programmes (relative to fixed-target machines with their many secondary beams). At SPEAR, for example, only two experiments could be mounted simultaneously, one at each e^+e^- intersection point; at PETRA one or two

experiments could be mounted at a total of four intersection points. This restriction on collider programmes acted to encourage the new physics dominance of e^+e^- experiment – programme committees allocated their finite resources to the phenomena of the greatest contemporary interest. Also by virtue of the restricted facilities for e^+e^- experiment, review articles provide rather comprehensive coverage of programmes at individual machines: for SPEAR and DORIS, see Feldman and Perl (1977); for CESR, Franzini and Lee-Franzini (1982); for PETRA, The Mark J Collaboration (1980), Schopper (1981a) and Duinker (1982); and for PETRA and PEP, Branson (1982). The single component of these programmes not discussed in the text is tests of QED. These were routinely performed as each machine opened up a new kinematic range; no departures from QED were found.

Amongst the electron machines listed in Table 12.1, the SLAC linear electron accelerator (denoted simply as SLAC) stands out as the only machine supporting a programme of fixed-target experiment. This programme was ambiguous with respect to the old and new physics. Experiments with the high-energy primary electron beam were new-physics experiments throughout the 1970s: they aimed to investigate scaling, scaling violation, and parity violation by the weak neutral current. However, the SLAC accelerator also made available lower-energy secondary beams of hadrons, and in this respect the SLAC programme resembled that at the CERN PS discussed below.

4 The CERN SPS, for example, provided a total of 17 secondary beams: two neutrino beams, one each of electrons and muons, and 13 hadron beams (Madsen and Standley 1982, 1). More than 17 simultaneous experiments were possible at the SPS, since detectors could be placed in series in neutrino beams (other beams were significantly degraded by single detectors). Statistics on the numbers of simultaneous experiments possible at e^+e^- colliders were given in the previous note. The superior data-collection rates of fixed-target experiments relative to colliders have been discussed in Chapter 2.

5 According to Martin and Irvine (1983b, 28, Table 8) the proton laboratories CERN, Brookhaven and Fermilab contributed 46.5% of the total number of experimental HEP publications between 1969 and 1978. In the same period, the electron laboratories DESY and SLAC contributed 18.5% of the literature, but this divided in a ratio of 5 to 1 in favour of fixed-target as against collider publications (Martin and Irvine 1983b, 18, Table 3).

6 I should emphasise that in this chapter I am concerned to establish the existence of overall trends in experimental programmes rather than with priorities of experimental discoveries. Certainly, several of the trends at CERN originated with discoveries made elsewhere. But in this respect CERN was the 'hardest case' for the new physics, and the transformation of experimental practice at CERN thus offers the best exemplification of the thoroughgoing penetration of HEP experiment by the gauge-theory world-view.

Some comment is also required on the experimental programme at Serpukhov. In energy, the Serpukhov accelerator was bracketed by the

CERN PS and SPS, and it is reasonable to assume (in default of detailed information) that trends at Serpukhov are illustrated by those at CERN. However, the problems encountered by the Soviet experimental HEP programme discussed in Chapter 2, note 2, may have led to significant differences between East and West. I have no quantitative information on the Serpukhov programme, but it is clear that it had little major impact upon the overall development of HEP in the 1970s, and I will not discuss it further here.

7   One further incidental advantage in focusing upon CERN can be mentioned: the wide availability of data upon the CERN research programme. The primary source of such data for all machines is the record of experimental proposals (successful and unsuccessful) submitted to the relevant institutions. Such records are held in unpublished form at all laboratories. CERN, however, is unique in its wide distribution of details of its experimental programme. (This uniqueness is related to CERN's position as an international organisation, catering to many national sub-communities.)

8   The primary source for Figures 12.1 to 12.3 are the *CERN Annual Reports* for the years 1970 to 1980, augmented from 1975 onwards by the yearly publication *Experiments at CERN* (both available from CERN on request). Also useful is Van Hove and Jacob (1980). Each experiment listed as on the floor and devoted to a particular topic is counted as one unit (occasionally an experiment aimed to span old- and new-physics topics, and is counted as contributing one-half a unit to each). There is clearly a degree of arbitrariness to this counting procedure which assigns equal weight to all experiments irrespective of the effort and range of resources involved. Fine details and small scale fluctuations in Figures 12.1 to 12.3 should therefore not be taken seriously. The large scale trends at issue are, however, immune to changes in counting procedures.

9   One of the Directors-General of CERN noted in 1980, 'Most of the expenditure transfer between the PS and SPS has been due to a migration of physicists over the years from the PS to the SPS which was more attractive from the research point of view' (Adams 1980, 14). In the US, the experimental programme at the 31 GeV Brookhaven AGS was similarly denuded of manpower by the advent of the 500 GeV Fermilab PS. Irvine and Martin (1983b, 20) note that Brookhaven's 'productivity plummeted in the mid-1970s as American experimentalists migrated to the more powerful facility at Fermilab'. For quantitative data on publication rates from the CERN PS and SPS, the AGS and Fermilab, see Irvine and Martin (1983b, 18, Table 3).

10   At Brookhaven, where the AGS was the sole accelerator throughout the 1970s, neutrino experiments and investigations of charm using hadron beams continued to flourish alongside baryonium physics to the end of the decade. For an extensive review of the AGS programme around 1976, see Hahn, Rau and Wanderer (1977); for subsequent developments, see the annual reports *Brookhaven Highlights* (available from the National Technical Information Service, Springfield, Virginia).

11   Sources for this figure are those listed in note 8, together with the extensive reviews of the ISR programme given by Giacomelli and Jacob

(1979, 1981). Note that, in comparison with the PS and SPS, relatively few experiments were performed at the ISR. This was a reflection of the limited number of proton–proton intersection points at which ISR experiments could be mounted. The apparent decline of the ISR programme in the late 1970s does not signify a lack of enthusiasm on the part of HEP experimenters. The early 1970s saw many quick first-generation ISR experiments; later experiments were larger, more sophisticated and slower.

12   Not included in Figure 12.2 are four old-physics experiments in 1980. These were first-generation experiments relating to the use of the ISR to store beams other than protons. One experiment studied the soft-scattering of alpha-particle (helium nuclei) beams, and the remaining three studied proton–antiproton collisions. Filling the ISR with alpha particles was part of programme designed to explore the versatility of the ISR, while the use of an antiproton beam was a by-product of the SPS pp̄ collider project discussed below. The initial round of alpha-particle and pp̄ experiments aimed to investigate the gross (old physics) features of the relevant interactions in the ISR regime prior to subsequent studies of rarer (new physics) phenomena. Thus the reappearance of the old physics in these experiments did not signal any general resurgence of interest in this area.

13   The pre-1974 new particle experiments shown in Figure 12.2 were unsuccessful searches for hypothetical particles: isolated quarks and magnetic monopoles. Such searches were routine at new accelerators and colliders.

14   The apparent drop in the number of experiments coincident with the first year of SPS operation is an artefact of this method of presentation. Replotting Figure 12.3 uniformly in terms of approval dates makes little difference to its overall form. A similar comment applies to Figures 12.1 and 12.2.

15   Although old-physics experiments still constituted a recognisable tradition at the SPS in the late 1970s it is worth noting that it was sustained by inertia rather than theoretical context. Planning and preparation for HEP experiments typically stretch over several years, and the old-physics experiments performed in the late 1970s had been conceived and approved in an era more sympathetic to the old physics. That this was the case is illustrated clearly in old-physics reviews presented in the late 1970s and early 1980s. For example Ganguli and Roy (1980, 203–4) began their review of the Regge phenomenology of soft inclusive hadronic processes as follows:

> The Regge phenomenology of inclusive reactions, like the Regge phenomenology in general, has virtually come to a halt since last five years or so. However, there has been a slow but steady accumulation of inclusive reaction and correlation data at low p_T, over these years. Indeed, many of these experiments were designed with the purpose of Regge analysis and are qualitatively superior to the older experiments of their kind. The time lag is simply the gestation time of a present day electronics experiment, further stretched in some cases no doubt by the declining interest of the

field . . . [T]he emphasis [in this review] will be on the data, accumulated over the last 5–6 years, much of which have remained unused so far.

Reviewing experimental data on exclusive soft-scattering processes, Alan Irvine (1979, 621) made a similar point more concisely: 'In conclusion', he remarked,

the data from this latest (and probably the last) generation of two-body hadronic experiments is of very high quality indeed and provides many answers. It is unfortunate that few can remember what the questions were.

The decline of conventional resonance experiments was even more marked than that of high-energy soft-scattering physics. Reviewing 'Prospects in Baryon spectroscopy' in 1980, one of the experts, CERN experimenter M. Ferro-Luzzi (1981, 45) began:

[T]he general impression which I felt whilst going through the reminiscences assembled below is that of a case of arrested childhood. Serious work started coming out in the early seventies but never reached any significant level. After a desultory scratch over the surface of what is surely a mine of unexplored information, all important activity declined. Only a few uncoordinated ventures went on to dribble out tantalizing but inconclusive results.

Ferro-Luzzi ascribed the decline of resonance spectroscopy to three factors. These were: the lack of development of suitable detectors; 'the unstoppable trend towards higher and higher energy accelerators' since it was 'hard to see how an energy increase of beams and facilities could possibly benefit a type of physics inherently more easy to study at low energy'; and 'perhaps of greater consequence than any of the above causes, . . . an undeniable lack of interest from the physicists community in this field.'

All of the authors just quoted emphasised that significant targets remained for future resonance and soft-scattering experiments, and that appropriate detector technology already existed or could be developed. All were pessimistic that the programmes they outlined would be carried out, and their pessimism was justified. We have noted the rundown of the PS experimental programme in the late 1970s; the ISR was scheduled for closure in the early 1980s (see Section 12.3); and visions of future SPS fixed-target experiment were dominated by the new physics (see note 30 below).

16 For technical reviews of developments in experimental neutrino physics spanning these three areas, see Cline and Fry (1977), Musset and Vialle (1978), Barish (1978), Sciulli (1979) and Fisk and Sciulli (1982).

17 Developments at Fermilab during the 1970s paralleled those at the SPS: an old-physics dominated programme in the early 1970s gave way to the new physics during the November Revolution. Giacomelli, Greene and Sanford (1975) review the state of the Fermilab programme in the mid-1970s. Whitmore (1974, 1976) discusses the early extensive Fermilab investigations of high-energy soft-scattering using 30-inch and (later) 15-foot bubble chambers. Later developments can be followed from *Fermilab Annual Reports* (available from Fermilab). An overview

374

of the evolution of the Fermilab programme was given by the US High Energy Physics Advisory Panel (1980, 13):

> The physics program at Fermilab uses a 400 GeV proton synchrotron . . . Proton beams can be directed simultaneously onto several targets producing secondary beams of hadrons, muons, photons, and neutrinos. As many as 15 experiments can be run simultaneously resulting in a program of great diversity.
>
> In its early days (1972–1974) the program was dominated by the exploration of a new energy regime. Experiments ranged from extensions of hadronic physics at lower energies to searches for quarks and monopoles. As the machine matured, the main thrust of the experimental program focused on the study of hadronic substructure, the interaction of subnuclear constituents with one another, and the investigation of new forms of matter such as charm.
>
> A major discovery of the Fermilab physics program was that of the upsilon resonance, leading to its interpretation in terms of a fifth quark, the b quark. The substructure of nucleons has been studied by means of muon and neutrino scattering. The structure of pions has been studied by means of dimuon production in pion – nucleon collisions. Static properties of hadrons such as hyperon magnetic moments have been measured. This program has also revealed unexpected dynamical effects such as the polarization of hyperons produced at large transverse momentum. Numerous measurements on the production and properties of charmed particles have been carried out. Notable are direct measurements of the lifetime of charmed particles.

18 It should be noted that the rise of the new physics in experiment around the world was, in a sense, self-reinforcing. Access to major accelerators and colliders was controlled by programme committees composed of leading HEP experimenters and theorists (at CERN, the Scientific Policy Committee). Such committees received experimental proposals and adjudicated upon whether they should be allotted machine time. These adjudications reflected the communal consensus which, in the wake of the November Revolution, favoured the new physics. Various individual experimenters remained keen to pursue programmes of old-physics research, but found it hard to convince the relevant committees that these were worthwhile (private communications).

19 Combinatorial background was yet another reason often cited for the delay in discovering the D mesons at SPEAR.

20 The other routes used nuclear emulsions, semiconductor detectors or high pressure streamer chambers. All of these approaches are reviewed in Bellini *et al.* (1982). For the high resolution bubble-chamber approach see the contribution to this review from Montanet and Reucroft (1982); for a popular account, see Sutton (1982). I am grateful to Dr C.Fisher for a discussion on the CERN bubble-chamber programme.

21 For a technical description of LEBC and its optimisation for charm searches, see Benichou *et al* (1981).

22 Allison *et al.* (1980).
23 For details of the EHS and its use in conjunction with LEBC, see Aguilar-Benitez *et al.* (1983).
24 For a review of the Fermilab measurements, see Whitmore (1974, 1976).
25 Bellini *et al.* (1982, 72, Fig. 6). This event was first reported in Adeva *et al.* (1981).
26 The spatial resolution achieved in LEBC using classical optics was of the order of 40 μm. To improve upon this figure, and thus to make visible the very short tracks of particles with lifetimes down to 10^{-14} secs (the expected lifetime of naked-b states), CERN experimenters began to use lasers and holographic techniques of data recording. First steps in this direction were taken in 1980, using another small, purpose-built bubble chamber, BIBC (the Bern Infinitesimal Bubble Chamber). Future experiments were envisaged using laser optics and a muon trigger, designed to enrich the sample of semileptonic charm decays relative to conventional hadronic background. For details see Montanet and Reucroft (1982, 69–83).

It is interesting to note that the development of high-resolution bubble chambers brought about a symbiosis of HEP experimenters and laser/holography experts: see CERN (1982).
27 Jacob (1982, 3).
28 Montanet and Reucroft (1982, 81).
29 Fisher, private communication.
30 In 1979, the head of CERN's Experimental Physics Division outlined the present and future of the overall SPS fixed-target programme as follows (Gabathuler 1979, 72, 77):

> The main thrust of the SPS programme is to help us in broadening our understanding of the strong interactions, where within the present model of Quantum Chromodynamics (QCD) we assume that the quarks interact [according to the dictates of gauge theory] . . . [T]he fixed target programme under way at the SPS is broad-based in its physics aims, and the present range of beams and detectors is well matched to that programme. New detectors in the form of 'fine grain' detectors using calorimetric and semi-visual techniques will certainly be added to existing experiments to carry out well-defined physics objectives. The physics programme will have as its main theme a range of specified detailed experimental tests on existing theoretical prejudices, mainly centred on QCD.

For detailed discussions of both the CERN and Fermilab fixed-target programmes for the period 1984–1989, see Mannelli (1983).
31 Figures 12.5 and 12.6 are adapted from Figures 5.6, 5.7, 9.2, 9.3 and 14.2 of Giacomelli and Jacob (1979). Giacomelli and Jacob also provide details on the experiments in question. They summarise the overall development of the ISR programme as follows (1979, 13):

> At present, as experimentation looks at more subtle effects than first exploration did, detectors have increased in sophistication and size . . .
>
> Trends in research have changed much in time . . . In 1972 there was a wide front approach using relatively simple detectors to study

σ_{tot}, elastic and single inclusive reactions [i.e. old physics]. In 1975, almost all experiments involved correlations among particles and the research programme was already very specialized. In 1978, the specialization is even stronger with lepton pair production and large transverse momentum phenomena [i.e. new physics] primarily being studied. Because of the low cross-sections involved, new large sophisticated detectors capable of momentum measurement and particle identification over a wide solid angle had to be built.

32 Darriulat (1980, 23).
33 See Fabjan and Ludlam (1982, 371). These authors discuss the use of large multi-element detectors in hadronic experiments. They bring out very clearly the connections between detector design, the perceived importance of new-physics phenomena and the inherent limitations of new-physics detectors:

> The event rates at hadron machines, either fixed-target or colliding beams, can be enormously high. The collision rate at the ISR . . . is $\gtrsim 1\,\mathrm{MHz}$ . . . The power of these machines is not the high collision rate in itself, but the capability to produce useful numbers of very rare events over periods of months. A very high degree of trigger selectivity is required, rejecting most events seen by the detectors. Indeed, a large detector system with thousands of readout channels generally cannot record more than a few events per second.

Fabjan and Fischer (1980) give a detailed review of the development of electronic detectors through the 1970s, emphasising the parallel developments in signal processing and triggering arrangements.

34 Further aspects of the new-physics structuring of ISR detector design can be noted. First, the priority accorded to triggering requirements fostered the development of calorimeter detectors (see Fabjan and Ludlam 1982). These measured particle energies directly, by stopping particles and converting a fraction of their energy into an electronic signal. They could thus discriminate between high and low-p_T particles much more quickly than conventional detectors (in which momenta were computed from measurements of deflections in magnetic fields) and offered correspondingly faster triggering.

Other advantages of calorimeters over conventional electronic detectors were (a) they were directly sensitive to neutral as well as charged particles; (b) they could be used to discriminate between electrons, muons and hadrons; and (c) they were smaller and more economical for the detection of high-energy particles: to achieve a given momentum resolution, calorimeter dimensions had to be increased in proportion to the logarithm of the energy of the detected particles, while the dimensions of conventional detectors had to be increased in proportion to the square-root of particle momenta. This last factor made calorimeters particularly attractive for experiments at the ultra-high-energy colliders discussed in the next section.

The first ISR calorimeter detector came into use in 1976 (experiment R806). Built by a Brookhaven–CERN–Syracuse–Yale collaboration, it used liquid argon and lead plates to identify electrons and photons. In 1980 it was superceded by a much larger uranium/scintillator calori-

377

meter used as a hadron detector (R807). Built by a Brookhaven–CERN–Copenhagen Lund–Rutherford–Tel Aviv collaboration, the new calorimeter together with other electronic detectors formed the so-called Axial Field Spectrometer (AFS). Approved for construction in April 1977, the AFS was in every respect optimised as a new-physics facility (for a technical description, see Gordon et al. 1982). The electronic arrangements, for example, were such that it was possible to trigger on a whole variety of new-physics phenomena: jets, multileptons and so on (Gordon et al. 1982, 311–12). At a more basic level, the AFS was so called because it used an axial magnetic field (oriented along the beam axis) for momentum measurements. The advantage of this field arrangement was that uninteresting softly scattered charged particles tended to spiral around the field lines and remain within the beam pipes – thus escaping the detectors. Earlier ISR facilities had used different field configurations. The SFM, for example, used fields perpendicular to the beam axis. Such fields acted to pull softly scattered particles out of the beam pipes and into the detectors. This made sense when the SFM was first designed – in the old-physics era – but necessitated considerable adaptation for the new-physics purposes of the late 1970s.

For a semi-popular review of the structuring of collider detectors around the new-physics world-view, see Willis (1978).

35   Panofsky (1974, xi–xii).

36   Bjorken (1977, 971). HEP experimenter David Miller (1977, 288) commented that 'Bjorken mentioned something which particle physicists feel very strongly; the case for building the next generation of accelerators can be made more precisely and with more confidence now than has been possible for any previous generation of accelerators.'

37   In the standard model the masses of the W and Z were determined by θ_W. Using the value of θ_W extracted from neutrino experiments, the W and Z were expected to be found at 78 GeV and 89 GeV respectively. Another potentially significant feature of unified electroweak theory were the Higgs particles. Assuming that the IVB masses arose from the Higgs mechanism, at least one such particle should exist. However, no clear predictions for the Higgs sector could be obtained. In the standard model, for example, there was a single Higgs particle, but the only constraint set by theory upon its mass was that it should lie between 7 and 300 GeV. Thus, unlike the IVBs, the Higgs particle was not a well defined target for a goal-oriented programme of machine construction and experimental research.

38   See Rubbia, McIntyre and Cline (1977).

39   Two cooling techniques were available. In each, the hot antiprotons were stored in a ring. In the first, devised by Novosibirsk physicist Gersh Budker, the antiprotons were cooled by interaction with cold electrons. In the second, devised by CERN physicist Simon van der Meer, the antiprotons were cooled 'stochastically': fluctuations in the antiproton beam were measured at a given point in the storage ring and used to generate a cooling signal at the opposite point of the antiproton orbit. For a discussion of both techniques and references to the original literature, see Cline and Rubbia (1980). For an extended review of the

stochastic technique eventually adopted at CERN, see Mohl, Petrucci, Thorndahl and van der Meer (1980).

40 Many detailed estimates of IVB production rates and decay modes were made during the 1970s. For reviews of their implications for future experiments, see Quigg (1977) and Ellis, Gaillard, Girardi and Sorba (1982). It is relevant to note that these reviews came from Fermilab and CERN respectively, the two laboratories where preparations for pp̄ collider experiments were under way.

41 Van Hove (1976, 34). At CERN the work required was principally to build a small antiproton cooling ring with appropriate beam transfer lines, and to construct experimental halls at the SPS pp̄ interaction points. The expenditure was funded from within the CERN operating budget.

42 Wilson (1980, 39).

43 Although the pp̄ colliders were conceived as the route to the IVBS, there was no *necessity* that they should be operated as new-physics machines. They could, in principle, have been devoted to old-physics topics, as the ISR had been in the early 1970s. However, the detectors designed for collider experiments effectively enforced their new-physics use. At the CERN SPS, for example, two pp̄ interaction regions were envisaged, and enormous detectors were planned for both. Designated UA1 and UA2, each detector was of the new-physics type developed at the ISR in the late 1970s (discussed in Section 2 and note 34 above). Beyond the search for IVBS, these detectors were designed to explore the standard range of new-physics phenomena at collider energies: hadronic jets, lepton production, heavy quark states (including searches for the t quark) and so on. The UA1 detector was 10 metres long, 5×5 metres in cross-section, and weighed 2000 tons. It was designed and built by a collaboration of more than 100 physicists from 11 institutions in Europe and the US, led by Carlo Rubbia. The UA2 detector was somewhat smaller; the UA2 collaboration, led by Pierre Darriulat, had only 51 members drawn from 6 European institutions. Technical descriptions of the UA1 and UA2 experiments (and of several smaller experiments designed to operate alongside them) are given in *Experiments at CERN*. A popular account of these experiments, and of the development of collider physics at CERN and Fermilab, is given by Cline, Rubbia and van der Meer (1982).

44 The first positive findings from the UA1 and UA2 experiments were presented at the third 'Topical Workshop on Proton–Antiproton Collider Physics' held in Rome, 12–14 January 1983 (Rubbia 1983, Darriulat 1983; see also Arnison *et al.* 1983a and Banner *et al.* 1983) and were announced at a CERN press conference on 25 January 1983 (see *CERN Courier* 1983a, 44). On 27 January 1983, a one-hour programme devoted to the SPS collider project and its findings, *The Geneva Event*, was shown on British television, and many accounts followed in the popular and scientific press. At the time of writing (July 1983) the status of experimental IVB physics is this. In an experimental run in late 1982, around 10^9 pp̄ interactions were monitored by the UA1 and UA2 collaborations. By triggering on jets and leptons, each group reduced the total of recorded events to around 10^6 (for a description of the extremely

sophisticated electronic trigger arrangements, see *CERN Courier* 1983c). From these samples, the UA1 and UA2 experimenters succeeded in isolating respectively 6 and 4 candidate W-events. These were events in which a high p_T electron was observed, ascribed to the characteristic decay W→ev, where the neutrino escapes undetected. On the basis of their events, the UA1 group assigned the W a mass of around 80 GeV – in the expected region (see *CERN Courier* 1982a). In a subsequent run, 3×10^9 p$\bar{\text{p}}$ interactions were monitored, reduced to 2×10^6 interesting events by triggering. From these, the UA1 collaboration isolated 5 Z^0-candidates, identified by their characteristic decays to lepton pairs: 4 events of the form Z^0→e$^+$e$^-$, and 1 similar event involving muons: Z^0→$\mu^+\mu^-$ (see *New Scientist* 1983 and Arnison *et al.* 1983b). The mass assigned to the Z^0 was 95 GeV – again in the expected range.

The statistical significance of the few candidate events was not overwhelming – recall, for example the disappearing 6 GeV upsilon at Fermilab (Section 10.1) – but, as one leading HEP theorist wrote (Harari 1983, 52):

> The first direct evidence of W bosons was recently reported . . . In the next several years new particle accelerators and more sensitive detecting apparatus will test the remaining predictions of the model [standard electroweak theory plus QCD]. Most physicists are quite certain that they will be confirmed.

45 Two further developments connected with the CERN antiproton project can be noted here. These concern alternative uses of the cooled antiproton beam. Although the beam was designed to feed the SPS, it could be and was used to fill the ISR. The beam was ready before modifications to the SPS were complete, and the first CERN p$\bar{\text{p}}$ observations were made at the ISR (see Schopper 1981b, 14, and note 12 above). Also, the advent of a cooled antiproton beam revived interest in low-energy antiproton physics at CERN. From 1977 onwards plans were laid for the construction of LEAR (the Low-Energy Antiproton Ring). LEAR was intended to take cooled antiprotons and make them available as beams for fixed-target experiments at energies between 0.1 and 2 GeV. In general, LEAR experiments promised to increase the quality of data on low-energy antiproton interactions enormously: its beams would be at least a thousand times more intense than conventional antiproton beams and of extremely well defined momentum, making possible extremely accurate measurements. Possible uses for LEAR included topics in nuclear and atomic physics. The principal HEP interest in LEAR centred upon baryonium searches and precision measurements on charmonium states. More conventional attributes of p$\bar{\text{p}}$ interactions could also be studied.

LEAR was intended to enter operation in 1983; for its history, details of its construction and the experimental programme envisaged, see Gastaldi and Klapisch (1981) and Jacob (1980).

46 See note 43 above.

47 See Adams (1980, 25).

48 Richter (1976b).

49   I thank P. Darriulat, J. Mulvey and A. Zichichi for information on the planning of LEP.
50   ECFA (1977).
51   Quoted in Landshoff (1979, 93).
52   Teillac (1981).
53   See ECFA Working Group (1979).
54   See the reports of the CERN and ECFA study groups: for example, Camilleri *et al.* (1976), ECFA/CERN (1979) and Zichichi (1980).
55   Adams (1979, 14–16).
56   LEP Study Group (1979, 2).
57   LEP Study Group (1979, 1).
58   For the Higgs particles, see note 37 above. For a technical description of LEP in all its guises, see LEP Study Group (1979).
59   A brief comment is appropriate here on future planning for major facilities at laboratories other than CERN (see Tables 2.1 and 2.2). In the 1970s, four major projects were envisaged in the US: at Fermilab it was planned to produce 1 TeV proton beams for fixed-target experiments, and to do p̄p collider physics at 2 TeV; at Brookhaven, the Isabelle pp collider project was underway, aiming at an energy of 800 GeV with a high interaction rate; and at Stanford studies were under way for the Stanford Linear Collider (SLC) which would realise e^+e^- collisions at around 100 GeV. All of these were conceived as new-physics machines and, in the US as in Europe, the electroweak IVBs provided the reference point for detailed planning. As the authoritative US High Energy Physics Advisory Panel (1980, 8) put it, in their review of the overall US programme: 'The clear central goal is the direct observation of these particles [W^+, W^- and Z^0].' (For details of machine design and proposed detectors for Isabelle, see Hahn, Month and Rau 1977 and BNL 1977; for the SLC, see SLAC 1980b, 1982.)
       In Europe, construction of one further new machine besides LEP was under active consideration in the late 1970s. This was HERA, an electron–proton (ep) collider, to be built at DESY in Hamburg. The idea of this machine was to make possible ep experiments at much higher centre-of-mass energies than those available at fixed-target electron accelerators like SLAC. Uniquely amongst the major projects entertained in the late 1970s, ep colliders were expected to be rather poor sources of data on the IVBs. Nevertheless, they were envisaged as new-physics machines. Their physics interest lay in the exploration of characteristic QCD phenomena – scaling violations and jets – and in the search for new heavy quarks and leptons (as well as other exotic new-physics phenomena). The possibility of building such a machine was first considered at DESY in 1972 and 1973. A British proposal, EPIC, was forthcoming in 1976, and in 1978 a CERN ep machine, CHEEP, was discussed. CHEEP, however, was displaced at CERN by LEP, and two DESY proposals, PROPER and then HERA, came to the fore. (For details of HERA and access to the literature on the earlier proposals, see ECFA 1980.) In 1983 the German government supported the construction of HERA, subject to sufficient international support becoming available (see *CERN Courier* 1983d).

No detailed information is available on the remaining new accelerator listed in Table 2.1 – UNK, the 3 TcV PS proposed for the Soviet Union.

13

GRAND UNIFICATION

Ayatellis: The conclusion is clear. QCD is obviously correct and its miracles have been convincingly proven by the experimental data (or if not, it soon will be).

Under these circumstances, it no longer requires a tremendous act of faith to believe in QCD, and I must go seek some new more far-out cult to believe in. I have resolved that I will go and work on grand unified theories of the strong, weak and electromagnetic interactions. These theories clarify and create many amusing new mysteries. One of the miracles they predict is the ultimate instability of matter: protons should have a finite lifetime of $10^{30 \pm 2}$ years. This raises many interesting theological problems on which I shall meditate. For example:

IS GOD MADE OF PROTONS?

De Oracle: The Ayatellis hopes She is, so that he may replace Her on Her throne.[1]

This account has almost reached its conclusion. In the previous four chapters we have seen how, in the late 1970s, electroweak theory and QCD permeated HEP, structuring both the present and future of research practice. Consensus reigned that the world was built of quarks and leptons, interacting as gauge theorists said they should. But Bjorken's 'new orthodoxy' encompassed three theoretical elements, and we have so far discussed only two. Besides the standard electroweak model and QCD, Bjorken included an SU(5) 'grand unified' theory within his orthodox trinity of gauge theories, and grand unification is the last topic of this history.[2] Section 1 outlines the theoretical development of grand unified theories. Whilst popular amongst HEP theorists, these had little impact on contemporary HEP experiment. Instead, the HEP theorists entered into mutually rewarding relationships with cosmologists and with experimenters buried deep underground. The cosmic and subterranean connections are reviewed in Sections 2 and 3 respectively.

13.1 GUTs and HEP

With the 1978 arrival of the IPT approach to QCD many HEP theorists discerned the end of an era. There now existed respectable field theories of the electroweak and strong interactions, and calculational techniques adequate to the phenomenological application of those theories. The phenomenological gauge-theory traditions existed in a state of satisfying symbiosis with the dominant experimental traditions of HEP. Certainly the quark confinement problem was unsolved in QCD, but several nonperturbative approaches to the problem were being developed. Theorists who had remained in the perturbative mainstream were content to leave confinement to the nonperturbative experts, trusting them to contribute the last piece to the jigsaw. From this perspective, theoretical HEP, as it had been previously understood, was finished: the building blocks of nature had been identified and their properties understood. This conclusion was, in itself, intolerable: if particle physics was solved then there was nothing left for HEP theorists to do. Unwilling to pack their bags and leave the field, theorists began, therefore, to look around for a new 'far-out cult to believe in'. In the late 1970s, many made their offerings at the shrine of grand unification.[3]

Grand unified theories – less pretentiously known as GUTs – embraced within a single gauge theory the weak, electromagnetic and strong interactions, all of the forces of particle physics. The historical development of the 'grand synthesis' (in Gell-Mann's phrase) had an air of inevitability about it. In the early 1970s, theorists learned how to unify the gauge theories of electromagnetism and the weak interactions, using the Higgs trick à la Salam and Weinberg. In 1973, a candidate gauge theory for the strong interactions was proposed: QCD. It was only a matter of time before the action replay: a unification of QCD and electroweak theory in an extended group structure, with Higgs particles suitably chosen to reproduce the low-energy phenomena. The first example of such a theory, published in 1973, was co-authored by Jogesh Pati, of the University of Maryland, and Salam himself.[4] The second example, from Harvard gauge-theorists Howard Georgi and Sheldon Glashow, appeared in 1974.[5]

Both the Pati–Salam and Georgi–Glashow models assigned quarks and leptons to the fundamental representations of some large group. This underlying symmetry was then assumed to be spontaneously broken down to a product of the $su(2) \times u(1)$ symmetry of the electroweak interactions and either an $su(3) \times su(3)$ (Pati–Salam) or simply an $su(3)$ (Georgi–Glashow) symmetry of the strong interactions. Pati and Salam's choice of $su(3) \times su(3)$ for the strong interactions corresponded to their predilection for integer-charge quarks and unconfined colour; accordingly their unifying group was

SU(4) × SU(4). Georgi and Glashow favoured the conventional choice of fractionally-charged confined quarks, corresponding to the colour SU(3) of QCD, and their unifying group was SU(5). As QCD prospered during the 1970s, and as confinement itself became a topic for active research, so the Georgi–Glashow model came to be seen as the prototypical GUT. In what follows, I will discuss the overall lines of GUT development, specialising, where appropriate, to SU(5).

GUTs were modelled on the unified electroweak theories. In electroweak theories, an exact gauge invariance was spontaneously broken by a set of Higgs scalars, suitably chosen to give large masses to the IVBs while leaving the photon massless. GUTs were based on a larger gauge group, broken in such a way as to leave the photon and the eight gluons of QCD massless, whilst giving appropriate masses to the IVBs, $W^{\pm}$ and Z^0. Because of the group structure, however, GUTs had to incorporate more than just these twelve gauge vector bosons. SU(5), for instance, required a further twelve 'X-bosons'. The Higgs sector had to be suitably chosen to give these bosons extremely large masses, in order to effect the observed decoupling of strong and electroweak interactions at 'low' – i.e. contemporarily accessible – energies. Through this hierarchic symmetry breaking, the surplus vectors acquired a mass comparable with the 'unification mass', whilst the IVBs acquired masses of the soon-to-be-accessible order of 100 GeV. The unification mass in GUTs was the mass – equivalently, energy – at which the strong, weak and electromagnetic interactions attained comparable strength, just as in the electroweak theories the weak and electromagnetic interactions were estimated to become comparable at energies corresponding to the mass-range encompassing the $W^{\pm}$ and Z^0.

The first detailed attempt to estimate the unification mass in GUTs was published in 1974 by three Harvard theorists, Howard Georgi, Helen Quinn and Steven Weinberg.[6] Central to their argument was the use of renormalisation group techniques. As we saw in Chapter 7, it was conventional to discuss the properties of QCD in terms of an effective coupling constant, the magnitude of which varied according to the energy scale of the physical process under consideration. One could similarly define effective coupling constants for the weak and electromagnetic interactions (although these had no practical utility in the computation of 'low' energy processes). Georgi, Quinn and Weinberg used renormalisation group calculations to trace out the evolution of the effective coupling constants for the strong, weak and electromagnetic interactions as a function of energy. They found that in the Georgi–Glashow SU(5) model the three couplings became comparable at around 10^{15} GeV (Figure 13.1). This was the grand unification mass at which the unity of the three interactions should become manifest, and at which the X-bosons should be produced.

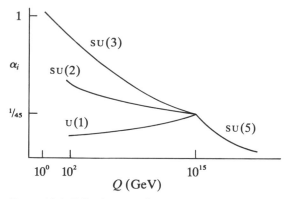

Figure 13.1. Effective coupling constants (α_i) as a function of energy scale (Q) for the standard electroweak model (SU(2) and U(1)) and QCD (SU(3)). Note that the three coupling constants become equal at an energy scale set by the grand unification mass, 10^{15} GeV.

What, then, were the phenomenological implications of GUTs? As far as mainstream HEP was concerned, the single major prediction of GUTs involved the electroweak mixing angle, θ_w. In electroweak theory θ_w was a free parameter, to be determined by experiment. This freedom arose from the fact that the standard model was invariant under gauge transformations specified by a product of two groups, SU(2) × U(1), rather than by a single group. In GUTs, the electroweak SU(2) × U(1) symmetry was embedded in a single larger group structure, and the arbitrariness of θ_w was removed. In a GUT, the value of θ_w could be calculated from simple group-theory arguments. In SU(5), for example, $\sin^2\theta_w$ was predicted to be $\frac{3}{8}$, as Georgi and Glashow noted in their original 1974 publication. This they regarded as the 'one easily testable prediction' of their theory.[7] However, more theoretical work at Harvard revealed that matters were not quite so simple. Only months after the first SU(5) publication, Georgi, Quinn and Weinberg noted that $\sin^2\theta_w$ was subject to renormalisation effects, just like the energy-dependent effective coupling constants of the theory. The prediction that $\sin^2\theta_w$ should equal $\frac{3}{8}$ applied only to measurements made at the grand unification energy. At lower energies the value of $\sin^2\theta_w$ would be renormalized downwards. Georgi, Quinn and Weinberg estimated that at low energies the measured value of $\sin^2\theta_w$ should be in the vicinity of 0.2 if su(5) were valid.[8]

This 1974 work represented only the outline of a calculation of renormalisation effects in GUTs, and in the outbreak of GUTs computation which took place from 1978 onwards the outline was filled in. There were threshold effects around intermediate boson masses, higher orders of perturbation theory and so on, to be included in the calculations. When this had been done at CERN, Harvard and elsewhere, and the latest measurements of the QCD parameter Λ had been inserted into the equations, the prediction for $\sin^2\theta_w$ was found not to have changed much. In 1979, the accepted theoretical estimate of $\sin^2\theta_w$ in the su(5) model was 0.20 ± 0.01, to be compared with an experimental value of 0.230 ± 0.015. There was a two-standard-deviation discrepancy between theory and experiment. As usual, this discrepancy was treated as an important result rather than a serious problem. Theorists could use it as an input to the construction of more complicated GUTs, while experimenters could cite it as evidence that more precise measurements were needed of $\sin^2\theta_w$.[9]

Besides this single point of contact, though, GUTs offered little to the accelerator-based HEP experimenter.[10] The problem lay with the X-bosons. The Xs were the grand unified analogues of the electroweak IVBs but, unlike the IVBs, they were located at unattainably high masses. While it was technologically feasible to attain the 100 GeV or so required to produce IVBs, the 10^{15} GeV needed to produce an X was beyond the range of any conceivable accelerator – it was estimated that a conventional proton synchrotron of sufficient energy would have a radius comparable with that of the solar system. Funding for a project of such magnitude being unavailable, even to the HEP community, theorists were obliged to look for more subtle applications of their new toy. These applications led in the two different directions discussed in turn in the following sections. One response to the astronomical size of the unification mass was to go where the energy was – the Big Bang at the start of the universe. This led to a social and conceptual unification of HEP and cosmology. The alternative response mirrored the tactics of the electroweak unification game, and consisted in the search for vestiges of grand unification in low-energy phenomena. Just as observed weak-interaction phenomena could be seen as a reflection of the exchange of virtual, high-mass IVBs, so virtual X-boson exchange could be expected to produce characteristic superweak phenomena in the terrestrial environment. The most striking such phenomenon was proton decay, discussed in Section 13.3 below.

13.2 GUTs and Cosmology

The most direct prediction of GUTs was the existence of X-bosons. But with masses of around 10^{15} times that of the proton it was clear that it was impossible to produce such particles in the laboratory. Physicists could only imagine one scenario in which the Xs would actually be produced as physical particles: the Big Bang, as cosmologists called it, the singularity at the beginning of the universe. According to Big Bang cosmology, the universe began as an infinitesimal domain of infinite energy density, which exploded and expanded to form the cosmos as we know it today.[11] The initially infinite energy density corresponded to an infinite temperature, but as the universe expanded it cooled. For some short time (around 10^{-35} secs) after the Big Bang, cosmologists expected that the thermal energies of the particles within the primordial glob would be sufficient for the production of X-bosons. Here, then, was a 'laboratory' in which the physical effects of the existence of X-bosons might be seen.[12]

What might those effects be? Theorists followed the lead offered by Motohiko Yoshimura of Tohoku University, Japan. In 1978 Yoshimura combined two of the attributes of GUTs and came up with a novel cosmological perspective. The attributes in question were baryon-number nonconservation and CP-violation, and the prediction was a baryon–antibaryon asymmetry in the contemporary universe.[13] Let us take baryon-number nonconservation first. Before the advent of GUTs it was taken for granted that baryon number was conserved. On this view a baryon could never turn into an antibaryon, or vice versa: the number of baryons in the universe minus the number of antibaryons was supposed to be a quantity fixed for eternity. This belief was routinely acceptable to particle physicists, but did cause something of a problem for cosmologists. As they had begun to emphasise by the mid-1970s, astrophysical observations suggested that there were no sizeable concentrations of antimatter anywhere within the universe.[14] Extrapolating back in time, this implied that at the time of the Big Bang there had been a very small numerical excess of baryons over antibaryons (most of the baryons and antibaryons annihilating one another as time drew on). Cosmologists had no idea why a small initial baryon excess should have existed: it would have been much tidier if the baryon number of the universe had been exactly zero.[15]

This was where GUTs came in. In the first instance, the X-bosons provided a mechanism for baryon-number nonconservation. In GUTs, the fundamental particles, quarks and leptons, were placed within unified families. And just as the IVBs of the weak interactions mediated transitions between leptons and between quarks of different flavours, and the gluons of QCD mediated transitions between quarks of different colours, so the X-bosons mediated transitions *between quarks and leptons.* The two hitherto distinct forms of matter were now intermingled. A single quark could, for example, transform via X-boson exchange into a pair of antiquarks and a lepton, with a net change of one unit in baryon number (Figure 13.2). Thus the X-bosons promised a way in which an initially symmetric, zero baryon number, universe could evolve from the Big Bang into the non-symmetric universe in which (cosmologists tell us) we presently find ourselves. However, the X-bosons could not do it alone. According to the standard lore of Big Bang cosmology,[16] for every X-boson active in the first 10^{-35} seconds in decreasing the net baryon number, there should be an anti-X busily and exactly compensating for it. To circumvent this conclusion, Yoshimura looked to CP-violation. As a theoretical proposition, CP-violation was a direct response to asymmetries seen between the decays of kaons and their antiparticles.[17] And, when incorporated in GUTs, CP-violation could be argued to lead to similarly asymmetric decay rates between the X-bosons and their antiparticles. Thus, as the universe cooled down after the Big Bang, it might well be that an initially symmetric population of Xs and anti-Xs could evolve into an asymmetric population of quarks and antiquarks, and hence to the accumulation of a baryon excess.

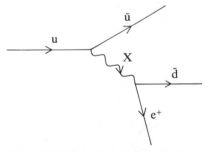

Figure 13.2. Baryon-number (B) changing transition mediated by an X-boson: the u quark has $B = +\frac{1}{3}$; the ū and d̄ antiquarks have $B = -\frac{1}{3}$; the positron has $B = 0$.

GUTs provided the framework for novel discussions of the very early stages of the universe, in which the X-bosons were produced and then decayed leaving an oversupply of baryons. HEP theorists lost no time in making this point,[18] and the cosmological community were quick to respond. GUTs offered cosmologists a new perspective on the old problem of the cosmic baryon excess, and guaranteed them a novel audience of particle physicists for their attempted solutions. According to the crude calculations of particle physicists, the prototypical grand unified theory, Georgi–Glashow's su(5) model, probably was not good enough to explain astrophysical observations. Su(5) straightforwardly predicted that the ratio of baryons to photons in the present universe was in the region of 10^{-16} to 10^{-20}, much smaller than the accepted value of 10^{-8} to 10^{-10}. However, detailed numerical computation of the evolution of the early universe was highly complex, and as John Ellis remarked in 1979: 'to get a precise number for [the baryon-photon ratio] will probably require 'quarkosynthesis' computer calculations for the thermodynamics and interaction rates fully as complicated as the old nuclear physics stellar evolution calculations of a previous physics generation'.[19] Here was an open invitation to astrophysicists with the right kind of resources and, as *Science* reported in February 1981, 'In the past few years, cosmologists concerned with the early universe have embraced the grand unified theories of particle interaction almost as enthusiastically as have the physicists who invented them': 'I think this is the most exciting thing in cosmology in the last 10 years', said Los Alamos astrophysicist Edward Kolb, an expert in computer modelling of particle dynamics in the early universe, 'now we're on the verge of understanding where the baryons themselves come from'.[20] Kolb himself joined three other astrophysicists at the University of Chicago in setting up the computer programmes required to model the early evolution of a grand unified universe. At Caltech, another centre of astrophysical modelling expertise, theorists did the same. The Chicago and Caltech models led to similar conclusions: HEP theorists were right in suspecting that su(5) was phenomenologically inadequate to generate a sufficient baryon asymmetry. The way out was to make things more complicated – go to larger unifying groups, introduce new heavy quarks and leptons, and, for really big asymmetries (baryon/photon ratios up to 10^{-3}) lots of massive Higgs bosons.[21]

The baryon-number asymmetry of the universe was possibly the most direct interface between cosmology and HEP in the late 1970s, but there were at least two others which should be mentioned. Detailed modelling of nucleosynthesis in the Big Bang enabled astrophysicists to assert that there were, at most, four distinct species of neutrinos – a number of great interest to theorists in the construction of realistic GUTs.[22] And the problem of the universe's 'missing mass' invited the speculation that neutrinos were not truly massless particles – a speculation with theoretical and observational conse-quences in both HEP and cosmology.[23] It is interesting to note that the massive neutrino solution to the missing-mass problem, like the preferred GUTs solution to the baryon-asymmetry problem, indicated that SU(5) was not, as Georgi and Glashow had once suggested, 'the gauge group of the world'.[24] This conclusion clashed with the classic requirement of theoretical simplicity, since SU(5) was acknowledged to be the minimal unifying group. It also clashed with the practice of HEP theorists, who had worked hard to explore the detailed implications of the SU(5) model. However, little effort was devoted to its rejection. Elaboration of GUTs based on groups other than SU(5) clearly represented a possible new dimension of theoretical practice,[25] but, more importantly, SU(5) was perceived to be almost a danger to the future of HEP. The danger, stressed from 1978 onwards by Bjorken, Glashow and other gauge-theory prophets, was that SU(5) implied a vast 'desert' stretching in energy from 10^2 GeV to 10^{15} GeV.[26] At projected LEP energies, around 100 GeV, new physics would emerge in the shape of the intermediate vector bosons of the electroweak interactions. Thereafter, according to SU(5), increasing energies would reveal no new phenomena until the X-boson region was reached at 10^{15} GeV. Not only was belief in the ability of SU(5) to extrapolate in energy over thirteen orders of magnitude extraordi-nary conceptual arrogance; it also offered HEP experimenters noth-ing – there were no targets to aim at in the desert – and it provided the perfect justification for building no new accelerators beyond the contemporary generation. For these reasons the HEP community were more than happy to entertain more complicated GUTs scenarios than that of SU(5). Theorists were well-equipped to work out more complicated gauge-theory structures, and experimentalists and acce-lerator-builders were pleased to listen. Thus cosmology gave back as much as it received from HEP and by 1980 this mutually reinforcing cycle was set to continue.[27]

13.3 GUTs and Proton Decay

In 1929, Hermann Weyl postulated the conservation of baryon number in order to account for the stability of the proton, and '[d]uring most of the half-century since Weyl . . . nearly everybody believed in it as an absolute truth'.[28] The stability of the proton, the lowest mass baryon, seemed intuitively obvious to physicists. Protons and neutrons made up atomic nuclei, and if they could decay to non-baryonic matter, then atoms throughout the cosmos would fall apart. Along with the disappearance of atoms would go the universe as we know it.

In the Big Bang cosmology discussed in the previous section, the universe was deemed to have existed for a finite length of time, around 10^{10} years. Thus, strictly speaking, the observation of protons in the present-day environment implied only that their lifetime was greater than 10^{10} years. But for practical purposes the difference between 10^{10} years and infinity seemed hardly worth making. This conclusion was reinforced in 1954 by the pioneering experiment of Frederick Reines and Clyde L. Cowan, Jr. Reines, Cowan and their collaborators set up a large tank of liquid scintillator adjacent to a nuclear reactor at Hanford, and observed occasional flashes corresponding to the production of an electron in charged-current neutrino events. Thus was the neutrino experimentally discovered; but Reines, Cowan and Brookhaven theorist Maurice Goldhaber also realised that these observations could be used to set a limit on the proton lifetime. Disintegrations of protons (or neutrons) within the scintillator would also lead to characteristic flashes from decay products, and from the absence of such flashes they were able to set a lower limit on the proton lifetime of 10^{20} to 10^{22} years (depending on how many theoretical assumptions one was willing to make in interpreting the data).[29] Reines and others persevered with this kind of neutrino experiment, using larger and larger quantities of scintillator as their detector, and in the mid-1970s the most severe lower limit on the proton lifetime – around 10^{30} years – came from Reines' work. His team had installed a 20-ton array of liquid scintillator detectors 3.2 kilometres underground (for shielding from cosmic rays) in a gold mine near Johannesburg, South Africa, and during observations taken from 1965 to 1974 they had seen no evidence for proton decay.[30]

All of these experiments with massive scintillation counters were primarily conceived as laboratories for neutrino physics, rather than with proton decay in mind: as the *New Scientist* reported in 1980:

> Reines emphasises that in this [the Johannesburg gold mine] and other experiments the limits which emerged for the nucleon lifetime were incidental to other work. Indeed, until 1974 there was virtually no theoretical guide or incentive to go out and look for nucleon decay. Only then, according to Reines, did the theories 'begin to come alive', although not sufficiently at first for the appropriate agencies to be persuaded to fund experiments.[31]

1974 was the year of the Georgi–Glashow su(5) GUT. And the theories began to come alive for Reines because GUTs predicted proton decay – exactly the kind of phenomena he had the resources to observe. We saw in the previous section that the X-bosons in GUTs mediated baryon-number nonconserving interactions, and vestiges of such interactions were expected to be manifest at low energies. Figure 13.3, for example, shows one X-boson mediated process through which a proton could decay to a positron plus mesons.[32]

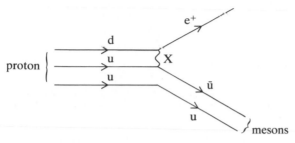

Figure 13.3. Proton decay to a positron plus mesons, mediated by X-boson exchange.

In their original papers Pati and Salam, and Georgi and Glashow, noted that proton decay was a rather direct consequence of grand unification, but adopted a distinctly tentative tone.[33] Only in 1978 and 1979 when GUTs became a theoretical industry did theorists speak more confidently. Because the X-bosons were extremely massive the expected lifetime of the proton was extremely long. One of the earliest detailed estimates came from a group of CERN theorists, who obtained a value of 10^{37} to 10^{38} years. 'Disappointingly' they commented, 'this would be too long for an observational test to be practicable in the forseeable future'.[34] However, more sophisticated consideration of the mass of the superheavy bosons soon brought the estimate down to the 'temptingly short' lifetime of $10^{31\pm2}$ years.[35] From the experimental point of view this was a very nice value: it was sandwiched between Reines' lower limit of 10^{30} years and the upper limit which could feasibly be measured, 10^{34} years.[36] Theorists had more good news for experimenters. Further calculations in the su(5) model, reported in 1979, indicated that protons (and neutrons bound in nuclei) would decay primarily to final states containing a positron (rather than a muon) and that a sizeable fraction of these decays would be to two-body final states: $p \rightarrow e^+\pi^0$ and $n \rightarrow e^+\pi^-$.[37] Such estimates had an important bearing on the feasibility of experimental measurements, the prominence of two-body decay modes offering a potentially clean and easily recognisable signal. As John Ellis remarked in 1979: 'Clearly more theoretical work is needed, particularly on estimating exclusive hadronic decay modes, but there is no need for experimentalists to drag their feet'.[38]

393

The outburst of GUTs theorising on proton decay created a climate wherein novel programmes of experiment could flourish. The predictive phenomenology of GUTs presented experimenters with relatively well defined goals, and the extent of theoretical involvement guaranteed would-be proton-decay searchers a sympathetic hearing from funding agencies. The way ahead was clear: to construct apparatus like that with which Reines had established the existing lower limit on the proton lifetime, but larger (and more expensive). In December 1978, a one-day seminar on proton stability was held at the University of Wisconsin, where the GUTs predictions and several experimental proposals were reviewed.[39] Many more such meetings and workshops followed and, one-by-one, proposed experiments became actual.

These experiments were all conceived along much the same lines. If the proton had a lifetime of, say, 10^{31} years, and if one monitored a total of 10^{31} protons (and neutrons) for a year, then one could expect to observe a single decay. 10^{31} protons weigh about 200 tons, so most experimenters aimed to assemble this much or more material, and to detect decays as efficiently as possible. A common factor in all of the proposals was to perform the experiments deep underground, in order to block out as many as possible background events caused by cosmic rays.[40] In finer details experimenters split on national lines – US physicists favouring the search for proton decays in water, whilst European, Indian and Japanese physicists put their faith in iron.[41] One of the first and largest experiments to be funded (by the US Department of Energy) was proposed by an Irvine–Michigan–Brookhaven collaboration led by proton decay experts Reines (Irvine) and Goldhaber (Brookhaven). They set about installing a tank containing 10,000 tons of very pure water (5×10^{32} nucleons) 2000 feet underground in the Morton Salt Mine east of Cleveland, Ohio. A Harvard–Wisconsin–Purdue collaboration, also funded by the DoE, mounted a similar but smaller experiment, aiming to monitor 1000 tons of water 2000 feet down a silver mine near Salt Lake City in Utah. The monitoring system in each was based upon conventional HEP techniques. Large arrays of photomultiplier tubes were set up to detect and amplify the flashes of radiation produced in water by the fast-moving products of proton decays. In the salt mine experiment the photomultiplier tubes surrounded the exterior of the tank; in the silver mine, the tubes were distributed within the water itself – an arrangement which the experimenters there considered to be more efficient. The third US group into the field was a team of physicists from the University of Pennsylvania led by Kenneth Lande. This group already had a 200-ton assembly of water/photomultiplier modules surrounding a neutrino experiment 5000 feet down the Homestake gold mine in South Dakota. The original objective of the Pennsylvania experimenters was to look for bursts of neutrinos from supernovae,[42] but in 1980 the Department of Energy agreed to finance the installation of an additional 600 tons of water and detectors to improve their chances of observing proton decays.

In Europe, progress was more stately. By 1980 no experiments had got under way. But two groups had come forward with definite proposals. From Italy came the Frascati–Milan–Turin NUSEX (nucleon stability experiment) concept. In collaboration with experimenters from CERN, the Italian physicists aimed to instal a 150-ton multilayer sandwich of iron plates and gas-filled detectors in the Mont Blanc Tunnel, 3000 metres below the summit. A French team planned to instal an iron/scintillator sandwich in the Fréjus tunnel under the Alps. In 1981 the Italian government approved plans to build an underground laboratory below the Gran Sasso mountain, which would be open to the international physics community, and which could accommodate 10,000-ton detectors. In 1981, developments were also under way in Japan, where a team of experimenters were constructing a 1000-ton detector instrumented with specially-developed phototubes, and planned to have it operational 1000 metres below ground by April 1982.

The first data, though, came from none of these experiments. Experimentation had been in progress 7000 feet down the Kolar gold mine in southern India for many years, and it was there that the first claim to have seen proton decay originated. An Indian–Japanese collaboration of physicists (from the Tata Institute, Osaka City University and the University of Tokyo) announced their preliminary findings in April 1981. They had been monitoring 140 tons of iron for 131 days and had observed around 200 events. Almost all of these could be ascribed to the passage through their apparatus of muons, generated by neutrino interactions in the overlying rock, but the experimenters concluded that three events could not be so explained. These three events were tentatively regarded as candidate proton decays, implying that the proton lifetime was somewhat less than 10^{31} years.[43]

Thus by the early 1980s proton decay was a well established field of experiment. Deep mines and tunnels around the world had experienced an influx of experimental physicists, anxiously monitoring large quantities of apparently inert matter at considerable expense. Positive findings had been reported, and already the next generation of more sophisticated proton-decay detectors were under active consideration.[44] Particle physics had escaped from the accelerator laboratory, and grand unified theory was incarnate below ground. Where men had once staked their future on society's need for minerals or transport, many physicists now staked their future on the chance that once in every 10^{31} years a proton would go pop.[45]

13.4 Gauge Theory Supreme

With GUTs we have come to the end of this history of particle physics. In the four preceding chapters we saw how, in the wake of the November Revolution, the twin gauge theories of the electroweak and strong interactions came to dominate the present and future of conventional HEP. In this chapter we have seen how gauge theory, in the shape of GUTs, further strengthened its grip upon HEP while extending its influence beyond the boundaries of that specialty. In Section 2 above, we saw GUTs and the universe become one, in a world-view which mediated the symbiosis of HEP theorists and cosmologists. In Section 3, we saw GUTs spilling out of HEP theory departments over the surface of our planet, collecting at the bottom of mineshafts and clogging important road tunnels, incarnate in kiloton water and iron detectors. Here too the GUTs world-view served to mediate a social symbiosis, the practice of GUTs theorists and non-accelerator experimenters generating mutually reinforcing contexts.

By 1980, Bjorken's gauge-theory 'new orthodoxy' – electroweak theory, QCD and GUTs – had come to fill the universe and to command space and time. At one extreme, physicists looked to GUTs for an understanding of the structure and evolution of the cosmos. At the other, they looked to the standard electroweak model and QCD for an understanding of yesterday's data and for a guide to tomorrow's experimental strategy. In between, investment decisions for the big machines of the coming decade were structured around the gauge-theory world-view. In their daily practice, and in their planning for the future, particle theorists and accelerator experimenters were joined by astrophysicists and searchers deep underground in testifying that their world was the world of gauge theory. They had built it, and it was impossible to imagine that they would surrender it.

NOTES AND REFERENCES

1   De Rújula, Ellis *et al.* (1982, 670).
2   For popular reviews of grand unification, see Weinberg (1974a), 't Hooft (1980) and Georgi (1981).
3   The late 1970s saw the rapid growth of several theoretical traditions which took the validity of electroweak theory and QCD for granted, and

which attempted to go beyond them. I focus upon grand unification here because it was the approach which had the most striking, direct and wide-ranging phenomenological applications, but a few words on the other popular theoretical approaches are appropriate. Besides grand unification, traditions devoted to 'subquarks', 'technicolour' and 'supersymmetry' all flourished in theoretical HEP in the late 1970s and early 1980s. Each displayed clearly the role of modelling in theory construction. The subquark tradition replayed the theme of composite-ness in a new key, asserting that quarks and leptons were themselves composites of more fundamental entities (see Chapter 10, note 23). Technicolour theories combined a different version of the compositeness theme with resources from gauge theory, and asserted that the Higgs particles responsible for spontaneous symmetry breaking were fermion composites bound together by a QCD-like but superstrong 'technicolour' force (for a review, see Farhi and Susskind 1981). Supersymmetry grew out of theoretical investigations of dual resonance models (see Section 3.4) and postulated the existence of symmetry relations between the otherwise distinct families of fermions and bosons (for a technical review, see Fayet and Ferrara 1977). In the late 1970s, interest focused upon the search for a supersymmetric quantum theory of gravitation (supergravity) and its unification with the supersymmetric grand unified theories (SUSY GUTs) of particle physics. The phenomenological implications of SUSY GUTs covered a similar range to those of the conventional GUTs discussed below, but were different in detail and more varied (for theoretical and experimental discussions of the implications of SUSY GUTs, see Nanopoulos, Savoy-Navarro and Tao 1982; for a popular account, see Waldrop 1983b).

4   Pati and Salam (1973a). This paper was, in a sense, anticipated by the 1972 work of three Lawrence Berkeley Laboratory theorists, Bars, Halpern and Yoshimura (1973). These authors also argued that all elementary particle interactions were mediated by gauge fields in a unified framework. However, they used the Higgs mechanism to give masses to the gauge vectors of the *strong interactions*: the gluons, in this model, became the hadronic vector mesons. As we have seen, this route to the understanding of the strong interactions lost its appeal after the discovery of the asymptotic freedom of gauge theory with massless gluons and the formulation of QCD.

5   Georgi and Glashow (1974).

6   Georgi, Quinn and Weinberg (1974).

7   Georgi and Glashow (1974, 440).

8   Georgi, Quinn and Weinberg (1974).

9   For access to the literature on calculations of $\sin^2\theta_W$ in various GUTs, see Weinberg (1980, note 58) and Ellis (1979, 942–3).

10 See the discussion of neutrino oscillations, note 45 below.

11 For a review of Big Bang cosmology, see Weinberg (1972b).

12 One of the most determined advocates of the interrelation between particle physics and cosmology was the Soviet theorist Ya. B. Zeldovich. See Zeldovich (1970) for an early expression of this perspective. For popular general accounts of the developing relations between HEP

theory and cosmology in the context of the Big Bang model, see Turner and Schramm (1979) and Ellis and Nanopoulos (1983); for technical accounts and reviews of the literature, see Steigman (1979), Wilcek (1981) and Dolgov and Zeldovich (1981).

13   Yoshimura (1978). For Yoshimura's early involvement with GUTs, see note 4 above. A similar mechanism to Yoshimura's had been proposed many years earlier by the dissident Soviet theorist Andrei Sakharov (1967). Sakharov's work predated the new physics and went largely unnoticed until Yoshimura's paper aroused the interest of GUTs theorists.

For popular accounts of the GUTs explanation for the cosmological baryon excess, see Wilcek (1980) and Waldrop (1981b).

14   See Steigman (1976).

15   One line of astrophysical argument denied that the present-day baryon excess existed, and asserted that the universe contained equal amounts of matter and antimatter separated by vast empty tracts. This symmetric cosmology was, however, generally considered to be in conflict with astrophysical observations (see Steigman 1976).

16   For a pre-GUTs popular review of the HEP theorist's version of the early evolution of the universe, see Weinberg (1977b).

17   As noted in Section 10.1, the embodiment of CP-violation in gauge theories had been first envisaged in 1973, in Kobayashi and Maskawa's 6-quark version of the standard model. It is significant to note that the work discussed here on the cosmological baryon excess was the first major tradition of novel theorising to embody CP-violation since the phenomenon was discovered by Fitch and Cronin in 1964. Perhaps this explains why Fitch and Cronin came to share the Nobel Prize in 1980 (and not earlier).

18   For an early review and references to the original literature, see Ellis (1979, 946).

19   Ellis (1979, 946).

20   Waldrop (1981b, 803).

21   Waldrop (1981b, 805).

22   See Olive, Schramm and Steigman (1978). Standard Big Bang lore plus detailed computer calculations linked the current helium abundance of the universe to the number of neutrino species (and to the number of species of 'hyperweak' particles: these were particles interacting even more weakly than neutrinos – their existence was predicted in many GUTs).

23   Ever since the invention of neutrinos in the early 1930s, particle physicists had routinely assumed them to be exactly massless particles. In the late 1970s, though, theorists began to suspect otherwise. There were several reasons for this. In cosmology a second puzzle analogous to that of the baryon-number asymmetry had come to a head over the preceding years. This was the puzzle of the 'missing mass' of the universe. Observations on the dynamics of astrophysical formations, from individual galaxies to giant clusters, led to estimates of their total masses much larger than observations of their luminous components suggested. The universe appeared to have a sizeable component of 'dark

matter' (for a popular account of the relevant astrophysics, see Rubin 1983). An obvious conjecture was that the dark matter was hadronic, but the cosmic abundance of helium appeared, through calculations of nucleosynthesis in the Big Bang, to rule this out. These calculations, however, did not apply to neutrino production in the Big Bang and thus, if neutrinos had even a very small mass (a few eV), the dark matter problem might be solved. Here cosmologists and particle physicists once more found common ground. In the basic SU(5) GUT, neutrinos were strictly massless, but in models based on larger groups, such as SO(10), neutrinos acquired masses in the range 1 to 100 eV. Interest in massive neutrinos further increased in 1980, when positive results were forthcoming from laboratory experiments (see note 45 below). For technical reviews of the GUTs and cosmological considerations pertaining to neutrino masses, see Marciano (1981) and Frampton and Vogel (1982); for a popular account, see Waldrop (1981a).

Besides massive neutrinos, a second candidate for the missing mass of the universe emerged from GUTs in the early 1980s: magnetic monopoles (see Chapter 7, note 50). In GUTs, these appeared as complex superheavy entities, with masses of around 10^{16} GeV. Like the somewhat lighter X-bosons, there was little prospect of producing monopoles in the laboratory but, again, they could be expected to have been produced in profusion in the very early stages of the Big Bang. Unfortunately, in the conventional GUTs/Big Bang picture, far too many monopoles would have been formed, in conflict with contemporary astrophysical data. This discrepancy fostered the construction of new cosmological and astrophysical models designed to reconcile particle theorists and cosmologists. The most popular such model was the so-called 'inflationary universe', in which monopole production was heavily suppressed (and other cosmological problems also solved). For a review of the literature on grand unified monopoles and cosmology, see Carrigan and Trower (1983a); for a popular account of the inflationary universe, see Waldrop (1983a). Like massive neutrinos, monopoles were also seen to have implications for terrestrial experiment: see note 45 below.

24 Georgi and Glashow (1974, 438).

25 A pool of basic resources for the construction of more complicated GUTs already existed. One of the first responses of theorists to the early work of Pati–Salam and Georgi–Glashow had been the investigation of alternative unifying groups. For example, at Caltech, Harald Fritzsch and Peter Minkowski (1975) explored the groups $SU_n \times SU_n$ ($n = 8, 12, 16$) and SO_n ($n = 10, 14$). F. Gürsey (at Yale) and P. Sikivie (at the University of Maryland) recommended one of the so-called 'exceptional' groups: Gürsey and Sikivie (1976), see also Ramond (1976). In an attempt to encompass the exotic phenomena then appearing in neutrino experiments (trimuons etc.) three Japanese physicists Inoue, Kakuto and Nakano (1977) produced a mutant SU(6) GUT. Gell-Mann, Ramond and Slansky (1978) offered a long review of grand unified schemes based upon the unitary groups, SU_n; the orthogonal groups, SO_n; the symplectic groups SP_{2n}; and the exceptional groups, G_2, F_4, E_6, E_7, and E_8.

26 Bjorken's (1979) 'The New Orthodoxy: How Can it Fail?' was entirely

devoted to experimentally interesting variants of the standard gauge-theory picture. Glashow, who had played such a central role in first creating and then advertising the elements of the new orthodoxy, had come by 1979 to a rather jaundiced perspective on the extent of his success, and gave several talks emphasising the importance of considering the alternatives to the standard picture. See Glashow (1980, 1981).

27 In subsequent years, the interdisciplinary symbiosis between HEP theory and cosmology received increasing institutional recognition. In the US, a theoretical astrophysics group was created alongside the HEP theory group at Fermilab (funded by NASA, at a cost of $500,000 over three years: see Arbeiter, 1982, and Ellis and Nanopoulos 1983, 213). And in Europe, for example, a joint European Southern Observatory–CERN Symposium on astrophysics, cosmology and particle theory was scheduled for 5 days in late 1983 (Ellis and Nanopoulos 1982, 215).

28 Goldhaber and Sulak (1981, 215).

29 Reines, Cowan and Goldhaber (1954).

30 Reines and Crouch (1974), Learned, Reines and Soni (1979). For a review of the history of experimental measurements on the proton lifetime, see Goldhaber, Langacker and Slansky (1980).

31 Sutton (1980, 1016).

32 For popular accounts of the relation between GUTs and proton decay, see Goldhaber, Langacker and Slansky (1980) and Weinberg (1981). For technical accounts and reviews of the literature, see Goldhaber and Sulak (1981), Primakoff and Rosen (1981) and Langacker (1981). Some idea of the intensity with which the HEP community took up GUTs can be gained from Langacker's 200 page review of 'Grand Unified Theories and Proton Decay'. This review, received for publication in November 1981, cites around 800 references (mainly theoretical): of these, the vast majority are dated 1978 or later.

33 In their first paper on grand unification, Pati and Salam (1973a) relegated consideration of proton decay to an appendix, where they concluded that the lifetime of the proton should be in the range 10^{15} to 10^{19} secs – some 16 to 20 orders of magnitude *less* than the contemporary *lower limit* from experiment. They did, though, return to the question some six months later. In a paper entitled 'Is Baryon Number Conserved?' (Pati and Salam 1973b) they concluded that, if free quarks were sufficiently massive, their model could be reconciled with contemporary data. Georgi and Glashow (1974) discussed proton decay in the penultimate paragraph of their paper, and their final footnote indicated that proton instability could be avoided in SU(5) at a theoretical price:

> A naive calculation [of the proton decay rate] indicates that the vector boson [X] mass must be greater that 10^{15} GeV $\simeq 10^{-9}$g! Let the reader who finds this hard to swallow double the number of fermion states and put quarks and leptons in different (but equivalent) thirty-dimensional representations. He must introduce both weakly interacting quarks and strongly interacting leptons. Now quark number is conserved modulo two and the proton is stable. The deuteron decays via the exchange of four superweak vector bosons, but this is not a serious problem.

Several authors subsequently devoted their efforts to contriving GUTs in which the proton was absolutely stable (Yoshimura 1977, Abud, Buccella, Ruegg and Savoy, 1977, Langacker, Segrè and Weldon 1978, Gell-Mann, Ramond and Slansky 1978). But this line of theorising lost its allure in 1978 when Yoshimura tied baryon-number violating reactions into cosmology, and GUTs calculations indicated that proton decay might be experimentally detectable (see below).

34 Buras, Ellis, Gaillard and Nanopoulos (1978, 67).

35 Goldman and Ross (1979, 1980), Marciano (1979). The quoted phrase is from Ellis (1979, 945).

36 The upper limit was set by the expected rate of background neutrino events in the proposed experiments (see note 40 below).

37 For early calculations see Jarlskog and Ynduráin (1979) and Machacek (1979).

38 Ellis (1979, 946).

39 Cline (1978).

40 The cosmic ray background arose as follows. At sea-level there is an appreciable flux of cosmic-ray muons. In large detectors they induce an event rate much higher than that expected for proton decays. However, muons are absorbed when they pass through matter. Thus, by burying their apparatus underground, experimenters sought to use the earth as a 'muon shield' and to reduce the rate of background muon events below that expected for proton decay. Unfortunately, the cosmic ray flux also includes neutrinos. These can also induce background events (recall that Reines' early experiments were directed at observing neutrinos). Furthermore, neutrinos are absorbed very weakly by matter. Thus burying detectors underground did not suffice to reduce neutrino background, and this source of error set the upper limit on the sensitivity of proton decay searches.

41 One principal constraint on the choice of material was that it be cheap and available in large quantities: hence the choice of water and iron. For descriptions and status reports on the experiments discussed below, see Sutton (1980), *Scientific American* (1980), *CERN Courier* (1981a, b), Weinberg (1981) and Goldhaber and Sulak (1981).

42 For a review of the use of large underground neutrino detectors to gather information of astrophysical significance, see Lande (1979).

43 The announcement of these findings was made at the Second Workshop on Grand Unification held at the University of Michigan at the end of April 1981. The claim was tempered by the fact that the estimated proton lifetime was close to the existing lower limit, and also because in each of three reported events one particle track reached the edge of the instrumented volume, leading to some ambiguity as to whether the events should be regarded as originating *within* the detector (i.e. proton decay) or outside (i.e. cosmic ray background). See *CERN Courier* (1981b). In 1982, the NUSEX collaboration reported a single candidate proton-decay event, again corresponding to a proton lifetime of around 10^{31} years (Battistoni *et al.* 1982). However, in 1983 the Irvine–Michigan–Brookhaven collaboration reported that in 80 days of data-taking they had found no proton-decay candidates. They concluded that the

lower limit on the proton lifetime was $6.5.10^{31}$ years, and suggested that the positive results elsewhere should be ascribed to cosmic-ray background (*The Times* Science Report, 26 Jan. 1983).

44 See the report on the September 1982 Los Alamos Workshop on underground experimentation: *CERN Courier* (1983c). The new generation of detectors were planned to be fine-grained, and hence capable of distinguishing between different proton decay modes (earlier detectors aimed primarily at establishing the existence of the phenomenon).

45 As mentioned in note 23 above, in the early 1980s two other classes of GUTs phenomena besides proton decay were believed to be accessible to terrestrial experiment: massive neutrinos and magnetic monopoles. Two classes of experiments related to the problem of neutrino masses. The former were nuclear-physics-style experiments which attempted a direct measurement of neutrino mass by very precise observations of nuclear β-decay. The second class relied upon the theoretical argument that if neutrinos had non-zero mass then 'neutrino oscillation' might be observed. Here the idea was that an initially pure beam of, say, electron-neutrinos would in time convert into a mixed beam containing electron-, muon- and tau-neutrinos. Such changes should be observable both in passive neutrino experiments (like the underground detectors mentioned in connection with proton decay) and in conventional accelerator neutrino experiments. Early indications from direct measurements and from searches for neutrino oscillations pointed to non-zero neutrino masses. With succeeding experiments conflicting data emerged, but again new traditions of novel experimentation had been founded (at CERN, for example, a low energy PS neutrino beam was purpose-built for neutrino oscillation experiments, and came into use in 1983: see Experimental Physics Division 1983). And, as one theorist put it: 'With the possible exception of proton decay, the establishment of a nonzero rest mass for neutrinos appears now as the most important and significant goal of laboratory experiments concerning elementary particles' (Frampton, in Frampton and Vogel 1982, 353). For reviews of the theory of neutrino oscillations and of experimental attempts to measure neutrino masses, see Bilenky and Pontecorvo (1978), Marciano (1981) and Frampton and Vogel (1982).

The superheavy magnetic monopoles predicted by GUTs were not expected to be produced in laboratory experiments, but it did appear possible that they might be found in the terrestrial environment. Great excitement prevailed when Blas Cabrera (1982) reported detection of a particle carrying the appropriate magnetic charge in a bench-top experiment at Stanford University. Many experimenters set about devising novel large-scale monopole detectors. Further excitement ensued when HEP theorists argued that monopoles could catalyse proton-decay, offering the proton-decay experimenters yet another phenomenon to look for. For a review of the literature in this area see Carrigan and Trower (1983a); for a popular account of monopole-catalysis of proton decay, see Waldrop (1982); and for expert discussions on observational and theoretical implications of superheavy magnetic monopoles see Carrigan and Trower (1983b).

14

PRODUCING A WORLD

Once there was a time when Coulomb's law, Ampère's law, Biot and
Savart's law and other laws were known but not yet dynamically
integrated. It is a stirring hope that the four forces will eventually
be part of a more unified theory. But there is presently no clear
road to that goal. . . .
   The state of particle physics . . . is a state not unlike the one in
a symphony hall a while before the start of the concert. On the
podium one will see some but not yet all of the musicians. They
are tuning up. Short brilliant passages are heard on some
instruments; improvisations elsewhere; some wrong notes too.
There is a sense of anticipation for the moment when the
symphony starts.   ABRAHAM PAIS, 1968[1]

If we avoid the fate of the builders of the Tower of Babel, then I see,
close at hand, another marvelous dream – a unified theory of the
hadrons . . .   MURRAY GELL-MANN, 1972[2]

The world is not directly given to us, we have to catch it through the
medium of traditions.   PAUL FEYERABEND, 1978[3]

The proposed [electroweak] models are a bonanza for experimentalists.   MURRAY GELL-MANN, 1972[4]

14.1   The Dynamics of Practice

The aim of this history of HEP has been to analyse the establishment
of the new-physics world-view. I have sought to explain how particle
physicists came to believe that the world was built from quarks and
leptons and that the interactions of these fundamental entities were
described by gauge theory. In conclusion, I will review the general
themes of my analysis and discuss some significant aspects of it.
   In Chapter 1, I outlined an archetypal 'scientist's account' of the
development of the new physics. This was very simple. It was based
upon the assertion that experimental facts obliged particle physicists
to adopt the new-physics package of beliefs: the observed hadron
spectrum implied the validity of the basic quark concept; the

403

observation of scaling in lepton–hadron scattering supported first the quark–parton model and later QCD; the discovery of the weak neutral current confirmed the intuition of electroweak gauge theorists; and so on. Throughout, experimental 'facts' served as an independent check upon theorising. In Chapter 1 I also outlined the philosophical and historiographical objections to the scientist's account. Philosophically, the key objection was that it obscured the ever-present role of scientific judgments in the research process – judgments relating to whether particular observation reports should be accepted as facts or rejected, and to whether particular theories should be regarded as acceptable candidate explanations of a given range of observations. I noted that the scientist's account factored such judgments out of consideration by adopting a stance of retrospective realism. Having decided upon how the natural world really is, those data which supported this image were granted the status of natural facts, and the theories which constituted the chosen world-view were presented as intrinsically plausible. The historiographical objection to such retrospective realism is obvious. If one is interested in how a scientific world-view is constructed, reference to its finished form is circularly self-defeating; the explanation of a genuine decision cannot be found in a statement of what that decision was.

I have approached the history of HEP with these deficiencies of the scientist's account in mind. My first objective has been to demonstrate the intervention of judgments throughout the developments at issue. Thus, for example, in discussing key experimental discoveries I have pointed to the potential for legitimate dissent, taking the discovery of the weak neutral current as a set-piece for detailed analysis. Similarly I have noted that even when accepted fields of 'facts' existed, a plurality of theories could be advanced for their explanation. None of these theories ever fitted the pertinent facts exactly: particle physicists were continually obliged to choose which theories to elaborate further and which to abandon. These choices concerning experimental data and theoretical plausibility had an irreducible character. Historically, particle physicists never seem to have been *obliged* to make the decisions they did; philosophically, it seems unlikely that literal obligation could ever arise. This is an important point because the choices which were made *produced the world of the new physics*, its phenomena and its theoretical entities. As we saw in most detail in the discussion of the neutral current discovery, the existence or nonexistence of pertinent natural phenomena was a product of irreducible scientific judgments. Similarly, the adoption of the theoretical apparatus of gauge theory entailed the view that misfits between predictions and data (and deductive lacunae) indicated grounds for further elaboration of the theory rather than for its rejection: again, an unforced and irreducible choice.

An exploration of the structure of scientific judgment has therefore been a central preoccupation of this account. I have sought to analyse why scientists should choose to accept the existence of one phenomenon rather than another, and why one theory rather than another should be regarded as a candidate for elaboration in the face of empirical misfits. For understanding, I have looked to the relation between particular choices and the contexts in which they were made. This involved the recognition that scientists are genuine agents: doers as well as thinkers, constructors as well as observers. My argument has been that judgments are best understood as situated within a continuing flow of practice – the day-to-day business of scientific research. Within that flow, judgments are seen to have implications for future practice: opportunities for research come into being or vanish according to choices concerning the existence of natural phenomena or the validity of theories. Thus, for example, acceptance of the Gargamelle group's novel interpretative practice implied at once the existence of the weak neutral current *and* the existence of new fields for experimental and theoretical research. The explanation of the neutral current in terms of electroweak gauge theory again pointed ahead to new areas of experimentation and theorising. And, I have argued, perceptions of those opportunities were structured by the resources available for their exploitation: the resources of hardware and expertise distributed amongst the experimental HEP community, and the stock of theoretical expertise stored in the theoretical HEP community. Scientific choice is in principle irreducible and open, but historically, options are foreclosed according to the opportunities perceived for future practice.

At the micro-level of individual research this seems an unsurprising and uncontentious observation. If scientists continually chose to put an intractable gloss on experimental data by a determined emphasis on the problematic aspects of every single empirical claim, what would emerge would have little in common with science as we know it. That individual scientists should seek to make irreducible judgments in a constructive rather than destructive fashion does not seem a radical conclusion. But at the macro-level of communal practice, some interesting and possibly less obvious conclusions can be drawn from the history of HEP.

14.2 Traditions and Symbiosis

In principle, the decisions which produce the world are free and unconstrained. They could be made at random, each scientist choosing by the toss of a coin at each decision point what stance to adopt. Instead, of course, we have seen that within HEP scientific judgments displayed a social coherence. Groups of scientists held to common agreements over phenomena and theories. The world of HEP was *socially* produced. The structure of social production is the theme of this section.

My analysis of the social coherence of scientific judgments has focused upon research traditions and their symbiosis. My suggestion has been that the history of HEP should be seen in terms of a fluctuating pattern of research traditions, wherein coherent sets of judgments were structured by *shared* sets of research resources. I have further argued that such traditions of experiment and theory from time to time existed in a state of symbiosis: the products of various traditions provided justification and subject matter for one another in a mutually supportive fashion. These symbioses of theory and experiment were both producers and consumers of world-views: particular views of the phenomenal world and associated theoretical entities were the medium of symbiosis, more refined views were its product. Thus my analysis of the establishment of the quark–gauge theory world-view has taken the form of an analysis of the growth to dominance of the new-physics traditions of HEP theory and experiment.

I will discuss the overall characteristics of both the new- and old-physics traditions in the following section, but it is appropriate to pause here to examine more closely the theoretical traditions of the new physics. On the 'scientist's account', the process of theory development is of little interest: theory choice is determined by the experimental facts, and the origins of candidate theories are irrelevant. But if the facts themselves are the products of judgments structured within a theoretical context, then the origins of that context become a central focus of interest. If explanatory theories are implicated in the constitution of empirical facts and phenomena, then theory cannot be absolutely constrained by experiment. Theory development must have, to some extent, a life of its own.[5]

In my account I have analysed this semi-autonomous process of theory development in terms of the dynamics of theoretical practice. I have argued that both the construction and elaboration of theories entail the recycling of theoretical resources from established areas of science to the research front. At both micro- and macro-levels the process is one of modelling or analogy.[6] Two great analogies were at the heart of the conceptual development of the new physics: drawing upon the well-oiled theoretical machinery of atomic and nuclear physics, hadrons were represented as composites of more fundamental entities – quarks; and theories of the weak and strong forces were modelled upon the quantum field theory of electromagnetism, QED. Neither of these analogical transplantations were made *en masse*. They were accomplished progressively and opportunistically. As schematised in Table 14.1, subsets of the established lore on composite systems and QED were recycled in limited research domains. The limited range of those transcriptions makes it analytically possible to circumscribe more closely the expertise of the theorists involved, and hence to associate the large scale process of theory development with the dynamics of individual theoretical practice.

Let me stress what is important in the above analysis of theory development. In the 'scientist's account', theory choice is absolutely constrained by experiment and the genesis of theories is regarded as unimportant. In contrast, I have argued that theory development is semi-autonomous, and only partially constrained by symbiosis with experiment. Since theory is the means of conceptualisation of natural phenomena, and provides the framework in which empirical facts are stabilised, an understanding of its internal dynamics is crucial to an understanding of the overall production of scientific knowledge. In the history of HEP, analogical transposition of shared resources was at the heart of theory development. Analogy was not one option amongst many (as some authors have argued)[7]; it was the basis of all that transpired. Without analogy, there would have been no new physics.

14.3   Incommensurability

The primary concern of this account has been with the social production of the new-physics world-view. I have argued that the phenomena and theoretical entities of that world-view were sustained by a coherent set of judgments – judgments which served, at the same time, to sustain the symbiosis of a constellation of research traditions. The dynamics of the new-physics traditions were, in turn, structured by the culturally specific resources available for their elaboration. Thus, the new-physics world-view was itself culturally specific.[8] In closing, I want to emphasise the significance of this observation by reference to the concept of incommensurability.

Arguments over the possibility of incommensurability have been endemic to the philosophy of science since the work of Thomas Kuhn in the 1960s.[9] Kuhn's argument was that if scientific knowledge were a cultural product then different scientific communities (separated in space or time) would inhabit different worlds. They would recognise

Table 14.1. Analogical structure of quark models and gauge theories (resource→construct).

Composite Systems→Quarks	Quantum Electrodynamics→Gauge Theory
spin→su(2) isospin symmetry	QED + su(2) →gauge theory
atoms (atomic and nuclear spectroscopy) + su(3)→ constituent quarks	hadron spectroscopy + gauge theory→su(3) Eightfold Way symmetry
Exclusion Principle + quarks→Han–Nambu colour	
atoms (impulse approximation) + scaling→quark-partons	gauge theory + coloured quark-partons→QCD
	many-body systems→renormalisation group
	QCD + renormalisation group→asymptotic freedom
	QCD + constituent quarks→QCD spectroscopy
positronium + QCD→charmonium	QCD + perturbative QED techniques→hard scattering, IPT analysis
QED (superconductivity)→spontaneous symmetry breaking (SSB)	
	gauge theory + SSB→unified electroweak theory
	electroweak theory + QCD→GUTs

408

the existence of different natural phenomena and explain their properties in terms of different theoretical entities. One striking consequence of this hypothesis is that the theories proper to different worlds would be immune to the kind of testing envisaged in the 'scientist's account'; they would be, in philosophical language, incommensurable. The reason for this is that each theory would appear tenable in its own phenomenal domain, but false or irrelevant outside it. There would be no realm of extra-cultural facts against which the empirical adequacy of different theories could be impartially measured. This is in many ways a disturbing conclusion which many authors have sought to reject. Numerous attempts to legislate against the very possibility of incommensurability in science are to be found in the philosophical literature.[10] But the history of HEP suggests that such attempts are misguided. I have analysed the emergence of the new physics against the background of the old, and manifestations of incommensurability are stamped across the transition between the two regimes. The old and new physics constituted, in Kuhn's sense, distinct and disjoint worlds.

To elaborate upon this observation, it is convenient to make a distinction between two forms of incommensurability. First, incommensurability is visible in the history of HEP at the *local* level of the performance and interpretation of individual experiments. The neutral current discovery can stand as an example of this. In the 1960s era of neutrino experiment, physicists adopted interpretative practices which made the neutral current nonexistent, in accordance with contemporary versions of the $V - A$ theory. With the 1970s explosion of work on electroweak gauge theory, a new set of interpretative practices were adopted by neutrino experimenters. These practices brought the neutral current into being, supporting the prototypical Weinberg–Salam unified model. Thus the 1960s and 1970s constellations of neutrino experiment and weak-interaction theory were incommensurable: the old and new theories of the weak interaction were each confirmed in its own phenomenal domain and were each disconfirmed outside it. To choose between the theories of the different eras required a simultaneous choice between the different interpretative practices in neutrino physics which determined the existence or nonexistence of the neutral current. The latter choice was irreducible; it cannot be explained by the comparisons between predictions and data which were its consequence.

I have pointed to many instances of local incommensurability in discussing the key discoveries in the history of the new physics. Besides the neutral current, other new-physics phenomena were also brought into existence by appropriate 'tuning' of interpretative practices. The discovery of scaling, for example, involved the acceptance of a potentially challengeable procedure for calculating radiative corrections. The existence of the naked-charm D mesons was established only via a departure from routine interpretative practice. The treatment meted out to the Washington and Oxford atomic-physics experiments also falls into this category. The Washington–Oxford results on parity-violation in atoms would have been uncontroversial in the no-neutral-current world of the old physics; in the new-physics era of electroweak gauge theory, they were simply unacceptable. All of these instances point to the local construction of a self-contained world of harmonious phenomena and theory, and hence to incommensurability, actual or potential.

Local incommensurability is often a subtle matter, requiring detailed analysis of individual experiments to bring it out. There is, though, a second *global* aspect to the incommensurability of the old and new physics which is not at all subtle. It concerns the gross rather than the fine differences between the experimental practices of the two regimes. We have seen that the phenomena of the old and new physics were disjoint. The old physics focused upon the most common processes encountered in the HEP laboratory: resonance production at low energies, soft-scattering at high energies. The new physics instead emphasised rare phenomena: the weak neutral current, hard-scattering, and so on. These differences between the phenomena of interest were reflected in gross differences of experimental practice. The very deployment of hardware in the laboratory was radically transformed in the shift from the old to the new regime. Old-physics experiments deployed beams and detectors in a common-sense fashion. They aimed to map out and study high-rate processes using hadron beams and targets. In so doing, retrospectively speaking, they smothered the rare phenomena of interest to the new physics. New-physics experiments achieved the opposite effect by artfully seeking out rare and precious events. Novel classes of experiment came to the fore: electron–positron annihilation experiments at colliders, fixed-target experiments using lepton rather than hadron beams, even non-accelerator experiments performed underground. Hadron-beam experiments did play a role in the new physics, but now detectors were contrived to filter out old-physics processes, recording only the one event in a million of interest to gauge theorists. Thus the disjuncture between the phenomenal worlds of the old and new physics was enforced not only in local interpretative practices but also globally on the laboratory floor, at the most basic technical level. New-physics phenomena were invisible by default in mainline old-physics experiments. Old-physics phenomena were invisible by construction in new-physics experiments.

The interlinked differences between the natural phenomena, explanatory theories and experimental strategies of the old and new physics are summarised in Table 14.2. They constitute the grounds for a simple and direct diagnosis of incommensurability. The theories of the two eras were well matched to the respective phenomenal worlds exhibited by contemporary experiments. Conversely, the old-physics theories had nothing to say on the rare phenomena of the new physics, and the new-physics theories had nothing to say on the common phenomena of the old physics. If the gauge-theory tool-kit of the late 1970s had been transmitted back in time to the mid-1960s, its phenomenological utility would have been nil, just as the phenomenological utility of the accumulated theoretical wisdom of the old physics was nil in the experimental context of the late 1970s. Each phenomenological world was, then, part of a self-contained, self-referential package of theoretical and experimental practice. To attempt to choose between old- and new-physics theories on the basis of a common set of phenomena was impossible: the theories were integral parts of different worlds, and they were incommensurable.[11]

One last issue remains to be discussed. On Kuhn's general account, the transition between incommensurable world-views usually involves acrimony and the exercise of institutional power. This seems plausible. If world-views are sustained by self-referential constellations of research traditions, rational argument alone is unlikely to dislodge entrenched habits of thought and action. But the transition from the old to the new physics within HEP showed little evidence of acrimony or duress. The prevailing climate within HEP during the 1970s was one of mutual congratulation rather than recrimination.[12] This departure from the Kuhnian picture was, I would argue, a contingent matter. The history of HEP can be understood as a communal search for a congenial world: a world which made social sense, and in which practice could be socially organised. At different times, different worlds of natural phenomena and explanatory entities were constructed. As it happened, the world of the old physics was conceptually and socially fragmented. Traditions organised around different phenomena generated little support for one another. Low-energy resonance physics, for example, was largely distinct from the study of high-energy soft-scattering processes: the practitioners of each acknowledged the existence of the other but went, in the main, their own separate ways. With the advent of the new physics, the conceptual unification of forces was accompanied by a social unification of practice. The quark–gauge theory world-view was at the heart of a community-wide symbiosis of experiment and theory.

Table 14.2. The old physics vs the new physics. Listed are the phenomena of primary interest in both eras, the appropriate experimental configurations and the preferred explanatory theories. The experimental configurations are classified according to the types of beam used, high energy (HE) or low energy (LE) and fixed target (FT) or colliding beam (CB). Hadron beam experiments are further subdivided into two categories: those using conventional detectors (CD) to investigate dominant processes, and those using purpose-built special detectors (SD) to investigate rare phenomena.

Phenomenon	Experimental configuration	Theory
Old physics		
Hadron resonances	LE hadron beams, FT, CD	CQM
Soft scattering	HE hadron beams, FT, CD	Regge
Hadronic weak decays	LE hadron beams, FT, CD	V–A theory
New physics		
Weak neutral current	HE neutrino beams, FT	Electroweak theory
New particles (c- and b-quark states)	HE e$^+$e$^-$, CB HE neutrino beams, FT HE hadron beams, FT and CB, SD	QCD (spectroscopic)/ electroweak theory
New particles (baryonium)	LE hadron beams, FT, CD	QCD (spectroscopic)
Hard scattering/jets	HE e$^+$e$^-$, CB HE lepton beams, FT HE hadron beams, FT and CB, SD	QCD (IPT)
Proton decay	Non-accelerator, passive underground detectors	GUTS

Research within new-physics traditions generated mutually reinforcing contexts throughout HEP. Oversimplifying only slightly, one can say that in the transition for the old to the new physics everyone gained and no-one lost.[13] Even outsiders to HEP – cosmologists, astrophysicists, underground experimenters and others – shared in the mutual benefits. It is often assumed that science is a zero-sum game where for every winner there is a loser, and hence that 'revolutionary' transitions between different worlds must be deeply marked by discord. The history of HEP does not support this view.

To summarise, the overall conclusions to be drawn from the history of HEP are these. The quark–gauge theory picture of elementary particles should be seen as a culturally specific product. The theoretical entities of the new physics, and the natural phenomena which pointed to their existence, were the joint products of a historical process – a process which culminated in a communally congenial representation of reality. I have analysed that process in terms of the conditioning of scientific judgments by the dynamics of practice. I offered a simple model for that dynamics – 'opportunism in context' was the slogan – and I have shown how it can illuminate major historical developments. The model is so simple that the query might legitimately arise: how could science be otherwise; what else might one expect? But the 'scientist's account' is deeply rooted in common-sense intuitions about the world and our knowledge of it. Many people do expect more of science than the production of a world congenial to social understanding and future practice. Consider the following quotation:

> Twentieth-century science has a grand and impressive story to tell. Anyone framing a view of the world has to take account of what it has to say . . . It is a non-trivial fact about the world that we can understand it and that mathematics provides the perfect language for physical science: that, in a word, science is possible at all.[14]

Such assertions about science are commonplace in our culture. In many circles they are taken to be incontestable. But the history of HEP suggests that they are mistaken. It is *unproblematic* that scientists produce accounts of the world that they find comprehensible: given their cultural resources, only singular incompetence could have prevented members of the HEP community producing an understandable version of reality at any point in their history. And, given their extensive training in sophisticated mathematical techniques, the preponderance of mathematics in particle physicists' accounts of reality is no more hard to explain than the fondness of ethnic groups for their native language. On the view advocated in this chapter, there is no obligation upon anyone framing a view of the world to take account of what twentieth-century science has to say. The particle physicists of the late 1970s were themselves quite happy to abandon most of the phenomenal world and much of the explanatory framework which they had constructed in the previous decade. There is no reason for outsiders to show the present HEP world-view any more respect. In certain contexts, such as foundational studies in the philosophy of science, it may be profitable to pay close attention to contemporary scientific beliefs. In other contexts, to listen too closely to scientists may be simply to stifle the imagination. World-views are cultural products; there is no need to be intimidated by them. In the irreverent words of a scientist from a bygone age: 'The lofty simplicity of nature all too often rests upon the unlofty simplicity of the one who thinks he sees it.'[15]

NOTES AND REFERENCES

1 Pais (1968, 24, 28).
2 Gell-Mann (1972, 338).
3 Feyerabend (1978, 34).
4 Gell-Mann (1972, 336).
5 At issue here is Reichenbach's distinction between the contexts of discovery and justification (see Reichenbach 1938). This distinction is central to much philosophical thinking on science. It asserts a clean separation between theory testing (the context of justification) and theory construction (the context of discovery). Theory testing, philosophers argue, is (or should be) amenable to explication according to canons of formal logic. It is an impersonal, ahistorical, culture-neutral process, and hence the proper locus for philosophical enquiry. Theory construction, on the other hand, is held to be immune to philosophical explication. It is seen as essentially private and personal, and hence to be relegated to the realms of psychology. Typically, the construction of major theories is ascribed to the quirks of scientific genius. My argument is that a clean separation between Reichenbach's two contexts is untenable both philosphically and historically. The distinction dissolves when one recognises the existence of irreducible judgments in the research process. These judgments tie data and explanatory theories together and imply an essential linkage between the dynamics of theory development (the context of discovery) and theory testing (the context of justification).

   For a similar critique of Reichenbach's distinction, see Knorr-Cetina (1981, 84). For a critique of psychological analyses of theory construction, which leads to an image of scientific discovery similar to that which I have advocated, see Brannigan (1981).
6 For access to the relevant literature on analogy in science, see the works cited in Chapter 1, note 12.
7 See Crane (1980b).
8 The easiest way to demonstrate this is counterfactually. Suppose, for example, that HEP theorists had approached their work without the stock of established analytical resources relating to composite atomic and nuclear systems; or that HEP experimenters had not had electron–positron colliders available to them. It is not hard to imagine either circumstance, and each would surely have led to the production of a different world-view from that of the new physics.
9 The work which brought the problem of incommensurability to the centre of philosophical attention was Kuhn's *The Structure of Scientific Revolutions* (first published in 1962 and reprinted with an important Postscript in 1970). Since Kuhn's work, the debate over incommensura-

bility has largely continued at an abstract philosophical level. Collins and Pinch (1982), however, organise their study of paranormal science around the idea of incommensurability, and their analysis parallels that given here.

10   The seminal collection of philosophical refutations of Kuhn is Lakatos and Musgrave (1970).

11   Depending upon the precise definition of incommensurability, it could be argued that the incommensurability of the old and new physics was partial rather than total. Data on hadronic resonances, for example, were held first to support the original CQM (old physics) and then to support the QCD-rejuvenated CQM (new physics). Two comments should be made on this. Historically, the areas of overlap between the old and new physics were small. Practice devoted to resonance physics was a minor component of new-physics research; few theorists and fewer experimentalists worked in this area in the late 1970s. Philosophically, whether the incommensurability was total or partial is not important. The primary thrust of my analysis is that the old and new physics were socially produced. Incommensurability is derivative upon social production, and I discuss it here just to emphasise the more striking implications of social production.

12   The general enthusiasm for, and lack of opposition to, the transition from the old to the new physics is evident throughout the technical and popular writings of particle physicists in the 1970s. Of course, some members of the HEP community were more enthusiastic than others: see the following note.

13   This is an oversimplification. Certain groups did find the transition difficult from the old physics to the new. Amongst experimenters, bubble-chamber physicists, for example, saw their expertise in danger of devaluation by the advent of the new physics. Bubble-chamber work had been at the forefront of old-physics strong-interaction experiment, but there was no obvious equivalent role for bubble chambers in the new physics. One response to this dilemma was to retrain in electronic techniques. Another was to adapt bubble-chamber technology to the needs of the new physics (as discussed in Chapter 12.2). Amongst theorists, exponents of the Regge approach were particularly hard hit when HEP experimenters lost interest in hadronic soft-scattering. Expertise in the techniques of Regge analysis found no significant place in new-physics theorising. Many Regge theorists retrained in field theory. A few, led by Chew, remained faithful to the hard-line bootstrap programme (even Chew expressed his pleasure when quark-like structures were found in a novel formulation of the bootstrap: see Chew and Rosenzweig 1978). Others sought to relate Regge theory and QCD, asserting the continued importance of the former while acknowledging the fundamental supremacy of the latter (see Collins and Martin 1982).

14   Polkinghorne (1983). Professor Polkinghorne was for many years a leading HEP theorist at the University of Cambridge. In 1979 he left physics for the Church.

15   This quotation is attributed to the eighteenth-century electrician Georg Cristoph Lichtenberg in Heilbron (1979, 448).

BIBLIOGRAPHY

Titles of technical journals have been abbreviated as follows:

AJP	*American Journal of Physics*
AP	*Annals of Physics*
ARNS	*Annual Review of Nuclear Science*
ARNPS	*Annual Review of Nuclear and Particle Science*
CMP	*Communications in Mathematical Physics*
CNPP	*Comments on Nuclear and Particle Physics*
JETP	*Soviet Physics. Journal of Experimental and Theoretical Physics (Translation)*
JETPL	*JETP Letters*
JMP	*Journal of Mathematical Physics*
LNC	*Lettere al Nuovo Cimento*
NC	*Nuovo Cimento*
NIM	*Nuclear Instruments and Methods*
NP	*Nuclear Physics*
Phys. Rept.	*Physics Reports*
PL	*Physics Letters*
PPNP	*Progress in Particle and Nuclear Physics*
PR	*Physical Review*
PRL	*PR Letters*
PTP	*Progress of Theoretical Physics*
RMP	*Reviews of Modern Physics*
RPP	*Reports on Progress in Physics*
SHEP	*Surveys in High Energy Physics*
SJNP	*Soviet Journal of Nuclear Physics (Translation)*
TMP	*Theoretical and Mathematical Physics*
ZFP	*Zeitschrift für Physik*

Abarbanel, H.D.I. (1976). 'Diffraction Scattering of Hadrons: The Theoretical Outlook'. *RMP*, *48*, 435–65.

Abers, E., Zachariasen, F. and Zemach, C. (1963). 'Origin of Internal Symmetries'. *PR*, *132*, 1831–6.

Abrams, G.S. *et al.* (1974). 'Discovery of a Second Narrow Resonance'. *PRL*, *33*, 1453–4.

Abud, M., Buccella, F., Ruegg, H. and Savoy, C.A. (1977). 'A New Unified Theory with Right-Handed Currents and Proton Stability'. *PL*, *67B*, 313–15.

Adair, R.K. and Fowler, E.C. (1963). *Strange Particles*. New York and London: Wiley Interscience.

Bibliography

Adams, J.B. (1979). 'Particle Accelerator Developments at CERN in 1979'. In *CERN Annual Report 1979*, 11–16.

Adams, J.B. (1980). 'The Development of CERN, 1970 to 1980'. In *CERN Annual Report 1980*, 13–25.

Adeva, B. *et al.* (1981). 'Observation of a Fully Reconstructed $D^0\bar{D}^0$ Pair with Long Proper Lifetimes in a High Resolution Hydrogen Bubble Chamber and the European Hybrid Spectrometer'. *PL, 102B*, 285–90.

Adler, S.L. (1965). 'Sum-Rules for Axial-Vector Coupling-Constant Renormalization in Beta Decay'. *PR, 140*, B736–B747.

Adler, S.L. (1966). 'Sum Rules Giving Tests of Local Current Commutation Relations in High-Energy Neutrino Reactions'. *PR, 143*, 1144–55.

Adler, S.L. (1969). 'Axial Vector Vertex in Spinor Electrodynamics'. *PR, 177*, 2426–38.

Adler, S.L. and Dashen, R.F. (1968). *Current Algebras and Applications to Particle Physics*. New York and Amsterdam: W.A.Benjamin.

Aguilar-Benitez, M. *et al.* (1983). 'The European Hybrid Spectrometer – A Facility to Study Multihadron Events in High Energy Interactions'. *NIM, 205*, 79–97.

Alguard, M.J. *et al.* (1976). 'Elastic Scattering of Polarized Electrons by Polarized Protons'. *PRL, 37*, 1258–61.

Alguard, M.J. *et al.* (1978). 'Deep-Inelastic e-p Asymmetry Measurements and Comparison with the Bjorken Sum Rule and Models of Proton Spin Structure'. *PRL, 41*, 70–3.

Ali, A. *et al.* (1980). 'A QCD Analysis of the High Energy e^+e^- Data from PETRA'. *PL, 93B*, 155–60.

Allison, W. *et al.* (1980). 'Direct Evidence for Associated Charm Production in 340 GeV πp Interactions', *PL, 93B*, 509–16.

Alper, B. *et al.* (1973). 'Production of High Transverse Momentum Particles in pp Collisions in the Central Region at the CERN ISR'. *PL, 40B*, 521–9.

Alpher, R.A., Fiske, M.D. and Porter, B.F. (1980). 'Physics Manpower: Present and Future'. *Physics Today, 33*(1), 44–53.

Altarelli, G. (1982). 'Partons in Quantum Chromodynamics'. *Phys. Rept., 81*, 1–129.

Alvarez, L.W. (1970). 'Recent Developments in Particle Physics'. In M. Conversi (ed.), *Evolution of Particle Physics*, 1–49. New York and London: Academic Press.

Amaldi, E. (1977). 'First International Collaborations between Western European Countries after World War II in the Field of High Energy Physics'. In Weiner (1977), 326–51.

Amaldi, E. (1981). 'The Bruno Touschek Legacy'. CERN Yellow Report 81-19.

'Amaldi Report' (1963). 'Report of the Working Party on the European High Energy Accelerator Programme'. CERN preprint CERN/FA/WP/23/ Rev. 3.

Amati, D., Bacry, H., Nuyts, J. and Prentki, J. (1964). 'SU4 and Strong Interactions'. *PL, 11*, 190–2.

Anderson, P.W. (1958). 'Random Phase Approximation in the Theory of Superconductivity'. *PR, 112*, 1900–16.

417

Bibliography

Anderson, P.W. (1963). 'Plasmons, Gauge Invariance and Mass'. *PR*, *130*, 439–42.

Andersson, B., Gustafson, G. and Sjostrand, T. (1980). 'A Three-Dimensional Model for Quark and Gluon Jets'. *ZFP*, *6*, 235–40.

Appelquist, T., Barnett, R.M. and Lane, K. (1978). 'Charm and Beyond'. *ARNPS*, *28*, 387–499.

Appelquist, T., De Rújula, A., Politzer, H.D. and Glashow, S.L. (1975). 'Spectroscopy of the New Mesons'. *PRL*, *34*, 365–9.

Appelquist, T. and Politzer, H.D. (1975). 'Heavy Quarks and e^+e^- Annihilation'. *PRL*, *34*, 43–5.

Arbeiter, L. (1982). 'Astrophysics Laboratory: Theoretical Gain'. *Nature*, *299*, 770.

Argento, A. *et al.* (1983). 'Electroweak Asymmetry in Deep Inelastic Muon-Nucleon Scattering'. *PL*, *120B*, 245–50.

Arnison, G. *et al.* (1983a). 'Experimental Observation of Isolated Large Transverse Energy Electrons with Associated Missing Energy at $\sqrt{s} = 540$ GeV'. *PL*, *122B*, 103–16.

Arnison, G. *et al.* (1983b). 'Experimental Observation of Lepton Pairs of Invariant Mass around 95 GeV/c^2 at the CERN SPS Collider'. *PL*, *126B*, 398–410.

Atwood, W. *et al.* (1972). 'Experimental Test for an Electromagnetic Axial-Vector Current of Hadrons in Inelastic Scattering of Polarized Electrons'. SLAC Proposal No. 95, unpublished.

Aubert, B. *et al.* (1974a). 'Further Observation of Muonless Neutrino-Induced Inelastic Interactions'. *PRL*, *32*, 1454–7.

Aubert, B. *et al.* (1974b). 'Measurements of Rates for Muonless Deep Inelastic Neutrino and Antineutrino Interactions'. *PRL*, *32*, 1457–60.

Aubert, B. *et al.* (1974c). 'Scaling-Variable Distributions in High-Energy Inelastic Neutrino Interactions'. *PRL*, *33*, 984–7.

Aubert, J.J. *et al.* (1974). 'Observation of a Heavy Particle J'. *PRL*, *33*, 1404–5.

Augustin, J.-E. *et al.* (1974). 'Discovery of a Narrow Resonance in e^+e^- Annihilation'. *PRL*, *33*, 1406–7.

Aurenche, J.P. and Paton, J.E. (1976). 'High-Energy Hadron Collisions: A Point of View'. *RPP*, *39*, 175–216.

Bacci, C. *et al.* (1974). 'Preliminary Result of Frascati (ADONE) on the Nature of a New 3.1 GeV Particle Produced in e^+e^- Annihilation'. *PRL*, *33*, 1408–10.

Bacci, C. and Salvini, G. (eds) (1983). *Proceedings of the Third Topical Workshop on Proton-Antiproton Collider Physics*, 12–14 January 1983, Rome. CERN Yellow Report 83-04.

Bacry, H., Nuyts, J. and Van Hove, L. (1964). 'Basic SU₃ Triplets with Integral Charge and Unit Baryon Number'. *PL*, *9*, 279–80.

Baird, P.E.G. *et al.* (1976). 'Search for Parity Non-Conserving Optical Rotation in Atomic Bismuth'. *Nature*, *264*, 528–9.

Baird, P.E.G. *et al.* (1977). 'Search for Parity-Nonconserving Optical Rotation in Atomic Bismuth'. *PRL*, *39*, 798–801.

Ballam, J. (1967). 'SLAC: The Program'. *Physics Today*, *20*(4), 43–52.

418

Bibliography

Ballam, J. and Watt, R.D. (1977). 'Hybrid Bubble Chamber Systems'. *ARNS*, *27*, 75–138.

Banner, M. *et al.* (1973). 'Large Transverse Momentum Particle Production at 90° in pp Collisions at the ISR'. *PL*, *44B*, 537–40.

Banner, M. *et al.* (1983). 'Observation of Single Isolated Electrons of High Transverse Momentum with Missing Transverse Energy at the CERN pp̄ Collider'. *PL*, *122B*, 476–85.

Barboni, E.J. (1977). *Functional Differentiation and Technological Specialization in a Specialty in High Energy Physics: The Case of Weak Interactions of Elementary Particles*. Cornell University PhD Thesis, unpublished.

Bardeen, J., Cooper, L.N. and Schrieffer, J.R. (1957). 'Microscopic Theory of Superconductivity'. *PR*, *106*, 162–4.

Bardeen, W.A., Fritzsch, H. and Gell-Mann, M. (1973). 'Light Cone Current Algebra, π^0 Decay and e^+e^- Annihilation'. In Gatto (1973), 139–53.

Barger, V. (1974). 'Reaction Mechanisms at High Energy'. In Smith (1974), I-193–227.

Barish, B.C. (1978). 'Experimental Aspects of High Energy Neutrino Physics'. *Phys. Rept.*, *39*, 279–360.

Barish, B.C. *et al.* (1974). 'Gauge-Theory Heavy Muons: An Experimental Search'. *PRL*, *32*, 1387–90.

Barish, B.C. *et al.* (1975). 'Neutral Currents in High-Energy Neutrino Collisions: An Experimental Search'. *PRL*, *34*, 538–41.

Barish, B.C. *et al.* (1976). 'Investigations of Neutrino Interactions with Two Muons in the Final State'. *PRL*, *36*, 939–41.

Barish, B.C. *et al.* (1977). 'Observation of Trimuon Production by Neutrinos'. *PRL*, *38*, 577–80.

Barkov, L.M. and Zolotorev, M.S. (1978). 'Observation of Parity Nonconservation in Atomic Transitions'. *JETPL*, *27*, 357–61.

Barkov, L.M. and Zolotorev, M.S. (1979). 'Parity Violation in Atomic Bismuth'. *PL*, *85B*, 308–13.

Barnes, B. (1977). *Interests and the Growth of Knowledge*. London: Routledge and Kegan Paul.

Barnes, B. (1982). *T.S.Kuhn and Social Science*. London: Macmillan.

Barnes, V.E. *et al.* (1964). 'Observation of a Hyperon with Strangeness Minus Three'. *PRL*, *12*, 204–6.

Bars, I., Halpern, M.B. and Yoshimura, M. (1973). 'Unified Gauge Theories of Hadrons and Leptons'. *PR*, *D7*, 1233–51.

Bartoli, B. *et al.* (1970). 'Multiple Particle Production from e^+e^- Interactions at C.M. Energies between 1.6 and 2 GeV'. *NC*, *70A*, 615–31.

Barton, M.Q. (1961). *Catalogue of High Energy Accelerators*. Brookhaven National Laboratory Report BNL 683 (T-230).

Battistoni, G. *et al.* (1982). 'Fully Contained Events in the Mont Blanc Nucleon Decay Detector'. *PL*, *118B*, 461–5.

Baum, G. *et al.* (1971). 'Measurement of Asymmetry in Deep Inelastic Scattering of Polarized Electrons by Polarized Protons'. SLAC Proposal No. 80, unpublished.

Baum, G. *et al.* (1975). 'A Test of Parity Violations in the Inelastic Scattering of Polarized Electrons at the Level of the Weak Interactions'. SLAC Proposal No. 122, unpublished.

419

Becchi, C. and Morpurgo, G. (1965a). 'Vanishing of the E2 Part of the $N^*_{33} \to N + \gamma$ Amplitude in the Non-Relativistic Quark Model of "Elementary" Particles'. *PL, 17*, 352–4.

Becchi, C. and Morpurgo, G. (1965b). 'Test of the Nonrelativistic Quark Model for "Elementary" Particles: Radiative Decays of Vector Mesons'. *PR, 140B*, 687–90.

Belavin, A.A., Polyakov, A.M., Schwartz, A. and Tyupkin, Y. (1975). 'Pseudoparticle Solutions of the Yang–Mills Equations'. *PL, 59B*, 85–7.

Bell, J.S. (1967). 'Current Algebra and Gauge Invariance'. *NC, 50*, 129–34.

Bell, J.S. and Jackiw, R. (1969). 'A PCAC Puzzle: $\pi^0 \to \gamma\gamma$ in the σ-Model'. *NC, 51*, 47–61.

Bell, J.S., Løvseth, J. and Veltman, M. (1963). 'CERN Neutrino Experiments: Conclusions'. In Bernadini and Puppi (1963), 584–90.

Bell, J.S. and Veltman, M. (1963a). 'Intermediate Boson Production by Neutrinos'. *PL, 5*, 94–6.

Bell, J.S. and Veltman, M. (1963b). 'Polarisation of Vector Bosons Produced by Neutrinos'. *PL, 5*, 151–2.

Bellini, G. *et al.* (1982). 'Lifetime Measurements in the 10^{-13}s Range'. *Phys. Rept., 83*, 1–106.

Benichou, J.L. *et al.* (1981). 'A Rapid Cycling Hydrogen Bubble Chamber with High Spatial Resolution for Visualizing Charmed Particle Decays'. *NIM, 190*, 487–512.

Benvenuti, A. *et al.* (1974). 'Observation of Muonless Neutrino-Induced Inelastic Interactions'. *PRL, 32*, 800–3.

Benvenuti, A. *et al.* (1975a). 'Observation of New-Particle Production by High-Energy Neutrinos'. *PRL, 34*, 419–22.

Benvenuti, A. *et al.* (1975b). 'Invariant-Mass Distributions from Inelastic v and $\bar{v}$ Interactions'. *PRL, 34*, 597–600.

Benvenuti, A. *et al.* (1975c). 'Further Observation of Dimuon Production by Neutrinos'. *PRL, 35*, 1199–202.

Benvenuti, A. *et al.* (1975d). 'Dimuons Produced by Antineutrinos'. *PRL, 35*, 1249–52.

Benvenuti, A. *et al.* (1976a). 'Further Data on the High-y Anomaly in Inelastic Neutrino Scattering'. *PRL, 36*, 1478–82.

Benvenuti, A. *et al.* (1976b). 'Measurement of the Ratio $\sigma_c(\bar{v}_\mu + N \to \mu^+ + X)/\sigma_c(v_\mu + N \to \mu^- + X)$ at High Energy'. *PRL, 37*, 189–92.

Benvenuti, A. *et al.* (1976c). 'Evidence for Parity Nonconservation in the Weak Neutral Current'. *PRL, 37*, 1039–42.

Benvenuti, A. *et al.* (1977a). 'Observation of a New Process with Trimuon Production by High-Energy Neutrinos'. *PRL, 38*, 1110–13.

Benvenuti, A. *et al.* (1977b). 'Characteristics of Neutrino-Produced Dimuon and Trimuon Events as Evidence for New Physics at the Lepton Vertex'. *PRL, 38*, 1183–6.

Berger, Ch. *et al.* (1978). 'Observation of a Narrow Resonance Formed in $e^+ e^-$ Annihilation at 9.46 GeV'. *PL, 76B*, 243–5.

Berman, S.M., Bjorken, J.D. and Kogut, J.B. (1971). 'Inclusive Processes at High Transverse Momentum'. *PR, D4*, 3388–418.

Bibliography

Berman, S.M. and Jacob, M. (1970). 'Connection Between Inelastic Proton–Proton Reactions and Deep Inelastic Scattering'. *PRL*, *25*, 1683–6.

Bernadini, C. (1978) 'The Story of AdA', *Scientia*, *113*, 27–44.

Bernadini, G. and Puppi, G.P. (eds) (1963). *Proceedings of the Sienna International Conference on Elementary Particles*, 30 Sept.–5 Oct. 1963, Sienna, Italy. Bologna: Società Italiana di Fisica.

Bethe, H.A. and de Hoffmann, F. (1956). *Mesons and Fields*, Vol. II. Evanston: Row, Peterson.

Bienlein, J.K. *et al*. (1964). 'Spark Chamber Study of High-Energy Neutrino Interactions'. *PL*, *13*, 80–91.

Bienlein, J.K. *et al*. (1978). 'Observation of a Narrow Resonance at 10.02 GeV in e^+e^- Annihilations'. *PL*, *78B*, 360–3.

Bilenky, S.M. and Pontecorvo, B. (1978). 'Lepton Mixing and Neutrino Oscillations'. *Phys. Rept.*, *41*, 225–61.

Bingham, H.H. *et al*. (1963). 'CERN Neutrino Experiment – Preliminary Bubble Chamber Results'. In Bernadini and Puppi (1963), 555–70.

Bjorken, J.D. (n.d.). 'Inelastic Electron (and Muon) Scattering at High Energies and Forward Angles'. Unpublished.

Bjorken, J.D. (1966a). 'Inequality for Electron and Muon Scattering from Nucleons'. *PRL*, *16*, 408.

Bjorken, J.D. (1966b). 'Applications of the Chiral $U(6) \otimes U(6)$ Algebra of Current Densities'. *PR*, *148*, 1467–78.

Bjorken, J.D. (1967). 'Inequality for Backward Electron- and Muon-Nucleon Scattering at High Momentum Transfer'. *PR*, *163*, 1767–9.

Bjorken, J.D. (1969). 'Asymptotic Sum Rules at Infinite Momentum'. *PR*, *179*, 1547–53.

Bjorken, J.D. (1974). 'A Theorist's View of e^+e^- Annihilation'. In Rollnik and Pfeil (1974), 25–47.

Bjorken, J.D. (1975). 'Symposium Summary and Prognosis'. In Kirk (1975), 987–1002.

Bjorken, J.D. (1977). 'Future Accelerators: Physics Issues'. In Gutbrod (1977), 960–1000.

Bjorken, J.D. (1979). 'The New Orthodoxy: How Can It Fail?' In *Neutrino '79: Proceedings of the International Conference on Neutrinos, Weak Interactions and Cosmology*, 18–22 June 1979, Bergen, Norway, 9–19.

Bjorken, J.D. and Brodsky, S.J. (1970). 'Statistical Model for Electron-Positron Annihilation into Hadrons'. *PR*, *D1*, 1416–20.

Bjorken, J.D. and Drell, S.D. (1964). *Relativistic Quantum Mechanics*. New York: McGraw-Hill.

Bjorken, J.D. and Drell, S.D. (1965). *Relativistic Quantum Fields*. New York: McGraw-Hill.

Bjorken, J.D. and Glashow, S.L. (1964). 'Elementary Particles and SU(4)'. *PL*, *11*, 255–7.

Bjorken, J.D. and Nauenberg, M. (1968). 'Current Algebra'. *ARNS*, *18*, 229–68.

Bjorken, J.D. and Paschos, E.A. (1969). 'Inelastic Electron-Proton and γ-Proton Scattering and the Structure of the Nucleon'. *PR*, *185*, 1975–82.

Bibliography

Bjorken, J.D. and Paschos, E.A. (1970). 'High-Energy Inelastic Neutrino-Nucleon Interactions'. *PR*, *D1*, 3151–60.

Blankenbecler, R., Brodsky, S.J. and Gunion, J.F. (1972). 'Composite Theory of Large Angle Scattering and New Tests of Parton Concepts'. *PL*, *39B*, 649–53.

Blewett, M.H. (1967). 'Characteristics of Typical Accelerators'. *ARNS*, *17*, 427–68.

Blietschau, J. *et al.* (1976). 'Observation of Muon-Neutrino Reactions Producing a Positron and a Strange Particle'. *PL*, *60B*, 207–10.

Block, M.M. *et al.* (1964). 'Neutrino Interactions in the CERN Heavy Liquid Bubble Chamber'. *PL*, *12*, 281–5.

Bloom, E.D. and Feldman, G.J. (1982). 'Quarkonium'. *Scientific American*, *246*(5), 42–53.

Bloom, E.D. *et al.* (1969). 'High Energy Inelastic e-p Scattering at 6° and 10°'. *PRL*, *23*, 930–4.

Bloor, D. (1976). *Knowledge and Social Imagery*. London: Routledge and Kegan Paul.

Bludman, S.A. (1958). 'On the Universal Fermi Interaction'. *NC*, *9*, 433–44.

BNL (1977). *Proceedings of the 1977 Isabelle Summer Workshop*, July 18–29, 1977, Brookhaven National Laboratory. BNL Report 50721.

Bogoliubov, N.N. and Shirkov, D.V. (1959). *Introduction to the Theory of Quantized Fields*. New York: Interscience.

Bogoliubov, N.N., Tolmachev, V.V. and Shirkov, D.V. (1958). *A New Method in the Theory of Superconductivity*. Moscow: Academy of Sciences of USSR.

Bogoliubov *et al.* (eds.) (1976). *Proceedings of the 18th International Conference on High Energy Physics*, Tbilisi, USSR, July 1976.

Bouchiat, C., Iliopoulos, J. and Meyer, Ph. (1972). 'An Anomaly-Free Version of Weinberg's Model'. *PL*, *38B*, 519–23.

Boulware, D. (1970). 'Renormalizeability of Massive Non-Abelian Gauge Fields: A Functional Integral Approach'. *AP*, *56*, 140–71.

Boyarski, A. *et al.* (1975). 'Limits on Charmed Meson Production in e^+e^- Annihilation at 4.8 GeV Centre-of-Mass Energy'. *PRL*, 35, 196–9.

Brandt, R.A. (1969). 'Asymptotic Behavior of Electroproduction Structure Functions'. *PRL*, *22*, 1149–51.

Brannigan, A. (1981). *The Social Basis of Scientific Discoveries*. Cambridge: Cambridge University Press.

Branson, J.G. (1982). 'Review of High Energy e^+e^- Physics'. Lectures at the International School of Subnuclear Physics, Erice, 1982. To be published.

Braunschweig, E. *et al.* (1974). 'A Measurement of Large Angle e^+e^- Scattering at the 3100 MeV Resonance'. *PL*, *53B*, 393–6.

Brézin, E., Gervais, J-L. and Toulouse, G. (eds) (1980). *Common Trends in Particle and Condensed Matter Physics*: Proceedings of Les Houches Winter Advanced Study Institute, February 1980. *Phys. Rept.*, *67*, 1–199.

Briedenbach, M. *et al.* (1969). 'Observed Behavior of Highly Inelastic Electron-Proton Scattering'. *PRL*, *23*, 935–99.

Broad, W.J. (1982). 'A Requiem for Isabelle'. *Science*, *216*, 158.

422

Bibliography

Brodsky, S.J. and Drell, S.D. (1970). 'The Present Status of Quantum Electrodynamics'. *ARNS*, *20*, 147–94.

Brodsky, S. and Karl, G. (1976). 'Parity Violation in Atoms'. *Comments on Atomic and Molecular Physics*, *5*, 63–9.

Bromberg, C. *et al.* (1977). 'Observation of the Production of Jets of Particles at High Transverse Momentum and Comparison with Inclusive Single-Particle Reactions'. *PRL*, *38*, 1447–50.

Bromberg, C. *et al.* (1978). 'Production of Jets and Single Particles at High p_T in 200 GeV Hadron-Beryllium Collisions'. *NP*, *B134*, 189–214.

Brower, R.C. and Ellis, J. (1972). 'Double-Regge Boundary Condition for Limiting Fragmentation'. *PR*, *D5*, 2253–7.

Brower, R.C. and Ellis, J. (1974). 'An Asymptotically Free Regge Field Theory'. *PL*, *51B*, 242–6.

Brower, R.C., Ellis, J., Savit, R. and Zakrzewski, J.R. (1975). 'Reggeon Field Theory on a Lattice: A Formulation'. *NP*, *B94*, 460–76.

Brower, R.C., Ellis, J., Schmidt, M.G. and Weis, J.H. (1977a). 'Hadron Scattering in Two-Dimensional QCD, 1: Formalism and Leading Order Calculations'. *NP*, *B128*, 131–74.

Brower, R.C., Ellis, J., Schmidt, M.G. and Weis, J.H. (1977b). 'Hadron Scattering in Two-Dimensional QCD, 2: Second-Order Calculations, Multi-Regge and Inclusive Reactions'. *NP*, *B128*, 175–203.

Brown, L.M. (1978). 'The Idea of the Neutrino'. *Physics Today*, *31*(9), 23–8.

Brown, L.M. (1981). 'Yukawa's Prediction of the Meson'. *Centaurus*, *25*, 71–132.

Brown, L.M. and Hoddeson, L. (1982). 'The Birth of Elementary-Particle Physics'. *Physics Today*, *35*(4), 36–43.

Brown, L.M. and Hoddeson, L. (eds) (1983). *The Birth of Particle Physics*. New York: Cambridge University Press.

Bucksbaum, P., Commins, E. and Hunter, L. (1981). 'New Observation of Parity Nonconservation in Atomic Thallium'. *PRL*, *46*, 640–3.

Budagov, I. *et al.* (1969). 'Measurement of Structure Factors in Inelastic Neutrino Scattering'. *PL*, *30B*, 365–8.

Buras, A.J. (1980). 'Asymptotic Freedom in Deep Inelastic Processes in the Leading Order and Beyond'. *RMP*, *52*, 199–276.

Buras, A.J., Ellis, J., Gaillard, M.K. and Nanopoulos, D.V. (1978). 'Aspects of the Grand Unification of Strong, Weak and Electromagnetic Interactions'. *NP*, *B135*, 66–92.

Burhop, E.H.S. *et al.* (1976). 'Observation of a Likely Example of the Decay of a Charmed Particle Produced in a High Energy Neutrino Inter-action'. *PL*, *65B*, 299–304.

Büsser, F.W. *et al.* (1973). 'Observation of π^0 Mesons with Large Transverse Momentum in High Energy pp Collisions'. *PL*, *46B*, 471–6.

Cabibbo, N. (1963). 'Unitary Symmetry and Leptonic Decays'. *PRL*, *10*, 531–3.

Cabibbo, N., Parisi, G. and Testa, M. (1970). 'Hadron Production in e^+e^- Collisions'. *LNC*, *4*, 35–9.

Cabibbo, N. and Radicati, L.A. (1966). 'Sum Rule for the Isovector Magnetic Moment of the Proton'. *PL*, *19*, 697–9.

Cabrera, B. (1982). 'First Results from a Superconductive Detector for Moving Magnetic Monopoles'. *PRL*, *48*, 1378–81.

Cahier Technique No. 6 (1973). 'Gargamelle: La Chambre à Bulles à Liquides Lourds'. CERN Report CERN/PIO 73-15.

Cahn, R.N. and Gilman, F.J. (1978). 'Polarized-Electron-Nucleon Scattering in Gauge Theories of Weak and Electromagnetic Interactions'. *PR*, *D17*, 1313–22.

Callan, C.G., Jr. (1970). 'Broken Scale Invariance in Scalar Field Theory'. *PR*, *D2*, 1541–7.

Calmet, J., Narison, S., Perrottet, M. and de Rafael, E. (1977). 'The Anomalous Magnetic Moment of the Muon: A Review of the Theoretical Contributions'. *RMP*, *49*, 21–9.

Camilleri, L. (1979). 'A Study of High Mass e^+e^- Pairs Produced in p-p Collisions at the CERN ISR'. In Homma *et al.* (1979), 187–8.

Camilleri, L. *et al.* (1976). 'Physics with Very High Energy e^+e^- Colliding Beams'. CERN Yellow Report 76-18.

Capra, F. (1979). 'Quark Physics without Quarks: A Review of Recent Developments in S-Matrix Theory'. *AJP*, 47, 11–23.

Carrigan, R.A., Jr. and Trower, W.P. (1982). 'Superheavy Magnetic Monopoles'. *Scientific American*, *246*(4), 106–18.

Carrigan, R.A., Jr. and Trower, W.P. (1983a). 'Magnetic Monopoles: A Status Report'. Fermilab preprint Fermilab-Pub-83/31.

Carrigan, R.A., Jr. and Trower, W.P. (eds) (1983b). *Magnetic Monopoles*. New York and London: Plenum.

Carruthers, P. (1971). 'Broken Scale Invariance in Particle Physics'. *Phys. Rept.*, 1, 1–30.

Cassidy, D.C. (1981). 'Cosmic Ray Showers, High Energy Physics, and Quantum Field Theories: Programmatic Interactions in the 1930s'. *Historical Studies in the Physical Sciences*, *12*, 1–39.

Cazzoli, E.G. *et al.* (1975). 'Evidence for $\Delta S = -\Delta Q$ Currents or Charmed-Baryon Production by Neutrinos'. *PRL*, *34*, 1125–8.

CERN (1982). *Photonics Applied to Nuclear Physics: 1*, European Hybrid Spectrometer Workshop on Holography and High-Resolution Techniques, Strasbourg, 9–12 November 1981. CERN Yellow Report 82-10.

CERN Courier (1977a). 'Gersh Budker and Ben Lee'. *17*, 246–7.

CERN Courier (1977b). 'Upsilon Hunting'. *17*, 319–21.

CERN Courier (1978). 'The Upsilon at DORIS'. *18*, 202–4.

CERN Courier (1981a). 'Looking for Proton Decay'. *21*, 195–6.

CERN Courier (1981b). 'Hunting the Unstable Proton'. *21*, 253–4.

CERN Courier (1982a). 'CERN: First Results at 540 GeV Total Energy'. *22*, 3.

CERN Courier (1982b). 'LEP Authorization'. *22*, 20.

CERN Courier (1982c). 'The Tilting of LEP'. *22*, 61–2.

CERN Courier (1982d). 'Fermilab Moves on From 400 GeV'. *22*, 316–18.

CERN Courier (1983a). 'First Signs of the W'. *23*, 43–4.

CERN Courier (1983b). 'US Science Underground'. *23*, 49–51.

CERN Courier (1983c). 'Fast Work'. *23*, 82–5.

CERN Courier (1983d). 'HERA Clearer'. *23*, 90.

Bibliography

Chang, C. *et al.* (1975). 'Observed Deviations from Scale Invariance in High-Energy Muon Scattering'. *PRL, 35*, 901 4.

Chanowitz, M.S. (1981). 'Have We Seen Our First Glueball?' *PRL, 46*, 981 4.

Chanowitz, M.S. and Ellis, J. (1972). 'Canonical Anomalies and Broken Scale Invariance'. *PL, 40B*, 397–400.

Charpak, G. (1970). 'Evolution of the Automatic Spark Chamber'. *ARNS, 20*, 195–254.

Charpak, G. (1978). 'Multiwire and Drift Proportional Chambers'. *Physics Today, 31* (10), 23–31.

Chew, G.F. (1964a). *The Analytic S Matrix*. New York and Amsterdam: W.A.Benjamin.

Chew, G.F. (1964b). 'Elementary Particles?' *Physics Today, 17*(4), 30–4.

Chew, G.F. (1968). '"Bootstrap": A Scientific Idea?' *Science, 161*, 762–5.

Chew, G.F. (1970). 'Hadron Bootstrap: Triumph or Frustration?' *Physics Today, 23*(10), 23–8.

Chew, G.F. and Frautschi, S.C. (1961a). 'Principle of Equivalence for All Strongly Interacting Particles within the S-Matrix Formalism', *PRL, 8*, 394–7.

Chew, G.F. and Frautschi, S.C. (1961b). 'Dynamical Theory for Strong Interactions at Low Momentum Transfers but Arbitrary Energies'. *PR, 123*, 1478–86.

Chew, G.F., Frautschi, S.C. and Mandelstam, S. (1962). 'Regge Poles in π-π Scattering'. *PR, 126*, 1202–8.

Chew, G.F., Gell-Mann, M. and Rosenfeld, A.H. (1964). 'Strongly Interacting Particles'. *Scientific American, 210*(2), 74–93.

Chew, G.F. and Rosenzweig, C. (1978). 'Dual Topological Unitarization: An Ordered Approach to Hadron Theory'. *Phys. Rept., 41*, 263–327.

Chodos, A., Jaffe, R.L., Johnson, K., Thorn, C.B. and Weisskopf, V.F. (1974). 'New Extended Model of Hadrons'. *PR, D9*, 3471–95.

Christenson, J.H., Cronin, J.W., Fitch, V.L. and Turlay, R. (1964). 'Evidence for the 2π Decay of the K_2^0 Meson'. *PRL, 13*, 138–40.

Christenson, J.H. *et al.* (1970). 'Observation of Massive Muon Pairs in Hadron Collisions'. *PRL, 25*, 1523–6.

Christenson, J.H. *et al.* (1973). 'Observation of Muon Pairs in High-Energy Hadron Collisions'. *PR, D8*, 2016–34.

Cline, D. (ed.) (1978). *Proceedings of the Seminar on Proton Stability*, University of Wisconsin, 8 Dec. 1978. Unpublished; no page numbers.

Cline, D. and Fry, W.F. (1977). 'Neutrino Scattering and New-Particle Production'. *ARNS, 27*, 209–78.

Cline, D.B., Mann, A.K. and Rubbia, C. (1974). 'The Detection of Neutral Weak Currents'. *Scientific American, 231*(6), 108–19.

Cline, D.B., Mann, A.K. and Rubbia, C. (1976). 'The Search for New Families of Elementary Particles'. *Scientific American, 234*(1), 44–63.

Cline, D. and Rubbia, C. (1980). 'Antiproton-Proton Colliders and Intermediate Bosons'. *Physics Today, 33*(8), 44–52.

Cline, D.B., Rubbia, C. and van der Meer, S. (1982). 'The Search for Intermediate Vector Bosons'. *Scientific American, 246*(3), 38–49.

Close, F.E. (1976). 'Parity Violation in Atoms?' *Nature, 264*, 505–6.

Bibliography

Close, F.E. (1978). 'A Massive Particle Conference'. *Nature, 275*, 267.

Close, F.E. (1979). *An Introduction to Quarks and Partons*. London, New York and San Francisco: Academic Press.

Coleman, S. (1973). 'Dilations'. In A.Zichichi (ed.), *Properties of the Fundamental Interactions*, Proceedings of the Erice Summer School, 1973, 359–99. Bologna: Editrici Compositori Bologna.

Coleman, S. (1979). 'The 1979 Nobel Prize in Physics'. *Science, 206*, 1290–2.

Coleman, S. and Gross, D.J. (1973). 'Price of Asymptotic Freedom'. *PRL, 31*, 851–4.

Coleman, S. and Weinberg, E. (1973). 'Radiative Corrections as the Origin of Spontaneous Symmetry Breaking'. *PR, D7*, 1889–98.

Collins, G.B. (1953). 'Scintillation Counters'. *Scientific American, 189*(5), 36–41.

Collins, H.M. (1974). 'The TEA Set: Tacit Knowledge and Scientific Networks'. *Science Studies, 4*, 165–86.

Collins, H.M. (1975a). 'The Seven Sexes: A Study in the Sociology of a Phenomenon, or the Replication of Experiments in Physics'. *Sociology, 9*, 205–24.

Collins, H.M. (1975b). 'Building a TEA Laser: The Caprices of Communication'. *Social Studies of Science, 5*, 441–50.

Collins, H.M. (ed.) (1981a). *Knowledge and Controversy: Studies of Modern Natural Science*, Special Issue of *Social Studies of Science, 11*, 1–158.

Collins, H.M. (1981b). 'What is TRASP? The Radical Programme as Methodological Imperative'. *Philosophy of the Social Sciences, 11*, 215–24.

Collins, H.M. (1981c). 'Stages in the Empirical Programme of Relativism'. *Social Studies of Science, 11*, 3–10.

Collins, H.M. and Pinch, T.J. (1982). *Frames of Meaning: The Social Construction of Extraordinary Science*. London: Routledge and Kegan Paul.

Collins, P.D.B. (1971). 'Regge Theory and Particle Physics'. *Phys. Rept., 1*, 103–234.

Collins, P.D.B. and Martin, A.D. (1982). 'Hadron Reaction Mechanisms'. *RPP, 45*, 335–426.

Collins, P.D.B. and Squires, E.J. (1968). *Regge Poles in Particle Physics*, Springer Tracts in Modern Physics, Vol. 45. Berlin: Springer-Verlag.

Combley, F.H. (1979). '(g-2) Factors for Muon and Electron and the Consequences for QED'. *RPP, 42*, 1889–935.

Commins, E.D. and Bucksbaum, P.H. (1980). 'The Parity Non-Conserving Electron-Nucleon Interaction'. *ARNPS, 30*, 1–52.

Connolly, P.L. *et al.* (1963). 'Existence and Properties of the ϕ Meson'. *PRL, 10*, 371–4.

Conti, R. *et al.* (1979). 'Preliminary Observation of Parity Nonconservation in Atomic Thallium'. *PRL, 42*, 343–6.

Copley, L.A., Karl, G. and Obryk, E. (1969a). 'Backward Single Pion Photoproduction and the Symmetric Quark Model'. *PL, 29B*, 117–20.

Copley, L.A., Karl, G. and Obryk, E. (1969b). 'Single Pion Photoproduction in the Quark Model'. *NP, B13*, 303–19.

426

Bibliography

Corcoran, M.D. *et al.* (1978). 'Comparison of High-p_T Events Produced by Pions and Protons'. *PRL, 41*, 9–12.

Courant, H. *et al.* (1963). 'Determination of the Relative Σ-Λ Parity'. *PRL, 10*, 409 12.

Coward, D.H. *et al.* (1968). 'Electron-Proton Elastic Scattering at High Momentum Transfers'. *PRL, 20*, 292–5.

Crane, D. (1980a). 'An Exploratory Study of Kuhnian Paradigms in Theoretical High Energy Physics'. *Social Studies of Science, 10*, 23–54.

Crane, D. (1980b). 'Reply to Pickering'. *Social Studies of Science, 10*, 502–6.

Creutz, M. (1981). 'Roulette Wheels and Quark Confinement'. *CNPP, 10*, 163–73.

Cronin, J.W. (1981). 'CP Symmetry Violation: The Search for Its Origin'. *Science, 212*, 1221–8.

Cundy, D.C. *et al.* (1970). 'Upper Limits for Diagonal and Neutral Current Couplings in the CERN Neutrino Experiments'. *PL, 31B*, 478–80.

Cushing, J.T. (1982). 'Models and Methodologies in Current Theoretical High-Energy Physics'. *Synthese, 50*, 5–101.

Dalitz, R.H. (1966a). 'Resonant States and Strong Interactions'. In *Proceedings of the International Conference on Elementary Particles*, Oxford, 1965, 157–81. Chilton: Rutherford Laboratory.

Dalitz, R.H. (1966b). 'Quark Models for the "Elementary Particles" '. In C. DeWitt and M.Jacob (eds.), *High Energy Physics*, 251–324. New York: Gordon and Breach.

Dalitz, R.H. (1967). 'Symmetries and the Strong Interactions'. In *Proceedings of the 13th International Conference on High Energy Physics*, 215–34. Berkeley: University of California Press.

Dalitz, R.H. (1982). 'Quarks and the Light Hadrons'. *PPNP, 8*, 7–48.

Danby, G. *et al.* (1962). 'Observation of High-Energy Neutrino Reactions and the Existence of Two Kinds of Neutrinos'. *PRL, 9*, 36–44.

Darden, C.W. *et al.* (1978a). 'Observation of a Narrow Resonance at 9.46 GeV in Electron-Positron Annihilations'. *PL, 76B*, 246–8.

Darden, C.W. *et al.* (1978b). 'Evidence for a Narrow Resonance at 10.01 GeV in Electron-Positron Annihilations'. *PL, 78B*, 364–5.

Darriulat, P. (1980). 'Large Transverse Momentum Hadronic Processes'. *ARNPS, 30*, 159–210.

Darriulat, P. (1983). 'Preliminary Searches for Hadron Jets and for Large Transverse Momentum Electrons at the SPS Collider'. In Bacci and Salvini (1983), 190–236.

Davier, M. (1979). 'e^+e^- Physics: Heavy Quark Spectroscopy'. In EPS (1979), 191–219.

De Camp, D. (1982). 'Lepton Pairs Versus Drell–Yan Mechanism'. In Trower and Bellini (1982), 311–44.

Deden, H. *et al.* (1975a). 'Experimental Study of Structure Functions and Sum Rules in Charge-Changing Interactions of Neutrinos and Antineutrinos'. *NP, B85*, 269–88.

Deden, H. *et al.* (1975b). 'Strange Particle Production and Charmed Particle Search in the Gargamelle Neutrino Experiment'. *PL, 58B*, 361–6.

De Grand, T.A., Jaffe, R.L., Johnson, K. and Kiskis, J. (1975). 'Masses and Other Parameters of the Light Hadrons'. *PR, D12*, 2060–76.

427

Bibliography

Derrick, M. (1970). 'The Bubble-Chamber Technique: Recent Developments and Some Possibilities for the Future', *Progress in Nuclear Physics, 11,* 223–69.

De Rújula, A. (1968). 'Algebra de Corrientes Sobra la Capa de Masas y Difusión Mesón Barión'. PhD thesis, University of Madrid, unpublished.

De Rújula, A. (1971). 'An Elementary Introduction to Deep Inelastic Electroproduction and the Parton Model'. *Journal de Physique, 32,* C3, 77–86.

De Rújula, A. (1974a). 'Proton Magnetic Form Factor in Asymptotically Free Field Theories'. *PRL, 32,* 1143–5.

De Rújula, A. (1974b). 'Lepton Physics and Gauge Theories', In Smith (1974), IV-90-4.

De Rújula, A. (1976). 'Plenary Report on Theoretical Basis of New Particles'. In Bogoliubov *et al.* (1976), Vol. II, N111–27.

De Rújula, A. (1979). 'Quantum Chromo Dynamite'. In EPS (1979), 418–41.

De Rújula, A. and De Rafael, E. (1970). 'Unitarity Bounds to T-Odd Correlations in Neutrino Reactions'. *PL, 32B,* 495–8.

De Rújula, A., Ellis, J., Petronzio, R., Preparata, G. and Scott, W. (1982). 'Can One Tell QCD from a Hole in the Ground?: A Drama in Five Acts'. In Zichichi (1982), 567–673.

De Rújula, A., Georgi, H. and Glashow, S.L. (1975). 'Hadron Masses in a Gauge Theory'. *PR, D12,* 147–62.

De Rújula, A., Georgi, H. and Glashow, S.L. (1976). 'Is Charm Found?' *PRL, 37,* 398–401.

De Rújula, A., Georgi, H., Glashow, S.L. and Quinn, H. (1974). 'Fact and Fancy in Neutrino Physics'. *RMP, 46,* 391–407.

De Rújula, A., Georgi, H. and Politzer, H.D. (1974). 'The Breakdown of Scaling in Electron and Neutrino Scattering'. *PR, D10,* 2041–146.

De Rújula, A. and Glashow, S.L. (1975). 'Is Bound Charm Found?' *PRL, 34,* 46–9.

De Rújula, A., Lautrup, B. and Peterman, A. (1970). 'On Sixth-Order Corrections to the Anomalous Magnetic Moment of the Electron'. *PL, 33B,* 605–6.

Deutsch, M. (1951). 'Evidence for the Formation of Positronium in Gases'. *PR, 82,* 455–6.

Deutsch, M. (1958). 'Evidence and Inference in Nuclear Research'. *Daedalus* (Fall 1958), 88–98.

Deutsch, M. (1975). 'Particle Discovery at Brookhaven'. *Science, 189,* 750, 816.

DeWitt, B.S. (1964). 'Theory of Radiative Corrections for Non-Abelian Gauge Fields'. *PRL, 12,* 742–6.

DeWitt, B.S. (1967a). 'Quantum Theory of Gravity II. The Manifestly Covariant Theory'. *PR, 162,* 1195–239.

DeWitt, B.S. (1967b). 'Quantum Theory of Gravity III. Applications of the Covariant Theory'. *PR, 162,* 1239–56.

Dokshitzer, Yu. L., Dyakonov, D.I. and Troyan, S.I. (1979). 'Hard Processes in Quantum Chromodynamics'. *Phys. Rept., 58,* 269–395.

428

Bibliography

Dolgov, A.D. and Zeldovich, Ya. B. (1981). 'Cosmology and Elementary Particles'. *RMP*, *53*, 1–41.

Dombey, N. (1979). 'The Road to Unification'. *Nature*, *282*, 131–2.

Donoghue, J.F. (1982). 'Glueballs'. *CNPP*, *10*, 277–85.

Drell, S.D. (1975). 'Electron-Positron Annihilation and the New Particles'. *Scientific American*, *232*(6), 50–62.

Drell, S.D. (1977). 'Elementary Particle Physics'. *Daedalus*, *1*, 15–31.

Drell, S.D. (1978). 'When Is a Particle?' *Physics Today*, 31(6), 23–32.

Drell, S.D., Levy, D.J. and Yan, T.-M. (1969). 'A Theory of Deep Inelastic Lepton-Nucleon Scattering and Lepton Pair Annihilation Processes, III: Deep Inelastic Electron-Positron Annihilation'. *PR*, 1617–39.

Drell, S.D. and Walecka, J.D. (1964). 'Electrodynamic Processes with Nuclear Targets'. *AP*, *28*, 18–33.

Drell, S.D. and Yan, T.-M. (1970). 'Massive Lepton-Pair Production in Hadron-Hadron Collisions at High Energies'. *PRL*, *25*, 316–20.

Duhem, P. (1954). *The Aim and Structure of Physical Theory*. Princeton: Princeton University Press.

Duinker, P. (1982). 'Review of e^+e^- Physics at PETRA'. *RMP*, *54*, 325–87.

Duinker, P. and Luckey, D. (1980). 'In Search of Gluons'. *CNPP*, *9*, 123–40.

Dydak, F. (1979). 'Neutral Currents'. In EPS (1979), 25–49.

Dyson, F.J. (ed.) (1966). *Symmetry Groups in Nuclear and Particle Physics: A Lecture Note and Reprint Volume*. New York and Amsterdam: W.A. Benjamin.

Dyson, F.J. (1979). *Disturbing the Universe*. New York: Harper and Row.

ECFA (1977). 'Recommendation by Plenary ECFA to the Scientific Policy Committee'. ECFA Report 77/21, 25 May 1977.

ECFA (1980). *Study on the Proton-Electron Storage Ring Project HERA*. ECFA Report 80/42.

ECFA Working Group (1979). 'First Report on High Energy Physics Activities in the CERN Member States'. CERN preprint ECFA/RC/79/47.

ECFA/CERN (1979). *Proceedings of the LEP Summer Study*, Les Houches and CERN, 10–22 Sept. 1978. CERN Yellow Report 79-01, 2 Vols.

Eden, R.J., Landshoff, P.V., Olive, D.I. and Polkinghorne, J.C. (1966). *The Analytic S-Matrix*. Cambridge: University Press.

Eichten, E. and Gottfried, K. (1977). 'Heavy Quarks in e^+e^- Annihilation'. *PL*, *66B*, 286–90.

Eichten, E., Gottfried, K., Kinoshita, T., Kogut, J., Lane, K.D. and Yan, T.-M. (1975). 'Spectrum of Charmed Quark-Antiquark Bound States'. *PRL*, *34*, 369–72.

Eichten, E., Gottfried, K., Kinoshita, T., Lane, K.D. and Yan, T.-M. (1980).'Charmonium: Comparison with Experiment'. *PR*, *D21*, 203–33.

Eichten, T. *et al.* (1973a). 'Measurement of the Neutrino-Nucleon and Antineutrino-Nucleon Total Cross Sections'. *PL*, *46B*, 274–80.

Eichten, T. *et al.* (1973b). 'High Energy Electronic Neutrino (v_e) and Antineutrino ($\bar{v}_e$) Interactions'. *PL*, *46B*, 281–4.

Ellis, J. (1970). 'Scale and Chiral Symmetry Breaking'. *PL*, *33B*, 591–5.

Ellis, J. (1971). 'Approximate Scale and Chiral Invariance'. *Proceedings of*

Bibliography

the Conference on Fundamental Interactions at High Energy, Coral Gables, 20–22 Jan. 1971, Vol. 2, 77–98. New York: Gordon and Breach.

Ellis, J. (1974). 'Theoretical Ideas about $e^+e^-\to$ Hadrons at High Energies'. In Smith (1974), IV-20–35.

Ellis, J. (1979). 'su(5)'. In EPS (1979), 940–50.

Ellis, J. (1981). 'Gluons'. *CNPP, 9*, 153–68.

Ellis, J., Finkelstein, J., Frampton, P.H. and Jacob, M. (1971). 'Duality and Inclusive Phenomenology'. *PL, 35B*, 227–230.

Ellis, J., Gaillard, M.K., Girardi, G. and Sorba, P. (1982). 'Physics of Intermediate Vector Bosons'. *ARNPS, 32*, 443–97.

Ellis, J., Gaillard, M.K. and Nanopoulos, D.V. (1975). 'On the Weak Decays of High-Mass Hadrons'. *NP, B100*, 313–28.

Ellis, J., Gaillard, M.K. and Nanopoulos, D.V. (1979). 'Baryon Number Generation in Grand Unified Theories'. *PL, 80B*, 360–4.

Ellis, J., Gaillard, M.K. and Ross, G.G. (1976). 'Search for Gluons in e^+e^- Annihilation'. *NP, B111*, 253–71.

Ellis, J. and Jaffe, R.L. (1973). 'Scaling, Short Distances and the Light Cone'. Lectures at the U.C. Santa Cruz Summer School on Particle Physics. SLAC-PUB-1353.

Ellis, J. and Nanopoulos, D. (1983). 'Particle Physics and Cosmology'. *CERN Courier, 23*, 211–16.

Ellis, J. and Renner, B. (1969). '$\rho\to4\pi$ Decay from Current Algebra'. *NP, B13*, 108–12.

Ellis, J. and Sachrajda, C.T. (1980). 'Quantum Chromodynamics and Its Applications'. In Levy *et al.* (1980), 285–432.

Ellis, J., Weisz, P.H. and Zumino, B. (1971). 'The Dimension of Scale Symmetry Breaking'. *PL, 34B*, 91–4.

Ellis, S.D. and Stroynowski, R. (1977). 'Large p_T Physics: Data and the Constituent Models'. *RMP, 49*, 753–75.

Englert, F. and Brout, R. (1964). 'Broken Symmetry and the Mass of Gauge Vector Mesons'. *PRL, 13*, 321–3.

Englert, F., Brout, R. and Thiry, M.F. (1966). 'Vector Mesons in Presence of Broken Symmetry'. *NC, 43*, 244–57.

EPS (European Physical Society) (1979). *Proceedings of the International Conference on High Energy Physics*, Geneva, 27 June-4 July 1979. Geneva: CERN.

Experimental Physics Division (1983). 'PS Neutrino Oscillation Facility'. In *CERN Annual Report 1982*, 43–4.

Fabjan, C.W. and Fischer, H.G. (1980). 'Particle Detectors'. *RPP, 43*, 1003–63.

Fabjan, C.W. and Ludlam, T. (1982). 'Calorimetry in High-Energy Physics'. *ARNPS, 32*, 335–89.

Fadeev, L.D. (1969). 'The Feynman Integral for Singular Lagrangians'. *TMP, 1*, 1–13.

Fadeev, L.D. and Popov, V.N. (1967). 'Feynman Diagrams for the Yang–Mills Fields'. *PL, 25B*, 29–30.

Faiman, D. and Hendry, A.W. (1968). 'Harmonic-Oscillator Model for Baryons'. *PR, 173*, 1720–9.

430

Bibliography

Faiman, D. and Hendry, A.W. (1969). 'Electromagnetic Decays of Baryon Resonances in the Harmonic-Oscillator Model'. *PR, 180*, 1572–7.

Faissner, H., Reithler, H. and Zerwas, P. (eds.) (1977). *Proceedings of the International Neutrino Conference, Aachen, 1976*. Vieweg: Braunschweig.

Farhi, E. and Susskind, L. (1981). 'Technicolour'. *Phys. Rept., 74*, 277 321.

Farley, F.J.M. and Picasso, E. (1979). 'The Muon (g-2) Experiments'. *ARNPS, 29*, 243–82.

Fayet, P. and Ferrara, S. (1977). 'Supersymmetry'. *Phys. Rept., 32*, 249–334.

Feldman, G.J. and Perl, M.L. (1977). 'Recent Results in Electron-Positron Annihilation above 2 GeV'. *Phys. Rept., 33*, 285–365.

Feldman, G.J. *et al.* (1977). 'Observation of the Decay $D^{*+} \rightarrow D^0 \pi^+$'. *PRL, 38*, 1313–16.

Fermi, E. and Yang, C.N. (1949). 'Are Mesons Elementary Particles?' *PR, 76*, 1739–43.

Ferro-Luzzi, M. (1981). 'Prospects in Baryon Spectroscopy'. In Isgur (1981), 415–38.

Feyerabend, P.K. (1975). *Against Method*. London: New Left Books.

Feyerabend, P.K. (1978). *Science in a Free Society*. London: New Left Books.

Feynman, R.P. (1963). 'Quantum Theory of Gravitation'. *Acta Physica Polonica, 24*, 697–722.

Feynman, R.P. (1966). 'The Development of the Space-Time View of Quantum Electrodynamics'. *Science, 153*, 699–708.

Feynman, R.P. (1969a). 'Very High-Energy Collisions of Hadrons'. *PRL, 23*, 1415–17.

Feynman, R.P. (1969b). 'The Behavior of Hadron Collisions at Extreme Energies'. In C.N.Yang *et al.* (eds.), *High Energy Collisions*, 237–58. New York: Gordon and Breach.

Feynman, R.P. (1972). *Photon-Hadron Interactions*. Reading, MA: W.A. Benjamin.

Feynman, R.P. (1974). 'Structure of the Proton'. Address given at Dansk Ingeniørforening, Copenhagen, Denmark, 8 Oct. 1973. Reprinted in *Science, 183*, 601–10.

Feynman, R.P. and Gell-Mann, M. (1958). 'Theory of the Fermi Interaction'. *PR, 109*, 193–8.

Feynman, R.P. and Hibbs, A.R. (1965). *Quantum Mechanics and Path Integrals*. New York: McGraw-Hill.

Field, R.D. (1979). 'Dynamics of High Energy Reactions'. In Homma *et al.* (1979), 743–73.

Field, R.D. and Feynman, R.P. (1978). 'A Parametrization of the Properties of Quark Jets'. *NP, B136*, 1–76.

Fisk, H.E. and Sciulli, F. (1982). 'Charged-Current Neutrino Interactions'. *ARNPS, 32*, 499–573.

Fiske, M. (1979). *The Transition in Physics Doctoral Employment 1960–1990*. New York: American Physical Society.

Fitch, V.L. (1981). 'The Discovery of Charge Conjugation-Parity Asymmetry'. *Science, 212*, 989–93.

431

Bibliography

Fleck, L. (1979). *Genesis and Development of a Scientific Fact*. Chicago and London: University of Chicago Press.

Fowler, W.B. and Samios, N.P. (1964). 'The Omega-Minus Experiment'. *Scientific American, 211*(4), 36–45.

Fradkin, E.S. and Tyutin, I.V. (1970). 'S Matrix for Yang–Mills and Gravitational Fields'. *PR, D2,* 2841–57.

Frampton, P.H. (1974). *Dual Resonance Models*. New York and Amsterdam: W.A.Benjamin.

Frampton, P.H. and Vogel, P. (1982). 'Massive Neutrinos'. *Phys. Rept., 82,* 339–88.

Franklin, A. (1979). 'The Discovery and Nondiscovery of Parity Nonconservation'. *Studies in History and Philosophy of Science, 10,* 201–57.

Franklin, A. (1983). 'The Discovery and Acceptance of CP Violation'. *Historical Studies in the Physical Sciences, 13,* 207–38.

Franzinetti, C. (ed.) (1963). 'The 1963 NPA Seminar: The Neutrino Experiment'. CERN Yellow Report CERN 63-37.

Franzinetti, C. (1974). 'Total ν and $\bar{\nu}$ Cross Sections and Inelastic Processes'. In Rollnik and Pfeil (1974), 353–67.

Franzini, P. and Lee-Franzini, J. (1982). 'Upsilon Physics at CESR'. *Phys. Rept., 81,* 239–91.

Freedman, C.Z. and van Nieuwenhuizen, P. (1978). 'Supergravity and the Unification of the Laws of Physics'. *Scientific American, 238*(2), 126–43.

Friedman, J.I. and Kendall, H.W. (1972). 'Deep Inelastic Electron Scattering'. *ARNS, 22,* 203–54.

Frishman, Y. (1974). 'Light Cone and Short Distances'. *Phys. Rept., 13,* 1–52.

Fritzsch, H. and Gell-Mann, M. (1971a). 'Scale Invariance and the Light Cone'. In M.dal Cin, G.J.Iverson, and A.Pcrlmutter (eds), *Fundamental Interactions at High Energy*, Proceedings of the Coral Gables Conference, 20–22 January 1971, Miami, Florida. New York: Gordon and Breach, 1–42.

Fritzsch, H. and Gell-Mann, M. (1971b). 'Light Cone Current Algebra'. In E.Grotsman (ed.), *Proceedings of the International Conference on Duality and Symmetry in Hadron Physics,* 5–7 April 1971, Tel Aviv, 317–74. Jerusalem: Weizmann Science Press of Israel.

Fritzsch, H. and Gell-Mann, M. (1972). 'Current Algebra: Quarks and What Else?' In Jackson and Roberts (1972), Vol. 2, 135–65.

Fritzsch, H., Gell-Mann, M. and Leutwyler, H. (1973). 'Advantages of the Color Octet Gluon Picture'. *PL, 47B,* 365–8.

Fritzsch, H. and Minkowski, P. (1975). 'Unified Interactions of Leptons and Hadrons'. *AP, 93,* 193–266.

Fry, W.F. and Haidt, D. (1975). 'Calculation of the Neutron-Induced Background in the Gargamelle Neutral Current Search'. CERN Yellow Report 75-1.

Fubini, S. and Furlan, G. (1965). 'Renormalization Effects for Partially Conserved Currents'. *Physics, 1,* 229–47.

Gabathuler, E. (1979). 'The SPS Programme beyond 1982'. In Landshoff (1979), 72–7.

432

Bibliography

Gaillard, M.K. (1965). 'On the Possibility of CP Violation in the K→3π Decay'. *NC, 35,* 1225–30.

Gaillard, M.K. (1968). 'Contributions a l'Étude des Interactions Faible'. Thèse de Doctorat, University of Paris, Orsay, unpublished.

Gaillard, M.K. (1972). 'Theory of Weak Interactions (Phenomenology)'. In Jackson and Roberts (1972), Vol. 2, 239–49.

Gaillard, M.K. (1974). 'Gauge Theories and Weak Interactions'. In Smith (1974), III-76–80.

Gaillard, M.K. (1979). 'QCD Phenomenology'. In EPS (1979), 390–417.

Gaillard, M.K. and Lee, B.W. (1974a). 'Rare Decay Modes of the K-Mesons in Gauge Theories'. *PR, D10,* 897–916.

Gaillard, M.K. and Lee, B.W. (1974b). '$\Delta I = \frac{1}{2}$ Rule for Non-Leptonic Decays in Asymptotically Free Field Theories'. *PRL, 33,* 108–11.

Gaillard, M.K., Lee, B.W. and Rosner, J.L. (1975). 'Search for Charm'. *RMP, 47,* 277–310.

Galison, P. (1982). 'Theoretical Predispositions in Experimental Physics: Einstein and the Gyromagnetic Experiments'. *Historical Studies in the Physical Sciences, 12,* 285–323.

Galison, P. (1983). 'How the First Neutral Current Experiment Ended'. *RMP, 55,* 477–509.

Ganguli, S.N. and Roy, D.P. (1980). 'Regge Phenomenology of Inclusive Reactions'. *Phys. Rept., 67,* 201–395.

Garelick, D.A. *et al.* (1978). 'Confirmation of an Enhancement in the $\mu^+\mu^-$ Mass Spectrum at 9.5 GeV'. *PR, D18,* 945–7.

Garwin, E.L., Pierce, D.T. and Siegmann, H.C. (1974). 'Polarized Photoelectrons from Optically Magnetized Semiconductors'. *Helvetica Physica Acta, 47,* 393.

Gasiorowicz, S. and Rosner, J.L. (1981). 'Hadron Spectra and Quarks'. *AJP, 49,* 954–84.

Gastaldi, U. and Klapisch, R. (1981). 'The LEAR Project and Physics with Low-Energy Antiprotons'. In *From Nuclei to Particles.* Bologna: Soc. Italiana di Fisica, 426–503.

Gaston, J. (1973). *Originality and Competition in Science: A Study of the British High Energy Physics Community.* Chicago and London: University of Chicago Press.

Gatto, R. (ed.) (1973). *Scale and Conformal Symmetry in Hadron Physics.* New York: Wiley.

Gell-Mann, M. (1953). 'Isotopic Spin and New Unstable Particles'. *PR, 92,* 833–4.

Gell-Mann, M. (1956). 'The Interpretation of the New Particles as Displaced Charge Multiplets'. *NC, 4,* Suppl. 2, 848–66.

Gell-Mann, M. (1961). 'The Eightfold Way: A Theory of Strong Interaction Symmetry'. Caltech Synchrotron Laboratory Report CTSL-20. Reprinted in Gell-Mann and Ne'eman (1964), 11–57.

Gell-Mann, M. (1962a). 'Strange Particle Physics. Strong Interactions'. *Proceedings of the International Conference on High Energy Physics,* 805. Geneva: CERN.

Gell-Mann, M. (1962b). 'Symmetries of Baryons and Mesons'. *PR, 125,* 1067–84.

433

Bibliography

Gell-Mann, M. (1964a). 'A Schematic Model of Baryons and Mesons'. *PL, 8*, 214–15.

Gell-Mann, M. (1964b). 'The Symmetry Group of Vector and Axial Vector Currents'. *Physics, 1*, 63–75.

Gell-Mann, M. (1968). 'Summary of the Symposium'. In Svartholm (1968), 387–99.

Gell-Mann, M. (1972). 'General Status: Summary and Outlook'. In Jackson and Roberts (1972), Vol. 4, 333–56.

Gell-Mann, M. and Lévy, M. (1960). 'The Axial Vector Current in Beta Decay'. *NC, 16*, 705–25.

Gell-Mann, M. and Low, F.E. (1951). 'Bound States in Quantum Field Theory'. *PR, 84*, 350–4.

Gell-Mann, M. and Low, F.E. (1954). 'Quantum Electrodynamics at Small Distances'. *PR, 95*, 1300–12.

Gell-Mann, M. and Ne'eman, Y. (eds.) (1964). *The Eightfold Way*. New York and Amsterdam: W.A.Benjamin.

Gell-Mann, M., Ramond, P. and Slansky, R. (1978). 'Color Embeddings, Charge Assignments, and Proton Stability in Unified Gauge Theories'. *RMP, 50*, 721–44.

Gell-Mann, M. and Rosenbaum, E.P. (1957). 'Elementary Particles'. *Scientific American, 197*(1), 72–88.

Georgi, H. (1981). 'A Unified Theory of Elementary Particles and Forces'. *Scientific American, 244*(4), 40–55.

Georgi, H. and Glashow, S.L. (1972). 'Unified Weak and Electromagnetic Interactions without Neutral Currents'. *PRL, 28*, 1494–7.

Georgi, H. and Glashow, S.L. (1974). 'Unity of All Elementary-Particle Forces'. *PRL, 32*, 438–41.

Georgi, H. and Politzer, H.D. (1974). 'Electroproduction Scaling in an Asymptotically Free Theory of Strong Interactions'. *PR, D9*, 416–20.

Georgi, H., Quinn, H.R. and Weinberg, S. (1974). 'Hierarchy of Interactions in Unified Gauge Theories'. *PRL, 33*, 451–4.

Gervais, J.L. and Neveu, A. (eds.) (1976). 'Extended Systems in Field Theory'. Proceedings of the Meeting Held at Ecole Normale Supérieure, Paris, 16–21 June, 1975. *Phys. Rept., 23*, 237–74.

Giacomelli, G., Greene, A.F. and Sanford, J.R. (1975). 'A Survey of the Fermilab Research Program'. *Phys. Rept., 19*, 169–232.

Giacomelli, G. and Jacob, M. (1979). 'Physics at the CERN-ISR'. *Phys. Rept., 55*, 1–132. Reprinted in Jacob (1981), 217–348.

Giacomelli, G. and Jacob, M. (1981). 'Physics at the CERN-ISR (1980 Updating)'. In Jacob (1981), 349–68.

Gilbert, W. (1964). 'Broken Symmetries and Massless Particles'. *PRL, 12*, 713–14.

Gilman, F.J. (1972). 'Photoproduction and Electroproduction'. *Phys. Rept., 4*, 95–151.

Gilman, F.J. (1974). 'Deep Inelastic Scattering and the Structure of Hadrons'. In Smith (1974), III-149–71.

Gilman, F.J. and Schnitzer, H.J. (1966). 'Symmetry Predictions from Sum Rules without Saturation'. *PR, 150*, 1362–71.

434

Bibliography

Glashow, S.L. (1959). 'The Renormalizability of Vector Meson Interactions'. *NP*, *10*, 107–17.

Glashow, S.L. (1961). 'Partial Symmetries of Weak Interactions'. *NP*, *22*, 579–88.

Glashow, S.L. (1974). 'Charm: An Invention Awaits Discovery'. In D.A. Garelick (ed.), *Experimental Meson Spectroscopy 1974*, 387–92. New York: American Institute of Physics.

Glashow, S.L. (1975). 'Quarks with Color and Flavor'. *Scientific American*, *233*(4), 38–50.

Glashow, S.L. (1980). 'Towards a Unified Theory: Threads in a Tapestry'. *RMP*, *52*, 539–43.

Glashow, S.L. (1981). 'Old and New Directions in Elementary Particle Physics'. In Ne'eman (1981), 160–70.

Glashow, S.L. and Gell-Mann, M. (1961). 'Gauge Theories of Vector Particles'. *AP*, *15*, 437–60.

Glashow, S.L., Iliopoulos, J. and Maiani, L. (1970). 'Weak Interactions with Lepton-Hadron Symmetry'. *PR*, *D2*, 1285–92.

Glashow, S.L., Schnitzer, H.J. and Weinberg, S. (1967a). 'Sum Rules for the Spectral Functions of $su(3) \times su(3)$'. *PRL*, *19*, 139–42.

Glashow, S.L., Schnitzer, H.J. and Weinberg, S. (1967b). 'Convergent Calculation of Nonleptonic K Decay in the Intermediate-Boson Model'. *PRL*, *19*, 205–208.

Goldberg, H. and Ne'eman, Y. (1963). 'Baryon Charge and R-Inversion in the Octet Model'. *NC*, *27*, 1–5.

Goldberger, M.L. and Treiman, S.B. (1958). 'Decay of the Pi Meson'. *PR*, *110*, 1178–84.

Goldhaber, G. (1976). 'One Researcher's Personal Account'. *Adventures in Experimental Physics*, *5*, 131–40.

Goldhaber, G. (1977). 'The Case for Charmed Mesons'. *CNPP*, *7*, 97–105.

Goldhaber, G. and Pierre, F.M. (1976). 'Evidence for a $K^{\pm} \pi^{\mp}$ State at 1.87 GeV Obtained from the SPEAR Data from 3.9 to 4.6 GeV'. LBL Physics Notes, TG-269, 5 May 1976, unpublished.

Goldhaber, G.,Pierre, F.M. *et al.* (1976). 'Observation in e^+e^- Annihilation of a Narrow State at 1865 MeV/c^2 Decaying to $K\pi$ and $K\pi\pi\pi$'. *PRL*, *37*, 255–9.

Goldhaber, G. and Wiss, J.E. (1980). 'Charmed Mesons Produced in e^+e^- Annihilation'. *ARNPS*, *30*, 337–81.

Goldhaber, M. (1956). 'Compound Hypothesis for the Heavy Unstable Particles. II'. *PR*, *101*, 433–8.

Goldhaber, M., Langacker, P. and Slansky, R. (1980). 'Is the Proton Stable?' *Science*, *210*, 851–60.

Goldhaber, M. and Sulak, L.R. (1981). 'An Overview of Current Experiments in Search of Proton Decay'. *CNPP*, *10*, 215–25.

Goldman, T.J. and Ross, D.A. (1979). 'A New Estimate of the Proton Lifetime'. *PL*, *84B*, 208–10.

Goldman, T.J. and Ross, D.A. (1980). 'How Accurately Can We Estimate the Proton Lifetime in an $su(5)$ Grand Unified Model?' *NP*, *B171*, 273–300.

435

Bibliography

Goldsmith, M. and Shaw, E. (1977). *Europe's Giant Accelerator: The Story of the CERN 400 GeV Proton Synchrotron*. London: Taylor and Francis.

Goldstone, J. (1961). 'Field Theories with "Superconductor" Solutions'. *NC*, *19*, 154–64.

Goldstone, J., Salam, A. and Weinberg, S. (1962). 'Broken Symmetries'. *PR*, *127*, 965–70.

Gooding, D. (1982). 'Empiricism in Practice: Teleology, Economy and Observation in Faraday's Physics'. *Isis*, *73*, 46–67.

Gordon, H. *et al.* (1982). 'The Axial Field Spectrometer at the CERN ISR'. *NIM*, *196*, 303–13.

Gottfried, K. (1981). 'Are They the Hydrogen Atoms of Strong Interaction Physics?' *CNPP*, *9*, 141–52.

Goudsmit, S.A. (1976). 'Fifty Years of Spin: It Might as Well Be Spin'. *Physics Today*, *29*(6), 40–3.

Greenberg, D.S. (1971). *The Politics of Pure Science*. New York: New American Library.

Greenberg, O.W. (1964). 'Spin and Unitary-Spin Independence in a Paraquark Model of Baryons and Mesons'. *PRL*, *13*, 598–602.

Greenberg, O.W. (1978). 'Quarks'. *ARNPS*, *28*, 327–86.

Greenberg, O.W. (1982). 'Resource Letter Q-1: Quarks'. *AJP*, *50*, 1074–89.

Greenberg, O.W., Dell'Antonio, G.F. and Sudarshan, E.C.G. (1964). 'Parastatistics: Axiomatic Formulations, Connection with Spin and TCP Theorem for a General Field Theory'. In Gürsey (1964), 403–8.

Greenberg, O.W. and Messiah, A.M.L. (1964). 'Symmetrization Postulate and Its Experimental Foundation'. *PR*, *136*, B248–67.

Greenberg, O.W. and Messiah, A.M.L. (1965a). 'Selection Rules for Parafields and the Absence of Paraparticles in Nature'. *PR*, *138*, B1155–67.

Greenberg, O.W. and Messiah, A.M.L. (1965b). 'High-Order Limit of Para-Bose and Para-Fermi Fields'. *JMP*, *6*, 500–4.

Greenberg, O.W. and Nelson, C.A. (1977). 'Color Models of Hadrons'. *Phys. Rept.*, *32*, 69–121.

Gregory, B. (1970). 'Introduction'. *CERN Annual Report 1970*, 9–21. Geneva: CERN.

Gribov, V.N. and Lipatov, L.N. (1972). 'Deep Inelastic ep Scattering in Perturbation Theory'. *SJNP*, *15*, 438–50.

Gross, D.J. and Llewellyn Smith, C.H. (1969). 'High-Energy Neutrino-Nucleon Scattering, Current Algebra and Partons'. *NP*, *B14*, 337–47.

Gross, D.J., Pisarski, R.D. and Yaffe, L.G. (1981). 'QCD and Instantons at Finite Temperature'. *RMP*, *53*, 43–80.

Gross, D.J. and Wilcek, F. (1973a). 'Ultraviolet Behavior of Non-Abelian Gauge Theories'. *PRL*, *30*, 1343–6.

Gross, D.J. and Wilcek, F. (1973b). 'Asymptotically Free Gauge Theories. I'. *PR*, *D8*, 3633–52.

Gross, D.J. and Wilcek, F. (1974). 'Asymptotically Free Gauge Theories. II'. *PR*, *D9*, 980–93.

Guralnik, G.S., Hagen, C.R. and Kibble, T.W. (1964). 'Global Conservation Laws and Massless Particles'. *PRL*, *13*, 585–7.

Gürsey, F. (ed.) (1964). *Group Theoretical Concepts and Methods in Elementary Particle Physics*. New York: Gordon and Breach.

Bibliography

Gürsey, F., Pais, A. and Radicati, L.A. (1964). 'Spin and Unitary Spin Independence of Strong Interactions'. *PRL, 13*, 299–301.

Gürsey, F. and Radicati, L.A. (1964). 'Spin and Unitary Spin Independence'. *PRL, 13*, 173–5.

Gürsey, F. and Sikivie, P. (1976). 'E_7 as a Universal Group'. *PRL, 36*, 775–8.

Gutbrod, F. (ed.) (1977). *Proceedings of the 1977 International Symposium on Lepton and Photon Interactions at High Energies*, 25–31 August 1977, Hamburg. Hamburg: DESY.

Hafner, E.M. and Presswood, S. (1965). 'Strong Inference and Weak Interactions'. *Science, 149*, 503–10.

Hahn, H., Month, M. and Rau, R.R. (1977). 'Proton-Proton Intersecting Storage Accelerator Facility ISABELLE at the Brookhaven National Laboratory'. *RMP, 49*, 625–79.

Hahn, H., Rau, R.R. and Wanderer, P. (1977). 'A Survey of the High Energy Physics Program at Brookhaven National Laboratory'. *Phys. Rept., 29*, 85–151.

Halliday, I.G. (1974). 'Strong Interaction Dynamics'. In Smith (1974), I-229–42.

Halzen, F. (1979). 'Signatures of Chromodynamics in Hadron Collisions'. In Homma *et al.* (1979), 215–21.

Han, M.Y. and Nambu, Y. (1965). 'Three-Triplet Model with Double $SU(3)$ Symmetry'. *PR, 139B*, 1006–10.

Hansen, K. and Hoyer, P. (eds.). (1979). 'Jets in High Energy Collisions'. Proceedings of a Symposium at the Niels Bohr Institute/NORDITA, Copenhagen, 10–14 July 1978. *Physica Scripta, 19*(2), 69–202.

Hanson, G. *et al.* (1975). 'Evidence for Jet Structure in Hadron Production by e^+e^- Annihilation'. *PRL, 35*, 1609–12.

Hanson, N.R. (1958). *Patterns of Discovery: An Inquiry into the Conceptual Foundations of Science*. Cambridge and New York: Cambridge University Press.

Hara, Y. (1964). 'Unitary Triplets and the Eightfold Way'. *PR, 134*, B701–4.

Harari, H. (1978). 'Quarks and Leptons'. *Phys. Rept., 42*, 235–309.

Harari, H. (1983). 'The Structure of Quarks and Leptons'. *Scientific American, 248*(4), 48–60.

Harding, S.G. (ed.) (1976). *Can Theories Be Refuted? Essays on the Duhem – Quine Thesis*. Dordrecht and Boston: Reidel.

Hasenfratz, P. and Kuti, J. (1978). 'The Quark Bag Model'. *Phys. Rept., 40*, 75–179.

Hasert, F.J. *et al.* (1973a). 'Search for Elastic Muon-Neutrino Electron Scattering'. *PL, 46B*, 121–4.

Hasert, F.J. *et al.* (1973b). 'Observation of Neutrino-Like Interactions Without Muon or Electron in the Gargamelle Neutrino Experiment'. *PL, 46B*, 138–40.

Hasert, F.J. *et al.* (1974). 'Observation of Neutrino-Like Interactions Without Muon or Electron in the Gargamelle Neutrino Experiment'. *NP, B73*, 1–22.

Heilbron, J.L. (1979). *Electricity in the 17th and 18th Centuries: A Study of Early Modern Physics*. Berkeley, Los Angeles and London: University of California Press.

Bibliography

Heilbron, J.L., Seidel, R.W. and Wheaton, B.R. (1981). *Lawrence and His Laboratory: Nuclear Science at Berkeley 1931–1961*. Berkeley: Office for History of Science and Technology.

Hendry, A.W. and Lichtenberg, D.B. (1978). 'The Quark Model'. *RPP, 41*, 1707–80.

Herb, S.W. *et al.* (1977). 'Observation of a Dimuon Resonance at 9.5 GeV in 400-GeV Proton-Nucleus Collisions'. *PRL, 39*, 252–5.

Hermann, A. (1980). 'CERN History: Feasibility Study'. CERN Report CC/1384.

Hesse, M.B. (1974). *The Structure of Scientific Inference*. London: Macmillan.

Hey, A.J.G. (1979). 'Particle Systematics'. In EPS (1979), 523–46.

Higgs, P.W. (1964a). 'Broken Symmetries, Massless Particles and Gauge Fields'. *PL, 12*, 132–3.

Higgs, P.W. (1964b). 'Broken Symmetries and the Masses of Gauge Bosons'. *PRL, 13*, 508–9.

Higgs, P.W. (1966). 'Spontaneous Symmetry Breaking without Massless Bosons'. *PR, 145*, 1156–63.

High Energy Physics Advisory Panel (1980). *Report of the 1980 Subpanel on Review and Planning for the U.S. High Energy Physics Program*. Washington: US Department of Energy.

Hoddeson, L. (1983). 'Establishing Fermilab in the US and KEK in Japan: Nationalism and Internationalism in High Energy Accelerator Physics During the 1960s'. *Social Studies of Science, 13*, 1–48.

Holder, M. *et al.* (1977a). 'Is There a High-y Anomaly in Antineutrino Interactions?' *PRL, 39*, 433–6.

Holder, M. *et al.* (1977b). 'Observation of Trimuon Events Produced in Neutrino and Antineutrino Interactions'. *PL, 70B*, 393–5.

Hollister, J.H. *et al.* (1981). 'Measurement of Parity Nonconservation in Atomic Bismuth'. *PRL, 46*, 643–6.

Holton, G. (1978). 'Subelectrons, Presuppositions, and the Millikan-Ehrenhaft Dispute'. In his *The Scientific Imagination: Case Studies*, 25–83. Cambridge: Cambridge University Press.

Hom, D.C. *et al.* (1976). 'Observation of High-Mass Dilepton Pairs in Hadron Collisions at 400 GeV'. *PRL, 36*, 1236–9.

Homma, S., Kawaguchi, M. and Miyakawa, H. (eds) (1979). *Proceedings of the 19th International Conference on High Energy Physics*, Tokyo, 23–30 August, 1978. Tokyo: Physical Society of Japan.

Horgan, R.R. and Dalitz, R.H. (1973). 'Baryon Spectroscopy and the Quark Shell Model. I. The Framework, Basic Formulae and Matrix Elements'. *NP, B66*, 135–72.

Horn, D. (1972). 'Many-Particle Production'. *Phys. Rept., 4*, 1–66.

Howard, F.T. (ed.) (1967). *Proceedings of the 6th International Conference on High Energy Accelerators*, Cambridge, MA, 11–15 September 1967.

Hoyer, P. *et al.* (1979). 'Hadron Distributions in Quark Jets'. *NP, B151*, 389–98.

Hung, P.Q. and Sakurai, J.J. (1981). 'The Structure of Neutral Currents'. *ARNPS, 31*, 375–438.

Bibliography

Ikeda, M., Ogawa, S. and Ohnuki, Y. (1959). 'A Possible Symmetry in Sakata's Model for Bosons-Baryons Systems'. *PTP*, *22*, 715–24.

Iliopoulos, J. (1974). 'Progress in Gauge Theories'. In Smith (1974), III-89–114.

Innes, W.R. *et al.* (1977). 'Observation of Structure in the Υ Region'. *PRL*, *39*, 1240–2, 1640(E).

Inoue, K., Kakuto, A. and Nakano, Y. (1977). 'Unification of the Lepton-Quark World by the Gauge Group su(6)'. *PTP*, *58*, 630–9.

Ioffe, B.L. (1969). 'Space-Time Picture of Photon and Neutrino Scattering and Electroproduction Cross-Section Asymptotics'. *PL*, *30B*, 123–5.

Irvine, J. and Martin, B.R. (1983a). 'CERN: Past Performance and Future Prospects. II – The Scientific Performance of the CERN Accelerators'. Sussex University preprint, submitted to *Research Policy*.

Irvine, J. and Martin, B.R. (1983b). 'Basic Research in the East and West: A Comparison of the Scientific Performance of High-Energy Physics Accelerators'. Sussex University preprint, to appear in *Social Studies of Science*.

Irving, A.C. (1979). 'Hyperchange Exchange Reactions and Hyperon Resonance Production'. In EPS (1979), 616–22.

Isgur, N. and Karl, G. (1977). 'Hyperfine Interactions in Negative Parity Baryons'. *PL*, *72B*, 109–13.

Isgur, N. (ed.) (1981). *Baryon '80: Proceedings of the 4th International Conference on Baryon Resonances*, 14–16 July 1980, Toronto. Toronto: University of Toronto.

Ishikawa, K. (1983). 'Glueballs'. *Scientific American*, *247*(5), 122–35.

Jachim, A.J. (1975). *Science Policy Making in the United States and the Batavia Accelerator*. London and Amsterdam: Feffer and Simons.

Jackiw, R. (1972). 'Introducing Scale Symmetry'. *Physics Today*, *25*(1), 23–7.

Jackson, J.D. (1969). 'Models for High Energy Processes'. In von Dardel (1969), 63–108.

Jackson, J.D. and Roberts, A. (eds). (1972). *Proceedings of the XVI International Conference on High Energy Physics*. National Accelerator Laboratory, Batavia, Illinois, 6–13 September 1972. Batavia: National Accelerator Laboratory.

Jackson, J.D., Quigg, C. and Rosner, J.L. (1979). 'New Particles, Theoretical'. In Homma *et al.* (1979), 391–408.

Jacob, M. (1971). 'An Editing Experiment'. CERN preprint (8 Jan. 1971).

Jacob, M. (1974). 'Hadron Physics at ISR Energies'. CERN Yellow Report CERN 74–15.

Jacob, M. (1980). 'New Directions in Elementary Particle Physics – p$\bar{\text{p}}$ from Very Low to Very High Energies'. *SHEP*, *1*, 213–48.

Jacob, M. (1982). 'Editorial Note'. In Bellini *et al.* (1982), 3–4.

Jacob, M. (ed.) (1974). *Dual Theory. Physics Reports Reprint Series, Vol. I.* Amsterdam: North Holland.

Jacob, M. (ed.) (1981). *CERN 25 Years of Physics. Physics Reports Reprint Book Series, Vol. 4.* Amsterdam, New York and Oxford: North-Holland.

Bibliography

Jacob, M. and Landshoff, P.V. (1976). 'Trigger Bias in Large p_T Reactions'. *NP*, *B113*, 395–412.

Jacob, M. and Landshoff, P.V. (1978). 'Large Transerse Momentum and Jet Studies'. *Phys. Rept.*, *48*, 285–350.

Jacob, M. and Landshoff, P.V. (1980). 'The Inner Structure of the Proton'. *Scientific American*, *242*(3), 46–55.

Jaffe, R.L. (1977a). 'Multiquark Hadrons. I. Phenomenology of $Q^2\bar{Q}^2$ Mesons'. *PR*, *D15*, 267–80.

Jaffe, R.L. (1977b). 'Multiquark Hadrons. II. Methods'. *PR*, *D15*, 281–9.

Jaffe, R.L. and Johnson, K. (1977). 'Unconventional States of Confined Quarks and Gluons'. *PL*, *60B*, 201–4.

Jarlskog, C. and Ynduráin, F.J. (1979). 'Matter Instability in the $su(5)$ Unified Model of Strong, Weak and Electromagnetic Interactions'. *NP*, *B149*, 29–38.

Jentschke, W. (1972). 'Physics Results from 1972'. In *CERN Annual Report 1972*, 11–17.

Johnson, K.A. (1979). 'The Bag Model of Quark Confinement'. *Scientific American*, *241*(1), 100–9.

Jones, L.W. (1977). 'A Review of Quark Search Experiments'. *RMP*, *49*, 717–52.

Jones, M., Horgan, R.R. and Dalitz, R.H. (1977). 'Re-analysis of the Baryon Mass Spectrum Using the Quark Model'. *NP*, *B129*, 45–65.

Jungk, R. (1968). *The Big Machine*. New York: Charles Scribner's Sons.

Kabir, P.K. (1979). *The CP Puzzle: Strange Decays of the Neutral Kaon*. London and New York: Academic Press, 2nd ed.

Kalman, C.S. (1981). 'Subquark Structure'. *Canadian Journal of Physics*, *59*, 1774–9.

Karl, G. (1974). 'Note on My Professional History'. University of Guelph, unpublished.

Kemmer, N. (1982). 'Isospin'. *Journal de Physique*, *43*, C8-359–93.

Kendall, W.H. and Panofsky, W.K.H. 'The Structure of the Proton and Neutron'. *Scientific American*, *224*(6), 61–77.

Kenyon, I.R. (1982). 'The Drell–Yan Process'. *RPP*, *45*, 1261–315.

Kevles, D.J. (1978). *The Physicists*. New York: Alfred A.Knopf.

Kibble, T.W.B. (1967). 'Symmetry Breaking in Non-Abelian Gauge Theories'. *PR*, *155*, 1554–61.

Kiesling, C.M. (1979). 'Results on Charmonium from the Crystal Ball'. In EPS (1979), 293–303.

Kim, J.E., Langacker, P., Levine, M. and Williams, H.H. (1981). 'A Theoretical and Experimental Review of the Weak Neutral Current: A Determination of Its Structure and Limits on Deviations from the Minimal $su(2)_L \times u(1)$ Electroweak Theory'. *RMP*, *53*, 211–52.

Kinoshita, T. (1962). 'Mass Singularities of Feynman Amplitudes'. *JMP*, *3*, 650–77.

Kirk, W.T. (ed.) (1975). *Proceedings of the International Symposium on Lepton and Photon Interactions at High Energies*, Stanford, 21–27 August 1975. Stanford: SLAC.

Klein, A. and Lee, B.W. (1964). 'Does Spontaneous Breakdown of Symmetry Imply Zero-Mass Particles?' *PRL*, *12*, 266–8.

440

Bibliography

Kleinchnect, K. (1974). 'Weak Decays and CP Violation'. In Smith (1974), III-23–58.

Knapp, B. *et al.* (1975). 'Photoproduction of Narrow Resonances'. *PRL, 34*, 1040–3.

Knapp, B. *et al.* (1976). 'Observation of a Narrow Antibaryon State at 2.26 GeV/c²'. *PRL, 37*, 882–5.

Knorr, K.D. (1981). 'The Scientist as Analogical Reasoner: A Critique of the Metaphor Theory of Innovation'. In Knorr *et al.* (1981), 25–52.

Knorr, K.D., Krohn, R. and Whitley, R. (eds.) (1981). *The Social Process of Scientific Investigation. Sociology of the Sciences, Volume IV, 1980.* Dordrecht: Reidel.

Knorr-Cetina, K.D. (1981). *The Manufacture of Knowledge: An Essay in the Constructivist and Contextual Nature of Science.* Oxford: Pergamon.

Kobayashi, M. and Maskawa, K. (1973). 'CP-Violation in the Renormalizable Theory of Weak Interactions'. *PTP, 49*, 652–7.

Koester, D., Sullivan, D. and White, D.H. (1982). 'Theory Selection in Particle Physics: A Quantitative Case Study of the Evolution of Weak-Electromagnetic Unification Theory'. *Social Studies of Science, 12*, 73–100.

Kogut, J. and Susskind, L. (1973). 'The Parton Picture of Elementary Particles'. *Phys. Rept., 8*, 75–172.

Kokkedee, J.J.J. (1969). *The Quark Model.* New York and Amsterdam: W.A.Benjamin.

Komar, A. and Salam, A. (1960). 'Renormalization Problem for Vector Meson Theories'. *NP, 21*, 624–30.

Kowarski, L. (1977a). 'New Forms of Organisation in Physical Research after 1945'. In Weiner (1977), 370–401.

Kowarski, L. (1977b). 'Some Conclusions from CERN's History'. In Zichichi (1977), Part B, 1201–11.

Krammer, M. and Krasemann, H. (1979). 'Quarkonia'. In Proceedings of the Schladming School 1979, *Acta Phys. Austriaca, Suppl. 21*, 259–349.

Kuhn, T.S. (1970). *The Structure of Scientific Revolutions.* Chicago and London: Chicago University Press, 2nd ed.

Kursunoglu, B., Perlmutter, A. and Sakmar, A. (eds) (1965). *Symmetry Principles at High Energies,* 2nd Coral Gables Conference, 20–22 January 1965, Miami. San Francisco and London: W.H.Freeman.

Kuti, J. and Weisskopf, V.F. (1971). 'Inelastic Lepton-Nucleon Scattering and Lepton Pair Production in the Relativistic Quark-Parton Model'. *PR, D4*, 3418–39.

Lagarrigue, A., Musset, P. and Rousset, A. (1964). 'Projet de Chambre à Bulles à Liquides Lourds de 17m³'. Paris: Ecole Polytechnique, unpublished.

Lakatos, I. and Musgrave, A. (eds) (1970). *Criticism and the Growth of Knowledge.* Cambridge: Cambridge University Press.

Lande, K. (1979). 'Experimental Neutrino Physics'. *ARNPS, 29*, 395–410.

Landshoff, P.V. (1974). 'Large Transverse Momentum Reactions'. In Smith (1974), V-57–80.

Bibliography

Landshoff, P.V. (ed.) (1979). *LEP: Report of a Meeting Held at the Rutherford Laboratory on October 15, 1979*. Didcot: Rutherford Laboratory.

Landshoff, P.V. and Polkinghorne, J.C. (1972). 'Models for Hadronic and Leptonic Processes at High Energies'. *Phys. Rept.*, *5*, 1–55.

Langacker, P. (1981). 'Grand Unified Theories and Proton Decay'. *Phys. Rept.*, *72*, 185–385.

Langacker, P., Segrè, G. and Weldon, A. (1978). 'Absolute Proton Stability in Unified Models of Strong, Weak and Electromagnetic Interactions'. *PL*, *73B*, 87–90.

LaRue, G.S., Fairbank, W.M. and Hebard, A.F. (1977). 'Evidence for the Existence of Fractional Charge on Matter'. *PRL*, *38*, 1011–14.

LaRue, G.S., Fairbank, W.M. and Phillips, J.D. (1979). 'Further Evidence for Fractional Charge of $\frac{1}{3}$e on Matter'. *PRL*, *42*, 142–5, 1019(E).

LaRue, G.S., Phillips, J.D. and Fairbank, W.M. (1981). 'Observation of Fractional Charge of $(\frac{1}{3})$e on Matter'. *PRL*, *46*, 967–70.

Latour, B. and Woolgar, S. (1979). *Laboratory Life: The Social Construction of Scientific Facts*. Beverley Hills and London: Sage.

Lautrup, B.E., Peterman, A. and de Rafael, E. (1972). 'Recent Developments in the Comparison between Theory and Experiment in Quantum Electrodynamics'. *Phys. Rept.*, *3*, 193–260.

Learned, J., Reines, F. and Soni, A. (1979). 'Limits on Nonconservation of Baryon Number'. *PRL*, *43*, 907–10, 1626(E).

Lederman, L.M. (1976). 'Lepton Production in Hadron Collisions'. *Phys. Rept.*, *26*, 149–81.

Lederman L.M. (1978). 'The Upsilon Particle'. *Scientific American*, *239*(4), 60–8.

Lee, B.W. (1969). 'Renormalization of the σ-Model'. *NP*, *B9*, 649–72.

Lee, B.W. (1972a), 'Renormalizable Massive Vector-Meson Theory – Perturbation Theory of the Higgs Phenomenon'. *PR*, *D5*, 823–35.

Lee, B.W. (1972b). 'Perspectives on Theory of Weak Interactions'. In Jackson and Roberts (1972), Vol. 4, 249–305.

Lee, B.W. (1972c). 'Model of Weak and Electromagnetic Interactions'. *PR*, *D6*, 1188–90.

Lee, B.W. (1972d). 'The Process $v_\mu + p \rightarrow v_\mu + p + \pi^0$ in Weinberg's Model of Weak Interactions'. *PL*, *40B*, 420–2.

Lee, B.W. (1978). 'Development of Unified Gauge Theories – Retrospect'. In M. Jacob (ed.), *Gauge Theories and Neutrino Physics*, 148–53. Amsterdam, New York and Oxford: North-Holland.

Lee, B.W. and Zinn-Justin, J. (1972a). 'Spontaneously Broken Gauge Symmetries. I. Preliminaries'. *PR*, *D5*, 3121–37.

Lee, B.W. and Zinn-Justin, J. (1972b). 'Spontaneously Broken Gauge Symmetries. II. Perturbation Theory and Renormalization'. *PR*, *D5*, 3137–55.

Lee, B.W. and Zinn-Justin, J. (1972c). 'Spontaneously Broken Gauge Theories. III. Equivalence'. *PR*, *D5*, 3155–60.

Lee, T.D. (1972). 'A New High-Energy Scale'. *Physics Today*, *25*, 23–8.

Lee, T.D. and Nauenberg, M. (1964). 'Degenerate Systems and Mass Singularities'. *PR*, *133*, B1549–62.

Bibliography

Lee, T.D. and Yang, C.N. (1960a). 'Theoretical Discussions on Possible High-Energy Neutrino Experiments'. *PRL*, *4*, 307–11.

Lee, T.D. and Yang, C.N. (1960b). 'Implications of the Intermediate Boson Basis of the Weak Interactions: Existence of a Quartet of Intermediate Bosons and Their Dual Isotopic Spin Transformation Properties'. *PR*, *119*, 1410–19.

Lee, T.D. and Yang, C.N. (1962). 'Theory of Charged Vector Mesons Interacting with the Electromagnetic Field'. *PR*, *128*, 885–98.

Lee, W. (1972). 'Experimental Limits on the Neutral Current in the Semileptonic Processes'. *PL*, *40B*, 423–5.

LEP Study Group (1979). *Design Study of a 22 to 130 GeV e^+e^- Colliding Beam Machine (LEP)*. CERN 'Pink Book' ISR-LEP/79-33.

Levy, M. *et al.* (eds) (1980). *Quarks and Leptons*. Proceedings of the Cargèse Summer Institute, 9–29 July 1979. New York: Plenum.

Lewis, L.L. *et al.* (1977). 'Upper Limit on Parity-Nonconserving Optical Rotation in Atomic Bismuth'. *PRL*, *39*, 795–8.

Lichtenberg, D.B. and Rosen, S.P. (eds.) (1980). *Developments in the Quark Theory of Hadrons, A Reprint Collection. Volume I: 1964–1978*. Nonantum, MA: Hadronic Press.

Lipkin, H.J. (1973). 'Quarks for Pedestrians'. *Phys. Rept.*, *8*, 173–268.

Lipkin, H.J. (1982). 'The Successes and Failures of the Constituent Quark Model'. Fermilab preprint 82/82-THY (Nov. 1982).

Lipkin, H.J., Rubinstein, H.R. and Stern, H. (1967). 'Strong and Weak Decays with Meson Emission in the Quark Model'. *PR*, *161*, 1502–4.

Litt, J. (ed.) (1979). *Nimrod the 7 GeV Proton Synchrotron*. Didcot: Rutherford Laboratory.

Livingston, M.S. (1980). 'Early History of Particle Accelerators'. *Advances in Electronics and Electron Physics*, *50*, 1–88.

Llewellyn Smith, C.H. (1970). 'Current-Algebra Sum Rules Suggested by the Parton Model'. *NP*, *B17*, 277–88.

Llewellyn Smith, C.H. (1971). 'Inelastic Lepton Scattering in Gluon Models'. *PR*, *D4*, 2392–7.

Llewellyn Smith, C.H. (1972). 'Neutrino Reactions at Accelerator Energies'. *Phys. Rept.*, *3*, 261–379.

Llewellyn Smith, C.H. (1974). 'Unified Models of Weak and Electromagnetic Interactions'. In Rollnik and Pfeil (1974), 449–65.

Llewellyn Smith, C.H. (1978). 'Jets and QCD'. *Acta Physica Austriaca, Suppl.* *19*, 331–61.

Lock, W.O. (1975). 'A History of the Collaboration between the European Organisation for Nuclear Research (CERN) and the Joint Institute for Nuclear Research (JINR), and with Soviet Research Institutes in the USSR 1955–1970'. CERN Yellow Report 75-7.

Lock, W.O. (1981). 'Origins and Evolution of the Collaboration between CERN and the People's Republic of China'. CERN Yellow Report 81-14.

Lubkin, G.B. (1980). '21-Term Series Yields Critical Exponents'. *Physics Today*, *33*(11), 18–20.

Machacek, M. (1979). 'The Decay Modes of the Proton'. *NP*, *B159*, 37–55.

MacKenzie, D.A. (1981). *Statistics in Britain, 1865–1930: The Social*

Bibliography

Construction of Scientific Knowledge. Edinburgh: Edinburgh University Press.

Madsen, J.H.B. and Standley, P.H. (1980). *Catalogue of High-Energy Accelerators*. Geneva: CERN.

Maki, Z. and Ohnuki, Y. (1964). 'Quartet Scheme for Elementary Particles'. *PTP, 32*, 144–58.

Mandelstam, S. (1968a). 'Feynman Rules for Electromagnetic and Yang-Mills Fields from the Gauge-Independent Field-Theoretic Formalism'. *PR, 175*, 1580–603.

Mandelstam, S. (1968b). 'Feynman Rules for the Gravitational Field from the Coordinate-Independent Field-Theoretic Formalism'. *PR, 175*, 1604–23.

Mannelli, I. (1979). 'Electron Pairs Production at the ISR'. In Homma *et al.* (1979), 189–91.

Mannelli, I. (ed.) (1983). *Proceedings of the Workshop on SPS Fixed-Target Physics in the Years 1984–1989*, 6–10 December 1982, CERN. CERN Yellow Report 83-02, 2 Vols.

Mar, J. *et al.* (1968). 'A Comparison of Electron-Proton and Positron-Proton Elastic Scattering at Four-Momentum Transfers up to 5.0 (GeV/c) Squared'. *PRL, 21*, 482–4.

Marciano, W.J. (1979). 'Weak Mixing Angle and Grand Unified Gauge Theories'. *PR, D20*, 274–88.

Marciano, W.J. (1981). 'Neutrino Masses: Theory and Experiment'. *CNPP, 9*, 169–82.

Marciano, W. and Pagels, H. (1978). 'Quantum Chromodynamics'. *Phys. Rept., 36*, 137–276.

Marshak, R.E. (1953). 'The Multiplicity of Particles'. *Scientific American, 186*(1), 22–7.

Martin, B.R. and Irvine, J. (1981). 'Internal Criteria for Scientific Choice: An Evaluation of Research in High-Energy Physics Using Electron Accelerators'. *Minerva, 19*, 408–32.

Martin, B.R. and Irvine, J. (1983). 'CERN Past Performance and Future Prospects. – CERN's Position in World High-Energy Physics'. Sussex University preprint, submitted to *Research Policy*.

Masterman, M. (1970). 'The Nature of a Paradigm'. In Lakatos and Musgrave (1970), 59–90.

McCusker, C.B.A. (1981). 'The Positive Results from Quark Searches'. University of Sidney preprint, unpublished.

McCusker, C.B.A. (1983). 'An Estimate of the Flux of Free Quarks in High Energy Cosmic Radiation'. *Australian Journal of Physics*, in press.

Melosh, H.J. (1974). 'Quarks: Currents and Constituents'. *PR, D9*, 1095–112.

Michael, C. (1979). 'Large Transverse Momentum and Large Mass Production in Hadronic Interactions'. *PPNP, 2*, 1–39.

Miller, C. *et al.* (1972). 'Inelastic Electron-Proton Scattering at Large Momentum Transfers and the Inelastic Structure Functions of the Proton'. *PR, D5*, 528–44.

Miller, D.J. (1977). 'Elementary Particles – a Rich Harvest'. *Nature, 269*, 286–8.

444

Bibliography

Miller, D.J. (1978a). 'Too Many Electron Neutrinos?' *Nature*, *272*, 205.

Miller, D.J. (1978b). 'More about the Beam Dump'. *Nature*, *272*, 668.

Mitra, A.N. and Ross, M.H. (1967). 'Meson-Baryon Couplings in a Quark Model'. *PR*, *158*, 1630–8.

Mohl, D., Petrucci, G., Thorndahl, L. and van der Meer, S. (1980). 'Physics and Technique of Stochastic Cooling'. *Phys. Rept.*, *58*, 73–119.

Montanet, L. (1979). ' "Narrow" Baryoniums – Experimental Situation'. In Nicolescu, Richard and Vinh Mau (1979), I-1–12.

Montanet, L. and Reucroft, S. (1982). 'High Resolution Bubble Chambers and the Observation of Short-Lived Particles'. In Bellini *et al.* (1982), 61–83.

Montanet, L., Rossi, G.C. and Veneziano, G. (1980). 'Baryonium Physics'. *Phys. Rept.*, *63*, 149–222.

Moorhouse, R.G. (1966). 'Photoproduction of N* Resonances in the Quark Model'. *PRL*, *16*, 772–4.

Moravcsik, M.J. and Noyes, H.P. (1961). 'Theories of Nucleon-Nucleon Elastic Scattering'. *ARNS*, *11*, 95–174.

Morgan, D. (1978). 'The Context of the Search for Axions'. *Nature*, *274*, 22–5.

Morpurgo, G. (1958). 'Inhibition of M1 γ Transitions with $\Delta T = 0$ in Selfconjugate Nuclei'. *PR*, *110*, 721–5.

Morpurgo, G. (1959). 'γ-Transitions between Corresponding States in Mirror Nuclei'. *PR*, *114*, 1075–80.

Morpurgo, G. (1961). 'Strong Interactions and Reactions of Hyperons and Heavy Mesons'. *ARNS*, *11*, 41–94.

Morpurgo, G. (1965). 'Is a Non-Relativistic Approximation Possible for the Internal Dynamics of "Elementary" Particles?' *Physics*, *2*, 95–105.

Morpurgo, G. (1970). 'A Short Guide to the Quark Model'. *ARNS*, *20*, 105–46.

Morrison, D.R.O. (1978). 'The Sociology of International Scientific Collaboration'. In R.Armenteros *et al.* (eds), *Physics from Friends: Papers Dedicated to C.Peyrou on His 60th Birthday*, 351–65. Geneva: Multi Office.

Morton, A.Q. (1982). *The Neutrino and Nuclear Physics, 1930–1940*. University of London PhD Thesis, unpublished.

Mukherji, V. (1974). 'A History of the Meson Theory of Nuclear Force from 1935 to 1942'. *Archive for the History of the Exact Sciences*, *13*, 27–102.

Musset, P. (1977). 'La Physique de Neutrino'. CERN reprint CERN/515-PU 77-07. Reprinted from *La Recherche*, *74*.

Musset, P. and Vialle, J.P. (1978). 'Neutrino Physics with Gargamelle'. *Phys. Rept.*, *39*, 1–130.

Muta, T. (1979). 'Deep Inelastic Scattering beyond the Leading Order in Asymptotically Free Gauge Theories'. In Homma *et al.* (1979), 234–6.

Myatt, G. (1982). 'Experimental Verification of the Salam–Weinberg Model'. *RPP*, *45*, 1–46.

Nadel, E. (1981). 'Citation and Co-Citation Indicators of a Phased Impact of the BCS Theory in the Physics of Superconductivity'. *Scientometrics*, *3*, 203–21.

Nakano, T. and Nishijima, K. (1953). 'Charge Independence for U-Particles'. *PTP*, *10*, 581–2.

Bibliography

Nambu, Y. (1965). 'Dynamical Symmetries and Fundamental Fields'. In Kursunoglu, Perlmutter and Sakmar (1965), 274–85.

Nambu, Y. (1976). 'The Confinement of Quarks'. *Scientific American*, *235*(5), 48–60.

Nambu, Y. and Jona-Lasinio, G. (1961a). 'A Dynamical Model of Elementary Particles Based upon an Analogy with Superconductivity. I'. *PR*, *122*, 345–58.

Nambu, Y. and Jona-Lasinio, G. (1961b). 'A Dynamical Model of Elementary Particles Based upon an Analogy with Superconductivity. II'. *PR*, *124*, 246–54.

Nambu, Y. and Sakurai, J.J. (1961). 'Odd $\Lambda\Sigma$ Parity and the Nature of the $\pi\Lambda\Sigma$ Coupling'. *PRL*, *6*, 377–80.

Nanopoulos, D.V., Savoy-Navarro, A. and Tao, Ch. (eds) (1982). *Supersymmetry versus Experiment Workshop*, 21–23 April 1982, CERN. CERN preprint TH 3311/EP.82/63-CERN.

Neal, R.B. (ed.) (1968). *The Stanford Two-Mile Accelerator*. New York: W.A. Benjamin.

Ne'eman, Y. (1961). 'Derivation of the Strong Interactions from a Gauge Invariance'. *NP*, *26*, 222–9.

Ne'eman, Y. (1974). 'Concrete Versus Abstract Theoretical Models'. In Y. Elkana (ed.), *The Interaction Between Science and Philosophy*, 1–25. Atlantic Highlands, NJ: Humanities Press.

Ne'eman, Y. (1978). 'Progress in the Physics of Particles and Fields'. *Physics Bulletin*, *29*, 422–4.

Ne'eman, Y. (ed.) (1981). *To Fulfill a Vision: Jerusalem Einstein Centennial Symposium, 1979*. Reading, MA: Addison-Wesley.

Ne'eman, Y. (1983). 'Patterns, Structure and Then Dynamics: Discovering Unitary Symmetry and Conceiving Quarks'. *Proceedings of the Israel Academy of Sciences and Humanities Section of Sciences, No. 21*, 1–26.

Newman, H. (1979). 'Measurement of High-Mass Muon Pairs at Very High Energy'. In Homma *et al.* (1979), 192–3.

New Scientist (1976a). 'Why is Charm so Rare?' *69*, 440.

New Scientist (1976b). 'Atomic Experiment Worries High Energy Physicists'. *72*, 654.

New Scientist (1979). 'Do Gluons Really Exist?' *83*, 709.

New Scientist (1983). 'Straight Tracks Point to the Z Particle'. *98*, 778.

Nguyen, H.K. *et al.* (1977). 'Spin Analysis of Charmed Mesons Produced in e^+e^- Annihilation'. *PRL*, *39*, 262–5.

Nicolescu, B., Richard, J.M. and Vinh Mau, R. (eds) (1979). *Proceedings of the Workshop on Baryonium and Other Unusual Hadron States*, 21–22 June 1979, Institut de Physique Nucléaire, Orsay, France.

Nishijima, K. (1954). 'Some Remarks on the Even-Odd Rule'. *PTP*, *12*, 107–8.

Novikov, V.A., Okun, L.B., Shifman, M.A., Vainshtein, A.I., Voloshin, M.B. and Zakharov, V.I. (1978). 'Charmonium and Gluons'. *Phys. Rept.*, *41*, 1–133.

Okubo, S. (1962). 'Note on Unitary Symmetry in Strong Interactions'. *PTP*, *27*, 949–66.

Okubo, S. (1963). 'ϕ-Meson and Unitary Symmetry Model'. *PL*, *5*, 165–8.

446

Olive, K.A., Schramm, D.N. and Steigman, D. (1978). 'Cosmological Constraints on Unification Models'. In Cline (1978).

O'Neill, G.K. (1962). 'The Spark Chamber'. *Scientific American*, *207*(2), 37–43.

O'Neill, G.K. (1966). 'Particle Storage Rings'. *Scientific American*, *215*(5), 107–16.

Pais, A. (1952). 'Some Remarks on the V-Particles'. *PR*, *86*, 663–72.

Pais, A. (1964). 'Implications of Spin-Unitary Spin Independence'. *PRL*, *13*, 175–7.

Pais, A. (1968). 'Twenty Years of Physics: Particles'. *Physics Today*, *21*(5), 24–8.

Pais, A. and Treiman, S. (1972). 'Neutral Current Effects in a Class of Gauge Field Theories'. *PR*, *D6*, 2700–3.

Palmer, R.B. (1973). 'A Calculation of Semileptonic Neutral Currents Assuming Partons and Weinberg's Renormalizable Theory'. *PL*, *46B*, 240–4.

Panofsky, W.K.H. (1968). 'Low q^2 Electrodynamics, Elastic and Inelastic Electron (and Muon) Scattering'. In J.Prentki and J.Steinberger (eds), *Proceedings of the 14th International Conference on High-Energy Physics*, Vienna, 28 Aug.–5 Sept. 1968, 23–39. Geneva: CERN.

Panofsky, W.K.H. (1974). 'Welcoming Remarks'. *Proceedings of the 9th International Conference on High Energy Accelerators*, Stanford, March 1974, xi–xiv.

Panofsky, W.K.H. (1975). 'Particle Discoveries at SLAC'. *Science*, *189*, 1045–6.

Particle Data Group (1977). 'New Particle Searches and Discoveries'. *PL*, *68B*, 1–30.

Particle Data Group (1982). *Review of Particle Properties*. Geneva: CERN. Reprinted from *PL*, *111B*, April 1982.

Paschos, E.A. and Wolfenstein, L. (1973). 'Tests for Neutral Currents in Neutrino Reactions'. *PR*, *D7*, 91–5.

Pati, J.C., Rajpoot, S. and Salam, A. (1977). 'Natural Left-Right Symmetry, Atomic Parity Experiments and Asymmetries in High Energy e^+e^- Collisions in PETRA and PEP Regions'. *PL*, *71B*, 387–91.

Pati, J.C. and Salam, A. (1973a). 'Unified Lepton-Hadron Symmetry and a Gauge Theory of the Basic Interactions'. *PR*, *D8*, 1240–51.

Pati, J.C. and Salam, A. (1973b). 'Is Baryon Number Conserved?' *PRL*, *31*, 661–4.

Pauli, W. (1941). 'Relativistic Field Theories of Elementary Particles'. *RMP*, *13*, 203–232.

Perkins, D.H. (1965). 'Neutrino Physics'. *Proceedings of the 1965 Easter School for Physicists Using the CERN PS and SC*, Bad Kreuznach, April 1–15, 1965. CERN Yellow Report 65-24, Vol. III, 65-93.

Perl, M.L. (1978). 'The Tau Heavy Lepton – A Recently Discovered Elementary Particle'. *Nature*, *275*, 273–8.

Perl, M.L. (1980). 'The Tau Lepton'. *ARNPS*, *30*, 299–335.

Perl, M.L. and Kirk, W.T. (1978). 'Heavy Leptons'. *Scientific American*, *238*(3), 50–7.

Bibliography

Perl, M.L. *et al.* (1975). 'Evidence for Anomalous Lepton Production in e⁺e⁻ Annihilation'. *PRL, 35,* 1489–92.

Perl, M.L. *et al.* (1976). 'Properties of Anomalous eμ Events Produced in e⁺e⁻ Annihilation'. *PL, 63B,* 466–70.

Peruzzi, I. *et al.* (1976). 'Observation of a Narrow Charged State at 1876 MeV/c² Decaying to an Exotic Combination of Kππ'. *PRL, 37,* 569–71.

Physics Survey Committee (1973). *Physics in Perspective: Student Edition: The Nature of Physics and the Subfields of Physics.* Washington: National Academy of Science/National Research Council.

Pickering, A.R. (1978). 'Model Choice and Cognitive Interests: A Case Study in Elementary Particle Physics'. Edinburgh, unpublished.

Pickering, A.R. (1980). 'Exemplars and Analogies: A Comment on Crane's Study of Kuhnian Paradigms in High Energy Physics' and 'Reply to Crane'. *Social Studies of Science, 10,* 497–502 and 507–8.

Pickering, A.R. (1981a). 'The Hunting of the Quark'. *Isis, 72,* 216–36.

Pickering, A.R. (1981b). 'Constraints on Controversy: The Case of the Magnetic Monopole', in Collins (1981), 63–93.

Pickering A.R. (1981c). 'The Role of Interests in High-Energy Physics: The Choice between Charm and Colour'. In Knorr *et al.* (1981), 107–38.

Pickering, A.R. (1984). 'Against Putting the Phenomena First: The Discovery of the Weak Neutral Current', to appear in *Studies in History and Philosophy of Science.*

Polanyi, M. (1973). *Personal Knowledge.* London: Routledge and Kegan Paul.

Politzer, H.D. (1973). 'Reliable Perturbative Results for Strong Interactions?' *PRL, 30,* 1346–9.

Politzer, H.D. (1974). 'Asymptotic Freedom: An Approach to the Strong Interactions'. *Phys. Rept., 14,* 129–80.

Politzer, H.D. (1977a). 'Gluon Corrections to Drell–Yan Processes'. *NP, B129,* 301–18.

Politzer, H.D. (1977b). 'QCD off the Light-Cone and the Demise of the Transverse Momentum Cut-Off'. *PL, 70B,* 430–2.

Polkinghorne, J.C. (1983). 'Quest for a Natural God'. *Times Higher Educational Supplement,* 10 June 1983, p. 24.

Polyakov, A.M. (1974). 'Particle Spectrum in Quantum Field Theory'. *JETPL, 20,* 194–5.

Pontecorvo, B. (1960). 'Electron and Muon Neutrinos'. *JETP, 37,* 1236–40.

Povh, B. (1979). 'Very Narrow States'. In EPS (1979), 604–9.

Prentki, J. and Zumino, B. (1972). 'Models of Weak and Electromagnetic Interactions'. *NP, B47,* 99–108.

Prescott, C.Y. *et al.* (1978). 'Parity Non-Conservation in Inelastic Electron Scattering'. *PL, 77B,* 347–52.

Prescott, C.Y. *et al.* (1979). 'Further Measurements of Parity Non-Conservation in Inelastic Electron Scattering'. *PL, 84B,* 524–8.

Primakoff, H. and Rosen, S.P. (1981). 'Baryon Number and Lepton Number Conservation Laws'. *ARNPS, 31,* 145–92.

Quigg, C. (1977). 'Production and Detection of Intermediate Vector Bosons and Heavy Leptons in pp and p̄p Collisions'. *RMP, 49,* 297–315.

Bibliography

Quigg, C. and Rosner, J.L. (1979). 'Quantum Mechanics with Applications to Quarkonium'. *Phys. Rept.*, *56*, 167–235.

Quine, W.V.O. (1964). 'Two Dogmas of Empiricism'. In his *From A Logical Point of View*. Cambridge, MA: Harvard University Press, 2nd ed.

Ramond, P. (1976). 'Unified Theory of Strong, Electromagnetic, and Weak Interactions Based on the Vector-Like Group E(7)'. *NP*, *B110*, 214–28.

Ramsay, N.F. (1968). 'Early History of Associated Universities and Brookhaven National Laboratory'. In *Brookhaven National Laboratory Lectures in Science: Vistas in Research, Vol. II*, 181–98. New York: Gordon and Breach.

Rebbi, C. (1974). 'Dual Models and Relativistic Quantum Strings'. *Phys. Rept.*, *12*, 1–73.

Rebbi, C. (1980). 'Monte Carlo Simulations of Lattice Gauge Theories'. In Brézin, Gervais and Toulouse (1980), 55–62.

Rebbi, C. (1983). 'The Lattice Theory of Quark Confinement'. *Scientific American*, *248*(2), 36–47.

Redhead, M.L.G. (1980). 'Some Philosophical Aspects of Particle Physics'. *Studies in History and Philosophy of Science*, *11*, 279–304.

Regge, T. (1959). 'Introduction to Complex Angular Momenta'. *NC*, *14*, 951–76.

Reichenbach, H. (1938). *Experience and Prediction*. Chicago: University of Chicago Press.

Reiff, J. and Veltman, M. (1969). 'Massive Yang–Mills Fields'. *NP*, *B13*, 545–64.

Reines, F. and Cowan, Jr., C.L. (1953). 'Detection of the Free Neutrino'. *PR*, *92*, 830–1.

Reines, F. and Cowan, Jr., C.L. (1956). 'Detection of the Free Neutrino – A Confirmation'. *Science*, *124*, 103–4.

Reines, F., Cowan, Jr., C.L. and Goldhaber, M. (1954). 'Conservation of the Number of Nucleons'. *PR*, *96*, 1157–8.

Reines, F. and Crouch, M.F. (1974). 'Baryon-Conservation Limit'. *PRL*, *32*, 493–5.

Renton, P. and Williams, W.S.C. (1981). 'Hadron Production in Lepton-Nucleon Scattering'. *ARNPS*, *31*, 193–230.

Reya, E. (1981). 'Perturbative Quantum Chromodynamics'. *Phys. Rept.*, *69*, 195–333.

Rice-Evans, P. (1974). *Spark, Streamer Proportional and Drift Chambers*. London: Richlieu.

Richter, B. (1974). 'e$^+$e$^-$ →Hadrons'. In Smith (1974), IV-37–55.

Richter, B. (1976a). 'One Researcher's Personal Account'. *Adventures in Experimental Physics*, *5*, 143–9.

Richter, B. (1976b). 'Very High Energy Electron-Positron Colliding Beams for the Study of the Weak Interactions'. *NIM*, *136*, 47–60.

Richter, B. (1977). 'From the Psi to Charm: The Experiments of 1975 and 1976'. *RMP*, *49*, 251–66.

Robinson, A.L. (1980). 'Budget Crunch Hits High Energy Physics'. *Science*, *209*, 577–80.

Robinson, A.L. (1982). 'Stanford Pulls off a Novel Accelerator'. *Science*, *216*, 1395–7.

Bibliography

Rollnik, H. and Pfeil, W. (eds) (1974). *Proceedings of the 6th International Symposium on Electron and Photon Interactions at High Energies*, Bonn, 27–31 August, 1973. Amsterdam and London: North Holland.

Rosenfeld, A.H. (1975). 'The Particle Data Group: Growth and Operations – Eighteen Years of Particle Physics'. *ARNS*, *25*, 555–98.

Rosner, J.L. (1968). 'Possibility of Baryon-Antibaryon Enhancements with Unusual Quantum Numbers'. *PRL*, *21*, 950–2, 1468(E).

Rosner, J.L. (1974a). 'Resonance Spectroscopy (Theory)'. In Smith (1974), II-171–99.

Rosner, J.L. (1974b). 'The Classification and Decays of Resonant Particles'. *Phys. Rept.*, *11*, 189–326.

Rosner, J.L. (1980). 'Resource Letter NP-1: New Particles'. *AJP*, *48*, 90–103.

Rubbia, C. (1974). 'Results from the Harvard, Pennsylvania, Wisconsin, FNAL Experiment'. In Smith (1974), IV-117-20.

Rubbia, C. (1983). 'Experimental Observation of Isolated Large Transverse Energy Electrons with Associated Missing Energy'. In Bacci and Salvini (1983), 123–89.

Rubbia, C., McIntyre, P. and Cline, D. (1977). 'Producing Massive Neutral Intermediate Vector Bosons with Existing Accelerators'. In Faissner, Reithler and Zerwas (1977), 683–7.

Rubin, V.C. (1983). 'Dark Matter in Spiral Galaxies'. *Scientific American*, *248*(6), 88–101.

Sakata, S. (1956). 'On a Composite Model for the New Particles'. *PTP*, *16*, 686–8.

Sakharov, A. (1967). 'Violation of CP Invariance, C Asymmetry, and Baryon Asymmetry of the Universe'. *JETPL*, *5*, 24–7.

Sakita, B. (1964). 'Supermultiplets of Elementary Particles'. *PR*, *136*, B1756–60.

Sakurai, J.J. (1960). 'Theory of Strong Interactions'. *AP*, *11*, 1–48.

Sakurai, J.J. (1978). 'Neutral Currents and Gauge Theories – Past, Present and Future'. In J.E.Lannutti and P.K.Williams (eds), *Current Trends in the Theory of Fields*. New York: American Institute of Physics, 38–80.

Salam, A. (1951a). 'Overlapping Divergences and the S-Matrix'. *PR*, *82*, 217–27.

Salam, A. (1951b). 'Divergent Integrals in Renormalizable Field Theories'. *PR*, *84*, 426–31.

Salam, A. (1962). 'Renormalizability of Gauge Theories'. *PR*, *127*, 331–4.

Salam, A. (1968). 'Weak and Electromagnetic Interactions'. In Svartholm (1968), 367–77.

Salam, A. (1979). 'A Gauge Appreciation of Developments in Particle Physics'. In EPS (1979), 853–90.

Salam, A. (1980). 'Gauge Unification of Fundamental Forces'. *RMP*, *52*, 525–38.

Salam, A. and Ward, J.C. (1959). 'Weak and Electromagnetic Interactions'. *NC*, *11*, 568–77.

Salam, A. and Ward, J.C. (1961). 'On a Gauge Theory of Elementary Interactions'. *NC*, *19*, 165–70.

Salam, A. and Ward, J.C. (1964). 'Electromagnetic and Weak Interactions'. *PL*, *13*, 168–71.

Bibliography

Sandars, P. (1977). 'Can Atoms Tell Left from Right?' *New Scientist*, *73*, 764–6.

Sanford, J.R. (1976). 'The Fermi National Accelerator Laboratory'. *ARNS*, *26*, 151–98.

Schilpp, P.A. (ed.) (1949). *Albert Einstein: Philosopher-Scientist*. Evanston: The Library of Living Philosophers.

Schon, D. (1969). *Invention and the Evolution of Ideas*. London: Tavistock.

Schopper, H. (1981a). 'Two Years of PETRA Operation'. *CNPP*, *10*, 33–54.

Schopper, H. (1981b). 'Introductory Review'. *CERN Annual Report 1981*, 13–22.

Schwarz, J.H. (1973). 'Dual Resonance Theory'. *Phys. Rept.*, *8*, 269–335.

Schwarz, J.H. (1975). 'Dual-Resonance Models of Elementary Particles'. *Scientific American*, *232*(2), 61–7.

Schwartz, M. (1960). 'Feasibility of Using High Energy Neutrinos to Study the Weak Interactions'. *PRL*, *4*, 306–7.

Schwinger, J. (1957). 'A Theory of the Fundamental Interactions'. *AP*, *2*, 407–34.

Schwinger, J. (ed.) (1958). *Quantum Electrodynamics*. New York: Dover.

Schwinger, J. (1959). 'Field Theory Commutators'. *PRL*, *3*, 296–7.

Schwinger, J. (1962a). 'Gauge Invariance and Mass'. *PR*, *125*, 397–8.

Schwinger, J. (1962b). 'Gauge Invariance and Mass. II'. *PR*, *128*, 2425–9.

Schwitters, R. (1975). 'Hadron Production at SPEAR'. In Kirk (1975), 5–24.

Schwitters, R. (1976). 'Plenary Report on the Validity of QED and Hadron Production in Electron-Positron Annihilation'. In Bogoliubov *et al.* (1976), Vol. II, B34–9.

Schwitters, R. (1977). 'Fundamental Particles with Charm'. *Scientific American*, *237*(4), 56–70.

Schwitters, R. *et al.* (1975). 'Azimuthal Asymmetry in Inclusive Hadron Production by e^+e^- Annihilation'. *PRL*, *35*, 1320–2.

Scientific American (1980). 'Waiting for Decay'. *243*(4), 66–70.

Sciulli, F. (1979). 'An Experimenter's History of Neutral Currents'. *PPNP*, *2*, 41–87.

Scott, W.G. (1979). 'Quark Distributions and Quark Jets from the CERN Bubble Chamber Neutrino Experiments'. In Hansen and Hoyer (1979), 184–90.

Segrè, E. (1980). *From X-Rays to Quarks: Modern Physicists and Their Discoveries*. San Francisco: W.H.Freeman.

Shapin, S. (1979). 'The Politics of Observation: Cerebral Anatomy and Social Interests in the Edinburgh Phrenology Disputes'. In R.Wallis (ed.), *On the Margins of Science: The Social Construction of Rejected Knowledge (Sociological Review Monograph 27)*, 139–78. Keele: University of Keele.

Shapin, S. (1982). 'History of Science and Its Sociological Reconstructions'. *History of Science*, *20*, 157–211.

Shaw, R. (1955). 'The Problem of Particle Types and Other Contributions to the Theory of Elementary Particles'. Cambridge University, PhD thesis, unpublished.

Shifman, M.A., Vainshtein, A.I., Voloshin, M.B. and Zakharov, V.I. (1978). 'η_c Puzzle in Quantum Chromodynamics'. *PL*, *77B*, 80–3.

451

Bibliography

Silvestrini, V. (1972). 'Electron-Positron Interactions'. In Jackson and Roberts (1972), Vol. 4, 1–40.

Sivers, D., Brodsky, S.J. and Blankenbecler, R. (1976). 'Large Transverse Momentum Processes'. *Phys. Rept.*, *23*, 1–121.

SLAC (1980a). 'SLAC Long Range Plans'. Stanford Linear Accelerator Center, unpublished.

SLAC (1980b). 'SLAC Linear Collider: Conceptual Design Report'. SLAC-Report-229.

SLAC (1982). *Proceedings of the SLC Workshop on Experimental Use of the SLAC Linear Collider.* SLAC-Report-247.

SLAC-MIT-CIT Collaboration (1966). 'Proposals for Initial Electron Scattering Experiments Using the SLAC Spectrometer Facilities'. SLAC Proposal No. 4, Jan. 1966, unpublished.

Slansky, R. (1974). 'High-Energy Hadron Production and Inclusive Reactions'. *Phys. Rept.*, *11*, 99–188.

Slavnov, A. and Fadeev, L. (1970). 'Massless and Massive Yang–Mills Fields'. *TMP*, *3*, 312–16.

Smith, J.R. (ed.) (1974). *Proceedings of the 17th International Conference on High Energy Physics*, London, July 1974. Rutherford Laboratory: Science Research Council.

Söding, P. (1979). 'Jet Analysis'. In EPS (1979), 271–81.

Söding, P. and Wolf. G. (1981). 'Experimental Evidence on QCD'. *ARNPS*, *31*, 231–93.

Spitzer, H.H. and Alexander, G. (1979). 'Search for Three-Jet Decay of the $\Upsilon(9.46)$'. In Homma *et al.* (1979), 259–62.

Squires, E.J. (1979). 'The Bag Model of Hadrons'. *RPP*, *42*, 1187–242.

Steigman, G. (1976). 'Observational Tests of Antimatter Cosmologies'. *Annual Review of Astronomy and Astrophysics*, *14*, 313–38.

Steigman, G. (1979). 'Cosmology Confronts Particle Physics'. *ARNPS*, *29*, 313–38.

Steinberger, J. (1949). 'On the Use of Subtraction Fields and the Lifetimes of Some Types of Meson Decay'. *PR*, *76*, 1180–6.

Sterman, G. (1976). 'Kinoshita's Theorem in Yang–Mills Theories'. *PR*, *D14*, 2123–5.

Sterman, G. and Weinberg, S. (1977). 'Jets from Quantum Chromodynamics'. *PRL*, *39*, 1436–9.

Stern, H. and Gaillard, M.K. (1973). 'Review of the $K_L \to \mu^+ \mu^-$ Puzzle'. *AP*, *76*, 580–606.

Steuwer, H. (ed.) (1979). *Nuclear Physics in Retrospect: Proceedings of a Symposium on the 1930s.* Minneapolis: University of Minneapolis Press.

Strauch, K. (1974). 'Electron Storage Ring Physics: Recent Experimental Results'. In Rollnik and Pfeil (1974), 1–24.

Stueckelberg, E.C.G. and Petermann, A. (1953). 'La Normalisation des Constantes dans la Theorie des Quanta'. *Helvetica Physica Acta*, *5*, 499–520.

Sucher, J. (1978). 'Magnetic Dipole Transitions in Atomic and Particle Physics: Ions and Psions'. *RPP*, *41*, 1781–838.

Sudarshan, E.C.G. and Marshak, R.E. (1958). 'Chirality Invariance and the Universal Fermi Interaction'. *PR*, *109*, 1860–2.

Bibliography

Sullivan, D., Barboni, E.J. and White, D.H. (1981). 'Problem Choice and the Sociology of Scientific Competition: An International Case Study in Particle Physics'. *Knowledge and Society: Studies in the Sociology of Culture Past and Present, 3,* 163–97.

Sullivan, D., Koester, D., White, D.H. and Kern, R. (1980). 'Understanding Rapid Theoretical Change in Particle Physics: A Month-by-Month Co-Citation Analysis'. *Scientometrics, 2,* 309–19.

Sullivan, D., White, D.H. and Barboni, E.J. (1977). 'The State of a Science: Indicators in the Specialty of Weak Interactions'. *Social Studies of Science, 7,* 167–200.

Sutton, C. (1980). 'Waiting for the End'. *New Scientist, 85,* 1016–19.

Sutton, C. (1982). 'Holography, Bubbles and Charm'. *New Scientist, 93,* 646–9.

Svartholm, N. (ed.) (1968). *Elementary Particle Theory: Relativistic Groups and Analyticity.* Stockholm: Almqvist and Wiksell.

Swatez, G.M. (1970). 'The Social Organization of a University Laboratory'. *Minerva, 8*(1), 37–58.

Symanzik, K. (1970). 'Small Distance Behaviour in Field Theory and Power Counting'. *CMP, 18,* 227–46.

Symanzik, K. (1973). 'A Field Theory with Computable Large-Momenta Behaviour'. *LNC, 6,* 77–80.

Taylor, J.C. (1976). *Gauge Theories of Weak Interactions.* Cambridge: Cambridge University Press.

Taylor, R.E. (1975). 'Inelastic Electron-Nucleon Scattering'. In Kirk (1975), 679–708.

Teillac, J. (1981). 'Foreword'. *CERN Annual Report 1981,* 3.

The Mark J Collaboration (1980). 'Physics with High Energy Electron-Positron Colliding Beams with the Mark J Detector'. *Phys. Rept., 63,* 337–91.

't Hooft, G. (1969). 'Anomale Eigenschappen van de Axiale Vectorskroom'. University of Utrecht, doctoral scriptie, unpublished.

't Hooft, G. (1971a). 'Renormalization of Massless Yang–Mills Fields'. *NP, B33,* 173–99.

't Hooft, G. (1971b). 'Renormalizable Lagrangians for Massive Yang–Mills Fields'. *NP, B35,* 167–88.

't Hooft, G. (1974a). 'A Planar Diagram Theory for Strong Interactions'. *NP, B72,* 461–73.

't Hooft, G. (1974b). 'A Two Dimensional Model for Mesons'. *NP, B75,* 461–70.

't Hooft, G. (1974c). 'Magnetic Monopoles in Unified Gauge Theories'. *NP, B79,* 276–84.

't Hooft, G. (1976). 'Computation of the Quantum Effects due to a Four-Dimensional Pseudoparticle'. *PR, D14,* 3432–50.

't Hooft, G. (1980). 'Gauge Theories of the Forces between Elementary Particles'. *Scientific American, 242*(6), 90–119.

't Hooft, G. and Veltman, M. (1972a). 'Regularization and Renormalization of Gauge Fields'. *NP, B44,* 189–213.

't Hooft, G. and Veltman, M. (1972b). 'Combinatorics of Gauge Fields'. *NP, B50,* 318–53.

453

Bibliography

Ting, S.C.C. (1975). 'Particle Discovery at Brookhaven'. *Science*, *189*, 750.

Ting, S.C.C. (1976). 'One Researcher's Personal Account'. *Adventures in Experimental Physics*, *5*, 115–27.

Ting, S.C.C. (1977). 'The Discovery of the J Particle: A Personal Recollection'. *RMP*, *49*, 235–49.

Trefil, J.S. (1980). *From Atoms to Quarks: An Introduction to the Strange World of Particle Physics*. London: Athlone Press.

Trenn, T.J. (1977). *The Self-Splitting Atom: The History of the Rutherford–Soddy Collaboration*. London: Taylor and Francis.

Tripp, R. (1979). 'In Search of Baryonium'. In Nicolescu, Richard and Vinh Mau (1979), II-1–20.

Trower, P.T. and Bellini, G. (eds) (1982). *Physics in Collision: High-Energy ee/ep/pp Interactions*. Proceedings of an International Conference on Physics in Collision, 28–31 May 1981, Blacksburg, Virginia. New York: Plenum.

Turner, M.S. and Schramm, D.N. (1979). 'Cosmology and Elementary Particle Physics'. *Physics Today*, *32*(9), 42–8.

Ueno, K. *et al.* (1979). 'Evidence for the Υ'' and a Search for New Narrow Resonances'. *PRL*, *42*, 486–9.

Uhlenbeck, G.E. (1976). 'Fifty Years of Spin: Personal Reminiscences'. *Physics Today*, *29*(6), 43–8.

Utiyama, R. (1956). 'Invariant Theoretical Interpretation of Interaction'. *PR*, *101*, 1597–607.

Van Dam, H. and Veltman, M. (1970). 'Massive and Mass-less Yang–Mills and Gravitational Fields'. *NP*, *B22*, 397–411.

Van der Velde, J.C. *et al.* (1979). 'Experimental Properties of Neutrino-Induced Jets'. In Hansen and Hoyer (1979), 173–8.

Van Hove, L. (1971). 'Particle Production in High Energy Hadron Collisions'. *Phys. Rept.*, *1*, 347–80.

Van Hove, L. (1976). 'Review of CERN Scientific Activities'. *CERN Annual Report 1976*, 19–34.

Van Hove, L. and Jacob, M. (1980). 'Highlights of 25 Years at CERN'. *Phys. Rept.*, *62*, 1–86. Reprinted in Jacob (1981), 13–98.

van Royen, R. and Weisskopf, V.F. (1967). 'Hadron Decay Processes and the Quark Model'. *NC*, *50*, 617–45, *51*, 583(E).

Veltman, M. (1963a). 'Higher Order Corrections to the Coherent Production of Vector Bosons in the Coulomb Field of a Nucleus'. *Physica*, *29*, 161–85.

Veltman, M. (1963b). 'Unitarity and Causality in a Renormalizable Field Theory with Unstable Particles'. *Physica*, *29*, 186–207.

Veltman, M. (1966). 'Divergence Conditions and Sum Rules'. *PRL*, *17*, 553–6.

Veltman, M. (1968a). 'Relations between the Practical Results of Current Algebra Techniques and the Originating Quark Model'. Lectures given at the Copenhagen Summer School, July 1968, unpublished.

Veltman, M. (1968b). 'Perturbation Theory of Massive Yang–Mills Fields'. *NP*, *B7*, 637–50.

Veltman, M. (1969). 'Massive Yang–Mills Fields'. In *Topical Conference on Weak Interactions*, 391–3. CERN Yellow Report 69-7.

Bibliography

Veltman, M. (1970). 'Generalized Ward Identities and Yang–Mills Fields'. *NP*, *B21*, 288–302.

Veltman, M. (1974). 'Gauge Field Theories'. In Rollnik and Pfeil (1974), 429–47.

Veneziano, G. (1968). 'Construction of a Crossing-Symmetric, Regge-Behaved Amplitude for Linearly Rising Trajectories'. *NC*, *57A*, 190–7.

Veneziano, G. (1974). 'An Introduction to Dual Models of Strong Interactions and their Physical Motivations'. *Phys. Rept.*, *9*, 199–242.

von Dardel, G. (cd.) (1969). *Proceedings of the Lund International Conference on Elementary Particles*, 25 June–1 July, 1969, Lund, Sweden. Lund: Berlingska Boktryckeriet.

von Krogh, J. *et al.* (1976). 'Observation of $\mu^- e^+ K_s^0$ Events Produced by a Neutrino Beam'. *PRL*, *36*, 710–3.

Wachsmuth, H. (1977). 'Accelerator Neutrino Physics'. Lectures given at the Herbstschule für Hochenergiephysik Maria Laach, Eifel, Germany, 14–24 September 1976. CERN reprint CERN/EP/PHYS 77-40.

Waldrop, M.M. (1981a). 'Massive Neutrinos: Masters of the Universe?' *Science*, *211*, 470–2.

Waldrop, M.M. (1981b). 'Matter, Matter, Everywhere. . . '. *Science*, *211*, 803–6.

Waldrop, M.M. (1982). 'Do Monopoles Catalyze Proton Decay?' *Science*, *218*, 274–5.

Waldrop, M.M. (1983a). 'The New Inflationary Universe'. *Science*, *219*, 375–7.

Waldrop, M.M. (1983b). 'Supersymmetry and Supergravity'. *Science*, *220*, 491–3.

Walgate, R. (1977). 'Would a Heavy Neutrino Do the Trick?' *New Scientist*, *73*, 766.

Ward, J.C. (1950). 'An Identity in Quantum Electrodynamics'. *PR*, *78*, 182.

Weinberg, S. (1967a). 'Precise Relations between the Spectra of Vector and Axial Vector Mesons'. *PRL*, *18*, 507–9.

Weinberg, S. (1967b). 'A Model of Leptons'. *PRL*, *19*, 1264–6.

Weinberg, S. (1971). 'Physical Processes in a Convergent Theory of the Weak and Electromagnetic Interactions'. *PRL*, *27*, 1688–91.

Weinberg, S. (1972a). 'Effects of a Neutral Intermediate Boson in Semi-leptonic Processes'. *PR*, *D5*, 1412–17.

Weinberg, S. (1972b). *Gravitation and Cosmology*. New York: Wiley.

Weinberg, S. (1973). 'Non-Abelian Gauge Theories of the Strong Interactions'. *PRL*, *31*, 494–7.

Weinberg, S. (1974a). 'Unified Theories of Elementary-Particle Interactions'. *Scientific American*, *231*(1), 50–9.

Weinberg, S. (1974b). 'Recent Progress in Gauge Theories of the Weak, Electromagnetic and Strong Interactions'. *RMP*, *46*, 255–77.

Weinberg, S. (1977a). 'The Search for Unity: Notes for a History of Quantum Field Theory'. *Daedalus*, *2*, 17–35.

Weinberg, S. (1977b). *The First Three Minutes*. London: Andre Deutsch.

Weinberg, S. (1979). 'Weak Interactions'. In Homma *et al.* (1979), 907–18.

Weinberg, S. (1980). 'Conceptual Foundations of the Unified Theory of Weak and Electromagnetic Interactions'. *RMP*, *52*, 515–23.

455

Bibliography

Weinberg, S. (1981). 'The Decay of the Proton'. *Scientific American*, *244*(6), 52–63.

Weiner, C. (ed.) (1977). *History of Twentieth Century Physics*. New York and London: Academic Press.

Weisberger, W.I. (1966). 'Unsubstracted Dispersion Relations and the Renormalization of the Weak Axial-Vector Coupling Constants'. *PR*, *143*, 1302–9.

Weisskopf, V.F. (1968). 'The Three Spectroscopies'. *Scientific American*, *218*(5), 15–29.

Weisskopf, V.F. (1978). 'What Happened in Physics in the Last Decade'. *Physics Bulletin*, *29*, 401–3.

Weisskopf, V.F. (1981). 'The Development of Field Theory in the Last 50 Years'. *Physics Today*, *34*(11), 69–85.

West, G.B. (1975). 'Electron Scattering from Atoms, Nuclei and Electrons'. *Phys. Rept.*, *18*, 263–323.

Weyl, H. (1931). *The Theory of Groups and Quantum Mechanics*. London: Methuen.

White, D.H. and Sullivan, D. (1979). 'Social Currents in Weak Interactions'. *Physics Today*, *32*(4), 40–7.

White, D.H., Sullivan, D. and Barboni, E.J. (1979). 'The Interdependence of Theory and Experiment in Revolutionary Science: The Case of Parity Violation'. *Social Studies of Science*, *9*, 303–27.

Whitmore, J. (1974). 'Experimental Results on Strong Interactions in the NAL Hydrogen Bubble Chamber'. *Phys. Rept.*, *10*, 273–373.

Whitmore, J. (1976). 'Multiparticle Production in the Fermilab Bubble Chambers'. *Phys. Rept.*, *27*, 187–213.

Wigner, E. (1937). 'On the Consequence of the Symmetry of the Nuclear Hamiltonian on the Spectroscopy of Nuclei'. *PR*, *51*, 106–19.

Wiik, B.H. (1975). 'Recent Results from DORIS'. In Kirk (1975), 69–96.

Wiik, B.H. (1976). 'Plenary Report on New Particle Production in e^+e^- Colliding Beams'. In Bogoliubov *et al.* (1976), Vol. II, N75–86.

Wilcek, F. (1980). 'The Cosmic Asymmetry between Matter and Antimatter'. *Scientific American*, *243*(6), 60–8.

Wilcek, F. (1981). 'Coming Attractions in SUMs and Cosmology'. *CNPP*, *10*, 175–85.

Wilcek, F. (1982). 'Quantum Chromodynamics: The Modern Theory of the Strong Interaction'. *ARNPS*, *32*, 177–209.

Williams, W.S.C. (1979). 'The Electromagnetic Interactions of Hadrons'. *RPP*, *42*, 1661–717.

Willis, W.J. (1978). 'The Large Spectrometers'. *Physics Today*, *31*(10), 32–9.

Wilson, K.G. (1969). 'Non-Lagrangian Models of Current Algebra'. *PR*, *179*, 1499–512.

Wilson, K.G. (1971a). 'The Renormalization Group and Critical Phenomena. I: Renormalization Group and Kadanoff Scaling Picture'. *PR*, *B4*, 3174–83.

Wilson, K.G. (1971b). 'The Renormalization Group and Critical Phenomena. II: Phase Space Cell Analysis of Critical Behavior'. *PR*, *B4*, 3184–205.

Bibliography

Wilson, K.G. (1971c). 'Renormalization Group and Strong Interactions'. *PR, D3*, 1818–46.

Wilson, K.G. (1972a). 'Feynman-Graph Approach for Critical Exponents'. *PRL, 28*, 548–51.

Wilson, K.G. (1972b). 'Renormalization of a Scalar Field in Strong Coupling'. *PR, D6*, 419–26.

Wilson, K.G. (1974). 'Confinement of Quarks'. *PR, D10*, 2445–59.

Wilson, K.G. (1979). 'Problems in Physics with Many Scales of Length'. *Scientific American, 241*(2), 140–57.

Wilson, K.G. and Fisher, M.E. (1972). 'Critical Exponents in 3.99 Dimensions'. *PRL, 28*, 240–3.

Wilson, K.G. and Kogut, J. (1974). 'The Renormalization Group and the ε Expansion'. *Phys. Rept., 12*, 75–200.

Wilson, R. (1977). 'From the Compton Effect to Quarks and Asymptotic Freedom'. *AJP, 45*, 1139–47.

Wilson, R.R. (1980). 'The Next Generation of Particle Accelerators'. *Scientific American, 242*(1), 26–41.

Witten, E. (1980). 'Quarks, Atoms and the 1/N Expansion'. *Physics Today, 33*(7), 38–43.

Yagoda, H. (1956). 'The Tracks of Nuclear Particles'. *Scientific American, 194*(5), 40–7.

Yamaguchi, Y. (1959). 'A Composite Theory of Elementary Particles'. *PTP, Suppl. 11*, 1–36.

Yan, T.M. (1976). 'The Parton Model'. *ARNS, 26*, 199–238.

Yang, C.N. and Mills, R. (1954a). 'Isotopic Spin Conservation and a Generalized Gauge Invariance'. *PR, 95*, 631.

Yang, C.N. and Mills, R. (1954b). 'Conservation of Isotopic Spin and Isotopic Gauge Invariance'. *PR, 96*, 191–5.

Yoshimura, M. (1977). 'Muon Number Nonconservation in a Unified Scheme of All Interactions'. *PTP, 58*, 972–7.

Yoshimura, M. (1978). 'Unified Gauge Theories and the Baryon Number of the Universe'. *PRL, 41*, 281–4, *42*, 746(E).

Young, E.C.M. (1967). 'High-Energy Neutrino Interactions'. CERN Yellow Report 67-12.

Zachariasen, F. and Zemach, C. (1962). 'Pion Resonances'. *PR, 128*, 849–58.

Zeldovich, Ya. B. (1970). 'The Universe as a Hot Laboratory for the Nuclear and Particle Physicist'. *Comments on Astrophysics and Space Physics, 2*, 12–40.

Zichichi, A. (ed.) (1977). *New Phenomena in Subnuclear Physics*. New York: Plenum Press.

Zichichi, A. (ed.) (1980). *ECFA-LEP Working Group 1979 Progress Report*. ECFA Report 79/39.

Zichichi, A. (ed.) (1982). *Pointlike Structures Inside and Outside Hadrons*. Proceedings, 17th International School of Subnuclear Physics, 31 July–11 August 1979, Erice, Italy. New York: Plenum.

Zweig, G. (1964a). 'An SU$_3$ Model for Strong Interaction Symmetry and Its Breaking'. CERN preprint 8182/TH401 (17 Jan. 1964).

457

Zweig, G. (1964b). 'An su₃ Model for Strong Interaction Symmetry and Its Breaking: II'. CERN preprint 8419/TH412 (21 Feb. 1964). Reprinted in Lichtenberg and Rosen (1980), 22–101.

Zweig, G. (1964c). 'Fractionally Charged Particles and su(6)'. In A.Zichichi (ed.), *Symmetries in Elementary Particle Physics*, Proceedings of the 'Ettore Majorana' International School of Physics, Erice, Italy, August 1964, 192–243. New York: Academic Press.

Zweig, G. (1981). 'Origins of the Quark Model'. In Isgur (1981), 439–59.

INDEX

The following abbreviations are used in the index:
f for figure, t for table and n for note. Italic page
numbers denote the main treatment of the subject.

ABJ anomaly, 203n98, 220-1
AFS (Axial Field Spectrometer), 378n34
AGS, *see* Brookhaven AGS
Abelian (and non-Abelian) group, 196n9
accelerators, 23-37, 348t, 364-6
aces, 85, 89-90
action-at-a-distance, 64
Adams, J. B., 368
Adler–Bell–Jackiw (ABJ) anomaly, 203n98, 220-1
Adler–Weisberger relation, 113
ADONE collider, *see* CNEN National Laboratory ADONE
agency in science, 7-8, 14, 405
Alternating Gradient Synchrotron (AGS), *see* Brookhaven AGS
Alvarez, L. W., 43n14, 44n25
analogy, x, 12, 19n12, 406-7, 408t
analyticity, 74
Anderson, C., 50
Anderson, P. W., 170
anomalous dimensions, 210-11
antielectrons, *see* positrons
antiparticles, 50
Appelquist, T., 263-5
Argonne National Laboratory, 34
  ZGS, 32t, 34
asymptotic freedom, 207-9, *213-15*, 223-5, 227-8, 230n48, 241, 243-4, 247, 263-4, 319, 321
atomic parity violation experiments, *see* Washington–Oxford experiments *and* Novosibirsk experiments
Axial Field Spectrometer (AFS), 378n34
axial vector quantities, 54

B mesons, 287-8

b quark, 286-8, 289, 290, 312, 318
BCS theory, 168, 197n26
BEBC (Big European Bubble Chamber), 322
BIBC (Bern Infinitesimal Bubble Chamber), 376n26
BNL, *see* Brookhaven National Laboratory
Bacry, H., 219
Bardeen–Cooper–Schrieffer (BCS) theory, 168, 197n26
Barkov, L. M., 299
baryonium, *312-15*, 336-7n10, 337-8n11, 350-1
baryons, 52, 57, 59f, 105t, 106, 107f
  number, 53
  number nonconservation, 388-93
beam cooling, 366, 378-9n39
beams, experimental, 23-8
Becchi, C., 100-1
Bell, J. S., 175
Berkeley Laboratory, *see* Lawrence Berkeley Laboratory
Bern Infinitesimal Bubble Chamber (BIBC), 376n26
beta decay, 68, 112
Bevatron, *see* Lawrence Berkeley Laboratory Bevatron
Big Bang, 387-90, 392, 398-9n23
Big European Bubble Chamber (BEBC), 322
Bjorken, J. D., *131-2*, 137-8, 140, 145, 153n13n14, 156n16, 231, 235-6, 238, 365, 383, 396, 399-400n26
Bjorken limit, 131, 141, 208-9
Blankenbecler, R., 333
Bloom, E. D., 276n26
Bludman, S. A., 166
Bogoliubov, N. N., 197n26

Bohr, N., 51, 68
bootstrap theory, *see* S-matrix
bosons, 52, 79n15
branching ratios, 106, 120n48
Brodsky, S. J., 333
broken symmetry, 59-60, 169
Brookhaven National Laboratory,
  33, 36, 94, 174, 273n9
  AGS, 32t, 34, 39, 49, 60, 144, 147-8,
  149, 253, 258, 283, 318, 331,
  348t, 372n9, 372-3n10
  Cosmotron, 32-3, 48, 49
  Isabelle, 37t, 40-1, 381n59
  see also Irvine–Michigan–Brookhaven collaboration
Brower, R. C., 249n31
bubble chambers, 25-6, 27, 43n14,
  44n25, 64-5, 355, 415n13
  high-resolution, 356-8
Budker, G., 378n39

c quark (charm), 184, 312
CEA, *see* Cambridge Electron
  Accelerator
CERN, 21, 34, 42n6. 89-90, 186,
  240-7, 291, 314, 319, 348-9,
  372n7, 393
  Gargamelle, 145-7, 157n57,
  159-60, 186-91, *192-3*, 195,
  205-6n115, 290, 293, 350, 405
  ISR, 37, 40, 125, 149-52, 238,
  288-9, 317-18, 331-5, 348t,
  351-2, 353, 354, *359-64*, 369,
  373n11n12, 376-7n31
  LEAR, 380-1n45
  LEP, 24, 37t, 39-41, 42n10, *367-70*
  LSR, 368
  membership of, 44n26
  PS, 32t, 34, 37, 40, 49, 58, 125, 144,
  318, 348t, *349-51*, 369, 372n9
  SPS, 32t, 35-7, 39, 41, 152, 227,
  288-9, 317-18, 322, 325-6, 348t,
  349, *352-3*, 354-5, 366, 367, 369,
  370, 372n9, 373-4n15, 376n30
CERN–Columbia–Rockefeller
  collaboration, 151f, 359, 360,
  361f
CERN–Dortmund–Heidelberg–
  Saclay (CDHS) collaboration,
  293, 294
CERN–Hamburg–Orsay–Vienna
  collaboration, 360, 362f
CESR, *see* Cornell Electron Storage
  Rings
CNEN National Laboratory, 38

CNEN National Laboratory— *contd*
  ADONE, 37t, 38, 148, 221, 255-6,
  273n7, 281
CP violation, 70, 289, 290, 389,
  398n17
CVC hypothesis, 112, 174
Cabibbo, N., 111
  Cabibbo angle, 123n71, 248n3
  Cabibbo–Radicati sum-rule, 113
Cabrera, B., 402n45
California Institute of Technology
  (Caltech), 116n1, 390
  see also SLAC–MIT(–Caltech)
  collaboration
Callan, C. G., Jr, 213, 214
calorimeter detectors, 333, 377-8n34
Cambridge Electron Accelerator
  (CEA), 32t, 35-6
  'Bypass' collider, 37t, 38, 221, 256
cascade, 48
centre-of-mass energy, 28
Chadwick, J., 47
charged currents (weak interactions), 182-3
charm, 4, 5, 17, 37, 39, 184-5,
  229n31, 238-40, 244, 246-7,
  253-73, 277-8n31, 280, 282,
  283, 318, 350-1
charmed pseudoscalar mesons,
  264-7, 272, 275-6n25, 276n26,
  312
charmonium model, *263-7*, 270,
  275-6n25, 276n26, 279n40, 286,
  302-3n12, 312
Chew, G. F., 73-5, 337n10, 415n13
chi mesons, 265-6, 275-6n25,
  276n26
Chicago cyclotron, 48
chiral symmetry, 242, 245, 247
Cline, D., 366
Close, F. E., 114, 296, 298, 309, 323
Coleman, S., 172, 213, 214, 215, 241,
  289, 291, 292, 294, 298
collaboration in experiment, 42n10
colliders, 27-8, *37-41*
colour, *216-22*, 233, 275-6n25
  symmetry SU(3), 207, 219-22,
  223-4, 233
Columbia–Stony Brook–Fermilab
  collaboration, 283
committee structure, 42-3n11,
  375n18
commutators (in field theory),
  109-12, 174, 222
composite quarks, 303-4n23, 397n3

confinement of quarks, *224-6*, 233, 265, 310, 312, 384
Conseil Européen pour la Recherche Nucléaire, *see* CERN
conservation laws, 50-60
conserved vector current (CVC) hypothesis, 112, 174
constituent interchange model (CIM), 333, 334
constituent quark model (CQM), *89-91*, 97-108, 115-16, 118n34, 120-1n53, 139-40, 216-17, 226, 228, 238, 266, 269, 312-13, *315-17*, 339-40n18
context of discovery / justification, 414n5
Cornell Electron Storage Rings (CESR)
  accelerator, 32t, 370n2
  CESR, 37t, 39, 287-8, 312, 348t, 368
cosmological baryon number excess, 388-91
cosmology, 387-91
Cosmotron, *see* Brookhaven Cosmotron
coulomb force, 265, 275n22
coupling (between particles), 62
coupling constant (strong interaction), 72, 134, 213, 226, 227, 230n50, 385-6
Cowan, C. L., Jr, 69, 392
Cronin, J. W., 81n47, 398n17
cross-sections, 49f
  defined, 28-9, 43n19
Crystal Ball experiment, 276n26
current algebra, *108-14*, 123n69, 155n27, 170, 240
current quarks, 115, 248n8

D mesons, 268-70
  discovery of, 261, 375n19
d quark, 86
DCI collider, 37t
DESY, 32t, 35, 39, 258, 368, 370n2
  DORIS, 37t, 39, 256, 260-1, 265-6, 267, 275-6n25, 282, 283, 287, 303n14, 312, 326, 348t
  HERA, 381-2n59
  PETRA, 37t, 39, 287, 288, 318, 319, 326, 328-30, 342n38, 348t, 371n3
Dalitz, R. H., 96-7, 118-19n35, 220, 340n21
Dardel, G. von, *see* Von Dardel, G.
Daresbury Laboratory NINA, 32t, 35
Darriulat, P., 335, 345-6n63, 379n43

deep inelastic electron scattering, 132-3, 136-9, 207, 208, 237, 321-5, 347
delta resonance, 48-9, 100-1
demography of HEP community, 21-2
De Rújula, A., *240-2*, 244, 248-9n13, 268-9
detectors, experimental, 23-6, 347, 377n33
Deutsch, M., 263
Deutsches Elektronen-Synchrotron, *see* DESY
dimensional analysis, 208
Dirac, P., 50
divergences (in field theory), 66, 176
DORIS, *see* DESY DORIS
Drell, S. D., 131, 147
Drell–Yan model, 148, 283, 331
dual resonance model, 77, 82n65, 275n22
duality, 77, 336n9
Dubna Laboratory, 34
  accelerator, 32t, 34-5
DuMond, J., 155n31
Dydak, F., 291-2, 299-300
Dyson, F. J., 67

ε-expansion, 212, 243
ECFA, 368-9
EHS, 356, 358
Eightfold Way symmetry SU(3), 46, 56-60, 78n2, 79n22, 85-7, 89, 91-2, 102, 109-11, 116n6, 164, 167, 169, 175, 234, 312
electromagnetic interaction, 3, 4, 5, 22, 46, 134, 159, 384
  see also electroweak unification *and* grand unification
electron–positron annihilation, *148-9*, 237-8, 244, 254-7, *286-8*, *326-30*, 348
electron-volt (eV), 44n22
electrons, ix, 3
electroproduction, 121n57
electroweak mixing angle, *see* Weinberg angle
electroweak theory, 4-5, 16, 17, 247-8n3
electroweak unification, 4, 159, *165-7*, 196n14, 235f, 237
Ellis, J., 240, *242-4*, 246-7, 248-9n13, 249n31, 256-7, 327, 390-3
emulsions, nuclear, 43n14
eta meson, 49

European Committee for Future Accelerators (ECFA), 368-9
European Hybrid Spectrometer (EHS), 356, 358
European Organisation for Nuclear Research, *see* CERN
Exclusion Principle, 216-18
exotic hadrons, 120-1n53, 313, 336n9

fallibility of experiment, 6, 9, 13-14, 20n14
Fermi, E., 68, 86
Fermi National Accelerator Laboratory, *see* Fermilab
Fermi theory, 68-71, 81n46, 165
Fermilab, 35, 36-7, 40, 152, 160, 186-7, 192-3, 227, 238, 245, 288-9, 290, 291, 317, 318, 319, 322, 325, 331, 333, 348t, 366-7, 374-5n17, 381n59
  accelerator, 32t, 35, 39, 41, 348t, 372n9
  Tevatron accelerator, 32t, 35
  Tevatron collider, 37t
  see also Columbia–Stony Brook–Fermilab collaboration *and* Harvard–Pennsylvania–Wisconsin–Fermilab collaboration
fermion, 52, 79n15
Ferro-Luzzi, M., 374n15
Feynman, R. P., 67, 70, 109, 112, 125, 132, 136-40, 147, 154n24, 155n31, 175
  Feynman diagrams, 67, 73, 134-5, 163
  Feynman integrals, 67, 201n68
  Feynman rules, 67, 70, 163, 173, 176
Field, R. D., 322-3
field theory, *see* quantum field theory
Fisher, C., 212, 355-6, 358
Fitch, V. L., 81n47, 398n17
flavour, 216, 261
forces, classification of, 3, 22
fragmentation functions, 330, 345n53
Frascati–Milan–Turin NUSEX (nucleon stability experiment) concept, 395, 401n43
Friedman, J. I., 127-31, 153n12
Fritzsch, H., 222-3

GIM mechanism, *184-5*, 187, 195, 203n98, 238, 246, 253-4, 267-8, 271

GUTs, *see* grand unification
Gaillard, M. K., 239-40, 244, *245-7*, 248-9n13, 261-2, 327
Ganguli, S. N., 373-4n15
Gargamelle, *see* CERN Gargamelle
Garwin, E. L., 306n47
gauge invariance, 160-2
gauge theory, x-xi, 4, 5, 12, 15, 16, 17, 18, 46, *159-95*, 231-5, 239, 240, 241-2, 270-1, 273
  Yang–Mills theory, *160-5*, 173, 174-7, 179-80, 196n8n9, 223
Gell-Mann, M., 57, 59-60, 70, 74, 85-9, 92, 93, 102, 108-12, 114, 115, 117n16, 122n62, 125, 164, 175, 194, 201n73, 207, 210, 211, 222-4, 239, 242, 403
Georgi, H., 186, 268-9, 384-7, 393
Georgi–Glashow model, 384-5, 386, 390
ghosts (in field theory), 176, 200n61
Gilman, F. J., 141, 155n27
Glashow, S. L., 164, 166-7, 184, 186, 198n42, 239, 241, 268-9, 277-8n31, 291, 300, 384, 393
  see also Georgi–Glashow model *and* GIM mechanism
glueballs, 338-9n13
gluons, 142-3, 146, 147, 151, 156n43, 165, 222-3, 320-1, 327-32, 343-4n51
Goldberger–Treiman relation, 112-13
Goldhaber, G., 258, 268, 273-4n11, 277-8n31
Goldhaber, M., 392, 394
Goldstone, J., 169-70
  Goldstone bosons, 170, 178
grand unification, 18, 20n17, 234-6, *383-96*
grand unification mass, 385-6
grand unified theories (GUTs), *see* grand unification
gravitational interaction, 3-4, 22, 164
Greenberg, O. W., 207, 218, 309, 317
Gregory, B., 145-6
Gribov, V. N., 243
Gross, D. J., 214, 215, 223, 224
group theory, 56
Gunion, J. F., 333
Gürsey, F., 94

HERA, *see* DESY HERA

H P W F collaboration, 187, 192-3,
204n105n107, 292-4, 296
hadrons, 4
couplings, 99-102
population explosion of, 33, 46,
47-50, 110
spectroscopy, 95-9, 226, 272,
310-17, 318
Halzen, F., 331-2
Han, M. Y., 207, 218, 219-20,
229n29
Han–Nambu model, 219-22
Hanson, G., 326
hard scattering, 30-1, 35, 37, 40, 125,
147-52, 238, 272, 330-5, 360
defined, 127
Harvard–Pennsylvania–Wisconsin–
Fermilab (H P W F) collaboration,
187, 192-3, 204n105n107, 292-4,
296
Harvard–Wisconsin–Purdue collaboration, 394
Heisenberg, W., 68, 80n32, 197n29
Hey, A. J. G., 309
hidden charm, 93, 261-6, 270
Higgs, P. W., 170-1
Higgs mechanism, 171, 179, 180,
198n39n41, 225, 233, 397n4
Higgs particle, 233, 292, 370,
378n37, 384, 397n3
high-resolution bubble chambers,
356-8
high-*y* anomaly, 292-3, 294, 298,
304n30
Hofstadter, R., 127
Hooft, G. 't, *see* 't Hooft, G.
Hughes, V., 305n47

I P T, 320-1, 331, 332, 334, 341n25
n26
I S R, *see* C E R N I S R
I V B S, *see* intermediate vector bosons
Iliopoulos, J., 184, 235
see also G I M mechanism
impulse approximation, 136
incommensurability, 407-14,
414-15n9, 415n11
infinite momentum frame, 123n77,
136
inflationary universe, 399n23
instantons, 226, 239n50
intermediate vector bosons (I V B S),
39-40, 41, 144, 148, 149-50, 165,
181-2, 232-4, 365-70, 387
discovery, 367

Intersecting Storage Rings collider,
see C E R N I S R
intuitive perturbation theory (I P T),
320-1, 331, 332, 334, 341n25n26
Irvine–Michigan–Brookhaven
collaboration, 394, 401-2n43
Isabelle, *see* Brookhaven Isabelle
Isgur, N., 316
isospin, 54-6, 72

J-psi particle, 39, 244, 253, 258-65,
268, 273n7n9, 275-6n25, 283,
312, 331
Jaffe, R. L., 243
Jentschke, W., 150
jets, 326-30, 333-4, 343n45, 343-4n51
Jona-Lasinio, G., 168-9
Joyce, J., 85
judgments, role in knowledge
production, 6-7, 8, 9, 13, 404-6,
407

K-star resonance, 49
K E K accelerator, 32t
K L N theorem, 341-2n26
kaons (K mesons), 48, 267
decay anomaly, 54, 69-70, 93f,
183-5, 245, 246
Karl, G., 316, 339-40n18
Kendall, H. W., 127-31, 139, 141,
153n12
Kibble, T. W. B., 172
Kinoshita–Lee–Nauenberg (K L N)
theorem, 341-2n26
Klein, A., 179
Kobayashi, M., 289
Kokkedee, J. J. J., 91, 97, 99, 103
Kolar gold mine experiment, 395
Kolb, E., 390
Kramers, H. A., 67
Kuhn, T. S., 18n6, 407-9, 411,
414-15n9
Kuti–Weisskopf model, 143

L A S S, *see* S L A C L A S S
L B L, *see* Lawrence Berkeley
Laboratory
L E A R, *see* C E R N L E A R
L E B C, 355-8, 376n26
L E P, *see* C E R N L E P
L P T H, 332-5, 345-6n63, 360
L S R, *see* C E R N L S R
Lagrangian (in field theory), 61-3
Lamb, W., 66
Lamb shift, 66-7

lambda, 48
  decay anomaly, 54
Lande, K., 394-5
Large Electron–Positron collider,
  see CERN LEP
Large Storage Rings, *see* CERN
  LSR
large transverse momentum hadron
  (LPTH) production, 332-5,
  345-6n63, 360
Latour, B., 18-19n6
lattice theory, 197n28, 226, 230n50,
  249n31, 275n22, 338-9n13
Lawrence, E. O., 33
Lawrence Berkeley Laboratory
  (LBL), 43n14, 44n25
  Bevatron, 32t, 33-4, 49, 50
  see also SLAC–LBL collaboration
Lederman, L. M., 148, 150, 283-6
Lee, B. W., 179-80, 186, 201n81,
  202n86, 245-6, 261-2
Lee, T. D., 69
leptons, 4
  number, 53
  -pair production, 147-8, 330-2
Leutwyler, H., 223
Lévy, M., 112
Lexan Bubble Chamber (LEBC),
  355-8, 376n26
light-cone algebra, 113-14, 155n27,
  222-3, 237, 244, 340n23
Liverpool–MIT–Orsay–Scandinavian
  collaboration, 360, 362f
London Conference, July 1974, 234,
  236-8, 241, 244, 246, 256
Lorentz group, 94-5
Low, F. E., 109
Low-Energy Antiproton Ring, *see*
  CERN LEAR

MIT bag model, 336-7n10
McIntyre, P., 366
magnetic monopoles, 226, 230n50,
  399n23, 402n45
Maiani, L., 184
  see also GIM mechanism
Marshak, R. E., 47, 70
Maskawa, K., 289
mass hierarchy problem,
  303-4n23
Maxwell, J. C., 4, 61, 67, 160
mesons, 52, 57, 58f, 104t, 106
Messiah, A. M. L., 218
Miller, D. J., 297
Millikan, R., 88

Mills, R. L., 159, 162
  see also gauge theory, Yang–Mills
  theory
mixing (of multiplets), 57-8
Mo, L., 153n12
modelling, *see* analogy
moments (of structure functions),
  322-3
monopoles, magnetic, 226, 230n50,
  399n23, 402n45
Morpurgo, G., 94-5, 100-1,
  117-18n32, 229-30n33
multiplets, 55-9
muons, 47-8, 303-4n23
  see also trimuons
Muta, T., 325

NINA, *see* Daresbury Laboratory
  NINA
NUSEX (nucleon stability experiment) concept, 395, 401n43
naked charm, 262, 266, *267-9*, 355-8
Nambu, Y., 168-9, 207, 218-20,
  229n29n31, 275n22, 313
  see also Han–Nambu model
National Accelerator Laboratory
  (US), *see* Fermilab
Ne'eman, Y., 57, 86, 116n9, 164, 167
neutral current, 4, 5, 7, 14, 16, 17, 37,
  159-60, *180-95*, 237, 238, 293,
  404, 405, 409
neutrinos, 36, 48, 68-9
  experiments: charm, 267, 271,
  276-7n28; hard-scattering, 318;
  neutral currents, *185-95*, 409;
  PS, 350-1; scaling, 36, 144-5;
  standard model, 254, 280,
  290-302
  massive, 391, 398-9n23, 402n45
  oscillation, 402n45
neutrons, ix, 4, 47
  background (in neutrino
  experiments), *188-90*, 191,
  206n17
new orthodoxy, 234-6, 238, 300, 383,
  396, 399-400n26
new particles, 17, 121n55, 238, 253-
  4, *258-61*, 270, 278-9n40, *312*
new physics, 15, 16, 30-2, 34, 35,
  36-7, 39, 40, 41, *125-247*, 248n6,
  315, 317, 335, 349-54, 359, 369,
  370, 375n18, 403, 404, 410-12
Nimrod, *see* Rutherford Laboratory
  Nimrod
Nishijima, K., 54

November Revolution, 17, 231, 240, 246, *253-4*, 267-73, 280, 288-9, 315, 327, 331, 365, 368
Novosibirsk experiments, 299-300
nuclear emulsions, 43n14
nucleon, 55
Nuyts, J., 219

OPE, 210-11, 242-3, 244, 319
Okubo, S., 59-60, 92
old physics, 15, 16, 30-2, 34-5, 40, 41, *46-108*, 231, 237, 248n6, 315, 317, 375n18, 410-12
Oleson, P., 275n22
omega-minus baryon, 49, 60, 79n28, 349-54, 370, 373n12n15, 86, 102
1/N expansion, 226, 230n50
operator product expansion (OPE), 210-11, 242-3, 244, 319
opportunism in context, 10-13, 239, 413

PCAC hypothesis, 112-13, 201n73, 220
PEP collider, *see* SLAC PEP
PMT, 26-7, 394
PS, *see* CERN PS
Pais, A., 54, 94, 403
Palmer, R. B., 205n115
Panofsky, W. K. H., 132, 139, 141, 365
parastatistics, 218-20
parity conservation/nonconservation (in weak interaction), 53-4, 69-70, 248n3
parity violation in electron scattering, *see* SLAC Experiment E122
partially conserved axial current (PCAC) hypothesis, 112-13, 201n73, 220
particles, classification of, 4
   see also new particles, real particles, strange particles, *and* virtual particles
partons, *see* quark-parton model
Paschos, E. A., 140
path integrals (in field theory), 177, 201n68
Pati, J. C., 384, 393
Pati–Salam model, 384-5
Pauli, W., 68, 160, 210, 216
Perl, M. L., 281-2
perturbation expansion/theory, 64, 72, 109, 134-5, 210

PETRA collider, *see* DESY PETRA
phase transitions, 212, 228n9, 249n31
phenomenology, *91-3*, 219, 227-8, 239, 245, 315
phi meson, 49
photomultiplier tube (PMT), 26-7, 394
photons, 52, 62-4
photoproduction, 106, 121n57
Pierre, F., 268
pions (pi mesons), 48, 151f
Politzer, H. D., 214-15, 223, 224-5, 241, 263-5, 341n26
population explosion (of hadrons), 33, 46, *47-50*, 110
positronium, 263
positrons, 32, 50
Povh, B., 314
Prentki, J., 186
Prescott, C. Y., 298, 304-6n47
propagator (in field theory), 61-2
protons, ix, 4
   decay, 393-6, 400-1n33, 401n40, 401-2n43
   structure, 126, 129, 134-6
Proton Synchrotron (CERN), *see* CERN PS
pseudoscalar quantities, 54
psi particle, *see* J-psi particle
psi-prime, 260, 265

quantum chromodynamics (QCD), 4, 5, 16, 17, 18, 36, 39, 125, *207-28*, 231-3, 234, 237-8, 254, 263-4, 270-1, 272, 279n40, 289, 290, *309-35*, 365, 369, 383, 384, 396
quantum electrodynamics (QED), x-xi, 12, 38, *61-5*, 108, 109-10, 122n61, 126-7, 132, 134, 232, 254-5, 258, 292, 320, 371n3, 407
quantum field theory, 46, 57, *60-73*, *107-8*, 159
quantum numbers, 50-60
quark-parton model, 16, 36, 40, 116, 125, *132-47*, 207-8, 209, 237, 238, 319-21, 323, 326-7, 331
quarkonium, 303n12
quarks
   derivation of name, 85
   non-observation in isolation, 4, 88-9
   perceived reality, x, *114-16*, 270
Quinn, H., 385-7

R, 221-2, 260, 278n36, 288, 319, 342n38
  mini *R*-crisis, 280-3
  R-crisis, *254-8*, 271
radiative corrections, 129-31, 153n12, 260
Radicati, L., 94
  see also Cabibbo–Radicati sum-rule
real particles, 65
realism, 7, 9, 14, 19n9
Regge, T., 75
  Regge models, 34, 114-15, 226, 228, 238
  Regge theory, 40, 47, 75-7, 78n2, 82n64, 108, 121-2n59, 373-4n15, 415n13
  Regge tradition, 34, 47, 237, 242, 279n42
Reggeon field theory, 243-4
Reichenbach, H., 414n5
Reiff, J., 177
Reines, F., 69, 392-4
Renner, B., 242
renormalisation/renormalisability (of gauge theory), *65-8, 173, 175-80*, 209
renormalisation group, *211-14*, 319, 385-7
research traditions, *see* traditions, research
resonances, 33, 34, 48-50, 78n8, 102-8, 114, 121-2n59, 128f, 237, 238, 315, 374n15
Retherford, R. C., 66
rho meson, 49
Richter, B., 38-9, 256, 258, 261, 343n42, 368
Rochester Conferences, 236, 261, 289, 322-5
Rosner, J. L., 246, 261-2
Ross, G. G., 327
Roy, D. P., 373-4n15
Rubbia, C., 366, 379n43
Rújula, A. de, *see* De Rújula, A.
Rutherford, E., 26, 138-9
Rutherford Laboratory, 34, 355
  Nimrod, 32t, 34

σ-model, 177-9, 201n73n78
S meson, 337-8n11
S-matrix, 33-4, 46-7, *73-8*, 90, 109, 122n60, 210
  bootstrap theory, 15, 33-4, 46, 74-5, 90, 337n10

s quark, 86
SFM, 360, 362f, 363f, 378n34
SLAC, 36, 126-7, 128-32, 227, 258, 273n9, 318, 321-2, 326, 347, 348t
  accelerator, 32t, 36, 38-39, 125, 152n5, 348t, 371n3
  collider, 37t, 39-40, 381n59
  experiment E122, 298-9, 301, 312n4, 304-6n47, 307-8n56
  LASS, 340n22
  PEP, 37t, 39, 288, 319, 326, 328, 348t
  SPEAR, 37t, 38-9, 244, 253, 256, 258-61, 266, 267-9, 273-4n11, 276n25, 281-2, 283, 287, 312, 328, 348t, 368, 371n3
SLAC–LBL collaboration, 256, 258, 268
SLAC–MIT(–Caltech) collaboration, 125-6, 127-32, 141, 152n7
SLC, *see* SLAC collider
SPS, *see* CERN SPS
SU(2), 56, 79n21
SU(2) × U(1) (electroweak theory), 166-7, 181, 234, 289, 297-8, 384, 386
SU(3), *see* colour symmetry *and* Eightfold Way symmetry
SU(4), symmetry in nuclear physics, 94, 96
  symmetry of hadrons, 184
SU(5), 234, 383, 385-7, 390-1
SU(6), *93-100*, 118n32
Sakata, S., 86, 116n5
  Sakata model, 86, 116n6, 230n35
Sakharov, A., 398n13
Sakurai, J. J., 164, 196n12
Salam, A., 167, 170, 171-2, 198n48, 291, 300, 308n57, 384, 393
  see also Pati–Salam model *and* Weinberg–Salam model
Sandars, P., 296, 297
saturation (of strong interaction), 219
scalar quantities, 54
scale invariance, *208-9*, 242-3, 244
scaling, 4, 5, 16, 36, 114, *125-32*, 137-40, 207, 238, 272, 410
  violation, *227-8, 321-6*, 331-2
scattering, *see* hard scattering *and* soft scattering
SCHOONSHIP, 199n58
Schwinger, J., 67, 165-6, 174, 177, 243
  Schwinger terms, 174

'scientist's account', 3-8, 18n3, 19n9,
    188, 193, 266, 301, 311, 403-4,
    413
scintillation counters, 26-7, 392
sea quarks, 142-3
Scrpukhov Laboratory, 34-5, 331,
    348t, 372n6
    UNK, 32t, 34-5, 382n59
Shaw, R., 167, 197n21
sigma baryon, 48
Silvestrini, V., 221
Sinclair, C., 306n47
soft scattering, 30-1, 34, 47, 77, 108,
    114, 237, 238, 374n15
    defined, 127
spark chambers, 27
SPEAR, *see* SLAC SPEAR
spin, 51-2, 93
spin-statistics theorem, 216-17
split-field magnet (SFM), 360, 362f,
    363f, 378n34
splitting (of masses within multi-
    plets), 55, 92, 96-9
spontaneous symmetry breaking,
    169-70, 178-80
standard model, *see* Weinberg–
    Salam model
Stanford Linear Accelerator Center,
    see SLAC
Stanford Linear Collider (SLC), *see*
    SLAC collider
Steinberger, J., 230n35
Sterman, G., 341-2n26
strange particles, 33
strangeness, 54-6, 72, 89, 185, 245
string model, 82n65, 275n22
strong interaction, 3, 4, 5, 22, 35, 46,
    71-3, 110, 132, 139-40, 235f,
    237, 238, 247, 384
    see also grand unification *and*
    quantum chromodynamics
strongly coupled field theory, 210,
    228n4
structure functions, 131,'133f, 137,
    142-7, 208, 330
subquarks, 396-7n3
Sudarshan, E. C. G., 70
sum-rules, 113, 123n77
Super Proton Synchrotron, *see* CERN
    SPS
superconductivity, 168-71
supersymmetry, 396-7n3
Symanzik, K., 213, 215, 229n19
symbiosis of research practice,
    10-11, 13, 14, 17, 18, *102-7*,

121-2n59, *193-5*, 267, 272-3,
    301-2, 310-11, 335-6n5, 396,
    406-7, 411-12
symmetry groups, 56
synchrotrons, defined, 23

t quark, 286, *288*, 289, 290, 318, 326,
    342n38
target, experimental, 23-6
tau lepton, *281-3*, 286, 288, 290
Taylor, R. E., 127-30, 304-5n47
technicolour, 396-7n3
Tevatron, *see* Fermilab Tevatron
't Hooft, G., 159, *177-80*, 202n87,
    214, 230n50, 243
Ting, S. C. C., 258-61, 273n9, 283-4
Tokyo Conference 1978, 289, 322-5
Tomonoga, S-I., 67
traditions, research, 8-13, 14-15, 16,
    17, *406-7*, 412
triggering (of detectors), 26, 359-60,
    364
trimuons, *292-4*, 298, 304n30
Tsai, P., 153n12
tuning (of experimental techniques),
    14, 20n16, 273-4n11, 409-10
two-neutrino hypothesis, 144, 247n1

u quark, 86
UA1 experiment, 379n43n44
UA2 experiment, 379n43n44
UNK accelerator, *see* Serpukhov
    Laboratory UNK
Uncertainty Principle, 62, 65, 69,
    78n8, 80n32, 225
unitarity, 71
units (mass/energy), 44n22
upsilons, *284-8*, 303n21, 312, 331,
    375n17

V − A theory, 69-70, 81n46, 109-10,
    112, 165, 181-4
VEPP-4 collider, 37t
valence quarks, 142, 146-7
Van Dam, H., 177
Van der Meer, S., 378-9n39
Van Hove, L., 174, 219, 367
vector dominance model, 196n12,
    221
vector mesons, 38, 54, 164, 175,
    258-61
vector quantities, 54
Veltman, M., *173-80*, 199n58n59
    n60, 200n61n63n64, 201n68,
    202n87

Index

Veneziano model, 77, 82n65, 275n22
vertex of interaction (in field
  theory), 62-4
virtual particles, 65
Von Dardel, G., 368

Walecka, J. D., 131
Ward, J. C., 167, 171
Washington–Oxford experiments,
  236, 292, *294-300,* 301-2,
  306-7n55, 308n57, 410
weak currents, 54, 109-10
weak interaction, 3, 4, 5, 22, 42n9,
  46, *68-71,* 110, 159, 165-6, 237,
  245, 253-4, 384, 409
  see also electroweak unification
  and grand unification
weak neutral current, *see* neutral
  current
Weinberg, E., 214
Weinberg, S., 66, 67-8, 73, 170, 171,
  180-1, 187, 198n41n42n48,
  202n87, 239, 241, 289, 291, 297,
  300, 341-2n26, 385-7
Weinberg angle, 172, 248n3, 291,
  386
Weinberg–Salam model, 167, *171-3,*
  180-4, 185, 187, 231, 232-3, 235,
  239, 253-4, 271, *280-302*
Weisskopf, V. F., 67, 143
  see also Kuti–Weisskopf model
Weyl, H., 195n4, 391

Wigner, E., 94
Wiik, B. H., 282, 328
Wilcck, F., 214, 215, 223, 224
Wilson, K. G., 155n31, 209-12, 214,
  230n50, 249n31, 275n22
Woolgar, S., 18-19n6

X bosons, 385-6, 387-9, 393

Yang, C. N., 69, 86, 159-64,
  196n5n6, 202n86
Yang–Mills theory, *see* gauge
  theory, Yang–Mills theory
Yerevan Laboratory, 32t, 35-6
Yoshimura, M., 388-9, 397n4,
  401n33
Young, E. C. M., 191, 192
Yukawa, H., 48, 71, 81n49

ZGS accelerator, *see* Argonne
  National Laboratory ZGS
Zeldovich, Y. B., 397-8n12
Zero Gradient Synchrotron (ZGS),
  see Argonne National
  Laboratory ZGS
zero-mass problem, 163-4
Zinn-Justin, J., 180
Zumino, B., 186
Zweig, G., 85-7, *89-93,* 102-3, 108-9,
  111, 114, 117n23, 125
Zweig rule, 92-3, 261-2, 263-4,
  313-14